Our Changing Views of Photons

Our Changing Views of Photons

A Tutorial Memoir

Bruce W. Shore

OXFORD

UNIVERSITY PRESS

OXFORD
UNIVERSITY PRESS

Great Clarendon Street, Oxford, OX2 6DP,
United Kingdom

Oxford University Press is a department of the University of Oxford.
It furthers the University's objective of excellence in research, scholarship,
and education by publishing worldwide. Oxford is a registered trade mark of
Oxford University Press in the UK and in certain other countries

First published 2020
Impression: 1

Published in the United States of America by Oxford University Press
198 Madison Avenue, New York, NY 10016, United States of America

British Library Cataloguing in Publication Data
Data available

Library of Congress Control Number: 2020938079

ISBN 978–0–19–886285–7

DOI: 10.1093/oso/9780198862857.001.0001

Printed and bound by
CPI Group (UK) Ltd, Croydon, CR0 4YY

Preface

Light surrounds us. It comes as sunlight, starlight and moonlight. It comes from lamps and lasers and fires, from television sets and control panels and dashboards of cars. It arrives through windows and optical fibers. Light is studied, very briefly, in high school classrooms where students play with lenses. More generally, as the science of Optics, the control and use of light is a major enterprise of academic study and industrial application; its practitioners connect through numerous professional organizations.

One very small but growing portion of that enterprise, known to its participants as *Quantum Optics*, has been my own dominant professional activity for many decades. The adjective "quantum" is more commonly expressed in this context with words "photon" and "photonics", widely found attached to organizations, conferences, publications and nationally funded initiatives. That descriptor originated in the early 20th century notion that light could be regarded as an assemblage of indivisible radiation increments, what might be called granules or corpuscles or particles—or photons.

Visible light, with its spectrum of colors, is but a small portion of a vastly broader range of electromagnetic radiations. X-rays assist dentists in identifying tooth cavities and physicians in identifying broken bones. Invisible infrared radiation in our kitchen microwave ovens heats our meals. Radar signals guide airplanes and give us weather information. In all of these radiation regimes the notion of photon applies, in principle.

Continuing theoretical advances and experimental implementations associated with photons have brought remarkable technologies, not only in traditional optics and imaging but in providing communication tools for Information Science and Quantum Information Processing (QIP) and for use in Cryptography and in Computer Science. No longer are questions of photon identity merely arcane academic exercises: There are now a growing number of institutional and commercial establishments, both national and international, that rely upon applying the contemporary concepts of photon that have replaced the early, more familiar textbook notions. From Google one can learn of Centres and Laboratories and Departments of Quantum Engineering throughout the world. In Germany there is a long-established Institute of Quantum Optics. In the UK there is National Quantum Technologies Programme that includes a Quantum Metrology Institute. In the US there are calls for a National Quantum Initiative, a National Photonics Initiative. (The name Quantum Optique is used by a Montreal maker of eyeglasses.) University-based research and industry-based development of devices that rely on quantum properties are regularly reported, and professional societies hold regular conferences devoted to application of Quantum Optics. Clearly, modern photons are moneymakers with career opportunities.

The views of photons Just as the very concept of "light" or radiation has innumerable manifestations—as I have indicated above—so too does the notion of photon. Ongoing discoveries and new experimental devices continue to force alteration of how best to regard, and work with, the concept of photon. In this Memoir I present an overview of some portions of physics (but by no means all) wherein the concept of photon occurs, and how the understanding (the <u>views</u>) of the physics community—and my own – has undergone revision over the years, from conflicting wave-particle interpretation onward. The book is, in part, a personal account (hence a *Memoir*) of the Quantum Optics research in which I have been particularly interested, and with which I have been involved, with many colleagues. But it also provides a far broader discussion of photons, as studied in other areas: Section 6.1 discusses Astronomical photons and Appendix A.19 presents the photon of the Standard Model of high-energy physics. Perfectly good photons can be found outside the boundaries of Quantum Optics.

This book The book-title words "... Changing Views ..." is intended to convey the central objective of explaining many (but by no means all) of the ways that researchers today regard the concept of "photon", and how that diversity of views differs from those held a century ago when quantum theory was just beginning to be developed. I offer a glimpse of ever-evolving diverse views, shared within several scientific communities (though with noted dissenters), distilled from fascinating published writings of numerous scientists and enlightening conversations with several of them. In common with many other intellectual activities, views that seemed quite clear at times have subsequently undergone significant change to fit results that continually enlarge our communal understanding of portions of physics united only by the sobriquet of "photon".

Although this goal necessarily requires mentioning past views, the book is not intended as a comprehensive history of a small part of science. For those whose interest lies therein, the internet provides numerous resources, as do some of my references.

The topics The present book is, in part, a personal account (hence a *Memoir*) of some of the research in which I have been involved, with many colleagues—a portion of atomic and optical physics now known as Quantum Optics. It offers lay readers a glimpse of scientific discovery—of how ideas become practical, as a small scientific community reconsiders its assumptions and offers the theoretical ideas that are then developed, revised, and adopted into technology for daily use.

The secondary portion of the book title, "... A Tutorial Memoir" sets a framework, an organization of topics and an approach, for the presentation. Although the book discusses a great many particulars of photons and their observations, there are far more potentially applicable topics that could fit those requirements than I can reasonably assemble, and the list continues to grow. But it also provides a far broader discussion of photons than is found in textbooks of quantum optics. In the first chapter it presents a conversational overview of the intellectual landscape into which the discussion of photons, and physics more generally, must fit—something of a contrast to the seemingly dry mathematics that underlies that landscape.

Because this book is offered as a Memoir it should not be regarded as a conventional textbook: I do not wish to compete with the many fine, detailed textbooks on quantum optics and quantum theory, nor the regularly appearing lengthy reviews of photon-related topics. It includes personal asides recalling my own progress toward understanding photons, beginning with graduate studies in the 1950s. Though its coverage of topics exceeds that of traditional textbooks (the realm of photons surpasses the coverage of any single textbook), it omits many topics found in them and it has little to say about actual experiments. There are few figures and no exercises for the student, and so I do not see it as a candidate for the sole textbook of a traditional class. It is a Memoir and a Tutorial, a useful description of changing views and a simple presentation of their bases.

The book structure I recognize that interest in photons includes readers that have a broad range of aptitudes and preparations. To include as many as possible the book organization is admittedly unusual.

In the first portion, some nine chapters, I have aimed to make the story of photons— those of today and those of an earlier day—understandable to not only readers who are anticipating making contributions to the research and the applications involved with the terms "photonics" and "photon", but also to those who, having read these labels in the press, are willing to take an interest in the subject despite having only taken rather rudimentary courses in mathematics and science. The exposition here aims, at the outset, to be conversational and untechnical, understandable by anyone with an interest in science. I have tried to make the unavoidable notions of quantum theory seem simpler and much more reasonable—and everyday ordinary—than they are often presented to be. These chapters categorize my own advancing views (paralleling those of the larger physics community), with few equations intruding into the narrative. Words, rather than formulas, tell this part of my story.

In writing this main body I have kept in mind conversations with my wife, my siblings, my offspring and grandchildren, and my Rotary-Club acquaintances—as well as with physics graduate students and tenured physics professors—with all of whom I have, at times, discussed brief portions of the exposition presented here, asking rhetorically: *What is a photon?*

For the benefit of students and others who want to understand the subject as a career prelude I have placed the actual mathematics—the most essential equations—into three tutorial appendices, leaving the preceding narrative largely unmarred by formulas, apart from a few side remarks. In these appendices I have collected the sort of self-contained supplementary material that I would like to find when I read a book written for the general public on a rather technical subject. Numerous Appendix-section citations punctuate the main narrative, giving links to more detailed explanations and thereby tying together the overall theme of photons. In keeping with the original intent of a conversational narrative, I have included only a few line-plots as figures.

The tutorials The three appendices, somewhat lengthier than the main narrative, provide a supplement of equations and mathematical support, a *tutorial* requiring minimal prerequisite apart from interest, that gives an idea of the world of quantum theory in which photons and other quantum systems are defined by working physicists. Any established scientist or engineer, perusing the first portion of the book to consider new opportunities, will appreciate the tutorial appendices that offer detailed underpinnings for basic quantum physics, radiation, and photons.

In writing these appendices I have drawn upon my lectures to graduate students and advanced undergraduates and my departmental colloquia to colleagues who were unfamiliar with quantum optics. I cover the basics that I think a newcomer should know to start work in quantum optics. My intent was to provide a fairly self-contained discussion fitting to a range of reader preparations. I hope that the presentation there of material well-known to my physicists friends will be regarded merely as a reminder of their working tools, needed as part of a logical foundation for presentation of less-known, or possibly even new, ideas. Conditioned by decades of professional activity, I have presented a rather idealized theoretician's view of experiments, but have been ever mindful of the importance of resting theory upon possible laboratory observation.

I have given considerable thought to how best to organize and present the technical material that would be a minimal support for the notions presented in the Memoir section—basically to summarize quantum theory, an impossible task. That was the aim. At various places the physics (and the conversational Memoir approach) presents opportunity to wander off into a vast world of mathematical concepts, not only those of vector spaces and groups but division algebras and topology. Such excursions have endless intellectual rewards, offering views of connections and organizations that might otherwise be unseen. However, their application to classical and quantum physics at the level of this Memoir requires only the briefest of presentations.

Unique features The breaking of the book into two parts (Memoir and tutorial appendices), with its implication that there are readers who fall between the two extremes just described, is admittedly unusual. Paragraphs above explain my justification.

In the introduction I have paid attention to careful definition of the many words that have particular technical meaning, and have kept in mind the usefulness of operational definitions. This introduces the philosophy of science as it applies to physics and to the concepts of photons—a desirable foundation for successful conversation..

A first appendix presents, very simply, the concepts of an abstract vector space, with homely examples that are readily understandable by adults or children, especially those who are familiar with data-base spread sheets. These concepts form the foundation of quantum theory, which is subsequently discussed.

As with other Memoirs, the present book has idiosyncrasies of the author. Most notably, I have included, on the opening page of each chapter, a cartoon I drew during my graduate-school training for my eventual career. These reflect the views of a grad student in the MIT of the 1950s, and are intended, not as guides for prospective students today, nor as sketches at all relevant to photons, but as whimsical (yes, light-hearted) asides for readers with long memories of their own studies, a little diversion from the often dull story of photons. Like the opening chapters of this Memoir, they collectively present a <u>view</u> set at a particular time, albeit one without any direct connection to that narrative on photons. At the end of the book a page shows four cartoons, created originally as annual Christmas greetings, that show the typical changing view toward the life of a graduate student during the advance toward the final degree: The concentration at first upon studies and examinations; the subsequent undertaking of research; and eventually the appreciation of the many intellectual and social opportunities that are available in any institution of education—a typical viewpoint change that has intellectual similarity (but no other connection) with the changing views of the physics community toward photons. They are placed together to emphasize their collective association with <u>changing views</u>, a phrase that appears in the title of this book, and in anticipation of the goal-depicting cartoon on the following final page of narrative. But with photons, *per se*, they have no deep connection.

Anticipated readership An obvious market for this Memoir will be readers of Scientific American and readers who have enjoyed popularizing books (see references below) on cosmology [Sin05,Sus08,Kra07b] , math [Cou41,Kli80,Der04,Ell15] , codes [Sin99] , navigation [Sob05] , risk [Ber96] , and contemporary technology, According to a reviewer, the principal market for this book is probably not the physics PhD student studying quantum optics or quantum technologies - there are many fine specialist texts for them. Rather, the potential readership is more varied, and might include, as the reviewer suggested:

(i) Scientists who are not specialists in quantum optics or photonics;
(ii) those interested in science education, public outreach and knowledge exchange;
(iii) young people (or others) with an interest in physics, being gifted the book;
(iv) philosophers of science, who will also appreciate the in-depth and carefully prepared arguments.

To this list I would add workers in photonics, high-energy physics, astrophysics , quantum engineering and other technology pursuits, who are interested in a broader view of their professional activities and want more than a few-page overview; A pastor interested in understanding the work of parishioners whose profession takes them to the forefront of physics research; A busy mother whose university major in liberal arts gave no time to science; her daughter whose viewing of TED talks and reading of NASA publicity on "space" has whetted an appetite for science. Makers of films and videos filled with special effects and writers of science fiction will benefit from reading this account of photons and their base in physics. Finally (another reviewer suggestion) it will also be of value to lecturers who have to teach Quantum Optics and introduce the ideas to new students.

Preface references

[Ber96]   Bernstein P.L. *Against the Gods. The Remarkable Story of Risk* (Wiley, New York, 1996).
[Cou41]   Courant R. and Robbins H. *What is Mathematics?* (Oxford Univ Press, Oxford, 1941).
[Der04]   Derbyshire J. *Prime Obsession. Bernhard Riemann and the Greatest Unsolved Problem in Mathematics* (Penguin Group, New York, 2004).
[Ell15]   Ellenberg J. *How Not to be Wrong. The Power of Mathematical Thinking* (Penguin, New York, 2015).
[Kli80]   Kline M. *Mathematics. The Loss of Certainty* (Oxford Univ Press, Oxford, 1980).
[Sin99]   Singh S. *The Code Book. The Evolution of Secrecy from Mary Queen of Scots to Quantum Cryptography.* (Doubleday, New York, 1999).
[Sin05]   Singh S. *Big Bang. The Origin of the Universe* (Harper Collins, New York, 2005).
[Sob05]   Sobel D. *Longitude: The True Story of a Lone Genius Who Solved the Greatest Scientific Problem of His Time* (Macmillan, 2005).
[Sus08]   Susskind L. *The Black Hole War* (Little, Brown, New York, 2008).

The cartoons

Following the suggestion of Professor Dr. Markus Roth of the Technical University of Darmstadt, I have adorned the chapter-starting pages with cartoons I drew in 1957, in my first year as a PhD student at the Massachusetts Institute of Technology (MIT), just learning about photons. These cartoons, clipped and enlarged from a poster, give a light-hearted glimpse of the preparation for a scientific profession as seen by one who was just beginning that process. They refer to a time before iPhones and iPads were ubiquitous, indispensible companions in classrooms. The paragraphs below provide brief explanations for what experiences at MIT underlay the cartoons.

Chapter 1 Introduction: *"The first view of grad school is rosy."*
The fanciful telescope here is not something that would function for viewing anything, but that is part of the point being made: One should approach any new adventure with a view through "rose-colored glasses" that do not show the less happy aspects of reality. In my personal experience the anticipation overlooked many challenges associated with coming from a small liberal arts, very coeducational, football-and-party centered college in the center of California (College of the Pacific, founded in 1851, was very much an institution of its times, where the President and the Chancellor were old family friends), to a large, urban, male-dominated technical institute, swarming with nerdy students and famous faculty – and snow.

Chapter 2 Basic background: *"The qualifying exam is the start."*
Entering graduate students had evidently supplied records of academic studies that satisfied an admission committee but all were nonetheless required to take a series of examinations on topics that they needed to master, either from past courses or by retaking undergraduate courses. My preparation was woefully inadequate in Thermo-dynamics, and so that became one of my first-year courses.

Chapter 3 Photons of Einstein: *"The teaching assistant is hard to meet."*
Portions of the undergraduate instruction dealt with homework assignments and these could sometimes be perplexing. Available to help, in principle, were graduate students titled "teaching assistant" who carried out these duties to help pay their education expenses. Like everyone with any academic responsibility these fellows (and perhaps rarely a gal) had their own work to look after. I had an appointment as a "laboratory assistant", a title that required me to wander around through a freshman chemistry lab session and try to answer questions from students who knew more than I did.

Chapter 4 Photons of Dirac: *"The professor seems pretty lofty."*
My graduate school was part of one of the most prestigious academic institutions in the US, and a person had to be recognized as being very clever indeed to be awarded a faculty position. Professors tended to be aware of the grandness of their titles, and undergraduate students tended to think of these fellows (no women in those days) as being aware of their lofty position. To be fair, I found that even department heads, authors of definitive textbooks, to be quite willing to give me some of their time, and those professors that I had closer contact with were extraordinarily friendly and kind. My fellow grad students found the same. We formed ties with professors that endured for their lifetimes.

Chapter 5 Photons as population changers: *"Grades can seem arbitrary."*
This rather defensive view was taken, primarily by undergraduates, to excuse poor test performance. The graduate courses I took paid little attention to grades, and professors routinely gave "A" grades. It was expected that, at that stage of one's life, the major portion of education would not be from classroom lectures and assignments. That has certainly been my experience.

Chapter 6 Message photons: *"Sometimes the course load seems unending."*
Coming from a leisurely small-school environment to the world of high academic expectations, on topics for which my coursework in physics had not prepared me well, I struggled to master the seemingly unending flow of unfamiliar concepts.

Chapter 7 Manipulating photons: *"Midnight oil is for burning before exams."*
Even in the more relaxed approach of my undergraduate college days, it was common for students to postpone all serious study until the night before an exam. I have seen that behavior in my own children, and I think it will remain a staple activity even as the internet replaces personal instruction.

Chapter 8 Summary: *"The major exam is the big one."*
At some point during the graduate years it was necessary to demonstrate to the departmental faculty that one had mastered enough of the basic subject matter of the discipline to be allowed to proceed to do the research that would result in a thesis and a PhD degree. This exam was pretty daunting.

Chapter 9 Finale: *"The oral exam can be stressful."*
All of us who progressed to writing a thesis were subjected to an oral exam, in which we stood before faculty members and answered whatever questions they wished to pose. This ordeal is common throughout academia in the US and Europe. In later years I have "sat in" on quite a few, mostly as a spectator.

Appendix A Atoms: *"Experimental apparatus can be mastered ... I was told."*
My thesis work, proposed in collaboration with a couple of professors, entailed what is known as "experimental" work with hardware, as contrasted with "theoretical" work with equation manipulation (pencil and paper in my day). I had to learn how to use a variety of devices, as small as Geiger counters and as large as a cyclotron. To some extent I did master their operation, and to a lesser extent I learned how to make minor repairs and how to find someone who could make larger repairs.

Appendix B Radiation: *"The toil of research can seem a swamp."*
This cartoon incorporates characters from a number of comic strips, including the Neanderthallic caveman Alley Oop of V.T. Hamlin, whose artwork was outstanding, the Pogo Possum of liberal cartoonist Walt Kelley, known for remarking "We have met the enemy and he is us".

Appendix C Coupled equations: *"The lack of research results may seem a desolate landscape."*
Well, the object of thesis work is to get some results, usually some numbers that satisfy some criteria (repeatability being a common one), and for an experimentalist this means that all of the equipment has to work for a suitable length of time. Until that happens one really can feel rather desolated.

Coda: *"The annual Xmas greetings."*
Four cartoons sent as Christmas cards give a fair picture of my changing outlook on life as a graduate student.

End page: *"The ultimate goal."*
The goal of all of this long effort is a PhD degree, granted over the signature of a faculty advisor, approved by the institution, and handed to the awardee at a suitable ceremony. Thereafter my mother took great pride in introducing me as "My son, *Doctor* Shore."

Contents

1

Introduction

The first view of grad school is rosy.

Among the questions that first began my own personal interest in the inanimate world of Nature were some very simple ones: What are atoms? What are electrons? What are photons? Those names have appeared for decades, even before my long life began, as part of the public face of Science. Today one hears of *Atomic* Energy (really meaning *nuclear* energy) as a possible route to continuing our national prosperity or *Atomic* Bombs (nuclear weapons) as an intended deterrent of national enemies. One reads of *Electronics*, and of professional societies whose very name incorporates that word. And there are organizations, magazines, conferences, and investment opportunities that are said to deal with *Photonics* (meaning the physics of light generation, transmission, manipulation, and detection; see page 6). Clearly atoms, electrons, and photons are part of contemporary culture, as is Energy, and so a modicum of understanding could be considered part of an education. Well might one ask: What are the entities that have given their names—atoms, electrons, photons—to this vast portion of an even vaster scientific enterprise?

These questions have led me, through studies of Chemistry and Physics and Applied Mathematics, to a succession of rewarding careers as a working Scientist. To some extent I have found answers that have satisfied my curiosity and have provided what I considered a consistent presentation of a very small part of the physical world, namely the nature of atoms (elementary increments of matter) and electrons (elementary increments of electricity). But the nature of photons (possible elementary increments of radiation) remains for me—and other physicists with whom I have spoken—something that has led to an ongoing, and still incomplete, path of discovery. This Memoir records changing views of the physics community about possible answers to the seemingly simple question, asked by many others [Scu72; Mar96b; Fin03; Mut03; Lou03; Ran05; Ben16]:

"What is a photon?"

Our Changing Views of Photons: A Tutorial Memoir. Bruce W. Shore, Oxford University Press (2020). © Bruce W. Shore.
DOI: 10.1093/oso/9780198862857.003.0001

Writing in a section with that title in a comprehensive 1990 text [Sho90] I began by saying (as quoted in [Bia06])

> The notion of [a] photon as a quantum of electromagnetic energy, like the notion of an atom as an elementary unit of matter, permits a number of interpretations.

The present Memoir revisits the answers given then and notes a variety of possible additional responses not anticipated at that time.

Foretaste. In describing phenomena of photons and quanta to nonexperts, physicists have, at times, built upon the following statements:

- *A packet of electromagnetic field having energy $h\nu$ is a <u>single photon</u>.*
- *All light is <u>photons</u>.*
- *At some basic level all electromagnetism is <u>quantum</u> electrodynamics (QED).*
- *Those aspects of light that can be made evident with tools available in the nineteenth century, and could have been understood then, is <u>classical</u>.*

In the course of this Memoir I will be examining the implications of such assertions, their limitations, and alternative views expressed by physicists.

1.1   Overview of the Memoir narrative

As the preface explained, this book has two parts: First, a conversational narrative (a Memoir) that is largely devoid of equations and technical details; Second, appendices that offer a tutorial on the mathematics and physics of quantum theory as it applies to the photons discussed in the narrative portion of the book.

This introductory chapter sets the stage for the narrative discussion by commenting on several basic concepts that are often taken for granted in expositions, but which I think are useful to make clear at the start of an intended nontechnical presentation of what is often regarded as a rather technical subject. It sets ground rules, so to speak, for the actual discussion that begins with the fundamentals, and the definitions of technical terms, to be presented in Chapter 2 and which then continues with the systematic appearance of photons in subsequent chapters.

The concept of a photon has undergone a notable evolution during the last century— even the last few decades. Like many others of my generation, I was introduced to the notion of "photons" as little packets of radiation that showered globs of energy onto atoms and molecules—a viewpoint that was associated with the names of Max Planck and Albert Einstein amongst others (see Chapter 3). I learned that those early researchers referred to these increments as radiation "quanta" and that the term "photon" appeared only much later – see page 62. When I gained needed mathematical tools I appreciated the view of Paul Dirac, who presented photons as incremental amounts of solutions to James Clerk Maxwell's equations of electromagnetism (see Chapter 4 and Appendix B).

This Memoir presents my own perspective on the changing views of photons that began in the research community of those days, and which continues today.

Well-established textbook approaches sufficed to describe radiation beams and to define traveling photons for generations of physicists—starting with Einstein. However, the quantum nature of the electromagnetic field, and its photons, were not immediately required in any of my work as a researcher in atomic physics or in the Laser Program at the Lawrence Livermore National Laboratory (LLNL), but the associated coherent excitation of atoms [Sho90; Sho11] sketched in Chapter 5, has been a foundation for the subsequent portion of Quantum Optics in which concepts of photons are essential. Notably, radiation confined by walls (cavity photons) offers alternative, definitive experimental evidence of radiation discreteness. Chapter 5 begins a discussion of that line of study.

I have learned how the photon, as a carrier of quantum information, has behavior that offers new protocols for such practical enterprises as Quantum Computing and secure communication; see Chapter 6. And advancing experimental techniques are allowing physical modifications of the physical entities referred to by experimenters as "photons"—changing their colors and their pulse shapes, even stopping and restoring them—all the while retaining their quantum heritage; see Chapter 7. These techniques have prompted those who carry out photon manipulation (and me) to re-examine how best, and most usefully, to define observable characteristics of radiation that mark particular samples as being constructed from photons—to be regarded as a quantum field rather than a classical field.

What was, in my youth, an esoteric academic discipline of Quantum Theory (later specialized as Quantum Optics) is beginning to prosper in more practical ways indexed by Google as Quantum Engineering and Quantum Technology. In such endeavors the quantum properties of photons have become matters of active research, by theorists and experimentalists alike. Increasingly, techniques of nonlinear quantum optics are making possible experiments in which "photons" within matter interact strongly, to form a many-body system of a sort hitherto limited to assemblages of electrons and ions. Chapter 8 presents an overview of some of the varied contemporary works in which the notion of photon makes an appearance.

It is that newer world of theory and experiment, well beyond what appears in my earlier book [Sho90], that has come to my attention in recent years, replacing the seemingly paradoxical historical world of textbook wave-particle photons, an ongoing journey of discovery that the present Memoir describes – a continuing, and evolving, personal interest in the long-standing question: "What is a photon?" I shall note some of the answers that physicists have given, and some of the recent technology that has made possible manipulations of single photons, thereby forcing a refinement of older, established answers to my question. Toward the end the question becomes: "What notions of photon are of use in experimental observations and established theory?"

References. The resources of internet-based Wikipedia offer much explanation for the technical terms I shall use. It is generally a good starting point. The twenty-five pages devoted by Wikipedia to "Photon" provide much discussion of both the history of the photon concept (what I am calling the photons of Planck, Einstein, and Bohr in Chapter 3) and the contemporary notions of photon as a gauge field associated with the Standard Model of particle physics (summarized in Appendix A.19). Wikipedia also offers some useful links to published material, as does Google Scholar. Nonetheless, as is customary in scholarly writings, I have included citations to many articles as insets, such as [Sho90; Sho11], that refer to alphabetized keys in the bibliography at the end of this Memoir . The relevant literature is far too vast to list more than a small portion. In my selection I have looked particularly for reviews and tutorials. For those things I have omitted, I would quote the words of Samuel Johnson in the preface of his great dictionary of 1755 [Joh55]:

> I may freely, without shame, leave some obscurities to happier industry, or future information.

1.2 Preliminaries: Defining terms

Definitions. Adults I recall, who hoped to diminish semantic-based quarrels amongst siblings, often relied on the mantra "define your terms". I think this is good advice when one starts along any journey of enquiry, because even commonplace words have a variety of meanings. The very simple word "up", for example, is famously used in such diverse expressions as "add up", "build up", "call up", "draw up", "end up", . . ., "wrap up", "zip up", and numerous others. The word "field" has different meanings to a farmer (or urban planner), to a baseball player, to a data-processor, to a biographer, or to an electrical engineer (or physicist). In the present book the technical word "polarization" must, unavoidably, appear with several different uses; Appendix B.1.7 presents the all-encompassing definition noted in the article [Ebe16b]. Certainly the word "photon" has been used with many meanings by different people.

Acknowledging this ambiguity property of language, I have tried to offer elementary definitions, with enlightening discussion, for the technical terms that must inevitably appear in discussions of technical concepts—of which the notion of photon is but one. In so doing I try to keep in mind the pragmatic notion of *"operational definition"*, associated with Percy Bridgman, in which the practitioner looks for some possible idealized measurement whose results define a technical word.[1] Notions of Epistemology (the theory of knowledge) and Ontology (the nature of being) and "objective reality" and whether tables and chairs (or electrons [Hou66; Fra97] or numbers [Cle16]) are "real", these I leave to philosophers.

Science. I was taught in grade school that Science is organized knowledge. True, but the knowledge base of Science is reproducible data, numbers obtained under controlled conditions. A list of aphorisms may be amusing, but it is not Science. Collected anecdotes may be Entertainment or Literature, but without controlled reproducibility they are not Science.[2]

The scientific method. To the goal of understanding the inanimate world, I aim to bring an approach patterned after the Scientific Method taught in schools to generations of pupils—procedures that involve not only logical *consistency* but *falsifiable* hypotheses and their experimental testing as well as exploratory *observations*. I like a comment made by neurologist and writer Oliver Sacks in a *New Yorker* note [Sac19]

---

[1] An initial start toward an operational definition of "light" might be that it is what comes from the sun and produces images in cameras. To ascribe an electromagnetic character to light requires further definition of waves and electrical phenomena. The hierarchy of linked operational definitions is seemingly endless.

[2] I acknowledge omitting from Science the many academic categories that organize study of human institutions, departmentalized under the title of Social Science.

[*Science*] moves cautiously and slowly, its insights checked by continual self-testing and experimentation.

Science as a puzzle. My leisure reading of 1920s-era English murder-mysteries ("whodunnits") has shown many examples in which a variety of ending scenarios seem to fit the assembled clues, until in the final paragraphs the sleuth explains how only one of these fits *all* of the clues. The pursuit of science, and the construction of models and viewpoints, might seem to have much in common with those narratives. However, although the experimental clues of science are never without error, the agreement between experiment and theory provided by quantum electrodynamics—the theory of electrons and photons—now routinely fits to more than nine digits, and so it is only the interpretation of these facts that has room for dispute.

Physics. More particularly, I approach scientific questions by way of Physics. For me that means setting aside questions of Value or Cost and Benefits or Elegance and Beauty [Hos18]—or politics or divinity or fantasy imagination—and dealing with *Concepts*—models of inanimate nature—that can be translated into mathematical terms—into tractable equations (see Section 2.1.2). These equations, in turn, allow one to obtain mathematical formulas with which to predict numerical values that are expected to result from particular measurements (*expectation values*). The divisions of traditional experimental physics that have most relevance to this Memoir are atomic, molecular, and optical physics, omitting astrophysics, biophysics, geophysics, and many other divisions that were absent from my schooling. Although I have myself carried out few experiments since my graduate-student days, I remain a confirmed believer in the need to tie theories to experiments. Logical consistency is necessary but not sufficient for a concept to be useful.

Photonics. The primary subject matter presented here could well be called "Photonics", a term that came into vogue in the 1960s, with the development of lasers, optical fibers, and semiconductor optical devices, as an optical analogy with the term "Electronics" used to express the controlled flow of electrical charge. It is now widely used to describe a range of topics that involve the controlled emission, transmission, amplification, detection, and modulation of light, with or without the requirement of discrete increments (photons). Its broad extent is well covered in the comprehensive textbook *Fundamentals of Photonics* now in its third edition [Sal19]. Photonics incorporates such subjects as optoelectronics, quantum electronics, and quantum optics. Lengthy reviews of some of these topics appear in the journals *Nature Photonics* and *Laser Focus World*. Subdivisions associated with particular applications now include healthcare photonics, biophotonics, nanophotonics, and astrophotonics (and even wearable photonics), each with an increasing literature.

The existence of the publication *Optics and Photonics News* and the peer-reviewed journal *Advances in Optics and Photonics* suggests that editors now distinguish between the categories of optics and photonics, although the articles appearing in those publications continue to appear in *Journal of Optics*, *Journal of Modern Optics*, *Optics Express*, and the serial *Progress in Optics*. But "The Optical Society" has yet to become "The Optical and Photonical Society".

Science and Technology. The result of a scientific enterprise is often an insight or a tool for further use. From scientific activity comes a variety of techniques and skills that have practical use in the production of goods and services—in *Technology*. Science and Technology are intimately connected but are at times distinguished, as in the institutional name Imperial College of Science, Technology, and Medicine, London.

Science and Engineering. In pursuing scientific activities a practitioner is traditionally aiming to determine the quantitative nature of observable phenomena, perhaps to test a hypothesis. This knowledge is put to practical use in *Engineering*, where the objective is customarily that of meeting some practical goal (say, the construction of a bridge or a gadget) with least expense. The notion of *optimization*, which is central to the work of an engineer, is less common in Science; but see the mention of optimal control in Section 5.8.

Science and Art. Physicists and others who engage in science look for organizable regularities, patterns that can be understood by simple rules—relationships often termed *laws of nature*—that allow reliable predictions and, in turn, practical application of their science. Artists may well be conscious of patterns—of sights and sounds—but Originality is typically more important than Reproducibility in the enterprises of Art. Success in both artistic and scientific endeavors is enhanced by appropriate aptitudes (and by good luck in choosing projects).

Alternatives to Science. Useful though Science may be, it is only one way to understand the world in which we find ourselves. Since the earliest days of recorded civilization people have found intellectually fulfilling paths toward contentment, or at least some reconciliation, with that world, some of which have involved Spirituality and Religion [Dys06]. Because these alternatives rest on different foundations than does Science as defined above, they need not conflict with the enterprise of Science that I shall be discussing.

Mathematics. Photons are part of physics, and physics relies on mathematics, so any understanding of photons must include some understanding of Mathematics; see Section 2.1 and Appendix A. But math and physics, though linked, do not stand on the same foundations. Mathematicians are free to propose any set of axioms, any rules for their symbols, from which they then deduce theorems and corollaries and lemmas—and display their ingenuity. It is only necessary that their structures—the rules of algebras and groups and so on—be free of hindering contradictions. By contrast, physics and other sciences are expected to conform to observations of nature, not just to be consistent within their mathematics. Sometimes mathematics and science converge (as will be true of what math I discuss), but this is not a necessary condition.

Simulation. In my student days the work in physical sciences—physics, chemistry, astronomy, and the like—fell into two categories[3]. First, there was the use (and construction and repair) of apparatus to make observations—to read dials or record oscilloscope traces or measure photographic images. Second, there was the use (and construction) of theory to treat observations—to deal with equations and their solutions. Academic

---

[3]Sometimes regarded as "practical know-how" and "book learning".

physicists typically identified themselves accordingly as *experimentalists* (using apparatus) or *theorists* (using equations). With the subsequent availability of electronic computers a third category became important: The *simulation* of nature through the automated construction of solutions to equations, usually differential equations (see Appendix A.1.2). It is the unification of these three approaches to science—theory, experiment, and simulation—that now engages researchers and writers about research.

The use of automatable computers has now largely displaced the laborious task of finding algebraic expressions for solutions to the differential equations that are intended to describe particular experimental or observational situations. The simulations involve several parts: The identification of particular equations; the design of algorithms and procedures to solve the equations (applied mathematics and mathematical physics); and the specification of initial conditions that pick out a particular case of the generality of all possible solutions to the given equations. If the assumed initial conditions—or the assumed equations of motion—are not a good fit to the actual portion of nature being treated, then the automated solution cannot be expected to agree with any observations. The need for good numerical input to the computer is encompassed in the oft-quoted admonition: *"garbage in, garbage out"* (GIGO).

Ambiguity. Although Science is widely regarded as being precise and exact, the language with which we must explain it is not. The English language is valued and praised for its ability to draw upon its roots in many past civilizations to create, with its epigrams, metaphors, aphorisms, and puns, an enormous variety of mental images, thereby bringing intellectual enlightenment, moral instruction, and even amusement to listeners and readers. When used in the presentation of scientific results the inherent ambiguity, praised by faculty in English Departments, remains. This is particularly evident in the numerous attempts to present the seemingly conflicting concepts of waves and particles; see Section 3.7. Resolving the difficulty requires careful attention to what the words "wave" and "particle" are to mean; see Section 2.7.

1.3 Models of physical phenomena

Observables. In the view presented above, Physics has the task of providing models suitable to particular sets of *observables*—things that can be measured and recorded as numbers. The diversity of phenomena of interest, and their observables, leads to a variety of rationales for the models. This variety, and the consequent equally valid theoretical constructs, lies at the very heart of modern physics – and the framework for understanding photons.

Refutable. Philosophers of Science have claimed that although there may be a long list of successful demonstrations of any scientific theory, a single experiment that does not fit the theory is sufficient to justify discarding it. These intellectuals therefore regard Science as built upon offerings of testable and falsifiable hypotheses, each capable of being refuted by an appropriate experiment [Gar01].

While this philosophy has merit, and underlies important tests of special relativity, general relativity, and quantum theory, the theory of photons discussed in this Memoir displays, with its wave-particle duality, examples where testability and refutability must be treated within a specific context. The words of Freeman Dyson, in his presentation of quantum electrodynamics [Dys49], are appropriate:

> ...the present treatment should be regarded as justified by its success in applications rather than by its theoretical derivation.

With this in mind, one needs to consider other criteria for judging a theory, notably contributors to its useful explanatory power.

Specialized. Just as a skilled craftsman has specialized tools for different purposes, using a hammer or a wrench depending on whether the fastening of objects is by nails or bolts, so too do physicists have different models for different phenomena.

For example: To model the effects of reservoir water flowing through outlet pipes or over spillways; To describe the flow of air over airplane wings, generating the lift needed for controlled flight; To predict the paths of explosion debris of a computer-generated entertainment; To design mirrors and lenses and optical fibers, and devise ways of communicating information between locations using beams of light; To design machines that carry goods and people or merely amuse them; and to the multitude of uses that continue to be found for laser-generated radiation in homes and laboratories.

Each of these very commonplace but economically important activities requires a different model of Nature and a different set of equations to be solved for different experimental observables. One might expect that hydrodynamic equations for use in modeling fluid flow would have little to say about the optical transparency of glass, but in fact there are many useful connections between the various models of nature, and a relatively few basic equations are to be found in all areas of everyday physics, cf. [Mor53; Men61].

Fundamental. Although atoms and electrons (and photons) are fundamental entities for some of the physics of everyday life, there has been a community of physicists who sought theories of underlying structures for all matter: Grand Unified Theories of Everything.[4] Whatever may be the future course of such researches, and of such concepts as quarks and gluons, of strings and branes, of dark matter and dark energy, whatever may be the observables for which such work ultimately offers quantitative predictions, the physics of atoms, electrons, and photons seen in most university and corporate laboratories proceeds independently, with equations that were completed nearly a century ago and experiments that continue to push technology and offer economic benefit. It is that limited portion of physics—Quantum Physics and Quantum Optics—to which this manuscript gives most attention, and in which my understanding of photons primarily takes place.

If one wishes to understand trees as biological objects then it is useful not only to look out over an unresolved forest but to approach individual examples. Similarly, for an understanding of photons as discrete entities it is desirable to supplement observations of vast crowds of photons with experiments that can examine their properties when found in small numbers, perhaps even as single photons. One must look for

---

[4]Meaning everything *inanimate*. Understanding of Life and Behavior lies outside this quest. But even without quarks, gluons and gravitons, work on Artificial Intelligence (AI) progresses towards simulation of human behavior.

characteristics of photons as individuals in weak samples of radiation, rather than industrial-laser cutting and welding or the extremely powerful multibeam laser systems constructed for research at National Laboratories and international collaborative facilities.

Systems and parts.   It often happens that a complicated object, say a self-driving automobile traveling along an interstate highway, can be regarded as a collection of independent parts—of distinct *Systems*—whose mechanical behavior can be treated separately, at least to a first approximation. The entire automobile as a whole has mass and kinetic energy of motion, is subject to air resistance and is powered by a motor (or motors) acting through wheels upon the surface of the roadway to provide propulsion. But inside the body of the auto are various parts that can be treated largely independently of the overall motion. The engine of a gasoline-powered car comes at once to mind, along with the power supplies. And notably there are the passengers. With windows closed they have negligible effect upon the vehicle performance, and their interactions amongst themselves are best treated, not by skills of engineering but by mastery of psychology. The passengers and the vehicle are treatable as separate (but interacting) systems.

So too are molecules and their constituent electrons and nuclei (to be defined in Section 2.2). Molecules are acknowledged to be comprised of distinct charged particles but these remain bound together, as a System. Their overall motion, expressed as that of a center of mass and an angular momentum (see Appendix A.1.3), is largely independent of the passenger-charges that move together as an aggregate. It is with the internal structure of composites such as atoms and molecules that the concept of photon has closest connection—changes to that structure can produce, or be altered by, radiation (see Section 2.8).

It is interesting to note that design issues associated with automobiles involve (at least) two often-conflicting viewpoints: That of the stylist and that of the engineer. In the physics world associated with radiation there is a similar viewpoint separation, of waves and particles as pictures for photons, a duality theme that appears throughout this Memoir (see Section 3.7).

But just as the automobile surroundings must be considered by its designers—the presence of other cars, the irregularities of the road surface, and the weather—so too must the surroundings of a molecule be considered for reliable detailed modeling of its behavior (see Section 5.1.3).

Simplicity. Skillful physicists often introduce models, and their associated equations, that are designed to match particular observations and to be no more complicated than is needed for that purpose. The fact that water is primarily molecules of an oxygen atom bound to two hydrogen atoms is irrelevant for the equations of hydrodynamics that are used to model fluid flows, although it undoubtedly has some relevance to modeling osmotic pressure and capillary action that raises water to the tops of Sequoia trees. It often happens that a variety of explanations are suggested for some natural phenomena thought to be governed by general rules propounded in physics and chemistry. To select amongst contending views it is often suggested that one should choose the explanation, consistent with existing data, that is simplest and requires the least elaboration of existing theories, a concept that is attributed [says Wikipedia], under a variety of labels (The Rule of Simplicity, the Principle of Parsimony, Occam's [sic] Razor), and to a variety of sources including Aristotle, Ptolemy, Thomas Aquinas, and William of Ockham [sic], as well as to the founders of quantum theory and numerous twentieth-century philosophers. Not that the involvement of fairies and leprechauns[5] is demonstrably wrong, just that they are unnecessary (and unreliable as forecasters).

Toy models. Many forms of model building have their uses, and their limits. At one extreme are the attempts to include all possible details, of wind resistance and friction and distributions of loosely held parts subject to heating and cooling. Such models require extensive sets of equations and supplementary data, and are intended to give very accurate predictions of complicated behavior. They are extensively used by engineers and planners.

At the other extreme are "toy models" based on extracting what are regarded as only those characteristics of a system that are most essential for some purpose. Frictionless planes, rigid bodies, point particles, and stationary, isolated two-state atoms are found there, with correspondingly simple equations that permit simple solutions. From such models one intends to gain intuition about simple aspects of complicated behavior. This Memoir offers numerous examples.

Quantifying physics. Words like "large" and "small" for lengths or "high" and "low" for intensity and strength, or "accurately" and "rapidly" all appear frequently in sciences. Their precise meaning varies with the particular objects subjected to study and with the tools available for measurement. Physicists are well known for choosing their measurement scales to produce numbers that do not differ much from unity.

Quantum physics. I have observed that, to a great many adults, the adjective Quantum brings even more apprehension than the noun Physics, and an even more overwhelming desire to be absent from exposure. But I have also observed that much of the content of quantum physics, though requiring some attention, involves little more than a shift in how our world is to be regarded and described. It requires nothing more than willingness to listen or read in order to understand how information we daily process can be regarded as examples of abstract vectors, as Section 2.5.2 and Appendix A.3 explain. It is only a small step further to bring the essential aspects of quantum physics into the conversation. A goal of the present Memoir is to make photons and atoms seem subjects that need not be dreaded by those who find real estate or stats of professional sports teams or English murder mysteries or even the weather to be topics worthy of ongoing conversation. These require some brain capacity but primarily they require the choice to take an interest. So it is also with quantum physics.

Quantum vs. classical. It is generally recognized by scientists—by physicists in particular—that the collection of rules and equations, with interpretation, that has been known since the 1920s as Quantum Theory, provides the basic formalism with which to describe the behavior of such things as atoms and inanimate matter in general. These

---

[5]Terms such as ghost image in optics [Str95; Erk10] and magic nucleon number [Bla52; Eva55], though using words that bring to mind fantasy fictions, are firmly based within physics.

are governed, at some microscopic scale, by Quantum Mechanics and its uncertainty principle (see Appendix A.4). It is therefore to be expected that radiation, being electromagnetic fields generated by atoms, must also, at some level, be governed by quantum theory. That view appears in much of the early twentieth-century discussion of radiation quanta (see Chapter 3).

However, each individual quantum system, each atom and electron, each molecule, that in an aggregate make up bulk matter, interacts with an environment that is not entirely controllable; see Section 5.1. Neighboring atoms bounce around with random, thermal motions that can only be followed statistically, on average. When such randomizing influences dominate the experience of every constituent of a quantum system it will suffice to use a description that makes almost no use of purely quantum characteristics. Under those conditions the matter, though constructed from atoms and electrons, exhibits properties that would have been familiar to nineteenth-century physicists, and much of the content of their textbooks requires no revision. Those earlier theories of matter and its motions—of machines and steam power, of motions of heavenly bodies and images formed by lenses and mirrors—has become known as *Classical Theory*.[6]

To observe non-classical properties (what one would call quantum properties) it is necessary that uncontrollable environmental influences (termed *decoherence*) be negligible; see Section 2.8.1. When that is the case, and we deal with a *coherent* system (of atoms and radiation) [Sho90; Sho13; Str17], then we should expect to model the behavior with quantum mechanics, and we should expect such modeling will suggest observable system characteristics that differ from what classical theory would predict. Appendix A.4 presents the essence of the needed modifications to classical theory.

In thinking about photons one needs to question how one can distinguish specifically quantum properties from the behavior that is expected to occur when quantum particles are seen in aggregates: What observable characteristics require quantum theory for explanation, and for which ones does classical theory suffice? It is with this view of the quantum-classical divide that we should consider such things as photons and single-photon quantum states. Appendix A.4 introduces the mathematical formalism that is commonly taken to provide a satisfactory answer.

Objective reality (OR). A reader of press reports will occasionally encounter mention of "*objective reality*" (OR) in connection with photons. To speak of OR is to consider objects that people report having recognized and handled and which therefore, in some sense, can be considered "real" even when nobody is actually touching or viewing them. It means (as the Walrus said to the Carpenter [Car71]),

> ... to talk of many things: Of shoes—and ships—and sealing wax, of cabbages and kings ...

But painters famously gloat over clouds that pilots of planes see as fog, and sommeliers gush over transitory olfactory sensations that novices hardly detect. Even a very solid elephant, when first encountered by a sight-impaired person, will seem a very different thing if the first touches are of the trunk or a tusk or an ear or a leg or the tail.

---

[6]According to Roy Glauber [Gla06] the word "classical" in physics means "Anything that we ... could have understood prior to the date of Planck's paper [on quanta], December 14, 1900".

What matters within the view of physics that I am following is our mental organization of observations into models that allow us to predict behavior—of inanimate objects or of institutions and social constructs. And it is well to agree on what observations we are including in our modeling—to rely upon operational definitions of our terms. Results of experiment can seem to be different from—at odds with—what particular views of OR seem to predict. The conflict between wavelike and particulate nature of light was an early example; see Section 3.7. The properties of correlations and entanglement amongst degrees of freedom offer newer examples; see Section 6.7 and Appendix C.8.5. Various discussions in this Memoir aim at offering theoretical approaches to reconciling these conflicting intuitions.

Special relativity. The theory of special relativity, with its curious predictions of time dilation and Fitzgerald contraction for moving but unaccelerated travelers and its use of intertwined four-dimensional space-time coordinates for Lorentz transformations between reference frames (see Section 8.3.1), is a purely classical theory, making no use of such concepts as uncertainty relations, expressed by eqn (2.12-1) or eqn (A.4-6), that are the hallmark of quantum theory. Although the quantum theory of atomic structure requires, for accurate results, the relativistic theory of electrons devised by Dirac [Dir28; Jau55; Fes58], much of atomic physics is suitably treated with nonrelativistic approximations—the assumption that all of the particles are moving much slower than the speed of light. Although they can also be considered separately, the two theories, special relativity and quantum theory, harmonize perfectly[7] in the conventional, fully-relativistic formalism of Quantum Electrodynamics (QED) [Dys49; Pow64; Bia75; Coh97].

Special relativity must be part of any quantitative description of reference frames that move with speed that is an appreciable fraction of the light speed c, as happens for any fast particle. It is therefore integral to the Standard Model of particle physics outlined in Appendix A.19. A free-space electromagnetic field moves always at this speed and so it is possible to regard a photon as a fully relativistic zero-mass particle; see Section 8.3.3. Much that is associated with the concept of photons, and with investigation of quantum optics and atomic physics, does not require the formalism of special relativity theory. However, there are a few places where the space-time symmetry of special relativity make explicit contact with notions of photons, most notably in group theoretical approaches that have become increasingly used; see Appendix A.18.3.

1.4 Caveats

The principal objective of this Memoir is to offer views of the physics of photons and how that landscape continues to change. I do not propose this as either a definitive technical treatise nor as a complete historical record.—it is a Memoir. Caveats for the reader who seeks those are therefore appropriate.

---

[7]It is with general relativity—quantum gravity—that quantum theory has significant challenges to be fitted [Sus08; Wil15].

Restrictions. At the outset I should mention that the energy quanta I have in mind are the sorts that are routinely used in university research labs and profit-making companies, based primarily upon microwave or optical radiation from masers and lasers, or are found in household heat sources—energy increments that are sufficient to give welcoming warmth or break apart atoms and molecules, even atomic nuclei, but are far from the astronomical and cosmological limit of destroying structures as large as a castle or a planet. I shall not be concerned with the photons associated with quantum gravity and general relativity [Roc13; Pla16], concepts discussed by cosmologists such as Stephen Hawking [Sus08], nor with the photons that appear along with other gauge bosons, leptons and quarks as constituents of unifying models of elementary particles and non-gravitational forces; Appendix A.19 offers an overview of the photons of that enterprise.

Science history. My own interests, to be presented in this Memoir , require some very specific equations, developed along with understanding and modeling over many decades—with origins from several centuries—and with contributions from many physicists. I have met a few of them who were active during my lifetime, and I have benefitted not only from their writings and public talks but also from conversations.

It is not my intention to provide a history of a portion of physical science, nor to trace the rather erratic course of discovery that led to our present understanding.[8] Historians of physics have done this very well, and have pointed out the mistakes to be found in textbook histories [Kra92]. But it is inevitable that some names of scientists appear in any discussion of science. Isaac Newton, Max Planck, Albert Einstein, Niels Bohr, Erwin Schrödinger, Werner Heisenberg and Paul Dirac are names, now associated indelibly with particular equations, that cannot be avoided in discussing the atomic physics that will be presented here. However, even a cursory delve into the subject will present names of others, not (yet) winners of the famous Nobel Prize, whose contributions have been of great importance to our present understanding—some from novel discoveries, others for providing insights. It would be impossible to list and credit them all.

There are two major difficulties with citing "original works". One is the often-challenging problem of assigning credit,[9] a problem I touch upon in mentioning how the name "photon" arose; see page 62. Historians demonstrably struggle over this.The discovery of electromagnetic radiation, and its application to communication, is an excellent example of the disputes over priorities, with dozens, if not hundreds, of justified claimants in addition to Heinrich Hertz and Guglielmo Marconi [Rab16]. There are difficulties in pinning down even the coining of the names electron, proton, and neutron and of collective excitations such as exciton, plasmon, phonon, etc. etc. [Wal70].

The second is that many of the sources that are traditionally cited are in journals that no longer exist, or are not readily available to the general public (see, for example, [Ste92]), or are written in a language (Russian, German, French, Italian, Persian) that

[8]Biagio Buonaural and Giuseppe Giuliani [Buo16] have written a fascinating history of the early years of photons, and the conflicting wave and particle interpretation, giving details of the work and reasoning of the scientists involved.

[9]See "List of scientific priority disputes" and "Nobel Prize controversies" in Wikipedia.

readers of this Memoir portion often do not know well enough to benefit from reading the original work. The articles certainly hold no interest for a lay person reading this Memoir. For purposes of a tutorial a reader is quite often better served by Wikipedia or by a reference to a good review article or textbook [McC67; Pai82; Kra02; Isa07].

On names and nomenclature.   In presenting ideas that rest on scientific discoveries I have followed the long-standing custom of assigning names of individuals to various physics concepts.[10] This Memoir is a story of ideas, not a resource of scientific biographies, and so those names are chosen for convenience in accord with established tradition, not always with historical justification. Even a brief reading of biographical literature will reveal remarkable stories of publications overlooked, of collaborations ignored, of credit wrongly assigned [Jac01; Ber05], of prizes unjustly denied, of recognition bypassed because of gender, and of glory given to men whose political views have become distasteful to many [Leo04; Bal14]. And I have learned that in physics, as in other endeavors, fame and celebrity are not reliable measures of accomplishment. Never mind. Those are stories for another occasion.

On wrong turns.   As a student I was often impatient during classroom expositions that followed a historical route and that noted various wrong turns taken by earlier chemists and physicists—for example the notion of phlogiston to explain combustion or the ether as the invisible carrier of light or the watermelon-seed model of atoms proposed by Joseph John ("J. J.") Thomson or the superseded quantization rules of the Old Quantum Theory of Bohr and Arnold Sommerfeld that preceded the present Quantum Theory as described in Appendix A.4. But now, figuratively donning a pedagogue hat, I find it impossible to avoid allusions to past concepts that, to some extent, remain part of commonly held views—the folklore of contemporary physics. Admittedly I cannot see how to present concepts of photons without acknowledging the notions of Planck and Bohr and Einstein (see Chapter 3) which, although somewhat outmoded, and even inconsistent with contemporary understanding, remain an inescapable part of what is commonly meant by photon. On reviewing my exposition, I understand why school curricula choose to repeatedly take up particular topics, giving increased clarification and detail with each visit. Repetition can enhance understanding. And it is always instructive to learn from mistakes—of others, if possible.

With those excuses I begin, in Chapter 2, by examining (and defining) some of the simple concepts that define those portions of mathematics and theoretical physics that have relevance to the notion of photons. The photons themselves arrive in Chapter 3.

---

[10]Equations and formulas, some of which are named after individuals, are quite commonly referred to as "laws" because they allow exact prediction of equation-related quantities. Thus one reads of Boltzmann's distribution law for energies, Coulomb's law of electrostatics, Newton's laws of motion, Planck's radiation law, the exponential decay law, the ideal-gas law, and innumerable others.

2

Basic background: Everyday physics and its math

The qualifying exam is the start.

A few concepts found in high-school courses of math and physics are probably indispensable in trying to explain how chemists and physicists have come to understand the inanimate objects with which they routinely deal in their professions. This chapter presents a brief overview of what I have found to be concepts of particular relevance to the stated quest for defining a photon.

Understanding of physics at any scale cannot proceed far without some supportive understanding of selected portions of mathematics (long held to be the language of science), primarily arithmetic and algebra, and so a brief review of portions of these disciplines appears in Section 2.1. (Appendix A.3 treats with more detail the portions of linear algebra and differential equations that are needed for putting quantum physics to work.)

Following this preliminary diversion, Section 2.2 overviews that portion of physics that became the fabric of my own professional career, dealing with the submicroscopic world of atoms and molecules. Regarded in the nineteenth century as somewhat hypothetical, these building blocks of matter are nowadays routinely isolated, and manipulated, as individuals. Complementing this discussion of particles, Section 2.3 presents a brief reminder of readily witnessed wave phenomena in bulk matter, to anticipate the discussion of electromagnetic waves in Section 2.8. Empty space—the vacuum—is not nowadays what it was imagined to be prior to the development of quantum theory. Section 2.4 notes the changed views.

Sections 2.5 - 2.6 present a qualitative introduction to some basic ideas of general physics: Forces, vectors, and energy. Section 2.7 then summarizes the (classical) equations with which one describes particle motions and electromagnetic wave phenomena. Section 2.8 begins a qualitative discussion of radiation, laying the groundwork for subsequent introduction of radiation quanta and photons in Appendix B.

Our Changing Views of Photons: A Tutorial Memoir. Bruce W. Shore, Oxford University Press (2020). © Bruce W. Shore.
DOI: 10.1093/oso/9780198862857.003.0002

A section on probability, 2.11, introduces notions that are essential for discussions of quantum theory. The concluding Section 2.12 introduces the notion of discrete quantum states, a concept that remains key to understanding both historical and contemporary photons.

2.1 Some mathematics

Mathematics appears in a variety of guises, not just as the memorization of arithmetic facts that once formed, together with reading and writing, the most essential core of children's education in the United States. And it is more than the geometry that fascinated the ancient Greeks and was taught, decades ago, to US high-school students. Following are some examples of other aspects of the vast subject of Mathematics, relevant to enquiry concerning photons.

> **Aside: Sets.** Mathematicians have attempted, with some success, to base their entire subject upon the notion of *set*, meaning a collection of distinct objects, tangible or imagined, whose defining characteristics allow them to be identifiable as being members (elements) of the collection [Dev12]. The objects might be as diverse as Platonic solids, odd integers, points in a plane, months whose names begin with J, or dining-room furniture. During the 1960s it was common for schools in the US to teach basic set theory to primary-grade students as part of the New Math. I missed that educational experiment, and my own children were not impressed. Consequently the language of set theory is not prevalent in this Memoir. The abstract vectors of Section 2.5.2 and Appendix A.3 will be recognized as examples of sets.

2.1.1 Mathematics of events and actions

All readers know that the ordering of sentences A and B in a story can make a significant difference. Automobile passengers know that the command B ("turn left") given before command A ("turn right"), denoted symbolically AB in the customary right-to-left reading of quantum optics (and some languages), will often result in a different destination than the commands ordered as BA. Craftsmen know that the actions of measurement A and cut B should always be performed in a definite order, BAA for the admonition "measure twice and cut once".

The written description of actions, and the symbols that represent them as a sequence, requires a formalism in which, unlike the arithmetic of number multiplication 2×4 and 4×2, ordering is important: The symbolic elements do not *commute*, AB is not the same as BA. In addition to the world of literature and navigation and craftsmanship, the world of quantum mechanics requires such a formalism, a mathematics (an *algebra*) of noncommuting elements. Appendix A.18 defines the basic mathematics (*group theory*) of actions or *operations*. Appendix A.3 (*vector spaces*) presents the underlying mathematics of the entities that are *acted upon* in describing quantum-mechanical objects such as electrons and photons. Neither of these is necessary for understanding the main text of this Memoir; they are part of the mathematical background presented tutorially in the appendices.

2.1.2   Mathematics of numbers and equations

Mathematics is an unavoidable part of daily life for all but a few of us. It may be only the viewing of a calendar or a digital watch, it may be the paying of a supermarket cashier or the forming of a betting strategy for some professional sports event. Sciences, whether dealing with touchable objects or only with organized data, cannot exist without the mathematics of numbers [Cle16], and it is therefore reasonable to begin this scientific exposition with that start.

Numbers.   In infancy I probably became aware of whole numbers, what mathematicians call the *Natural Numbers*, 1,2,3, ..., (a set they denote by $\mathbb{N}$) and the fact that there is no upper bound on them—you can always find a larger one. Only later, being introduced to the arithmetic operations of addition and multiplication (defining for mathematicians an *algebra*, see Appendix A.17), did I encounter the notion of a *Prime Number*, meaning a natural number that could not be written as a product of other natural numbers (numbers that are not prime are *composite*), a property that has made them crucial for encoding messages [Sin99; Nie00; Bar09].

The notion of *integer* (a set denoted $\mathbb{Z}$) adds to $\mathbb{N}$ the very valuable element of zero, needed to specify that the vacuum has zero photons, and the negative integers -1,-2,-3, ..., needed as labels on various discretized quantities (quantum numbers).

> **Aside: Powers of 10.** In expressing very large or very small numbers it is common to incorporate powers of 10, as in 10^3 for 1,000 or 10^{-3} for 0.001. The integer *exponent n* in the expression 10^n defines the position of the decimal point, with 10^{-n} meaning $1/10^n$ and 10^0 being just 1. This notation is used also for symbols: Sec^{-1} means 1 / sec (or per second). Defined exponents, with the prefixes used with them, include:[1]

	Large		*Small*
1	deka, da	-1	deci, d
2	hecto, h	-2	centi, c
3	kilo, k	-3	milli, m
6	mega, M	-6	micro, μ
9	giga, G	-9	nano, n
12	tera, T	-12	pico, p
15	peta, P	-15	femto, f
18	exa, E	-18	atto, a
21	zetta, Z	-21	zepto, z
24	yotta, Y	-24	yocto, y

Enveloping those numbers comes the category of *Real Numbers* ($\mathbb{R}$), positive and negative and zero, unbounded and typically regarded by physicists as possible results of

---

[1] One attometer is the sensitivity of the LIGO gravitational-wave detector. One yottameter is about 10^8 light years, roughly the distance between some galaxies. The *atomic unit of time*, $a_0/\alpha c$, is about 24 attoseconds; see eqn (A.2-5) in Appendix A.2. The age of the earth is around 140 petaseconds.

measurements. Physicists and other scientists use numbers as measures of quantity—the *variables* with which science deals— and it is those that I shall discuss. Mathematicians, being unhindered by any need to relate their objects to those of the world around us, struggled well into the twentieth century to find a satisfactory definition of real numbers.

When we select a pair of real numbers with which to specify the Cartesian (rectangular) x, y in a plane we deal with the two-dimensional set of independent reals $\mathbb{R}^2$. The unbounded coordinates in three dimensions (*Euclidean space*) form the set denoted $\mathbb{R}^3$. Very generally, a set of n independent, unrestricted reals is denoted $\mathbb{R}^n$.

On occasion we deal with subsets $\mathbb{S}^n$ of reals, a *locus* of points that satisfy in n dimensions a common distance, say r, from some central point. Thus the symbol $\mathbb{S}^1$ refers to points on the circumference of a circle while $\mathbb{S}^2$ refers to points on the two-dimensional surface of a three-dimensional sphere in Euclidean space, i.e. a subspace of $\mathbb{R}^3$; see the discussion of the Bloch sphere in Appendix A.14.2 and the Poincaré sphere in Appendix B.1.7.

When a measuring device has a digital output the measurement provides a special case of real numbers— a *Rational Number* (the ratio of two integers, forming the set $\mathbb{Q}$)— but other categories of number, such as *irrationals* (needed for solving algebraic equations, e.g. $\sqrt{2}$) and *transcendentals* (not solutions to algebraic equations, e.g. π) seldom need to be specifically identified by their restricting family characteristic.

Physicists frequently need to deal with the limit of $\mathbb{Q} \to \mathbb{R}$ obtained by considering the limit of indefinitely large denominators and correspondingly infinitesimal spacing of discrete rationals (see Appendix B.3.2). They also consider the replacement $\mathbb{R} \to \mathbb{Q}$ in order to discretize a continuum and replace an integral by a sum.

Number systems. As one contemplates seemingly conflicting aspects of photons, it is useful to keep in mind that, even with something as seeming mundane as numbers, there are various alternatives possible. Transactions in all our daily lives employ a decimal system of numbers, based on the ten symbols 1,2,3,4,5,6,7,8,9,0. But we also encounter other systems that use other symbols. Preface-page numbering, analog clock faces, volume numberings, and years of filming all frequently employ Roman numerals, I,V,X,L,C,M together with a position-sensitive system in which IV is not the same as VI and the doing of simple sums is an intellectual challenge formerly given to school children to keep them occupied.

Numerical strangeness; Paradox. Whenever we deal with numbers, or measurements, we encounter opportunity for strangeness or paradox—some surprising result of using logic.[2] Mathematicians, for amusement, like to point out that in the interval

---

[2]Paradox—a statement relying on logic that seems to contradict itself—enlarges our culture. In the operetta "The Pirates of Penzance" William Gilbert and Arthur Sullivan (G&S) provide the "most amusing paradox" of Frederick who was (mistakenly) obligated to serve as a pirate apprentice until his twenty-first birthday but, because he was born on 29 February in a leap year, he must serve on as a detested pirate for many years after having lived for 21 years of "reckoning in the usual way." The calendar-based paradox posed by G&S has a subsequent counterpart in the Einstein twin paradox, in which twins that travel by separate routes, involving different speeds, age at different rates, in accord with the timekeeping rules of special relativity; see Section 8.3.1 and [Lal01; Aha08].

between the integers 0 and 1 (or any other numerical interval) there are an unlimited count (infinite but countable) of rational numbers $\mathbb{Q}$, yet between any two of these, no matter how close together we may choose them, there are real numbers $\mathbb{R}$ that are not in this set: Though the set $\mathbb{Q}$ is infinite, in a denumerable way that accords with familiar counting of objects, the numbers of $\mathbb{R}$ are even more numerous—they form a *continuum*. It should be realized that even the most advanced digital computer must rely on discrete representation of its numbers: The size of its storage elements (its words) limits the number of digits in its numbers, thereby establishing a limit to the accuracy with which it can represent any number. Thus a number continuum is a mathematical construct akin to the frictionless plane or the perfect vacuum of a physicist.

This little mathematical curiosity of the continuum needs only mental exertion to appreciate, as does the seeming paradox that there are as many integers as there are odd integers (because they can be arranged with one-to-one matching). Other mathematical curiosities become evident when one considers integrals, such as the representation of signals by their frequency decomposition, as do electrical engineers. They find a relationship between frequency (or musical tone) and time duration of a signal, a purely mathematical characteristic associated with periodicity: You cannot specify, with absolute precision, both the frequency of a signal and an instant from the signal; frequency and time are complementary variables in such mathematics, formalized with Fourier transforms [Ebe77b]; see Appendix B.5.1.

Physicists encounter this sort of relationship in the experimentally-based position-velocity uncertainty relationship of Heisenberg, as discussed with eqn (2.12-1) and eqn (A.4-6), or the wave-particle duality relationships of Section 3.7. Physicists doing calculations get used to changing from $\mathbb{Q}$ to and from $\mathbb{R}$; see Appendix B.3.2. But they may find the change between wave and particle computations, at the heart of atomic physics and quantum optics, to seem strange, even mysterious.

The atomic hypothesis: Continuous matter or denumerable atoms?   Towards the end of the nineteenth century there arose a lively debate amongst physicists concerning the nature of atoms. Were they discrete and denumerable (as is the set $\mathbb{Q}$) or did they form a continuum (as does the set $\mathbb{R}$)? I can imagine that mechanical engineers, familiar with point-like centers of mass, would have been comfortable with Newton's equations for discrete masses (see Appendix A.1), while nautical and aeronautical engineers, dealing with flowing fluids, would prefer a continuum. It should be no surprise that a similar debate arose over the nature of radiation and its photons.

Symbols.   It was in high-school algebra class that I learned the power of using some letter symbol, say x, to stand in place of all possible numbers that might be used in some mathematical relationship. In my early FORTRAN programming, which used only upper case, the letters I, J, K, L, M, N were reserved for integers, leaving all other letters for use as the more general real numbers. In physics articles it is common to see i, j, k, l, m, n used to indicate numbers from the set $\mathbb{Z}$ as subscripts for denumerating instances; for example $E_1, E_2, \ldots E_n \ldots$ for a list of discrete energy values.

It is possible to discuss physics without using symbols; one does this in conversations that make no use of computer displays. However, to do so in writing, where contemporary software and hardware offer typesetting with a vast supply of fonts, poses a severe challenge to an author. I shall not abstain completely from the use of symbols, for they have become a seemingly indispensable part of physics.

Equations.   An equation is just a simple name for a relationship between variables, i.e. measurable quantities of interest. Quite commonly they consist of two sets of numbers or symbols separated by an equal sign (or other binary relationship symbol). Few adults escape viewing the famous formula of Einstein, $E = m\,c^2$, for the relationship between energy E, mass m, and light speed c, and all attendees of a class on physics will have seen Newton's formula[3] $\mathbf{F} = m\,\mathbf{a}$ for the relationship of force $\mathbf{F}$, mass m, and acceleration $\mathbf{a}$, see eqn (A.1-9). Formulas such as these, though absent from conversation, readily appear in writings, where their absence would seem an affectation.

Often a special notation indicates that a particular relationship, such as $i \equiv \sqrt{-1}$ amounts to a definition of one symbol, i on the left, with some expression, $\sqrt{-1}$, on the right. The inequality symbol, $\neq$, is used to indicate that two things are not equal, as in the relationship $AB \neq BA$ between noncommuting operations.

Some computer-programming languages use an equation structure to indicate an instruction for the moving of information from one location to another in storage. The equation $A = B + C$ is the FORTRAN instruction to add together the numbers stored in locations A and B and store the result in location C.

Even with the familiar symbols, an arithmetic may obey differing rules for addition and subtraction. Common alternatives are the analog-clockface system that leads to the addition rule $6 + 7 = 1$ and the binary system of electronic messages, with the rule $1 + 1 = 0$.

Often an equation presents on the left the symbol for some incremental change (say the velocity as the ratio of incremental position change dx to incremental time change dt) and on the right an expression for its functional relationship to dependent variables (say a position-dependent force). Such equations of change are examples of *differential equations* and, for pupils of my era, their systematic study awaited courses in Differential Calculus. Inevitably they appear in discussions of quantum physics, and they appear profusely in the appendices of this Memoir . Given suitable differential equations it is possible, in principle, to follow their instructions from a specified initial position at a specified initial time and be directed to all possible positions and time that are consistent with these rules. The organization of those instructions into functions form the solutions to the differential equations.

Limitations. The equations of physics present relationships between observable quantities. Inevitably there are limitations associated with application of these. It behooves us to appreciate these limitations—to recognize the assumptions and idealizations (frictionless planes, massless pulleys, point particles, empty surroundings, etc.)—that may negate their usefulness in particular instances.

Formulas, algorithms, and plans. In chemistry, a *formula* is a fixed set of symbols, such as H_2O or C_2H_6O, that describes the composition characteristics that distinguish between various compounds—that differentiate water from ethanol in this example.

---

[3]The boldface font indicates that force and acceleration are examples of vectors; see Section 2.5.

Organic compounds typically require elaborate diagrams of loops and lines for their definition, and these too could be considered formulas of a sort—a template that allows association of a name with a substance. But the formulas for nitroglycerine or gunpowder, by themselves, bring no information about how these substances can be formed into destructive devices, as would be explained in diagrams of an assemblage. In physics and math a formula is typically an equation expressing some mathematical relationship, such as $A = \pi r^2$ for the area of a circle A from its radius r, but it may just be some arrangement of symbols—the πr^2 part of the area equation.

Algorithm is a technical term for a *procedure*, typically for manipulating symbols or actions that then, from some starting material (or symbols or numbers) can produce a result that has desired characteristics. The area formula $A = \pi r^2$ presents an algorithm for obtaining A from r. A FORTRAN instruction is a written form of an elementary computer algorithm for moving numbers around in storage. Earlier generations of school children learned the algorithms for adding columns of multi-digit numbers (carries) and carrying out long division. Commonly the term is used for mathematical procedures, but it could apply to other activities. Students of organic chemistry are asked, on exams, to devise a synthesis procedure—an algorithm of sorts—for creating a compound defined by its formula. An experimentally-inclined physicist might try to explain the algorithm for producing a perfect cup of coffee; his formula for perfectly prepared coffee may have little to say, directly, about how it came to be so perfect.

The ambiguous noun "*plan*" is used with meanings similar to both those above: A retailer may compose a *plan* for an advertising campaign (an algorithm, not a formula, though it may include charts) and draw a (floor) *plan* for a revised store layout (which may include a timeline for actions). Conversely, in literary works one reads of "a formula for success" rather than the more accurate but less auspicious sounding "a plan for success". But it would be misleading to call Einstein's equation $E = mc^2$ the "formula for an atomic bomb" (as I was once told).

Complex numbers. Although measurements invariably produce real numbers, much of the mathematics associated with quantum theory (and hence with photons) is presented more simply by introducing the square root of minus one as a unit coordinate in a two-dimensional space (the *Complex Plane*) whose axes are, in one direction (horizontal), just real numbers, and in the perpendicular direction (vertical) are real numbers times the imaginary unit $i \equiv \sqrt{-1}$, defined by the requirement $i^2 = -1$. A complex number z (from the set denoted $\mathbb{C}$) therefore has the form $z = x + iy$, with real-valued x and y. The squared length z of such a two-dimensional vector is just the sum of the squares of its two components, $x^2 + y^2$, an example of the Pythagorean Theorem that students of trigonometry (and the diploma-endowed Scarecrow of Oz) learn for describing the long side of a right triangle:

> The square of the hypotenuse is equal to the sum of the squares of the other two sides.

Complex numbers are ubiquitous in physics, primarily for their mathematical convenience. In particular they occur as probability amplitudes, whose absolute squares are real-valued probabilities, and as electric-field amplitudes, whose absolute squares provide positive measures of radiation intensity and energy density (and hence of photon density). Appendix A.17 discusses various mathematical structures that generalize the notion of complex numbers—mathematical constructs (forms of algebras) sometimes termed *hypercomplex numbers*.

Aside: Magnitude and phase. Complex numbers have two parts, labeled real and imaginary. The two parts can also be expressed as a magnitude and a phase; see Appendix A.3.2. The *magnitude* of the complex number $x + iy$ is defined, for example, by the formula

$$|x + iy| = \sqrt{x^2 + y^2}. \tag{2.1-1}$$

The *phase* of a complex number z is the angle that specifies its direction on a line from the origin in the complex plane. It is the angle φ such that the real and imaginary components of z are obtained from the trigonometric relations defining sine (sin) and cosine (cos) for the two short sides of a right triangle having $|z|$ as hypotenuse (see Appendix A.1.1):

$$\text{real: } x = |z| \cos \varphi, \qquad \text{imaginary: } y = |z| \sin \varphi. \tag{2.1-2}$$

Physicists commonly express angles in radians, π rad $= 90°$, so $\sin \pi = 1$.

Variables and functions. Relationships between numbers form the basis for the equations with which physicists and others formulate their descriptions of Nature. For that purpose the equations are regarded as providing a connection between some quantities such as position and time that are to be chosen, and some other quantities, such as temperature or rainfall, that will then be found associated with that choice. The numbers involved are *Variables*: Either to be chosen (the *independent variables*, location and time) or to be found (the *dependent variables*, temperature and wind velocity). A mathematical *Function* is a rule for connecting the (resulting) values of dependent variables to the (chosen) values of the independent variables—a connection that I eventually learned is called by mathematicians a *mapping*.

In much of the physics literature a functional relationship of a single independent variable is exhibited typographically in the form $F(t)$ or $F(x)$, where t or x is the symbol being used for the independent variable (say, time or position) and F is the symbol for the dependent variable (say, temperature or stock price). To emphasize that the energy of Einstein's famous equation varies with mass one might write $E(m) = mc^2$. Appendix-defined examples from atomic physics include $\psi(x, y, z)$ for the complex-valued probability amplitude (a wavefunction) of finding a given electron at a position whose Cartesian coordinates are x, y, z (see Appendix A.5), and $\mathcal{E}(x, y, z, t)$ for the complex-valued amplitude of an electric field at time t and a position specified by coordinates x, y, z.

The range of values of the independent variable for which a function is defined is known as the *support* for the function. A pulse of temperature or electric-field magnitude that is modeled as occurring during only a finite time interval is said to have finite support. It often proves mathematically convenient to allow the interval to extend indefinitely, to become a pulse of infinite support.

The locus of points x that map $\mathbb{R}$ onto the function values $F(x)$ form a *curve*, possibly smooth but possibly with jumps and kinks, even *singular points* where $F(x)$

is undefined. The point at which two curves have the same value, $F_1(x) = F_2(x)$, is known as a *crossing*.

2.2   Particles: Elementary and structured

Particles, to physicists, are localized concentrations (usually small) of mass and other properties. They combine to produce the objects of everyday life, and they come in a great variety. Particles can be characterized, in part, by several *intrinsic* qualities (*attributes*), meaning characteristics that are unchangeable and independent of position and velocity. Examples include the electric charge of an electron, the chemical properties of an atom (or the number of its electrons), the color of a pool ball. *Spin*, associated with an intrinsic angular momentum, is another such attribute (see Section 2.2.8 and Appendix A.5). By contrast, characteristics such as the arrangement of electrons within an atom (governing atom size and shape) or of pool balls upon a pool table, are changeable and not intrinsic. Particles are regarded as *identical* if they have exactly the same set of intrinsic attributes.

Amongst the intrinsic attributes of everyday objects that are to be considered in identifying and using them is the pull of gravity (or resistance to force), quantified as *mass* [Adl87; Oku89; Roc05] and expressed in kilograms, kg.

> **Aside: Rest mass.** A stationary particle has an intrinsic measure of its resistance to motion (an *inertia*) quantified as a *rest mass* m_0. The special theory of relativity introduces changes of length and time measures as observed in different moving reference frames. Consequently a particle seen as moving steadily with speed v will have an apparent inertia (or relativistic mass) that increases without bound as v approaches c, the speed of light in vacuum—the ultimate limit of any particle speed; see eqn (A.1-11). It is rest mass that we list as an intrinsic characteristic of elementary particles.

2.2.1   Indivisibility

An important concept in physics and chemistry is the notion of *indivisibility*, and its ally, *elementarity*. Put operationally, this refers to the possibility, with given tools, to break apart some given object.

The notion of indivisibility very clearly depends upon the tools available for taking something apart and the decision to use them. When I gathered around our family dining table for a meal I recognized chairs as being basic units for sitting. These, along with plates and tableware, were present in unchanged numbers each day, and so they constituted the elementary constituents (particles) of meal support. But occasionally there would be a mishap, and a chair or plate might be broken into pieces. Discarded chairs, being wooden, could even be burned in a bonfire. The notion of indivisibility clearly depended on the conditions under which an object was being used.

Traffic engineers deal with motor vehicles as indivisible units, moving along discrete roadways. Air-traffic controllers deal with individual airplanes along approved flight paths. Ranchers deal with herds or flocks of individual animals pasturing during their growth. These and many other examples of temporarily indivisible units—structured particles—come readily to mind.

The notion of indivisibility occurs in chemistry for the definition of molecules and atoms. As long as only chemical processes are employed, these remain elementary units. Physicists use a variety of tools to break apart atoms, finding thereby a variety of more elementary units. The following paragraphs describe some of the particles that, for appropriate conditions, are treated as indivisible and elementary.

2.2.2   Atoms

It was common, in my school days, to read that "every schoolboy knows ..." and then to be presented with some fact from History or Science that, at the time of the writing, was part of the general education curriculum of boys—and of some girls. The notion of an atom (or a molecule) was one of those concepts. The idea was very simple, an example of what is called in German a *Gedankenexperiment* or thought experiment: you imagine cutting some chunk of Matter (a chair or a plate or a hammer) into smaller and smaller pieces. Eventually you will come to a situation when the pieces no longer carry the recognizable chemical properties of the larger object. The smallest piece of iron is an atom of iron. The smallest piece of a kitchen chair is a molecule of wood or plastic. This, in turn, can be broken into atoms of carbon and oxygen and hydrogen and nitrogen and other elements.

As every schoolchild used to know, there are nearly a hundred chemically distinguishable kinds of atoms (different chemical elements) found in nature. Charts listing these were once found displayed in classrooms, organized into the Periodic Table, starting with hydrogen and helium, elements 1 and 2, and ending with such heavy elements as uranium, element 92. Already in my elementary-school days we learned that the naturally occurring elements were being augmented by unstable, man-made elements (for example element 43, technetium, and, more recently, element 116, livermorium), but the Periodic Table remained our guide to the world of atoms which, when suitably combined into molecules and solids, provided all the inanimate objects of everyday life. We understood that every atom, every molecule, carried intrinsic chemical attributes (originating with the number of electrons and their structure) along with physical attributes of mass, size and shape. Isolated atoms might be treated as moving mass points but they carried intrinsic attributes as well.

Artificial atoms. Nowadays a variety of remarkable fabrication techniques and technology tools make possible the construction of submicroscopic (nanoscale) objects in which confined electrons mimic observable characteristics we attribute to the internal structure of traditional atoms and molecules. Often termed "artificial atoms" [Kas93; Kou01; Rei02; Cho10; Hon11; Lod15; Man16; Sal19] or "photonic nanostructures" they are available in many forms, including one-dimensional "quantum wires", two-dimensional "quantum wells", and three-dimensional "quantum dots", with properties that can be designed by an experimenter. The physics, and the mathematics, of such structures is very much what I learned in my years of professional concern with laser action upon traditional atoms—only the names have been changed. All of them serve equally well as examples of quantum-state systems that absorb, emit, and modify discrete increments of radiation (photons) in a coherent manner.

Subatomic entities; electrons and ions. We scholars also learned that the thought experiment of taking apart our everyday objects need not end with single atoms. These, we learned, are made from bits of electrically charged matter: Very light negatively charged *electrons*, removable from much heavier, positively charged atomic *nuclei* that are much smaller than the full atom (see Section 2.2.6 for a discussion of sizes). These electrons and nuclei are the basic particles whose intrinsic properties must be considered when one sets out to understand such things as the colors of stained glass windows or the strength of bricks and mortar or the combustion properties of gasoline. Their natures underlie the study of Chemistry and Materials Science, and the design of electronic devices that enable the creation of computer games and the sending of messages.

As commonly encountered, individual atoms and molecules have no electrical charge. They are electrically neutral because the number of their electrons exactly equals the number of positive charges in their nuclei. The removal of an electron from an atom or molecule[4] leaves a positively charged *ion*. If that freed electron becomes attached to a neutral particle the result is a negatively charged ion.

Individual atoms. The notion of breaking apart pieces of matter to obtain molecules and atoms as elementary units is not so far removed from household experiment. The simplest idea, say of grinding table salt into a fine powder, does not produce individual atoms of the sodium and chlorine that comprise the salt crystals, but when one pours the salt into water the molecular bonds between Na and Cl atoms break, and one has a salty solution in which individual atoms, carrying single electron charges as Na^+ and Cl^- ions, are to be found moving.

At least that is the picture that I was presented as a youngster. To some extent this picture of solvent action by water, breaking chemical bonds of solid compounds, remains valid, but the subject of Aqueous Chemistry—the study of water as a solvent [Stu95; Sch10; Hor12]—has brought a more complicated picture of the resulting liquid. Metal ions in aqueous solution carry with them an entourage of molecules from the fluid environment—a clothing of loosely held *ligands*[5] that depends on such conditions as how acidic or basic the environment is. Our glasses of dairy milk carry a variety of bovine-produced colloids. And seawater contains a vast assortment of minute organic and inorganic particles of sizes that range upward from a few aggregated atoms [Ben92; Guo97], all of which have their characterizing interactions (scattering and absorption) with photons.

As noted in Section 5.1.2, to best understand photons as individuals it is important that their source atoms be carefully prepared to prevent randomizing influences. For this purpose vapors have advantages over liquids, and so the individual atoms whose photon interactions are studied in Quantum Optics are often obtained from a dilute vapor. Conceptually the simplest procedure is to allow a heated gas to leak from a container as a particulate beam, available for manipulation that can capture and hold a single molecule, atom, or ion.

---

[4]It is relatively easy to remove an electron from an atom, thereby "splitting the atom" and creating an ion. A comb run through dry hair can do that. It is much more difficult to deliberately break apart an atomic nucleus.

[5]Photons traveling through matter become similarly dressed; see Appendix C.2.2 and C.4.2.

2.2.3   Electrons

Electrons, the indivisible bits of (negative) electricity, are the simplest of all contenders for the title of "*elementary particle*", meaning part of the stock from which all matter can be constructed. They are readily generated (from captivation in atoms) as *cathode rays* in evacuated containers, traveling in the space between two charged electrodes—from the negatively charged cathode to the positive anode. Flows of electrons occurred in the vacuum tubes of radio sets. Electron beams in the cathode-ray tubes of laboratory oscilloscopes and home television sets were directed by electric and magnetic fields towards designated target areas on the vacuum side of a transparent viewing screen, there to produce fluorescence from atoms, visible to human viewers from their side of the screen. This charged-particle motion through vacuum has become a model of elementary particle behavior; see Section 8.3.

Electrons can be stopped (by atoms) and extracted (from atoms) but cannot normally be otherwise created or lost: Electric charge is conserved under all conditions. Whatever may be their source, all electrons are identical, indistinguishable, and indivisible. It is natural to have electrons in mind when one contemplates indivisible elementary particles of radiation.

Within a metal there occur conduction electrons that are free to move throughout the solid, under the direction of applied voltages and localized structure. But in any sample of matter most of the electrons are bound to atomic nuclei, held by attractive Coulomb forces to form atoms and molecules or rigid structures.

> **Aside: Electron attributes.** The electron has three *intrinsic* (unalterable) attributes, electric charge e, rest mass m_e, and magnetic moment, whose values rank among the basic fundamental constants of nature (see Appendix B.1.1 for definition of SI electromagnetic units[6]):
>
> $$|e| \approx 1.60 \times 10^{-19} \text{ C}, \qquad m_e \approx 9.10 \times 10^{-31} \text{ kg}. \qquad (2.2\text{-}1)$$
>
> The intrinsic magnetic moment of a free electron differs very slightly from one *Bohr magneton*, defined as [here $\hbar$, called h-bar, is the reduced Planck constant (or Dirac constant), the elementary unit of angular momentum],
>
> $$\mu_B = \frac{e\hbar}{2m_e} \approx 9.27 \times 10^{-24} \text{ J/T, with } \hbar \equiv \frac{h}{2\pi} \approx 1.05 \times 10^{-35} \text{ J s/rad.} \quad (2.2\text{-}2)$$
>
> As Section 2.2.6 notes, electrons have no intrinsic size, although several combinations of fundamental constants associated with electrons have dimensions of length; see Appendix A.2.
>
> **Aside: Atomic (or Hartree) units.** In treating manipulations of quantum structures it is often useful to express charge, mass, and angular momentum as multiples of electron charge e, electron mass m_e and the reduced Planck constant $\hbar$, respectively. The resulting *atomic units* provide formulas that set $e = m_e = \hbar = 1$. The Bohr radius, a_0, provides the atomic unit of length, while the atomic unit

---

[6] Here C is coulomb, J is joule, T is tesla, kg is kilogram, s is second, and rad is radian.

of energy, the *Hartree energy* E^{AU}, is twice the ionization energy of an idealized hydrogen electron

$$a_0 = \frac{4\pi\epsilon_0\hbar^2}{m_e\, e^2} \approx 5.29 \times 10^{-11} \text{ m}, \qquad E^{AU} = \frac{e^2}{4\pi\epsilon_0\, a_0} \approx 27.21 \text{ eV}. \qquad (2.2\text{-}3)$$

In atomic units the speed of light is $c \approx 137$, the inverse of the fine structure constant α of eqn (A.2-3).

Aside: Fractional electrons. Electric currents are carried through gases and liquids by electrons and ions, bearing integer increments of the electron charge. In conducting solids (metals) both electrons and lack-of-electrons (*holes*) can carry the charges, whose voltage-driven motion is seen as currents, and is deflected by a magnetic field. But in two-dimensional semiconductors at very low temperatures, subject to a strong, static magnetic field, there can occur a collective fluid state wherein electrons and holes combine with magnetic flux to form charge-carrying *quasiparticles* (neither fermions or bosons, see Appendix A.8.3) that have fractional elementary charge, observable as the *fractional quantum Hall* (FQH) effect [Ait91; Sch98b].

2.2.4   Nuclei and their constituents

During the early years of the twentieth century physicists learned of two particles that account for the mass, charge, and volume of atomic nuclei: *Protons*, each carrying one unit of positive electric charge that exactly balances the negative charge of an electron, and electrically uncharged *neutrons*. The uncharged neutron is a constituent of the nuclei of all but the lightest hydrogen atoms.[7] Each of these (known collectively as *nucleons*) is some two thousand times heavier than an electron.

Aside: Nuclear masses. The rest masses of the proton and neutron are

$$m_p \approx 1836\, m_e \approx 1.008 \text{ u}, \qquad m_n \approx 1.007 \text{ u}, \qquad \text{u} \approx 1.66 \times 10^{-27} \text{ kg}. \quad (2.2\text{-}4)$$

Here the atomic mass unit, u, is $1/12$ the mass of a carbon ^{12}C atom.

The summed masses of constituent nucleons accounted well for the atomic weights of chemical elements, although later, more accurate, mass values revealed energies of binding interactions—the energy of nuclear fission.

Each proton and neutron has an intrinsic magnetic moment, smaller than that of an electron by the mass ratio m_e/m_p. These individual moments (vectors) combine with orbital motion of protons within the nucleus, to create an overall intrinsic nuclear magnetic moment (a vector) proportional to the nuclear spin [Eva55; Kop13]. In turn, the nuclear moment combines with the various magnetic moments of the electrons and their orbital currents to create a total atomic magnetic moment, again proportional to the total angular momentum of the electronic structure, as discussed in books on atomic structure [Con53; Bet57; Sla60; Hin67; Cow81; Sob92; Bra03; Dem10].

---

[7]The nuclei of (stable) deuterium and (unstable) tritium isotopes of hydrogen have one and two neutrons, respectively.

The names one encounters in reading of proposed particulate bits of matter—electron, proton, neutron, meson, and so on and on—originated with the notion that they had an *elementary* nature: They could not (normally) be further subdivided. Nowadays even the proverbial school attendee knows that, although electrons cannot be further divided (they are truly elementary particles), the atomic nuclei that chemists and atomic physicists treat as their elementary positively-charged particles, can, with sufficient effort and financial support, make evident an underlying structure based on neutrons and protons. These, we are told, are in turn built from quarks held together by gluons (see Appendix A.19).

But none of this sub-subatomic menagerie has immediate relevance to the workings of everyday objects. For their construction we need only electrons and nuclei. And for their understanding we need only the mathematics developed for the purpose in the early twentieth century, Quantum Theory; see Appendix A.4. It is within that world that the notion of photon, a companion to the electron, made its appearance.

2.2.5 Antimatter

Electrons and most naturally-occurring atomic nuclei (including the proton) are stable particles: Apart from radioactive nuclei they can remain unchanged indefinitely in isolation.

As physics experiments revealed during the mid-twentieth century, the particle constituents of the matter we encounter in everyday activities have *antimatter* counterparts that were first seen in cosmic rays and then in debris of target bombardments with particle accelerators. The positively-charged anti-electron (or *positron*) and the negatively-charged anti-proton, have opposite electric charge from (but identical mass as) their respective electron and proton matter partners.

The neutron too has an antiparticle, the uncharged antineutron. The photon has no electric charge and no mass, and so it is indistinguishable from its antiphoton.

Amongst the various elementary particles (leptons) that constitute the Standard Model of Particle Physics described in Appendix A.19 are three types of electrically neutral, almost-massless neutrinos. Though these are uncharged they have distinguishable antineutrinos—differing by *helicity* (the projection of spin onto the propagation axis).

Although neutrons can remain stable within nuclei, a neutron isolated outside of proximity to protons is unstable: It spontaneously decays into an electron, a proton, and an electron antineutrino (see Appendix A.19). The electron and proton charges exactly balance, thereby retaining overall conservation of electric charge in this decay.

The various particles of antimatter have exactly the same masses as their normal-matter counterparts, and so both sorts are affected equally by gravity and electromagnetism. However, particles of antimatter, upon encountering a matter counterpart, undergo mutual annihilation: Their masses are ultimately converted into radiation energy (photons) and neutrinos.[8] Conversely, a photon of sufficient energy can, in passing by an atom, become a particle-antiparticle pair. Such events are part of the realm of

---

[8] All electrons are identical and all positrons are identical, so any pairing will produce annihilation.

high-energy physics, not immediately relevant to the photon physics discussed in this
Memoir.

2.2.6 Particle sizes

It is important to know the sizes of objects you hope to fit into a room or a container, and we regularly make judgements of sizes (small, grande, big, huge, XXL)
even without that intent. The familiar units of feet and inches derive their usefulness
from their association with everyday objects—human bodies. To appreciate the world
of atoms and photons it is helpful to know how their sizes, however defined, compare
with everyday measures of size. Appendix A.2 discusses size measurement and presents
several values relevant to atoms, electrons, and photons based on constants of nature.
Here I summarize the qualitative details of that presentation.

> **Size estimates.** A simple measure of atom or molecule size comes from recog-
> nizing that in a liquid or solid the particles, though moving relative to each other,
> maintain contact. Thus if the mass density is ρ kg/m^3, and each particle has mass
> m kg/atom, then the volume occupied by one particle is
>
> $$\mathcal{V} = \frac{m\,[\text{kg/atom}]}{\rho\,[\text{kg/m}^3]}. \tag{2.2-5}$$
>
> Treating $\mathcal{V}$ as the volume of a sphere of radius r, one finds the particle radius r
> (a measure of size) from setting $\mathcal{V} = 4\pi r^3/3$.

Electrons have no intrinsic size, although several combinations of fundamental constants associated with electrons have dimensions of length; see Appendix A.2. Within
a metal, the conduction electrons are free to move throughout the solid. In other aggregates they occupy space around minuscule atomic nuclei but their electric-charge
(and mass) distribution is not uniform within that volume. The localization of electron charge within an atom is rather like a cloud: It has no sharp boundary and
is most concentrated near the nucleus, where the Coulomb attraction is largest, but
with notable overall nodal structure (places where the local mass and charge density
is negligibly small). Well outside the nucleus the mass density and charge density of
electrons within an atom falls exponentially with distance from the nucleus; the "size"
of an atom (or subatomic particle) therefore refers to a mean radius of mass or charge.
As atoms become heavier along the sequence displayed in charts of chemical families
their sizes vary in accord with chemical properties, but not in a monotonic way: Al-
kali atoms (e.g. Na) are appreciably larger than halide (e.g. Cl) or noble-gas (e.g. He)
atoms of comparable mass [DeV44].

 By contrast, the protons and neutrons that comprise the minute atomic nuclei
have basic intrinsic sizes (not sharp edges but mean radii): The radius of a proton is
around 1 fm (a femtometer or fermi, 10^{-15} m) while the radius of the first Bohr orbit
of hydrogen is approximately 0.5 Å $= 0.5 \times 10^5$ fm, see Appendix A.2. Nucleons are
not appreciably compressible—in first approximation they can be regarded as forming
a liquid. Thus as nuclei gain protons along the Periodic Table their volumes grow
in proportion to the number of their nucleons—their atomic weight. Nuclei generally

have shapes that are not spherical, as characterized by various charge-distribution multipole moments [Sto05; Kop13].

Well beyond an atom or nucleon its charge and mass can be treated as localized at a point-like center of mass. Any structure of the charge and current distribution appears as a series of multipoles; see Appendix B.2.6.

Although electrons occupy volumes that are defined by whatever confining forces may be present, several numbers with dimensions of length occur in theoretical treatments of electrons; see Appendix A.2 and Appendix B.8.2.

2.2.7   Radiations

Toward the end of the nineteenth century the terms "radiations" and "rays" found use in a variety of applications, and these still influence our discussions of the natural world. There were X-rays that could penetrate matter (subsequently found to be forms of electromagnetic waves). There were *cathode rays*, now known to be electrons. From radioactive solids came *alpha rays* (positively charged nuclei of helium devoid of any electronic cover), stopped by a sheet of paper, and *beta rays* (electrons at first, but subsequently also positrons), stopped by a few mm of aluminum, and massless, uncharged *gamma rays* (electromagnetic waves) that readily penetrated flesh and bone and were attenuated by sheets of lead. The form of radiation of most interest for the present exposition is visible light, now known to be electromagnetic waves, studied in the discipline of Optics. In the present Memoir the term *radiation*, with its photon increments, will mean exclusively electromagnetic fields. The examples of historical rays with nonzero rest mass are regarded here as particles.

2.2.8   Particle spin: Fermions and bosons

All of the common particles, whether elementary (an electron or photon) or composite (an atom), fall into two classes,[9] distinguished by their intrinsic spin:[10]

- *Fermions* have intrinsic spin of half-odd-integers. They include electrons, protons and neutrons, whose spin is $\frac{1}{2}$.
- *Bosons* have spin of 0 or positive integers, and include photons, whose spin is 1.

The distinction between fermions and bosons becomes pronounced when we consider constructs formed from multiple particles, such as a nucleus formed from nucleons or a molecule formed from atoms. The mathematical connection between spin and statistics (see Appendix A.8.3) and the consequent Pauli exclusion principle, prevents multiple fermions from occupying the same space, and thereby maintains the electronic structure of atoms. A composite formed from an odd number of fermions will be a fermion. An even number of fermions will be a boson.

---

[9]Situations occur when particle-like constructs do not fit these two choices; see the mention of "anyons", page 235.

[10]I distinguish between "spin", a dimensionless vector, and "spin angular momentum", commonly expressed in units of $\hbar$.

2.3 Aggregates: Fluids, flows, waves and granules

When we reverse the process of breaking apart material objects into their basic units—atoms and molecules—and instead assemble them to create liquids and solids we encounter collective properties of *aggregates* that are not to be found in the individual constituents. These are traditionally treated by idealizing matter as being continuously distributed within some defining volume—a continuum $\mathbb{R}^3$ of mass or electric-charge *density* (property per unit volume), without any specially noteworthy position, often without any specified boundary; see Appendix C.2.1. It is with such material that we find prototypes of wavelike behavior and so, to appreciate possible wavelike attributes of photons, we do well to review the equations found there.

Aggregate forms of matter. In my elementary-school education I was taught that gases, liquids, and solids were the three basic forms of *bulk matter* and that water could be found as any of these, depending on its temperature and atmospheric pressure. Much later I read of *plasmas* (ionized gases) as a fourth state of matter.[11] These categories appeared as organizings of physics (and its publications and academic subdivisions) into solid-state physics and plasma physics. Enlargement of the categories now include portions of matter that are soft and squishy. These are studied in condensed-matter physics and biophysics, and in departments of Biology. Section 2.3.5 notes more recently established categories.

Gases. A *gas* is imagined by physicists as a collection of many particles, each idealized as an infinitesimal concentration of mass and other attributes, to be treated as mathematical points distributed uniformly throughout a given volume. In the absence of external forces the gas particles move along straight-line paths, making occasional infinitesimally brief encounters with other particles, treated as *elastic* (momentum-transferring) and *inelastic* (energy-transferring) *collisions*. The mean time between collisions quantifies the duration of coherence of particle phases.

  A gas is usually regarded as a continuum of mass, momentum, and energy densities subject to an equation of state that relates the thermodynamic variables of pressure, temperature and volume. In particular, the averaged motion of the gas particles depends only on temperature; cooling of a gas requires that there be a mechanism for irreversibly transferring energy beyond the boundary of the gas—to a "bath", see Section 5.1.3. In accord with the precepts of quantum theory, no bulk matter can ever be entirely without some uncontrollable thermal energy, even when the thermodynamic temperature approaches its lowest value, zero degrees Kelvin.

Liquids. By definition, a given mass of liquid has a definite density and volume but no particular shape: When acted upon by gravity it readily takes the form of a container (although blobs of liquid freed from gravity will be shaped by surface tension). Like the molecules of a gas, the molecules of a liquid undergo continual random (thermal) motion conditioned by surrounding temperature. Though they are packed together closely they readily slip and slide past each other.

---

[11] The familiar substances steam, liquid water, and ice cubes are now regarded as just three *phases* of water, forms that are to be found in accord with pressure and temperature conditions. There are now some nineteen identified phases of the H_2O system, involving a variety of crystalline and amorphous fixed structures of ice [Gas18; Mil19b; Tse19; Kru19].

All liquids are held together by short-range forces between their molecules. These forces impose, but only briefly, a local structure akin to that of a solid, However, unlike the rigid structure of a solid, this orderliness is only in the immediate neighborhood of each molecule. The lack of long-range order is a defining characteristic of a liquid (or a glass).

The local, microscopic forces between molecules of a liquid produce a resistance to deformation, quantified macroscopically as *viscosity*—a frictional force between adjacent fluid increments that have different velocities.

Solids. The molecules of a solid, though subject to small, incessant thermal motions, are held by chemical bonds into a nearly-rigid stationary framework. The framework may exhibit a simple long-range regularity (as in a crystal) or may be quite irregular (as in glass). The defining characteristic of a solid is the (nearly) rigid structure of individual equilibrium positions of molecules, atoms or ions that form a relatively fixed lattice for electrons.

2.3.1   Fluids and flows

Aggregates of matter that can readily change shape—liquids and gases—are treated mathematically as *fluids*, a continuous distribution of mass, momentum, and kinetic energy whose parcels undergo collaborative motion—fluid flow. Streams and rivers gain their public popularity from their flows of water in bulk, while air flows drive windmills and sailboats. The equations that describe fluid flows, freely and past boundaries, form the topic of *hydrodynamics* [Lam45; Mil60; Lan87; Fal11]. (The specialized topic of *aerodynamics* provides a description of air flow over airplane wings.) All these modelings of nature form the topic of *fluid dynamics*—the treatment of a continuous distribution of properties including mass, momentum, and kinetic energy under the influence of bulk forces.

The basic equations of fluid dynamics express the time and space variation of a vector field, the velocity of an infinitesimal parcel of fluid. These equations derive, in part, from requiring that the motion of each fluid parcel should conserve mass and energy and respond appropriately (in accord with Newton's laws) to the imposition of forces: Those of pressure and the viscous forces between neighboring fluid parcels that move with different velocities. (The velocity gradient is referred to as a *strain rate*.) Additionally, the bulk matter is assumed to satisfy an *equation of state* relating thermodynamic variables of pressure, temperature, and volume. The resulting equation, the *Navier-Stokes equation* [Lam45; Mil60; Lan87; Fal11] [see eqn (A.1-47) in Appendix A.1.10] is a relationship—a partial differential equation (PDE)—between incremental changes in position and time whose solution describes distributions of fluid flow induced by pressure and viscosity. It is subject to constraints imposed by rigid boundaries of walls or wings or by boundaries between different fluids, such as the air-water interface bounding the upper surface of a sea.

Flows. The fluid flow may either follow simple paths (streamlines) along a watercourse or it may encircle a point, forming a *vortex*. An example of vortex motion is often visible in the flow of kitchen sink or bathtub water out through the drain. Along with wave motion, vortices are a characteristic of fluids, not evident in few-particle systems.

All real fluids are viscous; an idealization that has no resistance to deformation is known as an ideal or *inviscid* fluid. The dimensionless *Reynolds number*, the ratio of inertial forces to viscous forces, fixes the type of flow that will occur near a rigid boundary such as that of a water pipe or a ship hull. When viscous forces dominate, at low Reynolds number, the fluid motion is smooth (*laminar*), following simple flow lines. When inertial forces dominate, at high Reynolds number (typically several thousand, depending on the geometry), the motion is *turbulent* and irregular, producing vortices and other instabilities. The presence of viscosity leads ultimately to dissipation as heat; see page 35.

> **Aside: Photons as fluids.** In recent years I have appreciated that suitably crafted electromagnetic fields in free space (i.e. photons) can exhibit not only flows of energy and momentum but also vortices—distributed angular momentum; see page 98. In a vacuum, photons move freely with only negligible mutual interaction—the photon-photon scattering is there extremely slight. But when radiation passes into dense bulk matter the field can be strongly blended with matter characteristics, forming *quasiparticles* (polaritons) that carry the disturbances; see Appendix C.4. The possibility of frequent collisions between polaritons can lead to collective fluidlike behavior [Car13]; see Section 8.5.

2.3.2 Matter waves

All of the traditional forms of matter—solids, liquids, and gases—exhibit patterns of collective effects regarded as *wavelike* and characterized by minima (valleys or null-valued *nodes*) and maxima (*crests* or peak values or *antinodes*) that travel steadily (traveling waves) or are confined in space (standing waves).

Surface waves. As a vapor of diatomic water molecules condenses to form a mist of droplets and then a pan-filling liquid we are able to observe one of the most familiar properties of a liquid surface: Ripple patterns (waves) that move across the interface between the dense liquid and the overlying vapor. Viewings of water waves are a popular excuse for visiting shorelines. The waves, visible as crests and valleys of the water surface, are collective actions of the molecules that make up the bulk liquid.

From the equations of fluid dynamics it follows that at an interface between two different fluids, such as the air-water interface above a sea, there can occur surface waves. These are regular patterns of vertical and horizontal displacement of fluid elements, localized around the boundary, that transport energy along a propagation direction and which exhibit interference structure between colliding waves.

Such surface waves are also found in the equations of electromagnetism, associated with sharp boundaries between different dielectrics; see the topics in Appendix C.3.

Acoustic waves. Our ears, as well as such devices as microphones, respond to slight abrupt changes in air pressure that travel as acoustic waves. When these waves exhibit periodic variation of pressure they produce sensations of musical tones. Acoustic waves travel not only through gases but also through liquids and solids. They are generated by vibrating guitar strings, by gently struck wine glasses, and in many other ways. As molten iron solidifies the individual atoms of iron become confined between neighboring atoms. The atoms, though able to move slightly (a displacement from an equilibrium position), form a rigid macroscopic lattice. Collective motions of the atoms in an iron rod or a steel wire can be heard as sound waves, initiated by striking or plucking a source. The individual atoms of iron are not evident, and the solid can be regarded as a continuum of mass, lacking any granularity. Here too there is no evidence of waves when the sample of matter contains only a few atoms, although collective motions are used for understanding rotations of molecules and vibrations of their atomic parts.

All fluids are at least slightly compressible—the molecules can be pressed more closely together briefly. As a result of such small incremental changes in pressure there can occur *pressure waves*, moving at the local speed of sound.

Sound waves traveling through gases, liquids, and solids are examples of *longitudinal waves*: There occur small, compressional density changes (displacements of molecular constituents) along the direction of disturbance travel.

Solids and their waves. In solids there may occur not only compressional (longitudinal) waves but also transverse shear waves and torsional (twisting) waves, having displacements perpendicular to the direction the waves are traveling. Fluids, lacking the rigid structure of solids, do not exhibit transverse waves of shear and torsion. The mathematical description of all such waves requires parametrization of bulk properties (for example, mass density) but not the requirement of atoms. Nevertheless the collective behavior, of wave energy and momentum, can be regarded as comprising *phonons* as individual units (*quasiparticles*), akin to the photons of the electromagnetic field.

Solitary waves. Localized traveling disturbances, so-called solitary waves (*solitons*) occur in a variety of situations, most famously as destructive ocean-borne tsunamis. Solitary waves in channels have become a source of surfing activity that was long confined to coastlines [Fin18]; though the wave itself is distributed over a large distance, its action can be localized to that of an individual surfer.

Heat. The eventual fate of matter waves is as disorganized motions, first of wavelike collective actions (e.g. phonons) and finally as thermal jostling of the molecules that comprise the bulk matter. This eventual conclusion of organized waves constitutes heat. (The oft-mentioned "heat wave" of weather reports has no relationship to the waves described here.)

2.3.3 Electromagnetic waves

The electric and magnetic fields that jointly define electromagnetism (and photons) are governed by Maxwell's equations (see Appendix B and C), a set of PDEs that allow wavelike solutions. Waves of electromagnetic radiation, by contrast to waves of bulk matter, do not rely on any material medium; the notion of an all-pervading elastic "ether" as the carrier of these waves has long been disproven. Electromagnetic fields travel perfectly well through an idealized vacuum. In contrast to pressure waves of sound, elementary free-space radiation waves are treatable as *transverse*, meaning the fields that form the radiation, have their oscillatory variation in directions perpendicular to the wave-travel direction, as do shear waves of solids. Like all waves, these are spatially extended entities—a defining property of a field.

It is noteworthy that electromagnetic waves in dielectrics—or indeed, under conditions of tight-focusing or any spatial constraint—need not be transverse, and a variety of novel field structures involving longitudinal fields become possible [Bek11; Bli12; Bli14; Bli15; Tor15b; Bli17] in consequence of considering the nature of field momentum density (and angular momentum density) in a dielectric [Pad03; Pad04; Pfe07; Bar10; Bar10b; Bar10c; Mil10; Gri12].

2.3.4   Wave character

The familiar examples of traveling waves listed above all have one thing in common: They have some attribute that increases (to a peak value) and decreases (to a minimum, perhaps zero) over some characteristic distance (a *wavelength*) during some characteristic time (a period); see Section 2.7.3. That is, a wave is a *distributed disturbance*—along a line or on a surface or in a volume—that may have a characteristic length scale (of distance and time), but that cannot be localized completely, just as the beauty of a landscape painting cannot be localized more completely than the area enframed. In this respect wavelike phenomena are quite different from particle-like phenomena. Only with collective behavior in bulk matter are such waves evident, not with a few particles. Section 2.7 provides more details.

2.3.5   Granular character

In recent years the divisional organizing of professional physicists includes soft-matter physics, dealing with material whose properties differ from those of solids and viscous liquids. There has also come recognition that sand, sugar, salt, and other aggregates of *granules* [Jae92; Jae96; Ned05; Meh07; For08; Her13], can be poured and shaped by containers, but have collective characteristics markedly different from liquids. Sand and salt particles lack the interparticle attraction between water molecules and exhibit no surface tension.

   Aggregates of photons are not like any of these.

2.4   Free space: The Vacuum

It is common for physicists to imagine massive, moving objects, such as baseballs and rocket ships, as well as elementary particles such as electrons, to be replaced by hypothetical *point-particles* that have mass but no significant spatial extent around their center of mass (see Appendix A.1.2). Physicists also imagine space in the absence of objects of any kind: A *vacuum*, completely devoid of any matter or any particles—or even any radiation.[12] Outer space is a start, but even in the vast distances between stars or galaxies we have evidence that there are still a few atoms and molecules to be found. Certainly there is now known to be various forms of radiation supplementing visible pinpoints of starlight. And one reads increasingly often of "dark matter" [Tri88; Bar01; Kha02; Ber05b; Clo06; McD11; Cha15c; Aru17] and "dark energy" [Pee03; Aru17] that are now regarded seriously as entities to be considered when dealing with the vast scales of the universe; see Section 6.1. Section 5.1.2 presents an updated view of the quantum vacuum, a busy place that is far from devoid of matter and radiation.

---

[12]The *quantum vacuum* can be pictured in various equally valid ways; see page 335. In some pictures it is characterized by fluctuations in particle and photon numbers; see [Mil94].

The notion of Free Space, a complete vacuum, is an idealization, rather like the Frictionless Plane on which textbook-simplified motions of macroscopic objects might occur. It is in such a free-space background that one might imagine radiation traveling, at the speed of light c, between encounters with bits of matter (a viewpoint often associated with Richard Feynman, see Section 8.3). It is in such an empty void that one imagines a single atom to be placed, motionless, as it interacts with radiation.

Atoms and radiation in free space. Nowadays it has become possible to create situations very like this ideal: Single atoms or ions can be formed into beams in a vacuum chamber and slowed down, thereafter to be trapped in a cluster from which a single particle can be selected for study [Bla88; Phi98; Coh98; Jav06; Met12; Nev15]. Only relatively rarely will such a scrutinized particle be affected by unwanted collision with some contaminant particle or by any radiation other than that which an experimenter controls. So studies of single atoms or single ions, in otherwise empty space, are not the fantastic proposals they once were. Single atoms demonstrably exist and their internal electronic structure can be manipulated by radiation that an experimenter controls to a desired end, using techniques that are part of many physics-laboratory courses [Dem96; Dem98; Dem05] and whose theory I have explained in textbooks [Sho90; Sho11].

But what about radiation? Can it be regarded as made up of little increments of some sort, to be called *quanta* or photons? And if these exist, how are we to regard them and measure them, and with what mathematics are we to describe them?

These are questions that have occupied writings of many of the renowned physicists of the twentieth century. This Memoir describes my own encounter with those fundamental questions, and explains how my views (and those of the scientific community) were formed and have changed—and how our certainties have become less so as advancing technology has allowed more control and scrutiny of the weak radiation fields where individual photons are to be expected to be noticeable.

2.5 Forces and vectors

My formal introduction to the world of physics as a subject for study in classrooms came in a high school course in 1951. The presentation followed an expository route that had been in place for a century or so, and would continue for many more decades, a pattern that began with the notion of a *Force*—not, as has become part of entertainment culture, an intangible power for accomplishing good deeds ("the Force be with you" say the Jedi), but something measurable and reproducible.

2.5.1 Force

I learned that forces were not objects that could actually be seen, as could tables and chairs, but their effects could most certainly be observed. A push exerted by a playground bully would have had very definite consequences to me: I would find myself falling to the ground, perhaps acquiring an uncomfortable skinned elbow or knee. Forces caused playground swings to move. They caused the contact between bat and ball to advance the play of a baseball game. They moved chairs away from tables at mealtime. On and on the list could go. Forces, though not themselves visible, can definitely make themselves apparent. They are pushes and pulls that change the motion of an object, in a manner quantified by Newton's laws (see Appendix A.1.2).

2.5.2 Vectors

In high school I learned that there was a mathematics that went with the notion of force. It was very simple: A force has a *strength* (strong or weak) and a *direction*. If you were lifting a suitcase the direction was either up or down and the suitcase could be heavy or light. So your mathematics deals with magnitude (the strength) and direction. I learned that the technical term for this kind of object is *Vector*, something that has magnitude and direction. The simplest cases were those when the direction had only two possibilities, up and down, or left and right, or, when it was expressed as numbers along the real-number line, positive and negative.

We students learned about the mathematics of vectors primarily in order to deal with forces, and the trigonometry of resolving forces into components—two directions (sine and cosine components) for wind propelling a sailboat or for actions of gravity on ladders leaning against walls or, more generally, three directions in the space of objects completely free to move. And we considered corresponding vectors of position for the objects that the forces acted on, and for vectors that described changing positions—velocities (see Appendix A.1.1). Our vectors were (relatively) unconstrained in magnitude and direction, and were used to describe things like positions and velocities of everyday objects such as leaning ladders and baseballs and steam engines (which could still be seen in action). This was the world of Mechanics, regarded as a foundation of the larger study of Physics, but also a foundation for Mechanical Engineering—the stuff upon which a good career could be built; Vectors have not only intrinsic interest as a segment of mathematics but they can have economic value.

Discrete spaces. But we pupils did not give any thought to situations, such as describing the position of a checker on a checkerboard, that require only discrete whole numbers, integers that can, with some convention, specify which of the 64 discrete squares of a checkerboard is being considered. For the game of checkers the position vector of a piece has to be supplemented by a notion of the color, red or black. To treat a chess game the supplementary information must include not only the color, black or white, but the particular type of piece, pawn or bishop or king or queen or whatever. And the possible moves (the allowable discrete changes of position) of each piece can be described by a vector that leads from a starting square to a small set of possible ending squares. A checkerboard is an example of a flat (two-dimensional) space of finite size. Appendix A.3.1 discusses other everyday examples.

Abstract vectors. The velocities and forces discussed above are special examples of *Euclidean vectors* having three or fewer components and a number of relationships between these, notably how these numbers change with an alteration of position or orientation of coordinates. It is Euclidean vectors that mechanical engineers work with and that occur in equations of motion for particles—equations whose solution provides predictions of particle positions (see Appendix A.1.1).

Only after completing my public-school education, which in those days included instruction in arithmetic, algebra, geometry, and trigonometry but little else from the vast world of Mathematics, did I encounter the notion of (abstract) vectors as *ordered arrays of elements*—numbers or symbols or even words. In this view a five-dimensional vector is an ordered list of five elements, say the numerical values of measured temperature at five different weather stations on a particular morning. Applied to a family supper table the five elements of an abstract vector might be two spatial position coordinates, ordered as north-south and east-west, to which are appended the name of a piece of tableware that has those coordinates, the name of the family member by whom it is to be used, and a yes-no answer to whether it has been used at that meal and therefore requires washing. Such a mixed-element vector is typical of what programmers might deal with in providing instructions for a computer app or a robot.

Statevectors. As will be discussed in Section 2.12 and more fully in Appendix A.3, the mathematics of abstract-vector spaces provides a relatively straightforward approach to describing quantum systems. A particular system, characterized by its parts and their energies, is associated with a *statevector* in an abstract space. A task for those who wish a quantitative description for photon-changing alterations of a quantum system is to picture the motion of the statevector in its setting of a multidimensional abstract space. This activity is not so very different from following changes to weather maps on which are displayed daily inputs from a set of weather stations.

2.5.3   Forces from fields

My high-school physics studies of electricity and magnetism introduced me to the names of (Charles-Augustin) Coulomb, (Andrè-Marie) Ampére, and others, who had quantified the forces exerted by charges and currents on other charges and currents.

In principle these force laws allow one to evaluate the electric and magnetic forces that would act upon a small, idealized test charge, but in practice this evaluation is possible only for extremely simple, highly idealized systems of charges and currents. In general one must proceed as did Michael Faraday, who imagined the assorted charges and currents to be replaced by a spatial distribution of lines of force—an example of what physicists now call a *Field*. These space-filling force lines, of electric and magnetic fields, act as forces upon the test charges. I recall the fascination with which I played with iron filings on a sheet of paper above a magnet, as the filings patterns revealed magnetic field lines between what I learned were north and south poles.

The introduction of fields as a mathematical replacement for distributed charges and currents is an essential step in treating electromagnetic phenomena. It allows us to replace clumsy action-at-a–distance calculations with two separate and simpler calculations: First evaluating the field outside of a localized system of charges and currents, and then evaluating the effect of this field upon test charges. Fields, spread throughout space, rather than sources (localized moving charges), become the center of theoretical attention.

> **Aside: Mathematicians' notion of a field.** The quantities that physicists and engineers refer to as fields – typically a continuous distribution of force or energy – are examples of what mathematicians call *manifolds*. The typical mathematical approach to such things begins by defining a *topological space* in terms of a set of *points* along with a set of *neighborhoods* (or open sets) for each point. This

leads to defining properties such as *continuity* (smoothness and differentiability) and *connectedness* (gaps and cavities). For mathematicians a manifold is a differentiable topological space that at each point locally resembles an n-dimensional Euclidean space, $\mathbb{R}^n$. The everyday physical fields of fluids and electromagnetism are, amongst other things, three-dimensional differentiable manifolds for mathematicians.

Discontinuities. Physicists and engineers must often deal with idealizations in which there are abrupt (discontinuous) changes of field values with infinitesimal changes of position or time, or with *singular* field points at which field magnitude vanishes, so direction and phase have no definable value; see Section 4.1.4.

Superpositions. An essential characteristic of the fields considered by physicists is the possibility of *superposition*: We can add the values of two fields at a specified position and time, to produce a third field. It is this property of fields that is seen as interference between their waves and the production of node patterns. There is no counterpart in descriptions of classical particles, whether or not they are endowed with size.

Fundamental forces. Theoretical physicists like to say they have reduced the world of inanimate objects to the study of four fundamental force fields, from which all observable effects can be attributed. Two of these, known as the *weak interaction* responsible for radioactive beta decay and the *strong interaction* that holds atomic nuclei together despite the abundance of unbalanced positive charge of the protons, act only over subatomic distances and are of interest primarily to those whose research involves nuclear physics or to users of large charged-particle accelerators or to cosmologists who study the origin of the universe. They are considered to be important parts of any theory that attempts to explain Everything and their study has been rewarded by numerous Nobel Prizes over the decades. Appendix A.19 summarizes the Standard Model of particle physics that is regarded as the present frontier for such work. But these two interactions have negligible effect on household or university-laboratory experience.

Of the remaining two forces, both of them acting over indefinitely large distances, that of gravity is the most easily recognized. It is with us everywhere, unlimited in extent. We cannot shield against it nor alter its action—attraction between masses. On earth it dominates our daily activities: It keeps us standing and causes the flight of thrown balls to descend for possible capture.

The final force, that of electromagnetism, is responsible for all the non-gravitational forces we observe in our everyday lives. The chemical forces that draw atoms together into molecules, the forces of attraction between neutral molecules, these are all understandable results of quantum theory together with electrostatic interactions, as developed by chemists and described in their texts on atomic and molecular structure (see references in Section 9.2). There are two types of electric charge, positive and negative, and these are normally paired so that bulk matter is generally electrically neutral. For this reason the observed static fields outside bulk matter are usually magnetic rather than electric. It is with the extended, time-dependent electromagnetic field, able to break free, as radiation, from the moving charges that create it (see Section 2.8), that photons are to be found, and with which this exposition deals.

Although the strength of gravity and electrostatic forces both diminish with the same dependence on distance, their influence appears in different regimes. Atoms and molecules are held together entirely by electric and magnetic forces. The attraction of gravity has negligible effect in the immediate surroundings of atomic masses and in dealings with atomic and molecular physics it is justifiably neglected, apart from causing atoms to fall into boxes of radiation. (As noted by Feynman [Fey63], the gravitational attraction between two electrons is smaller by around 4×10^{42} than their electrostatic repulsion.) By contrast, stars, solar systems and galaxies are held together by gravity: Because opposed electrical charges occur in pairs, with resulting exact cancellation of charge seen at a distance, electric forces need not be taken into account for these structures. Magnetic forces are not so cancelled, and their effects are seen, for example, in the flow of ionized gases of the solar atmosphere.

Fields and quanta. Quantum theory provides a connection between the wavelike properties of a field and the discrete quanta that carry the field attributes such as mass and electric charge. There is an intimate connection between the range of a force and the mass of the quanta associated with the force field. Short-range forces, such as those of the weak and strong interaction, have quanta of finite rest mass: The larger the mass, the shorter the range. Their field quanta are the bosons of the Standard Model of particle physics, see Appendix A.19.[13]

The electromagnetic field in free space, by contrast, extends indefinitely, with magnitudes falling inversely with distance from the source. The waves move in vacuum with speed of light c and the quanta of electromagnetism—the photons—therefore have zero rest mass.

Effects of gravity are described by the general theory of relativity, in which gravitational forces are attributed to curvature of four-dimensional space-time induced by large masses and photons travel along mass-distorted free-space paths; see Section 6.1.2. There is presently no generally accepted quantum theory of gravity. The quanta of gravity (*gravitons*, spin-two bosons) remain to be observed, although its waves have been observed (with great difficulty).

2.6 Energy and heat

Although I realized as a schoolboy that forces were not visible, as were aggregates of particles, they certainly produced visible effects: They could cause motions of objects. Some forces were evidently steady but seemingly inactive: When you are standing on a structurally weak bench that breaks under your weight you learn that prior to the breakage there had been a force to keep you in place, but once that bench broke then the force of gravity, which had been there all along, was able to pull you steadily toward the ground, perhaps with uncomfortable consequences. But forces could also be deliberately transient: A push or pull might make its effect (some displacement) known but might cease thereafter.

---

[13]Particle physicists regard the Universe as comprising <u>matter</u> fields, having fermion quanta, and <u>force</u> fields, having boson quanta.

Work and energy.   Could the effects of a transient force be stored in some way? That question led our physics instruction to the notions of Work and Energy. We were taught that when a force moved an object through some distance (carrying a pail of water up a hill as Jack and Jill allegedly did, for example) the product of the two numbers (say a kilogram and meter) was to be regarded as *Work*, a technical term that we were not to confuse with Labor, which was what one had to do for payment or as a regular family chore—things like mowing a lawn or weeding the garden or just picking up toys. We schoolboys found it very amusing that these tasks, though they might be drudgery, were not Work as physicists knew that word. Work, for a physicist, meant a number, the product of force times distance. Labor, as economists use the term, is something with little connection to the Work as defined in physics, but has everything to do with daily life.

The interesting thing we scholars learned about Work is that it creates stored energy—Potential Energy—that can later be turned into energy of motion—Kinetic Energy. Physicists have formulas for these two sorts of energy.

> **Aside: Energy formulas; mass.** The kinetic energy of structureless particle of mass m is proportional to the square of its velocity v,
>
> $$E^{\mathrm{kin}} = \tfrac{1}{2}\, m\, v^2. \tag{2.6-1a}$$
>
> The potential energy is proportional to the distance moved by a force. A mass m raised a height h against the pull of gravity (acceleration constant g) has potential energy
>
> $$E^{\mathrm{pot}} = m\, g\, h. \tag{2.6-1b}$$
>
> The slope of the position dependence of potential energy—height or displacement from equilibrium—is a force. It is this force that causes balls to roll down hills, and causes a clock pendulum to oscillate; see Appendix A.1.4.

In each energy there is a proportionality constant that expresses the amount of material undergoing motion—what the textbooks called *mass* and denoted here by m.

Conservation of energy.   The neat thing about the notions of energy, from the viewpoint of a teacher constructing exam questions or a student seeking intended answers, is that it enables you to figure out how fast your dropped hammer will be going when it hits the ground below your ladder or how fast your car can coast at the bottom of a hill of some given height. Such problems are solved by invoking the principle that energy of different forms can be converted from one form to another – combining eqn (2.6-1a) with eqn (2.6-1b). This principle—Conservation of Energy – provided a major argument for the existence of photons, at least as energy increments: If atom energies are quantized (see Section 3.2) then so should be the energies carried to and from them by radiation fields. (This was one of the ideas associated with the photons of Einstein, see Section 3.4, and of Dirac, see Chapter 4.)

> **Aside: Energy units.** Workers in atomic and molecular physics often express energies in *electron volts*, eV. This is the energy gained, or lost, when a single electron moves through an electrostatic potential difference of one volt. In SI units [see eqn (A.1-13)] 1 eV is approximately 1.6×10^{-19} joule (J). Energies required to alter internal structures of atoms and molecules are typically a few eV; see Section 3.3. The energy required to photoionize a single hydrogen atom

is 13.6 eV. A photon having energy 1 eV would have an infrared wavelength 1240 nm and frequency 241.8 THz; see the table on page 73 and eqn (3.3-1).

Hamiltonians. When I gained the use of tools from differential calculus I learned that force can be expressed as the slope (directed spatial derivative) of a potential energy—balls roll down hills of potential energy. Energy thereby becomes a key director of motion; see Appendix A.1.4. The set of rules that relate energy to such measurable properties as mass, height and speed (for example eqn (2.6-1a)) form a *Hamiltonian*.[14] From the Hamiltonian for a particular collection of particles or other inanimate objects one can determine their possible motions and system changes—the falling of rocks and the creation of photons. Appendix A.1.6 provides details.

Friction and heat. But textbook calculations that rely on conservation of mechanical energy, if they are to be applied to real-world activities, have to deal with the reality of Friction Forces. These are what brings a sliding object to rest. These forces convert motional energy into *Heat*. Every child knows about friction heat, and about heat from fires and stoves. Heat is very detectable by our built-in body sensors. We learn very early in our lives not to touch the hot stove, and we recognize the pleasure we have when the home surroundings hold sufficient heat (as measured by temperature) to allow us to run barefoot outdoors. We know heat by the sweating it causes as the human body attempts evaporative cooling. We know *cold* as the absence of heat, and the need to prevent loss of body heat—accomplished by using layers of warm clothing. So heat is definitely something perceptible and measurable, like positions of objects (and potential energy) and like velocities of objects (and kinetic energy).

Temperature. As I was to learn from post-college instruction in *Thermodynamics*, heat is disorganized energy, only partially available for doing useful work. On a microscopic scale, heat is the amount of energy in random motion of atoms and molecules; see Appendix B.7. *Temperature* is the measure of the average kinetic energy of these constituents of matter.

> **Aside: Boltzmann's constant.** The numerical equivalence of an energy increment and a temperature increment (in degrees Kelvin, K) is given by the *Boltzmann constant* k_B:
>
> $$k_B = 1.380649 \times 10^{-23} \ \text{J/K}, \quad \frac{1}{k_B} = 11604.5 \ \text{K/eV}. \tag{2.6-2}$$
>
> In a gas of freely moving classical particles that are in thermal equilibrium with their surroundings every *degree of freedom* (DoF), meaning every independently specifiable, unconstrained coordinate in an equation of motion (see page 193), has an average kinetic energy of $k_B T/2$; see Appendix B.7.

Entropy. Energy that cannot be converted into useful work (in the technical sense of force times distance), and is therefore lost for practical use, is quantified as *Entropy*,

---

[14]Strictly speaking, a Hamiltonian is an expression of energy as a function of coordinates and momenta; see Appendix A.1.6. In practice, a Hamiltonian is a collection of relationships between energies and experimentally controlled parameters.

the amount of heat, as disorganized thermal motion, present in a sample of matter. The nineteenth century development of thermodynamics dealt with incremental changes of entropy as the ratio of incremental change of heat energy, during a reversible process, to the temperature T at which that change occurred. At a temperature of zero degrees Kelvin, the random thermal motion of matter constituents is minimal (it can never be completely absent), and this is taken as the zero point of entropy. Entropy need not be conserved: It can increase indefinitely. The notion of entropy, as a measure of randomness, is applicable not only to the statistical mechanics of particles (and photons) but to information [Sha48].

Chemical energy. The rather obvious exchange of mechanical energy between potential and kinetic forms has a counterpart in the chemical energy atoms have by virtue of their arrangement into molecules. As youngsters with Gilbert Chemistry Sets learned, certain sorts of atoms when brought together could form compounds with release of heat. Violent explosions could occur as atoms rearranged themselves. My first school course in Chemistry taught us about *exoergic* chemical changes that gave off energy and *endoergic* changes for which energy had to be supplied. Compounds were seen to be stores of Chemical Energy.

In college I learned that chemical changes occurred by breaking and remaking the electron-mediated chemical bonds that gave structure to molecules and solids. Only much later did I learn that energy changes of atoms and compounds could take place in non-damaging, reversible increments, induced by photons and discussed in Appendix A.10.2. These changes amount to reversible rearrangement of the internal structure of atoms and molecules. They are the changes ("coherent excitation") induced by laser radiation with which I have spent the last few decades, treated in expositions on coherent excitation [Sho90; Sho08; Sho11; Sho17]. Appendix A presents the mathematics used to describe these changes.

Energy sources: Atomic vs. nuclear. Prior to the 1950s almost all of our household energy came from the heat of chemical reactions—rearranging the electronic structure of hydrocarbon molecules and breaking them apart. The earliest domestic uses of fire by cave dwellers came from the rearrangement of wood molecules into smaller gaseous molecules. In time, power-line supplies of electrical energy came ultimately from combustion of coal or natural gas. All of these combustion-heat sources rely on rearrangement of electrons within atoms, and so they can be accurately called "atomic energy sources". The responsible molecules are broken apart (split) and their constituent atoms are rearranged, with release of stored chemical energy as heat. Studies of these processes are traditionally found in courses on Atomic Physics, Molecular Physics, and Chemical Physics.

The nuclei of the heaviest atoms, most notably uranium, will break apart into smaller "fission fragments" with release of energy and free neutrons when they absorb a free neutron from their surroundings. Such nuclear-fission reactions, and the heat energy obtained by "splitting the nucleus", are accurately called "nuclear energy sources" when they are controlled in nuclear reactors. They are studied in courses titled Nuclear Physics. Used militarily they are "nuclear weapons". The associated restructuring of the atoms that surround the newly created nuclei are but a minor portion of the energy changes. Ion accelerators such as cyclotrons are often called "atom smashers", although it is atomic nuclei that are the targets for their disruptive action.

Conservation laws. Energy is but one of the things that physicists list for conservation (in the absence of friction). The conservation of electrons (electric charge) underlies the analysis of electric circuits by electrical engineers. As embodied in *Noether's theorem* (see Appendix A.18.5), conserved quantities are often associated with some symmetry in the equations of motion. Although photons are not conserved, a number of other electromagnetic quantities are conserved [Lip64; Kib65; Cam12; Bar14].

2.7 Equations of change: Particles and fluids

The mathematics of Physics (and its dependent Engineering world of compromise and optimization) deals, in large measure, with equations that describe how Systems (collections of interacting parts, say linked pieces of a machine) change with time. I shall frequently mention *"equations of motion"*, meaning equations that govern the rate of change in some property, often the location of a localizable particle or a fluid mass. For solid objects the equations express how particle positions (and velocities) change when subject to forces; see Appendix A.1.

Other equations describe how forces acting on particles change with distance. The changes of interest are typically expressed as incremental changes to some observable (say position or velocity) resulting from some infinitesimal change (a differential) occurring in an independent variable (space or time), and so these are differential equations. Calculus is the broad branch of Mathematics that deals with their solution.

When the system we treat has motion in several directions or has several parts, each of which can affect others, we deal with sets of several interdependent (coupled) differential equations that must be solved simultaneously. A variety of techniques find use in obtaining solutions, either numerically or in algebraic expressions involving various well-studied functions bearing names of mathematicians. Appendix B.2 discusses several examples.

The lengthy literature on photons suggests that, under appropriate conditions, they can be regarded as particles or as waves. In finding appropriate equations for photons it is therefore useful to look not only to the physics of particles but also to the physics where waves are to be found—to fluids and their wave equations.

Rates of change. In discourse on motions there is always an unstated notion of "fast" and "slow". It is always necessary to place these adverbs into some context. Recollected extremes from childhood include "faster than greased lightning" and "slower than molasses in January". Applied to quantum-state change these distinctions are often termed impulsive or *diabatic* (fast) and quasistatic or *adiabatic* (slow); see Appendix A.13. In discussions of atomic processes a common reference interval is the lifetime of a given atomic state awaiting spontaneous emission, or the mean time between phase-interrupting collisions. In much of the discussion in this Memoir the changes to quantum systems are assumed to occur in a time interval that is shorter than a time that measures interruptions (decoherence time); see Section 5.1.

2.7.1   Particles: Newtonian mechanics

These (classical) equations describe, for example, how the gravitational force between two masses (or the Coulomb force between point charges) diminishes as they become increasingly far apart. The equations describe how the velocity of a massive object (the time derivative of a varying position) changes with time when forces act. For objects much larger than an atom (but smaller than the solar system) the common rule for calculating the rate of change in velocity $\mathbf{v}$ (i.e. the acceleration $\mathbf{a}$) is that this equals the ratio of force to mass, $\mathbf{a} = \mathbf{F}/m$. This equation of motion, known as one of Newton's laws, is at the heart of what is called (nonrelativistic) Classical Dynamics (see Appendix A.1) to distinguish it from motions over distances comparable to the size of atoms, where the behavior of electrons is governed by Quantum Mechanics. Underlying these equations is the assumption that the mass-endowed objects of interest are *localizable*—that it is possible to measure, and thereby define, positions for all the constituents of a system. Quite often these can be idealized as having mass localized at a specifiable position (the *center of mass*, a mathematical point) and having angular momentum centered at that point. This localization assumption is the essence of what we regard as a "particle". Appendix A.1 presents the needed equations.

Particulate light. Newton's laws of motion predict straight-line trajectories for particle motion when no forces act. In free space or uniform, transparent matter, unfocused light rays also follow straight-line paths. It is such ray paths that gave an early impetus for regarding light as corpuscular and which underlies lens design software today. But the use of discrete ray paths in optical design does not mean invoking the discrete photons of Planck, Einstein, and Dirac.

2.7.2   Wave equations

For descriptions of continuously distributed properties such as mass, velocity, electric charge, and force, the appropriate equations of motion are Partial Differential Equations (PDEs) whose two or more independent variables bring a connection between changes in time and changes over distance. Particularly noteworthy for eventual connection with photons are changes that appear as *waves*—patterns of regularity in some property value; see Section 2.3.2.

Maxwell equations for radiation. The essence of a particle is the notion of localizability. Radiation is rather different. The basic equations of electricity and magnetism, combining as electromagnetism (see Appendix B) are for electric and magnetic fields, envisioned as spatial distributions of forces that act on electric charges and which, in turn, are created by electric charges and currents. The fields themselves are, at every location and at every time, Euclidean vectors: They have a magnitude and a direction. And these vectors, one for the electric field and one for the magnetic field, may change over distance and with time. The changes, expressed as partial differential equations, are known as the Maxwell equations; see Appendices B.1 and C.2. They describe not only the static fields surrounding fixed point charges and permanent magnets but also radiation fields: Electric and magnetic fields that move through empty space at a constant velocity c, the velocity of light in free space. The basic Maxwell equations for fields in vacuum can be presented as four interconnected vector equations for six field-vector components. (Because they are partial differential equations, and because such things were regarded by curriculum writers as difficult mathematics, it was not until I was a graduate student that I saw the Maxwell equations in full.)

The four Maxwell equations are first-order equations, meaning that only first derivatives appear—three for spatial change, one for time change. For treatments of electromagnetic fields moving through space it is common to combine the four first-order equations into two equivalent second-order equations (i.e. equations involving second derivatives, see page 191), one for the electric field, another for the magnetic field (see Appendix B.1). These second-order PDEs are examples of *wave equations*: They have solutions that can describe either standing waves of stationary node patterns that form in enclosures or the traveling waves of moving nodal surfaces that appear as radiation beams and other constructs; see eqn (4.1-1).

Wave equations for electrons. A consequence of the uncertainty principle, re-garded as an expression of how Nature works, is the need to describe the behavior of electrons inside atoms as governed by a wave equation—an equation akin to those needed to describe electromagnetic radiation waves and known as the (time-independent) Schrödinger equation (see Appendix A.4.3). Its solutions are *wavefunctions*. The solutions to any wave equation shows patterns, in space and time, of places where the wave amplitude vanishes (a node) and places where the amplitude is a local maximum (an antinode). As quantum theory informs us, the electrons within atoms behave very much like an electric vapor, one that exhibits the sort of interference effects—nodes and antinodes—that are characteristic of waves in general—that occur in vibrating drumheads and tuning forks or that travel around obstacles. Section 4.1 discusses wave patterns expected from the charge distribution of electrons bound in atoms.

> **Aside: The de Broglie wavelength.** Any particle having momentum $\mathbf{p}$, rest mass m_0, moving with speed v in free space (far from any centers of attraction), has a wavelike behavior quantified by the *de Broglie wavelength* λ_{de},
>
> $$\lambda_{de} = \frac{h}{|\mathbf{p}|} = \frac{h}{m_0 v}\sqrt{1 - (v/c)^2}. \tag{2.7-1}$$
>
> This wavelength diminishes, and the particle becomes more localizable (pointlike) as its speed increases. For a particle with speed much less than the speed of light c this wavelength varies inversely with mass and velocity, $\lambda_{de} \approx h/m_0 v$, becoming large without limit as the particle slows. Such wavelike behavior is associated with the probability amplitude—the wavefunction—for any monoenergetic particle in free space; see Appendix A.5. Electrons are the lightest particles and so for given velocity their de Broglie wavelength is largest.

Wave-particle electrons. An electron can be localized as being within the small volume of an atom. If we do not enquire about its position more accurately than that, it can be idealized as is done in textbooks on classical motions: Its mathematical description is that of an electrically charged point particle, having a sharply defined position for use with Newton's laws of motion (suitably altered at high velocities to incorporate special relativity).

2.7.3 Wave attributes

A great variety of optical effects, studied as *Physical Optics*, involve phenomena that can be best understood by endowing light with wavelike properties: wavefronts of valleys (including null values—*nodes*) and crests (*antinodes*) that, in free space, move with constant speed c, the speed of light c in vacuum. When applied to optical radiation the wavelike properties refer to the electric and magnetic field vectors and to their squares, measured as radiation intensity, all of which are expressible as functions of space and time.

Wavelength and frequency. The distance between nodes, or antinodes, of any wavelike pattern is the *wavelength* λ. At any fixed location a passing wavefront of wavelength λ oscillates with *frequency* ν. Successive waves need not be identical; during an interval when the cycles repeat exactly the wave is *periodic*.

> **Aside: Units.** The unit of frequency ν, the *hertz* (Hz) is one cycle per second. The connection between frequency and wavelength of any disturbance is through a wave velocity. For light in free space this is the vacuum speed of light,[15] c:
>
> $$\nu = c/\lambda, \qquad c \approx 300 \times 10^6 \ \text{m/s}. \tag{2.7-2}$$
>
> **Aside: Angular frequency.** To avoid a plethora of 2π factors the mathematics of oscillatory behavior is commonly treated with *angular frequency* $\omega = 2\pi\nu$, measured in radians per second. A full cycle of rotation is 2π radians. Angular frequencies accompany reduced wavelengths $\lambdabar = \lambda/2\pi = c/\omega$.

Phase. Idealized, single-frequency (monochromatic) light is a periodically varying electromagnetic field: Its characteristics repeat regularly after a time interval (the wave period) $1/\nu$. During one period the field value will vary between its two extremes. Starting from a minimum (or any other arbitrary fiducial value), the subsequent time interval is quantified by a *phase*, varying from zero to 2π (radians) at the next repeat of the fiducial value (one period) . The moment when we start counting waves, say the first minimum value, fixes a phase for the enduring wave pattern: Further minima occur at times n/ν for integer n as long as the phase is unchanged.

The maintenance of constant phase, or a phase whose variation can be specified in closed form, makes the field *coherent*, and allows interference between field samples.

Interference. Amplitudes of any sort of wave are additive, so waves from different sources will, upon meeting, exhibit *interference*, producing weak or null intensities (destructive interference) when amplitudes cancel and peak intensities (constructive interference) when amplitudes reinforce each other. Wave properties include both interference and *diffraction*—the spreading of wavefronts that pass through a small opening in an opaque screen.

The existence of interference is demonstrated in electromagnetism and in quantum theory by observations on squares of amplitudes. In electromagnetic theory the amplitudes are electric and magnetic field magnitudes and the squares are radiation intensities; see Appendix B.1.1. In the quantum theory of atom structure the squares

---

[15]By international agreement c is defined to be exactly 299 792 458 m / s.

are probabilities and the amplitudes are wavefunctions or probability amplitudes; see Appendix A.3.

When optical intensity is a superposition of two or more competing field amplitudes their magnitudes can either add (constructive interference) to give a large (bright) intensity (maximal at an antinode) or subtract (destructive interference) to give a small (dark) intensity, perhaps zero (a node). The resulting bright-ark pattern of measurable intensities, in space or in time, is termed optical interference; see Figure 4.1 on page 97.

Quantum interference acts in the same way. When a probability is a superposition of two or more probability amplitudes these may contribute either constructively or destructively, to produce a pattern, in space or time, of varying high and low probabilities.

When traveling photons are invoked to treat examples of optical phenomena we consider superpositions of possible pathways leading from a source to a detector. If the pathways are indistinguishable then we add amplitudes before squaring. This allows interference. If pathways can be distinguished, even if only in principle, then we add separate squares. There is then no interference.

Interference, and the underlying superposition of different wave amplitudes, is the defining characteristic of waves, just as localization is considered the defining property of particles. It is therefore with measures of optical interference—in space or time—that we must look for evidence that optical photons have a wavelike nature. As was stressed by Glauber [Gla63; Gla95], what interferes in quantum theory are probability amplitudes (for specified events, such as electric field strength), not some notion of particle-like "photons". One might say[16] "photons follow from fields", not the reverse.

Energy quantization. Both the time-independent Schrödinger equation for the stationary distribution of an electron within an atom and the second-order Maxwell equations for free-space radiation have solutions that exhibit patterns of nodes and antinodes that is a defining characteristic of waves. Although the wave nature of radiation is evident in all the mathematics of the Maxwell equations, the wave properties of the electron, as given by the Schrödinger equation, come into play only when our concerns are with atomic-scale behavior. Then it becomes evident that something beyond classical mechanics is required. The most notable effect is that the energy of an electron, or groups of electrons, confined within an atom or molecule (or a nanoscale structure), can only take selected discrete values: The energies, combining potential and kinetic energy, form a discrete set and are said to be *quantized*.[17] This quantization is evidence that, although the electron has properties of a classical particle, it also has wavelike properties that invoke quantum theory—it is a *quantum particle*.

Not only are energies of electrons within atoms quantized, so too are the vibrations of atoms in molecules and solids, and the unhindered rotations of molecules. As older textbooks explained, the discreteness of these energies first became apparent in

[16]In a paraphrase of the maxim "form follows function" of architect Louis Sullivan.

[17]The second-order equations of electron distribution and radiation modes have discrete sets of solutions when boundary conditions enclose the field.

measurements of *specific heats*, in which energy exchanges with a thermal environment produce a temperature change. The transferred energy may be not only radiation but kinetic energy from particle collisions.

2.8   Light: Electromagnetic radiation

Sunlight is so ubiquitous in our lives that it seldom is directly noticed—except when it is absent or overly abundant. One of the welcoming characteristics of sunlight, and of other sources of radiation, is that it provides heat. It is that absorbed heat that helps popularize lying on warm sand at a seaside beach. As I learned eventually, visible light, whether from the sun outside or from a glowing lightbulb indoors, is a form of electromagnetic radiation, as are radio waves and X-rays. All of them carry, and can deposit, energy. And they all have wavelike characteristics—they are delocalized disturbances.

Electromagnetic waves: Optics. Already in the nineteenth century the unification of electricity and magnetism into *electromagnetism* by Maxwell provided the needed theoretical understanding for light as traveling waves of electromagnetic energy—of spatially distributed electric and magnetic field vectors; see Appendix B.1.1. Visible light was but one portion of the vast catalog of wavelengths that ranged from much longer radio waves and infrared (IR, the source of much solar heat) to shorter ultraviolet (UV, once termed "black light" because it was not visible to our eyes) and much shorter X-rays and gamma rays, and on without limit.

   Although it is common to regard all visible radiation as comprising photons, the sources of this radiation are remarkably varied, and the nature of the light, as recorded by its *spectrum* (the distribution of frequencies, responsible for the appearance of color), is accordingly also quite varied.

Elementary sources: Accelerated charges.   Any attempt to understand radiation must, at some point, deal with the possible microscopic mechanisms that generate it. Once it became recognized that radiation was a traveling electromagnetic disturbance, a phenomenon governed by Maxwell's equations, it was possible to find a description of radiation sources as motions of electrical charges.

   The simplest example, familiar to youngsters from demonstrations of electrostatics via shoe shuffling on carpets, is the electric field that surrounds a very localized electric charge (idealized as a mathematical point), say that of an electron. While the charge remains stationary the electric field, the Coulomb field, diminishes monotonically with distance, but when the charge undergoes acceleration, from whatever cause, the field falls more slowly: At large distances from the source charge it takes on characteristics we associate with traveling waves of radiation. Indeed,

> *all the radiation commonly encountered can be attributed to localized sources of accelerating charges.*

Appendix C.1 quantifies the connection between acceleration and radiation, for point charges. The force that causes the acceleration may come from encounters with other charged particles (in a hot gas) or it may come from an electromagnetic field, either static or oscillatory (i.e. radiation). The more abrupt is the acceleration the higher are the dominant Fourier frequency components of the associated radiation (and the more energetic are the photons).

The acceleration need not be steady; it can undergo reversal. When the source-motion reversals are periodic (as occurs with a harmonic oscillator, see Appendix A.1.7) the field carries this frequency and is idealized as a *monochromatic* (i.e. single color) traveling wave; see Appendix B.2.2. This simple model of a charge (say, an electron) oscillating around an equilibrium position and thereby radiating, guided much of the thinking about radiation sources that underlay the early proposals for discrete increments of radiation; The model, originated by Hendrik Antoon Lorentz and subsequently justified for a two-state quantum system, is discussed in Appendix A.14.4.

When the accelerating charges are bound within an atom, molecule, or other composite the radiated electromagnetic energy originates with loss of structural energy (potential and kinetic) of charges within their confinement—a change of rest energy (see Section 8.3.1) of the composite. Because states of internal motion exhibit identifiable quantum characteristics such radiation will also have measurable quantum character. Alternatively, the charges may be unbound, unconstrained in momentum and energy. The fields from such acceleration draw upon changes in unquantized kinetic energy.

Quantum theory does not alter the prediction that accelerating charged particles will radiate, but it imposes an important proviso: When the motion is constrained, as it is for electrons bound in an atom or molecule, then the system change can only take place between discrete quantum states. Furthermore, the state of lowest energy will not radiate; it is a stable state.

Not all acceleration produces radiation. Macroscopic objects that are charged but structured may create radiation fields in which there is phased destructive interference from different parts, so the overall external radiation is not seen [Abb85]. It is possible to design bulk matter so that local induced dipole moments contribute destructively to incident radiation, thereby rendering objects (nearly) invisible [Kra19].

Radiation sources can be categorized in two broad classes, *incoherent* (disorganized) and *coherent* (organized), distinguished by the arrangement of the accelerating charges within the source. Light signals emerging from these classes of sources maintain their coherence properties. During the time interval when a signal is coherent it is predictable. The irregularity of an incoherent signal prevents reliable prediction for times longer than its coherence time. The distinction between coherent and incoherent sources is important when considering the nature of the photons associated with the radiation. The following paragraphs define, and illustrate, the difference.

2.8.1　Coherence

As Newton first demonstrated by passing sunbeams through prisms, visible light, from whatever source, can be separated into a range of colors, each quantified by a specific frequency or wavelength. It is also possible to pass light through optical filters that transmit only a very small range of frequencies.

Idealized, single-frequency (monochromatic) light is a periodically varying electromagnetic field: Its characteristics repeat regularly after a time interval $1/\nu$. The moment when we start counting waves, say the first peak value, fixes a *phase* for the wave: Further peaks occur at times n/ν, for integer n, as long as the phase is unchanged.

Experience with actual light sources never meet this ideal periodicity, if only because the observations do not continue indefinitely; see Appendix B.5.1. The deviation from perfect periodicity—a perfectly coherent wave train—can occur for many reasons. Beams of frequency-filtered sunlight and beams from lasers differ qualitatively in their *coherence*—the time duration of perfect periodicity—and these differences are responsible for observable qualitative differences in how their light affects single atoms. In brief, laser light is able to produce *coherent excitation* [Sho90; Sho11; Sho17], see Section 5.4, whereas sunlight and incandescent light cannot produce such effects, see Section 5.2.

With incoherent sources it is possible to use filters that restrict the frequency to a narrow range (a narrow *bandwidth*), but it is not possible to eliminate all evidence of the underlying randomness of the many emitting electrons whose combined output produces the light. This remains evident in measurements of correlation between field values at different times; see Section 6.5.2 and Appendix B.5.3.

2.8.2 Incoherent sources

Amongst the most common readily available sources of visible light are the following. In each of these the motions of the elementary charges responsible for the radiation do not maintain any regular pattern and their individual fields combine irregularly. The bulk radiation is therefore disordered.

Sunlight. The most ancient of our light sources is undoubtedly that from the sun. As we now understand, this and other stellar sources is energy from a heated gas: Unconstrained collisions between atoms, ions and free electrons produce a very broad spectrum of frequencies (very close to a "blackbody" distribution) as energy alternates between kinetic energy of moving charges, potential energy of bound electrons, and traveling radiation.

Firelight. Controlled-burning fuel—a candle, a gas flame, a fireplace log—produces a flame of glowing gas whose radiation originates in the spontaneous emission of light from electrons of heated atoms and molecules undergoing chemical changes. The numerous constituent frequencies of the light are characteristic of the fuel composition.

Gas discharges. Several generations of experimental physicists made use of light that was generated by gas discharges: A glass tube that enclosed two voltage terminals, between which there was established a suitably high voltage. Electrons bound to gas molecules would, under the influence of the electric field, emerge from their binding molecule, accelerate under the influence of the electric field, and collide with other electrons (bound or free) to create an ionized vapor. The various processes would equilibrate, with some electrons becoming rebound and releasing their energy as radiation. The frequencies of the light were an indicator of the chemical composition of the gas.

Incandescence. Any heated metal will, as the temperature rises, be seen to glow; first red, then orange, then blue-white. The charges responsible for this light are electrons, confined to the metal. The light spectrum, like that of the sun, is very broad, following closely a blackbody distribution characterized by a temperature. Until recently the most common sort of household lighting apart from flames, dating back to invention by Thomas Edison, relied on the glow of a metal filament heated by electrical current inside an evacuated glass envelope—a light bulb. Thermal motion of electrons (in the filament) underlies the emission of radiation, termed *incandescent light.*

Fluorescence. A common alternative form of retail and workplace light comes from *fluorescent lamps.* The light in these originates with a glowing, electrically-driven gas discharge that produces ultraviolet (UV) light. The walls of the translucent glass enclosure are coated with compounds that absorb this light and reemit the energy, by spontaneous-emission fluorescence, in a range of other frequencies that partially fill out the visible spectrum. The accelerated charges are electrons, localized within atoms.

Phosphorescence. Some materials, having been exposed to light, will thereafter glow with delayed emission known as *phosphorescence.* The source atoms have been placed, during their exposure to illumination, into *metastable* excited states that decay only slowly, in the same way that radioactive sources emit their radiations. Here too, the motions of confined electrons produce the radiation.

Chemiluminescence. Various chemical reactions will produce products in states of electronic excitation that will subsequently emit light, for example glow sticks. This phenomenon is occasionally seen in nature as a will-o-the-wisp. When the process occurs in a living organism it is known as bioluminescence.

Light-emitting diodes. Nowadays much of the illumination we see originates with light-emitting diodes (LEDs) rather than the incandescent lightbulbs that once filled our household light sockets and automobile headlights. The light from these devices, like the incandescent or fluorescent sources that preceded them, is incoherent: It originates with spontaneous emission by electrons undergoing energy change in the spatial interface between n-type and p-type semiconductors, that is, a region between a solid whose structure makes available negative electrons (n-type) and one that has electron vacancies, or positive holes (p-type). What emerges is light of a frequency associated with a particular transition and spatially distributed throughout the transparent interface region. (White-light output is obtained either by incorporating three distinct frequency transitions or else converting a portion of blue light into yellow light by means of phosphors that absorb and reemit light.)

Bremsstrahlung. Any collision between charged particles involves an acceleration and hence the generation of radiation. The resulting radiation, known as *Bremsstrahlung* (German for "braking radiation"), occurs for inelastic scattering of any pair of charged particles. An example occurs when a free electron passes by an atomic nucleus: The closer is the approach the greater the acceleration and the more complete is the conversion of kinetic energy into radiation. A beam of electrons impinging on a metal target undergoes a variety of collisions and will emit a corresponding range of frequencies, up to the total incident kinetic energy of the electrons. Bremsstrahlung from beams of kilovolt electrons are the source of the continuum of frequencies found in X-ray tubes.

The stopping of electrons by solids also accompanies collisions that eject tightly bound electrons, leaving holes in the atomic structure. In refilling those vacancies the rebinding electrons emit discrete energies associated with the vacancies. These appear as *characteristic X-rays* on the continuous background.

The Bremsstrahlung occurring in thermal collisions in a hot plasma is known by astrophysicists as free-free radiation, between unquantized scattering states of free electrons. Free-bound radiation originates with an unbound electron and concludes with the electron in a discrete bound state, a process termed *radiative recombination*. The inverse of this process, *photoionization*, is known as bound-free radiation.

Proton-induced X-ray emission (PIXE). Although beams of protons in cyclotrons have traditionally been used to induce nuclear reactions (and were part of my thesis research [Sho61]), in recent years they have found new uses in non-destructive chemical analysis. A proton, passing a target atom, transfers kinetic energy to one of the tightly bound inner-shell electrons, expelling it from the atom. The filling of this vacancy by other electrons leads to emission of discrete-energy photons whose frequency provides a unique connection to the chemical element. The perfected technique, known as proton-induced X-ray emission (PIXE), is capable of determining the chemical composition of the ink used on individual letters of ancient manuscripts without harming or discoloring the document, a capability that was used in providing remarkable information about the printing procedures used by Johannes Gutenberg [Cah80; Cah81; Dav19]. The message conveyed by these freshly minted photons, about chemical composition of ink and paper, offers a vision of daily life from centuries past, told well by Margaret Davis [Dav19].

2.8.3   Coherent sources

The disorganized individual radiators of the sources listed above will, in turn, produce disorganized, incoherent radiation. We can expect them to produce single photons at random or uncorrelated crowds of photons. A number of radiation sources offer examples of radiation from organized groups of electrons, or from individual electrons. These sources have particular interest for studies of individual photons.

Radio sources.   The first class of controllably-generated coherent electromagnetic waves outside of the visible portion were those, originally called Herztian waves for their demonstrator Heinrich Hertz, in the spectral region now termed radio waves. They range from extremely low frequency (ELF) of 3 Hz to 30 Hz to tremendously high frequency (THF) 300 GHz to 3 THz. Radio waves are generated by oscillating electric currents—collective periodic accelerations of electrons—in an antenna, and are detected as comparable collective oscillations of electrons in a receiving antenna. For communication purposes the basic periodic charge oscillation (the *carrier frequency*) is given modulation, in amplitude, frequency or phase, in which the information is encoded.

Cyclotron and synchrotron radiation.   Charged particles that are constrained by static magnetic fields to move in circular orbits, as happens in cyclotrons, undergo centripetal acceleration that is accompanied by cyclotron radiation. When the particle speed approaches the velocity of light the required machine becomes a synchrotron, and the radiation is termed synchrotron radiation. Quantitative description of this type of radiation makes no use of photons, other than the name.

Cherenkov radiation. Charged particles in free space undergo abrupt deceleration when they strike the boundary of material where the speed of light is slower than the particle velocity. This deceleration produces Cherenkov (or Čerenkov) radiation, an electromagnetic version of a shock wave. The quantitative description of this process has no need for photons.

Atomic-transition radiation. The electrons bound together in an atom or molecule are forever in motion in orbits around atomic nuclei. In this motion they are constantly acted on by centripetal Coulomb forces that produce acceleration toward positively charged centers of attraction. This acceleration of charged particles can be regarded, in the viewpoint of radiation-reaction theory [Ack73; Sen73; Ack74; Mil75; Dal82; Ser86; DiP12; Bur14; Mil19], as producing the radiation seen as spontaneous emission from excited states of motion. In simple situations the emission from a single atom transition produces a single photon. In other situations a single atomic transition will produce two photons whose total energy is equal to the change of atom energy—a two-photon transition.

Single-atom sources of single photons. As lasers became commonplace tools in optics laboratories they made possible a variety of procedures in which single trapped atoms could be caused to emit single photons. Such sources require particular controls of the environment around the atom and particular control of its quantum state [Sho90; Sho08; Sho11; Sho13]; see Appendix A.10.2.

Lasers and masers. The word "*laser*" and the acronym LASER for light amplification by stimulated emission of radiation, refers to the mechanism responsible for the light: An initial spontaneous emission event, converting internal energy of an atom or molecule into radiation, acts to induce similar energy conversions as it travels through matter where such potential energy is present—a set of actions proposed by Einstein. Such radiation was first studied with microwaves, where the devices were known as *masers*. Unlike the incoherent sources listed above, laser radiation can be very directional (though it need not be [Hor12b]) and nearly monochromatic, at a frequency that is dictated in part by the originating transition [Scu66; Sar74; Sie86; Mil88; Sal19]. Section 5.3 summarizes the physics underlying maser and laser radiation.

Particle annihilation; Positron emission tomography (PET). Various processes act to produce anti-particles to the stable electrons and protons that comprise the matter of our surroundings. When an anti-particle encounters one of its particle counterparts, the two rest masses are converted into radiation energy in accord with the Einstein formula $E = mc^2$.

For example, one form of radioactivity alters nuclear structure by emitting a *positron*, the anti-particle to the electron, and turning a bound proton into a neutron. Just as all electrons are indistinguishable, so too are their antiparticles, and any pairing will result in annihilation of the pair. When positrons have little kinetic energy the energy of a single electron-positron annihilation appears as pair of gamma rays whose characteristics, taken with those of the positron, conserve energy, momentum and angular momentum. Particle annihilation of a single positron-electron pair produces a single pair of photons.

The positrons emitted from a radioactive isotope embedded in matter—say human tissue—travel only a very short distance before encountering an electron and converting to gamma rays. The source of these photons is therefore very close to the positron source, and so by imaging the photons it is possible to localize an isotopically labeled piece of tissue. This procedure is known as positron emission tomography (PET).

Quasistatic fields. As the carrier frequency ω of a traveling wave tends toward zero the wavelength tends to increase without bound (tends to infinity). Such fields are best treated as static or quasistatic (slowly varying) interactions. The Coulomb repulsion between electrons and the Coulomb attraction between electrons and nuclei is the primary example of such fields (but see page 242 for other examples). Their dominant effects, dealt with in nonrelativistic evaluations of atomic and molecular structure—the bare energies E_n—are commonly evaluated without the need of individual photons; but see Section 8.3.

2.8.4 Visualizing chaotic vs. coherent; Eddington's photons

In contemplating the two extremes of thermal light and laser light as sources of photons it is instructive to read a delightful description "The inside of a star" written in 1926 by Sir Arthur Eddington [Edd59] and reprinted in [Sho90]:

> The inside of a star is a hurly-burly of atoms, electrons and aether waves. We have to call to aid the most recent discoveries of atomic physics to follow the intricacies of the dance. We started to explore the inside of a star; we soon find ourselves exploring the inside of an atom. Try to picture the tumult! Disheveled atoms tear along at 50 miles a second with only a few tatters left of their elaborate cloaks of electrons torn from them in the scrimmage. The lost electrons are speeding a hundred times faster to find new resting-places. Look out! there is nearly a collision as an electron approaches an atomic nucleus; but putting on speed it sweeps around it in a sharp curve. A thousand narrow shaves happen to the electron in 10^{-10} of a second; sometimes there is a side-slip at the curve, but the electron still goes on with increased or decreased energy. Then comes a worse slip than usual; the electron is fairly caught and attached to the atom, and its career of freedom is at an end. But only for an instant. Barely has the atom arranged the new scalp on its girdle when a quantum of aether waves runs into it. With a great explosion the electron is off again for further adventures. Elsewhere two of the atoms are meeting full tilt and rebounding, with further disaster to their scanty remains of vesture.

> As we watch the scene we ask ourselves: Can this be the stately drama of stellar evolution? It is more like the jolly crockery-smashing turn of a music-hall. The knockabout comedy of atomic physics is not very considerate towards our aesthetic ideals; but it is all a question of time-scale. The motions of the electrons are as harmonious as those of the stars but in a different scale of space and time, and the music of the spheres is being played on a keyboard 50 octaves higher. To recover this elegance we must slow down the action, or alternatively accelerate our own wits; just as the slow motion film resolves the lusty blows of the prize-fighter into movements of extreme grace—and insipidity.

In 1990 I wrote the following counterpoint to that lively description of thermodynamic equilibrium [Sho90]: "Eddington's charming prose presents a vivid mental picture of the excitation processes occurring in most hot gases, although the word 'photon' has now replaced his phrase 'aether wave'. As a portrayal of the microscopic behavior in such a milieu, his words remain a reliable description today as they were when first written in 1926. But whereas an atom in Eddington's star must respond to frequent collisions with particles from chaotic surroundings, atoms exposed to laser light encounter primarily a procession of nearly identical photons. An atom encounters one of these photons far more frequently than it collides with other perturbers. Indeed, the encounters are so frequent that the discreteness of energy quanta becomes blurred. We can best view this situation as a scene dominated by a traveling electromagnetic wave. The periodic electric field of this wave forces sympathetic oscillations of the electrons. The result, coherent excitation, is a far more harmonious activity than Eddington imagined occurring inside a star. Eddington's words give no hint of the diversity of effects that occur under conditions of coherent excitation—the contemporary world of quantum optics. Were Eddington offering today a metaphor for coherent excitation he might suggest the following":

> Imagine a marching band on parade. In the distance we hear first the regular pulsations of the bass drum. Then the brasses are heard harmonizing a melody. Soon the full orchestration becomes apparent. As the players come into view we notice the precision of their march. Each left foot rises in unison, as if moved by some irresistible force—the feet of the musicians seem to be locked together with rigid but invisible bars. Soon even a few spectators can be see tapping feet in time to the music.

> As we watch this scene we ask ourselves: How is it that these individuals become, for a few moments, part of a musical machine? What force periodically propels each foot? We know the answer: It is the rhythm of the music, directed by the drum major, that provides this invisible bond between musicians. Each individual acts so as to fit coherently into a pattern produced by an unseen but audible carrier melody.

One must proceed with care in placing reliance upon such pictures and analogies. Unlike the marching band, which always has a fixed number of members during any performance, a laser beam does not have a precisely defined number of photons; see Appendix B.4.3. It has a large average number and so individuals are not readily discerned—they seem merely an organized crowd. Just as we know that a distant, indistinctly seen mountain is clothed with unresolved, individual trees, we know that radiation ultimately comprises photons. This Memoir aims to explain just what that assertion means—how we are to regard the photon trees of the radiation forest.

2.8.5   Pencil beams; Rays

Radiation can be formed, by means of mirrors and lenses and a succession of apertures (small openings) in opaque screens, into pencil-sized beams that travel, when unconstrained, along straight-line paths (at least to a first approximation) [Car74; Dav79; Sim16]. These are idealized as *light rays*, treated as mathematical lines. Radiation from a laser quite naturally forms a pencil beam, as does light from individual stars. The radiation itself, during passage from its source (e.g. the sun, a street lamp, or a laser pointer) is visible only by scattering from dust particles along an air path (or a path through some translucent medium such as milk) or from surfaces such as furniture or street pavement. Its arrival can be evidenced in many ways, including reflection and absorption (to become heat).

Contemporary designers of optical devices rely on modeling their assorted lenses and mirrors by software that carries out extensive ray tracing through optical paths— so-called *Geometric Optics*. Superficially their procedures resemble the use of what Newton regarded as light: Streams of colored "corpuscles" that followed straight-line paths in free space and bent paths upon entering or leaving transparent materials. This model offered a simple explanation for how lenses act and how a beam of white light became dispersed, by a prism or grating, into constituent beams of colored light.

Nowadays it is common to find optical beams traveling through optical fibers, where confining action of refractive-index variation guides narrow beams of light along curved paths, thereby eliminating the need for beam-directing mirrors in laboratories and allowing long-distance optical communications; see Appendix C.3.2. Here too one can imagine finding photons, as indivisible increments of energy.

Beam characteristics. Beams of laser radiation are found not only in optics laboratories but as pointers in lecture halls and scanning devices at retail checkouts. Such beams, as well as beams of incandescent radiation and the sunbeams with which Newton experimented, have three common attributes [Car74; Dav79; Sim16].

Direction: A pencil beam is characterized first by its direction—idealized as a three-component wavevector **k** that specifies the propagation direction of the electromagnetic-field wavefronts; see Appendix B.2.2.

Radiation which is confined to a narrow beam has a well-defined dominant wavevector, pointing along the beam axis (the longitudinal direction), but it must incorporate a superposition of many wavevectors in the transverse direction in order to produce the confinement; see Appendix B.2.2. It is therefore very different from an idealized simple *plane wave*, which extends indefinitely and uniformly in the transverse direction— termed *Nadelstrahlung* (needle radiation) by Einstein and others (the needle-like coordinates are in momentum space). Beams also contrast with *Kugelstrahlung* (literally "ball radiation") or spherical waves (multipole fields, see Appendix B.2.6).

Color: Next its specification requires some measure of its frequency composition— its colors;[18] see Appendix B.5.1. The frequencies may be spread over a broad range, as with sunlight, or they may be concentrated within a narrow range, as with laser radiation (idealized as *monochromatic*). The distribution of frequencies present in the intensity is the spectrum of the light. A monochromatic light wave having wavevector **k** has angular frequency $\omega = |\mathbf{k}/c|$. Thus color and propagation direction together require only three numerical parameters for complete specification.

---

[18]Color, as distinct from frequency composition, is associated with the physiology of vision. In particular any given color can be produced by many combinations of three primary frequencies; see Chapter 35, on "Color vision", of the Feynman Lectures on Physics Vol. 1, available through https://www.feynmanlectures.caltech.edu/info/

Phase: As with any wave, a beam of coherent radiation can be assigned a phase— a number that specifies the moment during any period that is to serve as the origin for counting waves. The phase of a single beam can be chosen arbitrarily, but the difference in phase between two beams is determinable by interference measurements if they are coherent.

Polarization: Finally, light beams may be *polarized*—linear, circular, or ellipti- cal. (Circularly polarized light is said to have a definite *helicity*: Positive helicity for right-circular polarization.) This attribute (demonstrable with polaroid sheets or sun- glasses) is the time-averaged direction of the electric-field vector, perpendicular to the propagation direction (see page 298). At every position and every instant the electric field has a definite direction, but on average there may be no preferred direction: Un- polarized light is an incoherent mixture of polarizations, a time average of randomly varying electric-field directions. Appendix B.1.7 discusses intermediate situations of *partial polarization* as quantified by Stokes parameters.[19]

2.8.6   Elaborate fields

In recent years both theory and experiment have greatly enlarged the variety of beam- like classical fields being considered. Appendix B.2.1 mentions some of these, with references. But beams are only one of the many examples of radiation fields that contemporary optical techniques can prepare. A review article lists the many ideas that are under active investigation [Aie15]. Each of these fields offers possibilities for defining photons.

2.9   Possible radiation granularity; Photons

Discreteness in daily life is by no means uncommon. Until recently all cash financial transactions had to make use of a discrete set of values for the bills and coins that are to be found in all traditional monetary systems.

One might ask, as did philosophers of ancient times who thought about the struc- ture of matter, whether there might be an ultimate granularity of radiation, as has been demonstrated for matter with its base of atoms: Might light rays possibly comprise some collection of massless elementary units—perhaps the corpuscles of Newton—just as matter is known to comprise collections of electrons and atomic nuclei? A stream of water flowing out of a faucet can be demonstrated to comprise discrete molecules. Can a flashlight beam somehow be demonstrated to comprise granules of light?

An affirmative response came in the early twentieth century. The hypothesized elementary units of radiation, the radiation Quanta, would eventually be termed Pho- tons. They would be counterparts to the electron, the elementary unit of negative electricity, and to the atom, that elementary bit of matter. Presumably whatever def- inition might be used, a photon, in common with an electron, cannot be further split into parts.

---

[19]The name of George Gabriel Stokes appears in three separate contexts in the present monograph: That of the Stokes vector and Stokes parameters used for describing polarization characteristics of an electromagnetic field, Appendix B.1.7; the name of one field of the pair that contribute to a two-photon Raman transition, Appendix A.15; and, in passing, as a namesake of the Navier-Stokes equation, eqn (A.1-47).

Particles and rest mass. However, there are important differences between radiation and electrons, or other bits of everyday matter. Most obviously the traditional particles that make up matter can not only move but they can be brought to rest. And at rest they have nonzero *rest mass*—the characteristic that gives numerical magnitude to potential and kinetic energy. Radiation, on the other hand, travels through free space always at a constant velocity, the speed of light c; see eqn (2.7-2). Unlike atoms and electrons, free-space radiation increments can have no rest mass. However, radiation traveling through bulk matter, whether transparent or opaque, will travel more slowly and will acquire other characteristics that differ fundamentally from radiation in a vacuum; see Appendix C.

The notion that radiation might be regarded as being composed of massless particle-like entities was for some time quite perplexing. George Darwin wrote in 1932 [Dar32]:

> ... it is quite unlike any known particle to come into existence and later to disappear without a trace.

Nowadays, with the possibility of experiments using very high-energy particle accelerators, and the relativistic theory of particle-antiparticle pair production, even this seemingly outrageous proposal has become an accepted part of nuclear chemistry and high-energy physics; see page 55. The "unlikely" has become commonplace.

Stopping photons. When radiation is stopped by some obstruction, be it an opaque surface or some small particle, it deposits energy—this is the heat energy of sunlight. Might the energy deposition (or its emission) occur in discrete increments—photons being absorbed and emitted?

This was a question that not only I, as a newcomer to physics, pondered, but one that had brought thoughtful examination and detailed expository writing by many illustrious physicists: Do photons exist? If so, how are they to be understood and treated with appropriate mathematics [Scu72; Lam95; Mar96; Coh98; Lig02; Mut03; Bia06c; Fri09]? After all, physics must eventually reduce Nature to mathematical expressions (equations of some sort) that will lead to predictions of measurable quantities. What are the equations for photons that supplement well-established equations for atoms and electrons? Only when we have such equations can we think we understand the associated physics of photons.

The following chapter begins the presentation of various lines of experiment interpretation that supported the notion of granular radiation. There were several, and they led to somewhat differing details of how the radiation granules were to be regarded.

Photon size. Massive particles such as atoms, and electrons bound to atoms, have a measurable extent—a size within which (most of) the mass and charge will be found; see Appendix A.2. Massless free-space photons are more difficult to pin down to any location.

For an idealized monochromatic field (one that extends indefinitely in space and time as a traveling wave or a standing wave) one might argue that localization could take place only near an antinode, over a distance set by the wavelength. When multiple frequencies are present there will occur interferences between waves, and localization along a standing-wave beam axis can become much more precise than a wavelength. This is the principle behind interferometry [Buc86; Pac93; Har96; Ber97; Cro09; Che16].

But the required periodicity that accompanies monochromatic waves then prevents identification of just which antinode is providing the localization. Superpositions of waves can improve the localization along a beam axis, but the ambiguity remains. Only when an atom or other bit of matter registers the absorption of a radiation increment is it possible to assign a position to the vanished photon—a position that is fixed by the localizable atom, not by the field (or photon). Appendix C.4.2 discusses issues related to localization in bulk matter.

Although localization is a defining characteristic of particles, light is regularly used to view small objects by means of lenses and light-sensitive surfaces. Monochromatic light that emerges into vacuum through an opening (an aperture) in an opaque surface will expand as spherical wave fronts. Along a single line of sight the waves crests move with the speed of light, separated in space by the wavelength, producing a steady intensity that is spread uniformly over small segments of a plane transverse (perpendicular) to the line of sight—there is usually no obvious nodal indicator of wavelength in the transverse direction (but see Appendix B.2 for examples of beams with transverse nodes). Nevertheless, the wavelike nature of light imposes long-established limitations upon the size of objects that can be distinguished using focusing optics. Traditional methods of optical observation set the limit of transverse localization of an unstructured beam as roughly a wavelength; see Appendix A.2.3. However, with the use of coherent radiation various techniques now exceed that classical resolution limit; see page 205.

Photon density. Although electrons have no intrinsic size, there is a limit on how many can be packed into any finite volume. Apart from the Coulomb repulsion, which can be offset with positive charges of nuclei, there is a fundamental limit, expressed by the Pauli *exclusion principle*, that prevents two electrons (or any two fermions) from having the same set of four quantum numbers and that thereby inhibits them from occupying the same space; see Appendix A.8.2. This principle leads electrons in atoms to cluster into inert central cores (shells), around which a few valence electrons move, and it leads electrons in metals to occupy momentum bands. But there is no such fundamental restriction on photons (or any bosons) and on the associated intensity of electromagnetic radiation—the density of electromagnetic energy. Indeed, bosons prefer to share a common quantum state, and there is no theoretical upper limit on the energy density or the focused intensity of an electromagnetic field in vacuum.

As radiation becomes increasingly intense, the Lorentz force becomes sufficiently strong to rip off bound electrons, setting them free beyond residual ions. This becomes evident when the electric field of the radiation exceeds the Coulomb field that binds the electrons. Pulses from industrial lasers used for welding and cutting routinely produce such intense radiation, well above the values used for controlled manipulation of quantum states by single photons.

Aside: Radiation scales. The significant scales of electric field magnitude $\mathcal{E}$ and radiation intensity I for overwhelming the electron-binding Coulomb forces are the atomic units of these quantities:[20]

$$\mathcal{E}^{AU} = \frac{1}{4\pi\epsilon_0} \frac{e}{(a_0)^2} \approx 5.14 \times 10^{11} \text{ V/m}, \tag{2.9-1a}$$

$$I^{AU} = \frac{(\alpha\hbar c)^2}{(a_0)^4} \approx 6.4 \times 10^{19} \text{ W/m}^2, \tag{2.9-1b}$$

where a_0 is the Bohr radius of eqn (A.2-5), the atomic unit of length. That is, a monochromatic wave of intensity 1 watt/m^2 creates an electric field whose magnitude $|\mathcal{E}|$ is about 5×10^{-7} atomic units of field strength.

When radiation fields become still more intense they eventually affect even the vacuum, and it is no longer possible to avoid considering the creation and annihilation of massive particle-antiparticle pairs; see eqn (8.5-1). These provide a "dressing" to the "bare" photons and the field equations become nonlinear.

The name "photon". The name "photon" is commonly credited to physical chemist Gilbert Lewis who, in 1926, used it to describe his notion of a particle-like quantum of radiation [Lew26]. As his article title, "The conservation of photons", makes clear, he had in mind discrete particles, of energy $h\nu$, momentum magnitude $h\nu/c$ and mass $h\nu/c^2$ that, together with atoms, obeyed conservation laws of energy, momentum and mass and which were stored in atoms awaiting release. The "photons" of contemporary physicists, which I describe in this Memoir, are not what Lewis had in mind. As Willis Lamb has stressed [Lam95] (see page 177), the unconventional ideas of Lewis (conservation of photons) failed to gain acceptance—they were demonstrably wrong, photons are not conserved—but his chosen name has prevailed.

Update: The photon name. Historian Helge Kragh [Kra14] has pointed out, in a posted but unpublished article cited by Wikipedia (in the wiki article "Photon") that the word "photon" was used prior to Lewis's 1926 article by three other authors, primarily for describing physiological effects of light: Leonard Troland (in 1916), John Joly (in 1921), René Wurmser (in 1924), and it was used in 1926 by Frithiof Wolfers. Kragh credits Arthur Compton for first citing Lewis and for using the term photon with a more lasting interpretation.

2.10   Angular momentum: Orbital and spin

A slowly moving (nonrelativistic) particle of mass m and velocity $\mathbf{v}$ (a vector) has *linear momentum* $\mathbf{p} = m\mathbf{v}$ (a vector in the direction of motion) and kinetic energy (a scalar) expressible as either

$$E^{\text{kin}} = \frac{m}{2}|\mathbf{v}|^2 \quad \text{or} \quad E^{\text{kin}} = \frac{1}{2m}|\mathbf{p}|^2. \tag{2.10-1}$$

In the absence of any force, both the kinetic energy and the linear momentum remain constant.

---

[20]The atomic unit of electric field magnitude is the electric field at a distance of one Bohr radius from one electron charge.

Around any chosen reference position, such as the center of a fixed star, a particle may have an angular momentum (a moment of linear momentum—a twist), a vector whose magnitude is proportional to its mass and its velocity magnitude or to its linear momentum magnitude, and whose direction (the twist axis) is perpendicular to these collinear vectors and to the time-varying vector **r** that points from the chosen reference position to the moving point particle; see eqn (A.1-15) in Appendix A.1.2. The changing values of the particle-position vector **r** trace out a path known as the particle *orbit* (not necessarily a closed path), and the angular momentum associated with that path is therefore termed *orbital angular momentum*. Although these expressions are most commonly found in discussions of bound-state positions, they apply to any trajectory.

When the particle twists, or turns about an axis, as does a toy top, a gyroscope, a bicycle wheel, or a typical planet, it has a second form of angular momentum, a vector independent of its velocity or its position, pointing along the rotation angle (for the earth, toward the north star) with a magnitude proportional to the particle mass and to the rate of rotation about its axis (its angular velocity). This rotation, or twist, is termed *spin*, and its angular momentum is therefore termed *spin angular momentum*, commonly expressed in units of $\hbar$. This contribution to a total angular momentum is independent of the particle trajectory: It is an intrinsic property of a planet whatever may be the planetary orbit.

By contrast with a particle, a field can have spatially distributed energy and momentum, quantified as energy density and momentum density. It can also carry two forms of angular momentum density, orbital and spin. Quantum theory associates with an electron or other massive particle a field (a wavefunction) of electron-charge density and of mass density. The angular momentum density of this electron field has both an orbital contribution, dependent on the velocity and kinetic energy density, and an intrinsic contribution, a spin angular-momentum density whose spatial integral, for a single electron, is $\hbar/2$. Although the name suggests rotation about a moving axis, its origin is attributable to a wave-nature distribution (in a wavefunction) of intrinsic angular momentum [Oha86]. The spatially integrated spin density of the electromagnetic field of a single photon is $\hbar$, twice that of a single electron. Photons, as the quanta of electromagnetic fields, are therefore said to have spin one, while electrons have spin one-half. This marks them as bosons and fermions, respectively (see Section 2.2.8).

2.11   Probabilities

Much of the literature of modern life mentions probabilities, and numerous specialized texts and courses discuss their broad reach [Cox46; Fel68; Bal70; Dav70; Gri12b]. The enterprises of economists and investors, of weather forecasters and sports fans, of merchants and farmers, of police enforcers and medical advisors, all make use of probabilities—if only implicitly and unconsciously—for their decisions. Few organized contemporary activities are free of reference to probability (or likelihood)—and from the enticement of financial gain from using probabilities to place bets or avoid loss. The essence of probability is expressed as:

> *Probabilities are basically numerical representations of our best guess at outcomes of future or ongoing events.*

For the events of everyday activities—of weather and traffic flows—such predictions are based on extrapolating data of past events. Their organization and use is now taught in courses entitled statistics, often part of the basic requirements for an undergraduate degree and administered by academic departments of Statistics. There the courses of interest for students of economics, medicine, and sociology often incorporate subjective notions of Value and Utility into procedures for decision making and Risk Management [Ber96].

In physics, and particularly quantum physics, probabilities are typically defined in terms of a set of measurements on a collection of identically prepared physical systems (or repeated observations of a single system). Expressed in words, the definition of probability used in Physics can be stated as:

> *The ratio of the number of events that are associated with some observable attribute x, to the totality of observed events, is taken to be the probability P(x) of attribute x.*

The probabilities that underlie quantum theory and statistical physics—and which therefore appear in this Memoir—do not rely on historical records of measurements but upon simple rules. They have much in common with odds calculated centuries ago by Renaissance gamblers for games of chance played with dice [Ber96].

All interests in probability and chance, whether in everyday activities or in quantum theory, have some common features, discussed below.

2.11.1   Events

The events that are to be described by probabilities—observations of measurable characteristics of some reproducible phenomenon—may be of two types (casinos combine both).

Serial: First, we can repeatedly observe a single system, for example seeing whether a coin flipped into the air falls heads or tails, the only acceptable two possibilities. Or we might have a box of marbles, differing only by color, say, either green or orange. We shake the box well and, without peeking at the color, draw one out.

Parallel: Alternatively, we may have many copies of identical systems, prepared identically. For example, the contents of a well-shaken basket of dice is tossed onto a flat surface and the distribution of die faces is noted. A team of coin flippers announce their collective results after a coin flip. A group of competitors are issued decks of cards to shuffle and deal, and the assorted hands are duly noted. A morning report arrives from a set of weather stations giving their air temperature.

So it is with observations of atoms. We can either deal with a large collection of identically prepared atoms (this is the traditional assumption) or we can somehow trap and hold a single atom and, after repeatedly restoring it (an essential step), make an observation of its properties. This procedure has become possible in recent years.

2.11.2   Evaluating probabilities; Classical and quantum

Within any enterprise that uses probabilities there are well established procedures for evaluating them.

For example, the probability that it will rain tomorrow is estimated from the ratio of past rainy days to the total number of days in the weather record (a matter of statistics). The possibility that a racehorse or a sports team will win is (usually) based on its past record.

Alternatively, one can define probabilities in terms of the known, or presumed, distribution of possible events. Unless you have reason to believe otherwise, you assume that each distinguishable possibility is equally likely. If a coin is well balanced so that neither side has an obvious advantage, we estimate the probability of heads to be the fraction $1/2$. In the marble-box example, if we know there are equal numbers of green and orange marbles in the box, and they are indistinguishable by touch, then our best guess is that in half the experiments of drawing out one marble its color will be green. Prior to drawing out a marble, but knowing the marble composition in the box, we estimate the probability of getting a green marble will be the fraction $1/2$. The probability for drawing an ace (one of four possibilities) from a well-shuffled deck of 52 playing cards is regarded as the ratio $4/52$ (a matter of *combinatorics*, an enumeration of possible ways of achieving a particular result).

Those probabilities deal with activities that are governed by everyday, classical physics. Quantum theory has its own special rules for calculating probabilities. Most notable is that all the real-valued probabilities are obtained as absolute squares of complex-valued *probability amplitudes*. It is with those amplitudes, leading to wavelike nodal structure associated with photons, that the several equations of quantum theory deals; see Appendices and A.3.3 and A.5.

2.11.3   Properties of probabilities

With either scenario—multiple systems or multiple trials with a single system—a probability is used to give a numerical estimate of our best guess of the results of many observations. It is with the prediction of such outcomes, taken in the limit of a very large number of trials, that probabilities occur. They generally say nothing about a single observation, only about an accumulation of a large number of observations, expressed as a fraction—a number between 0 and 1 inclusive.

Only if the probability is known to be 0 (meaning a result that is impossible) or 1 (meaning a mandatory, guaranteed result) have we any certainty about a single trial. It is the uncertainties of individual events that draws bettors to casinos. It is the numerical probabilities that keep the casinos in business.

One of the essential properties of any definition of probability is that the sum of the probabilities for all possible individual outcomes must be unity. For example, if one takes marbles from a box that contains only green and orange marbles, the probabilities of taking an orange one or taking a green one must sum to unity—the chosen marble can only be orange or green, given the underlying model. Probability cannot increase beyond complete certainty. (This intrinsic constraint does not affect the sportscasters who regularly exhort athletes to exert an effort in excess of 100%.)

Because of the way probabilities are defined they must be non-negative, real numbers, bounded by zero and unity. For persons of a mathematical inclination, they are mappings of event measures onto this small portion of the real-number line. For the examples above, of coins, cards, and marble colors, the mapping is to non-negative rational numbers.

As my education continued in graduate school, I learned how important probabilities are in any discussion of atoms and electrons—in the quantum mechanics with which their behavior is described. And I have found that concepts of photons rely also on probability concepts. Despite what Einstein hoped (in allegedly saying "God does not play dice")[21] our present physics of photons cannot proceed without encountering chance—meaning influences over which an experimenter has no control.

2.11.4   Random variables; Expectation values

The events with which probability deals can, for familiar examples, be represented by numbers associated with measurements or observations: The reading of a thermometer at a specified location; the value of a playing card drawn from a (presumably) well-shuffled deck; the published winning score of an athletic contest. (Measurements of qualities such as color can be given numerical values in accord with some scale.) The possible numerical outcomes of such measurements form a *sample space*, a set of numbers (say a discrete set $\mathcal{N}_1, \mathcal{N}_2, \ldots$) from which any given observation will select one. Unless the measurement outcome is predetermined, it will have some uncontrollable irregular variation (termed *random* or *stochastic*) of the successive values. A measurement provides an incidence, or particular value, of such a variable. It is the purpose of probability values to provide estimates of the distribution of any set of observation values. From whatever source they come, a given set of probabilities allows us to predict the most likely result of measuring a random variable. This *expectation value* (the numerical value predicted as the average of an indefinitely large number of cases) for some stochastic variable V is here denoted $\langle V \rangle$. For example, $\langle \mathcal{N} \rangle$ denotes the (theoretically) expected value of the (experimental) average defined in the simple example of eqn (2.11-1).

> **Aside: Mean and variance.** In any set of events governed by chance there are many measurable properties that can be of use in characterizing the underlying probability distribution. Typically it is possible to assign some single number to an event, say the summed values of reading of the upper faces of cast dice. From recording numerical values $\mathcal{N}_i$ of repeated events, $i = 1, 2, \ldots$ one can evaluate the average numerical value, or *mean*, denoted $\overline{\mathcal{N}}$, defined as the sum of all successively observed numbers divided by the number of events, a construction of the form
>
> $$\overline{\mathcal{N}} = \frac{\mathcal{N}_1 + \mathcal{N}_2 + \mathcal{N}_3 \cdots}{1 + 1 + 1 + \cdots}, \tag{2.11-1}$$
>
> where each measured numerator number $\mathcal{N}_i$ accompanies an addition of 1 to the denominator count. The expectation value is the mean value, $\langle \mathcal{N} \rangle = \overline{\mathcal{N}}$.

---

[21] To which Bohr is said to have replied: "Einstein, you cannot tell God how He is to run the world". [Isa07] page 609.

When dealing with quantities that may be positive or negative (as happens with oscillations) it is common to use root-mean-square (RMS) values (the *quadratic mean* or square root of the mean square) to quantify the distribution of values. For n numbers N_i the RMS obtains from the formula,

$$\mathcal{N}_{\mathrm{RMS}} = \sqrt{\frac{\mathcal{N}_1^2 + \mathcal{N}_2^2 + \cdots + \mathcal{N}_n^2}{n}} = \sqrt{\langle\mathcal{N}^2\rangle}. \qquad (2.11\text{-}2)$$

Such measures, along with such quantities as the most probable single result, have practical value as one ponders a probabilistic investment. But other characteristics are also important in making such decisions: How likely is it that a single event returns a number close to the average? A measure of the variation amongst a succession of events (and the uncertainty in the next one) is the *variance*,

$$\mathrm{var}(\mathcal{N}) = \frac{(\mathcal{N}_1 - \overline{\mathcal{N}})^2 + (\mathcal{N}_2 - \overline{\mathcal{N}})^2 + \cdots + (\mathcal{N}_n - \overline{\mathcal{N}})^2}{n} = \langle\mathcal{N}^2\rangle - \langle\mathcal{N}\rangle^2. \quad (2.11\text{-}3)$$

This number, the difference between the mean square and the square of the mean, expresses how sharply peaked around the mean value is the distribution of observed values. Such measures are important not only for risk managers of investment funds [Ber96] but for descriptions of quantum processes.

Uncertainty. Whenever we measure properties of some system that has distinguishable irregularities we encounter variation of the numbers acquired from a succession of measurements: Weather patterns at different times or different places; prices of different stocks at different times. The tabulated values may trend, on average, with a predictable pattern (night and day, north and south, desert or coastline) but they may fluctuate around an average. One common measure of the range of values around the mean is known to mathematicians as the variance. Physicists call it uncertainty.

The uncertainties of physics occur for three reasons. First, the measurements may be of uncontrollable phenomena such as the rainfall outside a particular workplace during the lunch hour. Such data is stochastic—predictable only on average. A different, second, sort of uncertainty occurs when we measure the tone (frequency) of a musical note: The accuracy of the measurement depends on its duration. A third class of uncertainty occurs when the data comes from simultaneous measurements of certain characteristics of a single quantum system—say, measurements of position and velocity (or momentum) of a very small particle. Each of these three types of uncertainty requires a distinct approach and has distinct results and interpretation. Most notable for discussions of photons are the unavoidable uncertainties in the mathematically different pairings of frequency and time (see Appendix B.5.1) and of position and momentum (see Appendix A.4.2).

2.11.5 Conditional probabilities

Often we wish to consider repeated observations under constrained conditions in which we exclude some of the possible outcomes. We might, for example, know that a card packet comprises only spades or only black cards. Under these conditions we can assign probabilities to the occurrence of specific combinations. These probabilities are *conditional*—they depend upon prior certainty concerning the nature of the sample space. The conditional probability $P(x|y)$ is the probability that, if the property y is known to hold, then the property x will be found. This conditional probability is the ratio of the number of systems that have both property x and property y, to the number of systems that have property y.

Uncertainty. The differing views of probability used by physicists and by others listed above have to do with the nature of the events being described. Whereas every electron, every proton, is identical and indistinguishable, such uniformity is not found in other domains of enquiry: Every person, every pet animal, every sporting event, every new day, is unique. Thus predictions of human, or weather, behavior stand on a somewhat different base than predictions applied to simple inanimate objects. Nonetheless, the general properties of probabilities noted above apply broadly to all dealings with events and measurements that involve uncertainty about outcomes.

However, significant differences occur between classical and quantum probabilities and their uncertainties. Whereas one might hope that improved tools might continually diminish errors (uncertainty) in a manufacturing process, and that more elaborate computer facilities might improve weather forecasting, all quantum systems are subject to the Heisenberg uncertainty principle that fixes a lower bound for many sorts of measurements and predictions.

2.12 Quantum states

Physics deals with information about physical systems—not only about what sorts of individual pieces and linkages may be present, what sort of atoms or molecules may be present, but how these constituents are distributed in space and how they are moving. Information about those basic variables—positions and velocities—defines a *state of motion*.

2.12.1 The uncertainty (indeterminacy) principle

Until the twentieth century brought quantum theory as the basic description of physical systems, there seemed no limit to the accuracy with which one could measure positions and velocities, thereby specifying a state of motion. But as even the most basic course in quantum theory informs students, the distinction between particle behavior governed by classical equations and behavior governed by quantum mechanics is set by *Heisenberg's Uncertainty Principle*: The product of uncertainty[22] in particle position Δ_x and the uncertainty in its nonrelativistic velocity Δ_v cannot be smaller than an amount that is inversely proportional to the mass of the particle, m.

> **Aside: Heisenberg uncertainty.** The Heisenberg uncertainly relationship is commonly expressed in terms of coordinate and momentum variables; see eqn (A.4-6). With momentum $p = mv$ the position-velocity relationship reads
>
> $$\Delta_x \times \Delta_v \geq \hbar/2m. \tag{2.12-1}$$

The proportionality constant in this uncertainty product is the reduced Planck constant $\hbar = h/2\pi$; its value essentially fixes the sizes of atoms. Only under special

---

[22] Appendix A.4.2 quantifies uncertainty.

circumstances, termed minimal uncertainty states, does the uncertainty product reach its minimum value; see Appendix B.4.4.

2.12.2 Defining a quantum state

The notion of information defined by particular instances (a *state*) of system behavior is found in the microscopic systems, such as atoms and electrons (and photons), governed by quantum theory. There the information, though constrained by the uncertainty principle, still provides a distinction between different examples of system behavior. To paraphrase Michael Raymer [Ray97],

> A *quantum state* can most simply be regarded as whatever is needed to specify the probabilities of measurement outcomes of all observable quantities pertaining to a physical system.

An experimental view, stated by Leonard Mandel [Man99], is:

> In an experiment the [quantum] state reflects not what is actually known about the system, but rather what is knowable, in principle, with the help of auxiliary measurements that do not disturb the original experiment.

In these views, which I adopt, a "quantum state" is not an actual object that you can touch or see, or push and pull, it is <u>information</u> about how some physical object is constructed—a building plan. (The term "blueprint" would have been used decades ago, when such objects were commonplace amongst draftsmen.)

There are two obvious logical concerns with this definition: How is the information to be acquired? This is known as the *Measurement Problem* [Wig63; Leg80; Roy89; Sch05; Van08; Whe14]. And: How is the information to be stored and retrieved for use? This question leads to consideration of lists and to notions of vectors and abstract spaces for them. (Other questions concern information manipulation.) Two approaches, defined more fully in Appendix A.3, find regular use.

Wavefunctions. When the information that defines a quantum state is presented as probabilities for finding a set of particles at a prescribed set of positions, the underlying probability amplitude is termed a *wavefunction* (see Appendix A.4.3). In keeping with Heisenberg's uncertainty constraint, such a function of coordinate variables says nothing about their velocities.

Statevectors. Apart from very simple idealizations, quantum systems have many degrees of freedom, each of which requires description by a probability amplitude. The resulting multidimensional wavefunction is too complicated to be viewed without simplification. When only discrete-energy states need to be considered, as happens when we treat photon creation and destruction, the abstract vector spaces defined in Section 2.5.2 offer a convenient way of organizing and working with the information that describes the quantum states. Appendix A.3 describes the abstract-vector spaces commonly used in describing quantum systems and their changes. As I shall be explaining there, such a space is a setting for a *statevector*: An abstract vector whose components are *probability amplitudes*; their absolute squares provide probabilities for observing a quantum system in one of its possible quantum states; see Appendix A.8. Under many conditions all of the possible information about a quantum system, and its possible changes, is embodied in such a statevector—an abstract vector that has as many dimensions as the number of discrete quantum states that may be required to fully define the system. (Under more general conditions the description requires a *density matrix*; See Appendix A.14.)

For the discussion of photons that I will be presenting in the narrative portion of this Memoir, little more than those general principles (of quantum states and probability amplitudes viewed as components of abstract vectors) are needed for dealing with the results of quantum theory. The view I shall take is that it is not necessary to speculate on what an electron or a photon "really is". It suffices to have equations that consistently predict its observable attributes and how it behaves under controlled, specified forces—results of experiments.[23] This is what quantum theory provides.

---

[23] This is a view that Feynman often expressed in his talks and books [Dud96].

3

The photons of Planck, Einstein, and Bohr

The teaching assistant is hard to meet.

Notions of discrete radiative increments (termed *quanta*) were proposed on a variety of grounds during the early decades of the twentieth century. This chapter discusses the predominant views of partisans in the long-standing controversy of whether radiation had a particulate, corpuscular nature analogous to the atoms of which matter was composed (and akin to the mathematicians' denumerable set $\mathbb{Q}$) or whether it was a purely wavelike phenomena (the mathematicians' continuum $\mathbb{R}$). The preceding chapter presented a basis for this decision with remarks on the general concept of particles (discrete, localizable entities) and waves (unlocalized entities exhibiting interference). Here I review the evidence of those century-old experiments—and then how their interpretation has been overturned.

Four primary lines of experiment were first taken as evidence of radiation quanta—what we now refer to as photons [Bia06c]: The work of Max Planck on the spectrum of thermal (blackbody) radiation (Section 3.1); the Bohr-Einstein explanation for spectral lines (Section 3.4); the Einstein theory of the photoelectric effect (Section 3.5); and the description of Compton scattering of radiation by free electrons (Section 3.6). Although these experiments were regarded at the time as definitive proof that discrete radiation increments existed, and they remain convincing justifications for regarding radiation as comprising photons, it turns out that all of these effects can be explained quantitatively without invoking photons; the notion of photons is sufficient but not necessary. Section 3.8 discusses revisionist alternatives, with references.

3.1 Thermal light: Planck quanta

Abraham Pais has written [Pai79] an instructive account of the early introduction of the notion of radiation as energy packets—light quanta (*Lichtquanten*) as they were first termed—and the subsequent notion of particle (corpuscular) characteristics of radiation—what we now call photons. I had not appreciated how outspoken was the opposition of the established community of physicists, devoted as they were to

Our Changing Views of Photons: A Tutorial Memoir. Bruce W. Shore, Oxford University Press (2020). © Bruce W. Shore.
DOI: 10.1093/oso/9780198862857.003.0003

Maxwell's equations for continuum fields, to the idea that radiation should be treated as having granular, particle-like attributes. Pais says [Pai79]:

> Never ... has the idea of a new particle met for so long with such almost total resistance ....

Historian Russel McCormmach put it [McC67]:

> Only very gradually did a largely reluctant community of physicists become reconciled to light quanta.

An article written for the Nobel Prize organization by Gösta Ekspong presents a detailed and instructive description of views on the dual wave-particle nature of light as reflected in the Nobel archives associated with the award of the physics prize; see the web site https://www.nobelprize.org/prizes/themes/. For example, on page 14 of his 1922 Nobel Prize lecture Bohr expressed his opposition to a particulate nature of radiation with the following words [Boh23]:

> In spite of its heuristic value the hypothesis of light quanta, which is quite irreconcilable with the so-called interference phenomena, is not able to throw light on the nature of radiation.

Hendrik Antoon Lorentz, renowned for his command of nineteenth century classical physics, maintained the view [Kox13]:

> This is all without doubt very striking, but nevertheless it seems to me that on closer inspection grave objections to the light quantum hypothesis arise.

The prevailing view was evidently that because light was known to be waves, therefore it must follow that light could not be particles. That was a seeming paradox that I too encountered. Its resolution is a major theme of the present Memoir. However, the world is not always one of "either - or ", it is often "both - and", and it is not difficult to find several common culinary examples of such seeming duality paradoxes: The dessert Baked Alaska, which is both cold (interior ice cream) and hot (exterior pastry crust) or the sweet-and-sour sauce served in Chinese restaurants.

Visible radiation; Colors. As is commonly demonstrated with classroom use of a prism through which a pencil beam of light passes, sunlight comprises a continuum of colors, the solar spectrum.

> **Aside: Wavelength and frequency units.** The sensation that we call color is dependent upon the physiology of the eye, but in keeping with the notion of light as a wave disturbance, monochromatic color can be expressed either as a frequency ν (expressed in hertz, Hz, i.e. cycles per second, or in terahertz, 1 THz $= 10^{12}$ Hz, or as a wavelength λ (expressed in nanometers, 1 nm $= 10^{-9}$ m or Ångströms, 1 Å $= 0.1$ nm $= 10^{-10}$ m): The product of these two numbers, $\nu\lambda$ (or $\omega\lambda$), is the speed of light in vacuum, the c of eqn (2.7-2).

The classroom separation of optical radiation into seven colors is attributed to Newton: Red, orange, yellow, green, blue, indigo, and violet, with its school-taught mnemonic ROY G BIV. There is no sharp demarcation between these colors, and various references present different boundaries. The following table, after [Woo61], gives nominal values for monochromatic colors [see eqn (3.3-1) for photon-energy relationships]:

Color:	*Violet*	*Blue*	*Green*	*Yellow*	*Red*
Wavelength (nm):	400	450	500	580	650
Frequency (THz):	750	670	600	520	460
Photon energy (eV):	3.1	2.8	2.5	2.1	1.9

The electromagnetic spectrum extends indefinitely toward small and large wavelengths (large and small frequencies). Various segments have been assigned labels, but these are to be regarded as rough definitions of particular spectral regions. At shorter wavelengths the radiation is termed ultraviolet (UV) and then X-ray and gamma ray. At longer wavelengths the radiation is infrared (IR) and subsequently microwave and radio. The following table[1] identifies a few specific wavelengths with their spectral regions. (see page 18 for prefixes E, P, f, p, etc.; gamma rays are further classified on page 135.):

Region	Wavelength	Frequency	Photon energy
Radio	100 km	3 kHz	12.4 peV
	100 m	3 MHz	12.4 neV
Microwave	1 m	300 MHz	1.24 μeV
	1 cm	30 GHz	124 μeV
Infrared	1 mm	300 GHz	1.24 meV
Visible	500 nm	600 THz	2.5 eV
Ultraviolet	1 μm	300 THz	1.24 eV
X-ray	1 nm	300 PHz	1.24 keV
Gamma ray	1 pm	300 EHz	1.24 MeV

Thermal radiation.   As textbooks recount, the first mathematical treatment of radiation quanta was in the writings of Max Planck, who was considering explanation for measurements of how radiation energy from incandescent thermal sources, such as sunlight, was distributed over the visible spectrum. Such light from a hot object was idealized as *Blackbody Radiation* because carbon-black surfaces absorb all colors equally and, in turn, emit all colors equally. There was a simple idealized source of such radiation: It was to be found inside an enclosure whose walls were maintained at a specified temperature T—what in German is called a *Hohlraum*. The walls would be in Thermodynamic Equilibrium (TE) with the enclosed radiation, which would be steady, uniform in space, isotropic in direction and unpolarized.[2] By allowing this radiation to be seen through a very small hole, one would see steady radiation in equilibrium with its surroundings and whose spectrum of colors would be that of a blackbody at temperature T.

---

[1]From Wikipedia "Electromagnetic spectrum" and "Radio spectrum".

[2]The notion of thermodynamic equilibrium, like that of a frictionless plane or a point particle or free space, is a common idealization that requires adjustment in practice.

The radiation from an electrically-heated light-bulb filament (incandescent light) or from glowing fireplace embers has a similar blackbody spectrum: A continuum of frequencies peaking at a value that increases (red to yellow to blue-white) with temperature.

3.1.1 Planck quanta

The distribution of Hohlraum energy with frequency—the spectrum of blackbody radiation—had been identified as an important challenge for theoretical physics. It was successfully treated by Planck, by Einstein, and by Satyendra Nath Bose [Bos24; Pai79] using ideas from statistical-mechanics underpinnings of thermodynamics; see Appendix B.7.

Planck's model of the thermal radiation from a Hohlraum (*Hohlraumstrahlung*) or incandescent source (see Appendix B.7.2) treated it as though each frequency came from a distinct, independent radiator in the enclosure wall, regarded as an example of an harmonic oscillator in which an electric charge vibrates under the influence of a restoring force and which, in vibrating, gains and loses energy at the frequency of vibration. To obtain an explanation for his successful but empirical formula for blackbody radiation [see eqn (B.7-3)] Planck treated the changes of wall-matter energy at frequency ν (in Hz) as occurring in discrete energy increments ΔE proportional ν,

$$\Delta E = h\nu, \qquad h \approx 6.626 \times 10^{-34} \text{ J s}. \tag{3.1-1}$$

The proportionality constant h, subsequently known as *Planck's constant*, has units that are the product of energy and time (frequency has units of inverse time); see also eqn (3.3-1a). (Discussions in quantum optics typically write $\Delta E = \hbar\omega$ rather than $\Delta E = h\nu$, where $\hbar = h/2\pi$ is the reduced Planck constant.)

The model of Planck, with its assumed discreteness of energy increments (quanta) taken and given to samples of heated matter, successfully predicted the spectrum of blackbody radiation. It made no reference to the particular sorts of atoms that were to comprise the enclosure walls: Everything was determined solely by the temperature. For the first time it gave some evidence that radiation, at least light from a thermal source, might have a granular character, in that it could be regarded as exchanging discrete quanta of energy with material walls. It was natural to assume that what emerged through the tiny observing pinhole was a steady stream of radiation quanta (or photons).

It was Einstein who followed through by proclaiming quantization of the radiation itself, as radiation quanta (the photons), and Bose who completed that approach by considering the radiation as being distinguished by phase-space elements, as would occur for particles—examples what are now called *bosons* (particles with zero or integer spin, see Appendix A.8.3) [Bos24; Pai79].

This picture—this model of Planck photons—was a part of how I was first introduced to the notion of photons, and it is often presented pedagogically for that purpose. But it is not sufficiently complete to be as useful as it needs to be, and its interpretation raises additional questions. Although the walls of the Hohlraum, being matter, were constructed of atoms, these were tied together by chemical bonds into a solid, and so they could not be expected to exhibit the atomic structure of free atoms.

Indeed, the valence electrons of atoms in metals are not localized. They are *conduction electrons* that move throughout the metal as electric currents, driven by voltages, and their lack of confinement means that their energies need not take only the discrete (quantized) values they take when each atom is isolated in free space. Only much later did I appreciate that characteristics of thermodynamic equilibrium, by its very nature, cannot depend upon any details of the material, and that Planck was therefore free to model this in any way he chose [Gla06]; see Appendix B.7.1.

I understood that Planck himself never fully accepted the notion that radiation itself had a corpuscular nature. To place his objection into perspective one should realize that until well after Planck's youth many prominent physicists regarded even the existence of atoms and molecules as an unproven hypothesis, a useful mathematical fiction. (The physicist Ernst Mach and the physical chemist Wilhelm Ostwald are said to be amongst the disbelievers of the molecules that Ludwig Boltzmann used in his statistical mechanics.)

3.1.2   Thermal averages and fluctuations

One approach to the subject of thermodynamics, a traditional staple of chemistry and physics studies, proceeds from considering the microscopic, atomic-scale, constituents of matter, a subject listed in college courses as *statistical mechanics* or statistical physics. Its roots trace back to the latter part of the nineteenth century and the writings of most of the celebrated physicists of that time. Einstein wrote several papers on aspects of the subject, papers which, along with his insightful discussions and correspondence, helped establish the revolutionary notion of quantized energies and the related notion of quantized radiation (and the existance of atoms).

Averages.   According to classical statistical mechanics energies of individual atomic and molecular motions should, when in thermodynamic equilibrium at temperature T, average to a multiple of $k_B T$ where k_B is the Boltzmann constant; see Appendix B.7. Underlying this deduction is the assumption that individual molecular energies were unconstrained and could take on a continuum of values. Observations of the specific heats of solids (the energy required to produce a temperature change) deviated markedly from the theory at low temperature. It was Einstein who first suggested that these observations could be understood if the vibrational kinetic energies of single solidified atoms were discrete, and that at low temperature the thermal energy $k_B T$ fell below the value needed to induce vibrations of a single molecule. Similarly, at low temperatures $k_B T$ was insufficient to excite various rotations and vibrations of individual gas molecules—an explanation for puzzling measurements of their specific heats. This theoretical break from conventional assumptions of statistical mechanics— the notion of quantized energies—was a major preparation for the eventual acceptance of quantum theory [Kle65].

Fluctuations.   Once the Planck formula, eqn (B.7-3), became established for the distribution of frequencies of radiation from a thermal source Einstein examined its prediction for expected variations (*fluctuations*) about the mean energy density for given temperature. He found this energy variance had two parts [Scu72; Pai79; Mil81; Mil19]. One part could be regarded as what would be expected if radiation were to consist of independently moving localizable quanta with energy $h\nu$. The other part obtains from considering unlocalized standing waves enclosed in a bounded box. Particles and waves.

Einstein further examined the fluctuations in radiation pressure (photon momentum) for a thermal source, as would be detected by an immersed plane mirror. Again the results comprised two terms, one interpretable as quanta of momenta $h\nu/c$ and the other a wave contribution [Pai79; Mil19].

These deductions of radiation granularity resulted from applying notions of statistical mechanics and thermodynamics to thermal radiation—to vast crowds of random photons. Historically the next important contribution to the notion of photons came from Einstein's explanation of the photoelectric effect or, as he put it, "the generation of cathode rays by the illumination of solid bodies".

But before introducing that source of evidence for photons in Section 3.5 it is instructive to consider another line of scientific investigation, one that was essential in establishing what became quantum theory and which offers definitions of photons that remain in use today: Spectroscopy [Col90; Dem96; Bro03; Mar06; Pav08; Tow13].

3.2   Spectroscopy; Photons as energy packets

The very practical subject of Spectroscopy—originally the interpretation of colors present in light from various sources—gained its present importance from work of Gustav Kirchoff and Robert Bunsen in the mid 1800s. Their spectroscopes that broke light into its component colors did so by first passing the light through a narrow vertical slit that served as a source for subsequent optical elements. Its colors were then dispersed horizontally by wavelength either by passing through a prism or reflecting from a grating, to form a horizontal spread of slit images that appeared as bright or dark vertical *lines—spectral lines*—in an otherwise featureless horizontal background of thermal radiation. Kirchoff and Bunsen observed that light passing through a relatively hot vapor, such as glowing hydrogen or sodium or mercury, would acquire discrete new frequencies—bright spectral lines. Light passing through a cold vapor would have discrete frequencies removed—dark lines.

Spectra as chemical identifiers. Kirchoff and Bunsen found that the discrete frequencies that were added or removed, as light was emitted or absorbed by a vapor, were characteristic of the particular chemical composition of the vapor, so that the pattern of spectral lines could be used to identify specific chemical elements that were present. Every chemical element, every compound, was associated with a unique set of spectral lines. Conversely, a given spectrum could be linked to a unique set of atoms and molecules, much as a fingerprint is associated with a unique individual. Today the principles first expounded by Kirchoff and Bunsen underlie the vast enterprise of quantitative spectroscopic chemical analysis, a welcome replacement to the chemical balance and white precipitates that were once part of the toolbox of analytic chemists (and my own undergraduate studies). Not only are the lists of frequencies of value in identifying constituents, but details of relative intensities of the lines provide information about the temperature and pressure of the source. Spectroscopists now rely not only on visible light but on all accessible portions of the electromagnetic spectrum for their sources.

Bohr and energy quantization. The physical explanation for the observed discreteness of the spectral frequencies came first from the work of Niels Bohr who, in devising what became known as the Bohr-Rutherford model of the hydrogen atom (a massive but very tiny positively charged nucleus, around which travels a negatively charged electron), proposed that isolated atoms in free space had quantized internal energies (energies associated with motions of their electrons). That view, subsequently elaborated during the development of quantum mechanics in the 1920s, was very simple: Atoms and molecules could exist only with certain discrete energies of their internal parts (quantized excitation energies). This was, and still is, one of the characteristics that distinguishes a quantum system from a classical system. That discreteness is now attributed to the mathematical properties of localized solutions to the time-independent Schrödinger equation; see Appendix A.5.

Radiation increments. To change the structural energy of an atom by action of radiation (as contrasted with changes induced by collisions with other particles) the radiation must carry off, or impart, the energy difference between the discrete energy states of the atom. This notion, of radiation energy $h\nu$ ($= \hbar\omega$) given or taken from a single isolated atom, gave a very simple explanation of the existence of discrete colors in the emission spectrum of a hot gas or the absorption spectrum of a cold gas: The frequencies (i.e. the discrete colors) originated with discrete energy differences, converted into frequency units with the multiplier Planck's constant h (or $\hbar$). That is, observed bright spectral lines originated with atoms that had emitted photons (emission lines) while dark lines of missing frequencies owed their existence to absorbed photons (absorption lines).

Bohr condition. It was natural to regard these notions of radiation being given up or acquired by atoms as changes to photon numbers in a radiation field: A *single photon* is what carries the radiation energy increment $\hbar\omega$ to or from a single isolated atom as it changes between stationary states of discrete allowed internal energies. The radiation frequency is determined by the atom energy change in accord with the rule

$$\omega = \omega_{12} \quad \text{with} \quad \hbar\omega_{12} \equiv |E_2 - E_1|, \tag{3.2-1}$$

for transitions between energies E_1 and E_2. The angular frequency ω_{12} defined by eqn (3.2-1) is the *Bohr (transition) frequency*; the equation expresses the Bohr condition that the photon frequency (a Bohr photon) should match the Bohr frequency.

The equality sign $=$ in eqn (3.2-1) should, more correctly, appear as the approximation symbol $\approx$ because absorption and emission can take place in a range of frequencies around the Bohr frequency. The breadth of this frequency distribution – the *spectral line width*—and the relative intensity of various lines, gives useful information about the environment of the radiative interaction [Bre61; Ber75; Van77; Sob12].

In this picture the change in the number of photons (the overall strength of the change to the spectrum) is set by the number of emitting or absorbing atoms, and the overall energy content of the radiation, available for heating or viewing, is the summed energy of individual photon increments. For a monochromatic beam this is the product of the photon number density and the energy of individual photons. As photon numbers increase, the effects of the radiation become more readily observed.

Radiation attenuation. When the frequency of steady illumination approaches that of a particular atomic transition (a particular Bohr frequency) the light intensity diminishes steadily with passage through the matter. The light is scattered from its path or is absorbed, to appear subsequently as fluorescence (of various frequencies) and heat (thermal motions of the absorber atoms). The absorption events are associated with temporary formation of an excited state of the atoms or other constituents of the matter through which the radiation propagates. Appendix B.8.1 presents details.

Momentum and angular momentum. Radiation carries not only energy (ability to heat) but momentum (ability to exert pressure). This follows from interpretation of Maxwell's equations, as textbooks show; see Appendix B.1. Absorption or scattering of radiation by any particle will convey momentum to the particle and will serve as a force to accelerate particle motion. It is from momentum-carrying photons that we picture these actions being instigated on atomic centers of mass.

Radiation may also (but need not) carry angular momentum (ability to exert a *torque* or twisting action); see Appendix B.2.3. With the full quantum theory of atomic structure came recognition that the stationary states of isolated atoms could be characterized by more than just their energies; atoms and molecules had shapes, and they could take discrete orientations in space. These states are associated with collective angular momentum of the constituent electrons and nuclei, specifically the magnitude of the total angular momentum and its projection on a chosen axis—known as the *quantization axis*—that one could choose arbitrarily. For such quantum states, discussed in texts on atomic and molecular structure (see Appendix A.7.2), the radiative transitions can add or subtract angular momentum, and so the photons must carry appropriate attributes beside energy.

3.3 Discrete energies of atoms

The observed patterns of spectral-line frequencies implies discretized regularities of the energies of the underlying quantum system. Historically these energies were associated with the internal structure of the atoms (or molecules or ions) that interacted with the radiation. Such discreteness of structural energy originates with the arrangement of system constituents, including rotation and vibration, not with unconstrained overall kinetic energy (although atoms and ions held in traps have discrete values of kinetic energy).[3] The first successful predictions of those discrete energies came from Bohr's modeling the behavior of the single electron of an isolated hydrogen atom. A variety of other examples come from structures of molecules.

> **Aside: Photon energy.** From the discrete energy differences of atoms, molecules, and other structures come the discrete energies of photons. Although the joule (J) is the SI unit of energy, it is more convenient to express energies of atom and molecule excitation in units of electron volt (eV). The needed formula, relating photon energy to frequency ν (in Hz = 1/s), is
>
> $$E^{\mathrm{phot}} = h\,\nu, \qquad h \approx 4.136 \times 10^{-15} \text{ eV/Hz}. \tag{3.3-1a}$$

---

[3] Discrete energy occurs with a degree of freedom that is constrained by boundary conditions.

The connection with wavelength is

$$E^{\text{phot}} = hc\,/\,\lambda, \qquad hc \approx 1.240 \times 10^{-6} \text{ eV m}. \qquad (3.3\text{-}1\text{b})$$

These formulas, taken with the energy values associated with classes of structural change, establish the various spectroscopic regimes in which photon evidence is to be found.

Orbital energy. The Coulomb force that binds an electron to a point-charge nucleus has the same dependence on particle separation distance as does the gravitational force between a planet and the much more massive sun, a force that leads to the elliptical Kepler orbits of planetary motion. As first courses in quantum mechanics now discuss, the resulting electron motion as part of the atom is restricted to a succession of discrete orbits identifiable by a *principal quantum number n*, a positive integer. The zero of energy of the hydrogen electron-nucleus system is conventionally taken to be the electron at infinite distance from the nucleus. As the electron approaches the nucleus the Coulomb attraction produces a negative binding energy, greatest in magnitude when $n = 1$. For hydrogen the area enclosed by an electron orbit increases as n^2 and the negative energy of binding to the nucleus increases inversely with n^2. The series of possible energies approach a limit as n increases; the area enclosed by the orbit becomes large without bound and the binding energy approaches zero from below. The acquisition of further energy carries the electron into an unbound quantum state, no longer constrained to discrete energy, as it travels away from its former captivation, leaving behind an ion.

Similar infinite sequences of energy levels (so-called *Rydberg series*[4]) occur not only in hydrogen but in all the chemical elements that have just a single, chemically (and optically) active electron, as do the *alkalis* Li, Na, K, Rb, and Cs. The remaining electrons of these atoms form *shells* of charge that are inactive in chemical processes such as molecule formation and are unaffected by optical light—their spectral properties are found in the X-ray regime. The electrons of other chemical elements act collectively and their energies form more complicated patterns, although these too exhibit Rydberg series associated with single-electron activity.

These spectral series, and other patterns formed by radiative transitions between electronic-structure states, are dominant features in the visible region of the electromagnetic spectrum.

Vibrational energy. The electrons of molecules also exhibit Rydberg series of discrete energies, but their collective motions, as atoms, become more important: Molecules have quantized energies in which their constituent atoms vibrate about an equilibrium arrangement with energy proportional to a positive, integer *vibrational quantum number* ν. Like electronic motions, these too form an infinite sequence, terminating with motions so vigorous that the constituents fly apart.

Vibrational energy levels are more closely spaced than electronic-structure levels. Their changes are seen as patterns in the infrared portion of the electromagnetic spectrum.

---

[4]For alkalies the Rydberg energies vary inversely as $(n - \delta)^2$ where n is a positive integer and δ, the *quantum defect*, is not an integer.

Rotational energy. Molecules also have a tumbling energy of rotation, identifiable by an angular momentum quantum number J that, for any particular molecule, may take an infinite number of non-negative values separated by unity. (For a given molecule the possible values start either from 0 or $\frac{1}{2}$.) The rotational energy is proportional to $J(J+1)$ and to a moment of inertia.

   Rotational energy levels are more closely spaced than vibrational levels and their changes are seen in microwave spectra. Combined changes of rotation and vibration (ro-vibrational) appear as bands of spectral lines in the infrared.

Orientation energy.   Still another form of quantized energy occurs with magnetic moments, either of orbital motion of electrons or of intrinsic (spin) angular momentum of nuclei and elementary particles. For angular momentum J there occur $2J+1$ discrete orientations, leading to $2J + 1$ discrete energies when the system is exposed to a magnetic field. (This field-induced energy increment is the *Zeeman shift*.) The relevant identifier is the *magnetic quantum number* M, taking $(2J+1)$ values between $-J$ and $+J$.

   Transitions between orientation states typically occur at radio frequencies. A particularly celebrated transition occurs between the two possible relative orientations of electron and nuclear spins of hydrogen atoms. This has a wavelength near 21 cm and a frequency near 1480 MHz (numbers known to more than a dozen decimal places). The transition is very weak—the excited state has a mean lifetime of ten million years—but it is observable in galactic clouds of neutral hydrogen and its observation by radio astronomers starting in the 1950s brought important information about the early universe.

Nuclear structure.   The protons and neutrons in an atomic nucleus are not frozen into static structures. Indeed, they have been regarded as a rotating liquid drop. Like the electrons of an atom or the atoms of a molecule, nucleons can take a variety of motions, both individually and collectively, characterized by various quantum numbers [Bla52; DeS74]. These excitations, which can be seen following beta decay of radioactive elements, are typically short lived and are associated with energy changes that are seen in transitions as gamma rays.

Doubly-excited states; Autoionization.   The simplest forms of photon-induced excitation are those in which the photon energy goes into a single electron, or a single degree of freedom. This is what happens with the hydrogen atom. But in systems with several electrons the energy may be shared amongst two bound electrons, which form a *doubly-excited* state. When the two confined electrons collide their shared energy may be transferred to a single electron, which then has enough kinetic energy to leave the atom spontaneously, a process known as *autoionization* [Fan61; Fan65; Sho67; Sho68; Pra05; Tem06; Tem13]. When autoionization is possible the absorption spectrum exhibits notable characteristics: A plot of intensity versus frequency is markedly asymmetric, a Fano profile.[5] [Fan61; Fan65; Sho68b; Lou92; Sta98] in which

---

[5]The spectral profiles have been called Beutler-Fano profiles in recognition of an experimental paper by Hans Beutler [Beu35] and a theoretical explanation by Ugo Fano [Fan35] It was Fano, in 1961, who first presented the analytic expression eqn (B.8-12b) now commonly used for the intensity distribution of autoionizing lines and other spectral structures in a photoionization continuum.

much of the spectral line absorbs but a portion can be transparent; see the discussion of resonance-scattering theory on page 363 of Appendix B.8.2. The inverse of autoionization, known as *dielectronic recombination* [Sho69], is a three-body process in which two free electrons collide with an ion to produce a doubly-excited atom state plus an outgoing photon.

There exist simple idealized models of all these motions, with energy expressions presented in most courses of quantum mechanics. The actual behaviors are typically combinations of all of these simplifications. Inevitably there is a state of lowest energy, termed the *ground state*, into which all systems tend as surrounding temperatures are lowered.

With all such models of orbital, rotational, and vibrational motion there occur infinite sequences of quantum states associated with each degree of freedom. However, several practical considerations diminish the useful number of quantum states. These limitations originate with the environment encountered by the atoms, molecules, and other quantum systems of actual experiments. Inevitably there are neighboring atoms. For solids the environment of any single atom is a more-or-less rigid framework of charge distributions, and so the idealization of infinitely large hydrogen-like orbital motion of electrons fails. For liquids this surrounding neighborhood is not static. For gases the environment resembles a succession of brief collisions that prevent continuation of a quantum state. In all of these situations the number of quantum states that need to be considered in treating adjustable system behavior becomes limited to a finite number, often only two or three. The number of *essential states* needed for a satisfactory description depends not only on the details of the experiment but upon the precision desired. Quite commonly only two states are considered [All87].

3.4 The Bohr-Einstein emission and absorption photons

Einstein used these notions in presenting a description of changes to internal excitation of atoms as induced by thermal radiation. Einstein proposed three mechanisms, each of which can be regarded as an incoherent change in photon number needed to maintain thermodynamic equilibrium: Absorption, spontaneous emission (occurring in a vacuum, with no photons present), and stimulated emission (induced by pre-existing radiation). Appendix A.16 discusses these. The differential equations that described these three processes under conditions of thermodynamic equilibrium proved satisfactory until maser and laser radiation become commonplace tools, requiring a major revision of the notion of how one must describe radiation-induced changes (e.g. the Jaynes-Cummings model of Appendix C.6)—and presenting new challenges for how photons should be regarded.

My own picture of photons as little packets of energy was based upon the ideas first presented by Planck, Einstein, and Bohr in the early years of the twentieth century, after Planck had proposed radiation quanta but before the full development of quantum theory as it came to be constructed by Heisenberg, Schrödinger, Dirac, and many others during the 1920s. That Planck-Einstein-Bohr view was very simple: Atoms and molecules could exist only with certain discrete energies of their internal parts (quantized excitation energies). Any exchange of energy with radiation must therefore involve discrete energy increments, $h\nu = \hbar\omega$, to be carried by the radiation as photons. This is a simple viewpoint illustrated schematically by eqn (5.1-1a) of Section 5.1.

The picture attributed to Planck, Bohr, and Einstein was for me, and many others, a satisfactory definition of a photon. The granularity of thermal radiation, as embodied in the successful Planck formula for the blackbody spectrum, seemed to fit with the quantization of energy levels of isolated atoms. This notion of a photon has two versions, discussed with more detail in Section 4.3:

> **Emission photon:** A photon can be regarded as a packet of radiation (a superposition of frequencies, see Appendix B.5.1 and discussion after eqn (B.5-4)) emitted from a single excited atom (or other quantum entity) as it undergoes a change to a discrete energy state of lower energy. The photon carries off that energy as radiation. (Photons may also, but need not, be carriers of momentum and angular momentum.) The emission can either be induced by an existing radiation field (called stimulated emission) or, like the radiative decay of an atomic nucleus, can occur spontaneously. Section 2.8 discussed some examples of radiation emission [Sha91; Bia93].

> **Absorption photon:** The alternative version is that a gain of internal energy (or momentum or angular momentum) by a single atom (or other entity) is caused by absorbing a single photon. The photon is thereby defined "postmortem" as having been localized at the atom. A gamma-ray photon, in being absorbed, may produce an audible "click" in a Geiger counter, a record of earlier photon presence. As Winnie the Pooh might have said to Piglet, "We know where the woozel was" (but it is not now there). Appendix B.8.1 discusses some models of radiation absorption from beams.

I have found that these two very traditional operational definitions of photons have given way to more nuanced views, as Chapters 5 - 7 will explain. As a postdoc I first learned how simple rules for quantizing electromagnetic fields, supplemented by the elegant mathematical theory of angular momentum (an example of a Lie algebra, see Appendix A.18.4), could explain quantitative details of the absorption and emission as originating with selection rules that depended on the orientation of individual atoms in the traveling-wave electric vector of a radiation field [Con53; Sob72; Sob92].

Einstein's notion of emission and absorption presumed that these process are independent. This was how I learned, from writings of MIT professors, to view nuclear reactions—a neutron or proton striking a nucleus would form a temporary *compound state* which would subsequently rid itself of that energy [Bla52; Fes58b; Fes62; Sho67].

However, this is not a valid assumption for elastic (energy-preserving) scattering of light by an atom or molecule: For such situations the incoming field creates an induced dipole moment with a magnitude and phase set by the incoming field; see Appendix C.4 and [Scu72]. This induced dipole, in turn, is the source of the outgoing field. It is only if there is a delay between incoming and outgoing field sufficiently long that the dipole moment loses phase memory that the notion of uncorrelated, independent re-emission has merit [Sho67]. As subsequent developments have shown, this is a description that applies fully only in the incoherent limit.

3.5   The photoelectric effect; The Einstein photon

When suitable metals are illuminated by radiation, free electrons (photoelectrons) emerge from the surface. This effect has been most commonly observed with ultra-violet radiation illuminating clean metal surfaces in vacuum. When the radiation is monochromatic, the photoelectrons are mono-energetic. It is significant that the energy of the outgoing electron does not depend upon the intensity of the radiation, only on its frequency. The experimentally observed relationship, quantified in the formula for the kinetic energy of the emerging electron,

$$E^{\text{kin}} = h\nu - E^{\text{bind}},\tag{3.5-1}$$

was given interpretation as an example of energy conservation by Einstein in his 1905 Nobel Prize winning paper "On a heuristic point of view about the creation and conversion of light" [Ein05; Aar65; Kra92].

Einstein's quanta. Einstein wrote (as translated by Dirk ter Haar [Ter67]):

> It seems to me that the observations on blackbody radiation, photoluminescence, the production of cathode rays by ultraviolet light and other phenomena involving the emission or conversion of light can be better understood on the assumption that the energy of light is distributed discontinuously in space. According to the assumption considered here, when a light ray starting from a point is propagated, the energy is not continuously distributed over an ever increasing volume, but it consists of a finite number of energy quanta, localized in space, which move without being divided and which can be absorbed or emitted only as a whole.

Einstein applied this view to obtain the formula of eqn (3.5-1) for the description of the photoelectric effect: He proposed that the energy given to each electron came from the energy $h\nu$ of a single quantum of radiation added to the binding energy E^{bind} that initially holds the electron into the surface (and known as the *work function*).

Although the Einstein picture of photonionization, expressed with the formula eqn (3.5-1), fitted well the experimental results of that time (but see [Kra92]), subsequent writers have pointed out that the photoelectric effect is by no means unequivocal proof that radiation is intrinsically granular. Section 8.6 presents these arguments.

Einstein later completed the picture of the particulate nature of radiation by noting that these quanta would carry not only energy in discrete increments of $h\nu$ but also linear momentum in discrete increments of magnitude $h\nu/c$. It is these field increments that subsequently became the particles known as photons. (As noted in Appendix B.3.7, standing-wave photons carry no linear momentum.)

Traveling electromagnetic plane waves may (but need not) be regarded as coherent superpositions of left- and right-circular polarizations. The photons of such circularly-polarized vector fields have an intrinsic angular momentum (or *spin*) of $\hbar = h/2\pi$: They are spin-one particles (*bosons*) that can exchange energy, momentum, and angular momentum with atoms.

> **Aside: Spin projection.** It is the *projection* of spin along the propagation axis that serves as a dynamical degree of freedom, measurable as polarization, not the

intrinsic spin, which is always unity (just as the photon electric charge is always zero).

3.5.1 Fanciful photons

Fig. 3.1 *Menzel photons.*

The concept of photons just presented led Donald Menzel (then director of the Harvard College Observatory and author of numerous technical articles on radiation passage through hot gases) to create a rather fanciful anthropomorphic picture of photons preparing to interact with an atomic electron, a picture that became part of the dust jacket for a 1968 textbook [Sho68]. His depiction of photons in Figure 3.1 is rather like the "raindrops" that Feynman used for their description [Mar96b] and the traveling globs of energy that have often been suggested for visualizing photons. In this whimsical cartoon the anthropomorphic photons are intended to represent the crowd of indistinguishable thermal photons, not the organized troops of photons that march from a laser—which had not yet become a common source of radiation.

3.6 Scattered photons: Doppler and Compton

The notion of particle scattering—of collisions between either elementary particles or of composite systems—made an appearance in my graduate-school studies. I found with this topic, known as scattering theory [Gol64; New66; Rod67; Sho67; Tay72; Wat07; Mis10; Fre18] , an elegant combination of mathematics and theoretical physics that, to this day, remains a significant part of fundamental physics as well as atomic and optical physics [Sho67]. The topic can be approached from a viewpoint of either particles colliding or of steadily flowing waves of definite frequency (and energy) intersecting. Appendix B.8.2 presents some of the formulas that quantify radiation-scattering events. The following paragraphs summarize two aspects of photon scattering.

3.6.1 Photons and Doppler shifts

Everyday experience with sound waves of approaching and receding sirens and horns bring examples of Doppler shifts in acoustic-wave frequencies. Such frequency shifts of moving sources are found in all wavelike phenomena. In particular, they affect the frequency of light emitted by a traveling atom or by any other moving radiation source [Lan51; Pan55; Jac75; Lan80; Pic14]. Measured frequency shifts of identified spectral lines from astronomical sources (red shifted when moving away) have given us our picture of the universe and its constituent assemblages of matter.

Although the Doppler effect is to be found with wave phenomena in general, it has also been placed within the context of discrete radiation quanta [Sch22; Dir24; Mic47; Str86; Giu13; Red13]. The affect appears in the theory of special relativity as examples of relativistic particles that have zero rest mass and therefore move in free space with speed *c*. An observer who is stationary in a laboratory reference frame will observe a radiation quantum emitted or absorbed by a moving atom to be Doppler shifted. Conversely, to produce excitation of a moving atom a laser in the laboratory rest frame must emit light with a carrier frequency that incorporates a Doppler shift.

The Maxwell equations of Appendix B.1.1 that define classical (non-quantum) electricity and magnetism predict that traveling waves of electromagnetism in free space carry both energy (thereby bringing heat) and momentum (thereby bringing pressure). Experiments upon beams of atoms (or molecules) often rely on radiation-induced velocity changes to produce cooling and trapping of individual atoms. In such situations the atoms undergo excitation in a discrete transition, interpreted as the absorption of a single photon, followed by spontaneous emission of a second photon. This photon-based description of energy change brings a concomitant change of atom momentum, also attributed to discrete quanta.

In many situations we deal with a collection of atoms whose Doppler shifts originate in random, thermal motions. Appendix A.11.2 discusses how these velocity distributions are to be treated in modeling effects of radiation.

3.6.2   Compton photons

The notion of a photon as a traveling entity, albeit one without rest mass, lends itself to considering collisions with particles of matter. The photon has no electric charge, but it can scatter from charged particles, such as an electron or an atomic nucleus. Observation of such scattering by Arthur Holly Compton starting in 1923 (work rewarded by the 1927 Nobel Prize in physics) did much to convince the physics community that radiation quanta existed. Generations of physicists were subsequently taught that, as Albert Messiah put it [Mes61]:

> The Compton effect represents another confirmation of the photon theory, and a refutation of the wave theory.

As Section 8.6 notes, passing time has altered this strict judgement, but it remains a useful view.

In the photon-based picture of the Compton effect (Compton scattering) a photon having definite energy (typically from an X-ray or gamma-ray source) and definite linear momentum (as part of a beam) collides with a relatively unconstrained charged particle (typically an electron). Conservation of discrete energy and discrete linear momentum (particle-like conditions) as these are given by collision to the charged particle then provides a condition on the wavelength of the scattered radiation as a function of scattered angle. For backscattering from an electron (i.e. reversal of the photon momentum and forward recoil of the electron) the change of wavelength (the Compton shift) is twice the *Compton wavelength*, see Appendix B.8.3.

Although Compton scattering by electrons can, in principle, be observed with arbitrarily weak radiation (a single photon), it requires either unbound electrons or photon energies that are much larger than the electron binding energy.

As noted in Appendix B.8.2, a variety of other scattering processes are observable, either with change of photon energy (inelastic scattering) or without change of energy (elastic scattering). Compton scattering is an example of inelastic scattering of a photon by a classical charged particle.

3.7   The wave-particle photon

The suggestion by Einstein that the photoelectric effect demonstrated the absorption of a discrete quantum of radiation—a photon—leads to the expectation that a quantum of radiation has been localized as a discrete event in space and time: A single atom, or other localized receptor, has absorbed a single photon. Given an appropriate recording device, such as a photographic sensor or a position-sensitive detector, one can imagine that, for a sufficiently weak beam of radiation, individual photons could be detected (but see Section 7.3 for a more detailed explanation of the photographic process). Their spatial distribution would, with continued exposure, build a pattern that corresponded to the intensity distribution of the radiation.

Such an experiment, of the interference pattern seen with light passing through two slits, was carried out by Geoffrey Taylor in 1909, using a photographic plate and a very weak light source, one in which the mean photon number was quite small [Tay09]. The exposure endured for many days, while Taylor vacationed, at the end of which the accumulated pattern of single-photon detections formed the expected interference pattern of unexposed nodes and bright antinodes. Further demonstrations of interference effects (wavelike behavior) in feeble light have been reported over the years [Dem27; Jan57; Rey69; Rue96; Dim08; Buo16; Qur16]. Section 3.8 notes a revisionist interpretation of the original Taylor work.

3.7.1   Wave-particle duality

From this and other similar experiments comes the recognition that an experimenter can find either wave properties (interference patterns and waveness) or discrete quanta (localization and particleness) properties of radiation, depending on what the experiment is designed to detect. There is said to be a *duality* of particle and wave attributes. A photon can justifiably be regarded, therefore, as having observable wavelike and particle-like characteristics. Put simply, as "being" either a wave or a particle.

However, an experiment need not make a clear distinction between these two extremes. Nowadays the wave-particle duality of radiation, like the particle-wave duality of electrons, is commonly regarded as residing in an experimenter's application of quantum theory to a choice of measurements. The fact that something (say an electron) requires, for full description, separate equations of motion for particle behavior and wave behavior, thereby introducing the so-called *wave-particle duality*, has posed a challenge for philosophers of science, and many physicists too, but it is how Nature seems to be.

A common contrary view is expressed by Stanford professor of theoretical physics Leonard Susskind [Sus08]:

...having two incompatible—even contradictory—theories of nature is intellectually intolerable ....

Similarly, MIT professor and Nobel laureate Frank Wilczek asks [Wil15]:

Is it not unnatural to separate our understanding of the world into parts that we do not seek to reconcile?

By contrast, Caltech professor and Nobel laureate Richard Feynman is quoted as saying [Dud96],

... I never think "This is what I like, this is what I don't like". I think "This is what it is, and this is what it isn't".

Feynman further said, in his Nobel Prize lecture [Fey66] (quoted by Schwinger [Sch89])

We are struck by the very large number of different physical viewpoints and widely different mathematical formulations that are all equivalent to one another.

Princeton professor Freeman Dyson, renowned for unifying the approaches to quantum electrodynamics of Feynman and Schwinger [Dys49], expressed this view as [Dys06]:

There is no such thing as a unique scientific version ... Science is a mosaic of partial and conflicting visions.

More confidently, Paul Dirac, Nobel laureate and the original formulator of quantum electrodynamics, wrote in 1927 [Dir27]:

There is thus a complete harmony between the wave and quantum descriptions of the interaction.

The two sorts of equations, and the wave and particle nature of the solutions, are appropriate for different sorts of observations of the electron, and we should expect the same duality with photons. Appropriate radiation measurements of weak fields exhibit discrete detection events, whereas other measurements reveal nodal interference patterns indicative of waves. What you see (or measure) is what you get (or must deal with in equations)—abbreviated WYSIWYG.

Pedagogical alternatives in NMR. For an example of reconciling seeming intellectual conflict of physics models, Appendix A.14.7 presents the alternative pedagogical approaches of two groups of professorial physicists in explaining the physics underlying their 1952 Nobel Prizes for developing what became nuclear magnetic resonance (NMR), using the language of quantum transitions (photons) at Harvard [Pur46] and of spinning tops at Stanford [Blo46; Blo46b; Blo46c].

Living with duality. In thinking of the mental challenge this dual view of the world can present, what comes to my mind is the common use of a twelve-hour clock dial to display a twenty-four-hour day. As I have found, to my chagrin, one can awaken in a darkened room, disoriented after traveling through multiple time zones, and not know whether to apply the daytime AM interpretation or the night-time PM interpretation to the seven o'clock reading of a timepiece, a dilemma that my observation of pedestrian and street traffic did not immediately resolve. In due course a falsified hypothesis (a wrong guess) removed the ambiguity.

Update: Waves and particles. The Dirac approach to defining photons, discussed in Chapter 4 and Appendix B, provides a relatively simple mathematical explanation for the wave-particle duality. The electric and magnetic fields are distributions in space with time variations. The mathematical expressions for these fields can be separated into a factor that depends on time and a factor (a field mode) that depends on position. The spatial mode contains the node structure that we associate with classical waves and interference. The spatial factor has the discreteness that we associate with counting photons as discrete increments of radiation.

Update: Particles or fields. Nowadays wave behavior (e.g. superpositions and interference patterns) is regarded as associated with a field equation, so the seeming conflict between particle and wave nature of photons is regarded as a contrast of particles and fields [Hob05; Hob13]. Discussions, even disputes, of how to regard particles (concentrations of mass and charge) and fields (extended distributions of force and energy) have not ceased amongst professional physicists, particularly those who seek understanding of how the universe is "really" constructed.

Feynman presented a seemingly particulate view with his famous diagrams that suggest an electron or photon appearing at some point, traveling through space and time, and finally vanishing at a second point; see Section 8.3. The paths and endpoints of those diagrams commonly refer to a momentum space appropriate to wave components, but Feynman also presented a detailed position-trajectory formalism appropriate to nonrelativistic point particles as basic entities [Fey48].

Historically, Faraday espoused a view of space-filling lines of force (fields) that originated with charges, currents, and magnetic poles. With the development of Quantum Field Theory (QFT), the fields are associated with discrete energy increments, *quanta*, that are interpreted as single particles. The entities Feynman dealt with are regarded, in QFT, as field quanta: Photons for the electromagnetic field of Maxwell's equations, electrons as quanta of the electron-positron quantum field of the Dirac equation.

The observation of interference between paths from two phase-related sources (e.g. two slits in a screen [Woo79; Sto94; Jon94; Bra04; Men12; Abo17]) has long been regarded as evidence for the wave nature of quantum entities: First for photons and then for electrons, atoms, and even molecules [Kni98; Nai03]. As published images demonstrate, weak fields display the random detector events associated with particles, while stronger fields exhibit the node structure expected with fields [End89; Dim08; Hob13]. Feynman, looking for particulate interpretation, regarded the interference as a fundamental puzzle of quantum theory. However, the discreteness of individual events seen in images is attributable to the spatial distinguishability of distinct detector elements—the resolution pixels.

Another defining characteristic of a field is its *nonlocal* nature: It has values not just at a single point but throughout a volume. The mathematical connections between points in space-time establishes the consistency of a quantum field with the requirements of special relativity. The group-theoretical classification of the particles for the Standard Model of Appendix A.19, including photons, is based upon field (wave) characteristics. It is notable that solutions to the free-space Maxwell equations (photons) are unable to describe spacelike mathematical points (particles) [Kel05; Haw07; Smi07; Bia09; Bia14].

3.7.2 Complementarity

The curious occurrence of both distributed, wavelike and localized, particle-like attributes of electrons, photons, and other quantum systems (sometimes referred to as *quantons*), given the name *complementarity* by Bohr, has provoked thoughtful discussion for nearly a century. In essence, any two measurable attributes are termed *complementary* if a precise measurement of either of them precludes any knowledge of the other; there is said to be a *duality* of the two attributes.

Although the duality between particles and waves has commonly been regarded as expressing mutually exclusive qualities (either-or choices), quantification of particle and wave natures offers examples in which there may be a continuum of these relative attributes. Studies of such systems have often been idealized as interfering near-monochromatic waves from two distinguishable sources (say, two slits): The occurrence of intensity maxima and minima (interference fringes) at a detector betokens wave phenomenon quantified by the *fringe visibility* $\mathcal{V}$, a number bounded in magnitude by 0 and 1. Alternative, complementary measurements can reveal which path the disturbance took to reach the detector—a *welcher-Weg* (which-way) determination that has interpretation as a beam or localizable particle trajectory [Rem91; Sto94; Bjo98; Dur98; Bra04; Men12]. A *distinguishability* variable $\mathcal{D}$, also bounded by 0 and 1, quantifies the confidence with which it is possible to determine this path localization. The tradeoff between the wave and particle-beam aspects of observables takes the form of a complementarity relation

$$\mathcal{D}^2 + \mathcal{V}^2 \leq 1. \tag{3.7-1}$$

Although eqn (3.7-1) has been viewed as an expression of this complementarity [Dur00; Col14], the customary Planck constant makes no appearance here, and studies are showing that such an expression as eqn (3.7-1) proves applicable not only to the description of quantons but also to paired degrees of freedom for classical optical fields [Qia15; Ebe16; Ebe16b; Qia16; Ebe17; Qua17; Qia18; Qia18b]. Appendix B.1.8 provides some basic background.

3.8 Revised views of Planck, Einstein, and Compton photons

During the century that followed the papers of Planck and Einstein an atomistic, granular view of radiation became an accepted counterpart to the atomistic view of matter: Discrete units of radiation were companions to discrete units of mass and charge. But even as successive generations dealt with these concepts there remained a shifting community of physicists who wondered whether radiation quanta—photons— were really necessary or if they were only a convenience in treating how radiation and matter interacted (see Section 8.6).

These questions have prompted ongoing re-examinations of experiments and their interpretation and a realization that many observations—but not all—do not require traditional, discrete radiation increments (photons) for their explanation, only a classical field and the quantum mechanics of atoms—a *semiclassical* model. Amongst the radiative effects whose qualitative behavior is no longer regarded as proof of photons are the blackbody spectrum, the photoelectric effect, the Compton effect, and the processes of absorption, stimulated emission, and spontaneous emission [Scu72; Kid89;

Gar08], although the semiclassical model does have limitations [Cla72; Cla74]. The opening chapter of [Gar08] shows how portions of the phenomena rely only on kinematics (conservation of energy and momentum). In particular they note that the well-established ways of modeling the photoelectric effect without quantizing the radiation field mean that the photons are not required to understand the famous wave-particle demonstration of G. I. Taylor in 1909 [Tay09], redone more recently [Dem27; Jan57; Rey69; Rue96; Dim08; Buo16; Qur16].

It is particularly with quantitative details of radiation effects, most notably fluctuations and correlations, that one finds evidence for photons [Ray90]. The subsections below summarized some of the revisionist views of photons. Section 8.8 summarizes the contemporary evidence that photon discreteness is still an important consideration.

Update: Planck photons.   In writing this Memoir I have read a variety of sources concerning the arguments Planck used to achieve the accepted formula for the frequency distribution of blackbody radiation and his views of the significance of the results. Amongst these writings were several by historians of science [Kuh84; Kuh87; Kra00]. They tend to propose a revisionist view of the traditional description I presented above. I recognize that my narrative is an example of what is commonly done in pedagogy: Invoking a name (Planck in the present case) to unify a particular formalism (in this case the statistical mechanics of thermal photons, discussed in numerous textbooks [Mor64]).

I have found general agreement that it was Einstein, in his paper on radiation [Ein05; Aar65], who most clearly argued that the Planck work demonstrated the existence of radiation quanta (later to be called photons). He argued, in his section on blackbody radiation, that the hypothetical oscillators or resonators that Planck assumed were interacting with radiation could be regarded as examples of the Lorentz atom discussed in Appendix A.14.4: Electrons bound to equilibrium positions by forces that increase linearly with separation. These electrons were affected by random collisions with free electrons and gas molecules to produce an equilibrium. Further invoking the statistical thermodynamics of Boltzmann, Einstein argued that the radiation itself behaved as an ensemble of discrete quanta, a step Planck avoided.

Update: Photoionization.   Although the notion of photons fits very nicely with understanding of the basic photoelectric effect, later writers, starting with Gregor Wentzel in 1927 [Wen27], have shown that the observed energy relationships need not originate with quantized radiation [Scu72]—it is sufficient that the electrons alone obey quantum mechanics. Indeed, the modeling of photoionization by a classical field is sometimes posed as a student exercise in the use of time-dependent perturbation theory and the derivation of Fermi's Famous Golden Rule for calculating transition rates, [see eqn (A.16-10) and eqn (C.5-11)]. In both descriptions, that of Einstein photons and that of semiclassical atomic theory (see Section 5.4), the radiation is taken to be thermal radiation.

Although the link of eqn (3.5-1) between electron kinetic energy and radiation frequency does not require photons, quantum theory predicts that electron ejection need not be delayed awaiting its accumulation of binding energy, as would be required by classical waves of radiation. Such measurements [Car89] were not available when physicists were testing Einstein's quantum views, and so they could not then serve as proof of photons. They now serve to link ejected electrons with the depth of their origin below the surface, as discussed in the following Update on photoelectrons. Subsequently, experiments involving simultaneous photoionization in two locations have shown that a full theory of all aspects of the photoelectric effect requires a quantized field [Cla74].

Update: Photoelectrons. The photoelectric effect discussed by Einstein is an extension to bulk matter of the photoionization to be seen when a single atom is illuminated by radiation whose photon energies exceed the binding energy of one or more of the bound electrons. The picture of photoelectric events has undergone significant refinement since Einstein first presented his photon description [Ber64; Ber64b; Mah70; Fei74; Pen76; Dam03; Paz15; Isi17]. The incident field must first enter the bulk matter, typically a metal. There it acts to impart energy and momentum to an electron that must subsequently travel back to the surface, from which it will, if it has sufficient energy and momentum, emerge into a vacuum. There its energy and momentum (angular distribution around the direction of the incident field) become measurable as photoemission spectra.

Within the metal there are several potential sources of electrons. The valence electrons of the atoms that coalesce to form the metal are merged into conduction bands of kinetic energy. These electrons can be ejected by ultraviolet light. Around the nuclei there remain localized electron states, of more sharply limited energy, most notably those of the most tightly bound K shell. Ejection of these electrons requires the energy of X-rays. When an electron is photoejected from such a state it leaves a vacancy that will refill from less tightly-bound occupied orbitals. That process expels energy either in the form of photons (observed as discrete-frequency *characteristic X-rays*) or as electrons (observed as discrete-energy *Auger electrons* [Dup09; Bur14b]).

This picture of photoionization in stages has led to treatment of photoemission as a scattering process involving an initial wavepacket of photons and a final, detected, free-electron wavepacket, with a succession of intermediate elastic and inelastic scatterings along the way [Mah70]. With the development of tools capable of resolving events within attoseconds (1 as $= 10^{-18}$ s) it has become possible to distinguish the different depths from which the electrons originate, i.e. their different travel times [Cav07]. Their angular distribution provides information about the momentum distribution of the conduction bands.

Update: Compton photons. The essence of the Compton effect is the conservation of kinetic energy and momentum between radiation, regarded as a brief impulse, and a free electron. It can be regarded in several ways. Viewed in the rest frame of the electron it is an example of elastic scattering, and the wavelength shift is an example of a Doppler shift [Nie66; Kan71; Coo77]. Schrödinger, in 1927, explained it without any reference to quantized radiation [Sch27; Dod83; Str86b; Giu13]. Peter Milonni has given a nice exposition of what the Compton effect implies, noting that, as with the photoelectric effect, once the electron is taken to obey a time-independent Schrödinger equation the energy constraints on the field must follow [Mil97].

Update: Spontaneous emission. It is often said that a fully quantum theory of radiation, i.e. QED, is needed to predict spontaneous emission—that a semiclassical theory is unable to predict this effect. This is not quite true. The traditional two-state system undergoing resonant electric-dipole excitation can be in a superposition of states 1 and 2. It will therefore have an electric-dipole transition moment that oscillates with the Bohr frequency [Ebe68; Scu72]. Any oscillating dipole will, according to classical theory, radiate. This is spontaneous emission, and it, along with other effects such as the Lamb shift, were part of the so-called *neoclassical theory* proposed and studied during the 1960s and 1970s; see Section 8.7. Unfortunately for that theory, its description of spontaneous emission (and other things) is not quite right. For example, a completely excited state, like a pure ground state, has no dipole moment, and so classically it will not radiate—it would be metastable.

3.9   Beyond emitted and absorbed quanta

The two viewpoints presented in Section 3.4, of a photon created by emission or a photon destroyed by absorption, each deal with discrete field increments, but their photons need not be exactly the same, as I came to understand (see Section 4.3.3). Nor do they exhaust the possible consistent ways of defining photons, as I shall note in Chapter 8. Nevertheless, they each seem to give an operational definition of a photon, a definition that seems to offer an experimental procedure for creating or destroying a single photon. These definitions aim to answer the very fundamental question: What is a photon?

But lurking behind this seemingly simple view of quantized radiation to match quantized energy levels of atoms was an apparent *non sequitur*: The quantization that was invoked by Planck had apparently nothing to do with the atomic structure proposed by Bohr that underlay the discreteness of isolated-atom energies and whose discrete emission and absorption frequencies were a unique fingerprint of each chemical element. And the quantization of energy levels by Bohr made no use of any granularity of the radiation field introduced by Planck. It might very well be that photons existed, but that possibility was not at all a certainty. Section 3.8 discusses some of the ways that the earlier presumed proofs of photons (those of Planck, Einstein, and Compton) have been given alternative interpretations that have no need for discrete field quanta.

As I subsequently realized, the results of Planck dealt with thermodynamic equilibrium, and under those highly idealized conditions all properties depend only on temperature, not at all on any model of the absorbing and emitting matter. To understand regimes that are far from thermodynamic equilibrium, as are the typical events in a laboratory, it is necessary to consider the equations that govern electron motion in atoms (the Schrödinger equation) and that govern electromagnetic fields (the Maxwell equations). Chapter 4 begins that discussion—after a brief digression.

3.10   Bohr

The two great scientific giants of the early development of what became known as Old Quantum Theory, the brief precursor to what became the Quantum Theory, were Bohr and Einstein. It was the Nobel-Prize-winning development by Bohr of the Bohr-Rutherford model of the hydrogen atom, giving a formula that explained the remarkable regularities of the discrete colors seen in the emission and absorption spectra of that simplest of atoms, that led to the important suggestion that the energies of atomic structure were discrete values – were quantized. It was Einstein who, soon thereafter, presented a successful description of how changes in the discrete energies of atoms were to be brought by thermal radiation: He proposed the three mechanisms of absorption, stimulated emission, and spontaneous emission that prevailed until the development of masers and lasers allowed the use of coherent radiation and brought a different view with different mathematics. Einstein's description of radiative processes relied on notions of radiation quanta, the discrete energy increments that had been introduced by Planck and which later were titled photons. He used these also in his explanation for the photoelectric effect, work that won him the Nobel Prize but which, as it has turned out, was not the definitive evidence for photons that he and others had thought at the time.

I never met Einstein but I once met Bohr. He spoke with me. The occasion was a visit he paid to MIT around 1959. As part of his activities he was to participate in one of the monthly Dinners at the Graduate House, the then residential accommodation for many of the unmarried graduate students. I was one of the residents at the time, pursuing my PhD, and it was my responsibility, as the elected President of the Graduate House Executive Committee, to host the speakers at these events. So I had to make the introduction of the after-dinner remarks of a man who, probably more than any other at that time and with that audience, "needed no introduction". No record remains of what either of us said. But in the informal mingling of the grad students prior to being seated for the meal, I had a chance to speak to Bohr. What does one say to such a giant? Surely not "Do you think the Red Sox might win the pennant this year and defeat the Babe Ruth Curse?" And surely not "What were your feelings about Einstein and his rejection of Quantum Theory?"

I was quite unprepared for this challenge, but somehow I managed a question, to which Bohr began to reply. I have learned that even his fellow Danes had great difficulty understanding the remarks of their famous countryman. For sure I did not understand a word, whatever the language he was speaking. But no matter. I spoke with *Bohr*. And Bohr spoke with *me*.

4

The photons of Dirac

The professor seems pretty lofty

The discrete packets of radiation energy proposed by Planck, Einstein, Bohr, and others provided a simple picture of the discrete energy interchanges that came to be associated with quantum theory. But the picture could be criticized for being rather *ad hoc*, not derived from the basic principles that had provided the very successful theory of atomic structure. It is Paul Dirac who holds credit for first placing radiation onto the quantum framework that had been used for particles and who first provided a consistent theory of radiation and matter. In so doing, he provided a definition of photon, explained in some detail in Appendix B.3, that has prevailed into today's world of quantum optics—but with revisions to be mentioned in Chapter 8.

To explain the Dirac view of discrete radiation increments (particles of radiation) I will begin by recalling ways of viewing wavelike properties of particles, namely electrons in atoms. This approach rather reverses the historical record, in which Duke Louis de Broglie argued from the obvious wave properties of light to suggest wave properties of electrons, but the discussion fits an outlook that has become commonplace in chemistry courses since then.

4.1 Modes: Electron orbitals and cavity radiation; Superpositions

Descriptions of mechanical motions often introduce a set of basic motions—stretching, bending, twisting—as useful coordinates for describing more complicated motions [Eyr44; Wil80; Jaf18]. These *normal modes* have counterparts throughout physics.

4.1.1 Discrete electron modes

It was during my high-school course in Chemistry that I first glimpsed pictures of the solutions to the (time-independent) *Schrödinger equation* applicable to the single electron of an isolated hydrogen atom. The pictures, shown in many elementary

Our Changing Views of Photons: A Tutorial Memoir. Bruce W. Shore, Oxford University Press (2020). © Bruce W. Shore.
DOI: 10.1093/oso/9780198862857.003.0004

chemistry and physics texts, presented the electron charge distribution to be expected in the discrete stationary states of motion—the Quantum States—for this simplest of stable atoms: A single electron moving under the influence of a single, much more massive, point charge. These pictures of electron-density clouds, known as electron *orbitals*, showed slices through three-dimensional patterns of increasing complexity—of an increasing number of nodes where no charge was to be found—as the energy of the electron increased from its minimum allowable value through a succession of discrete values, associated with orbitals of increasing size. These atomic-electron orbitals are often presented as the basic building blocks for constructing many-electron atoms. They generalize to molecular orbitals discussed in chemistry texts.

Electron orbitals. Pictures of electron orbitals, with their spectroscopy-inspired labels s, p, d, f (commemorating the so-called Sharp, Principal, Diffuse, and Fundamental series of spectral lines [Con53]), first sparked my interest in understanding quantum structures. The most tightly bound orbital, the $1s$, was a simple fuzzy ball of electric charge. The similarly fuzzy electron distribution of the first excited-state, the $2p$, was cut by a planar node into a dumbbell of unconnected charge.[1] Then came the successively more energetic $3d$ and $4f$ orbitals, of increasing size and nodal complexity, until the imagined limit when the electron could move to arbitrarily large distances from the nucleus and became unbound. The atom then became an ion and an unbound electron, whose kinetic energy was no longer required to take only discrete values.

Although a bound electron might be treatable as an inseparable part of an atom—as a particle localized within that small volume—pictures of electron orbitals made clear an underlying quantum-mechanical wavelike behavior beneath the particle behavior of classical mechanics: The small-scale behavior of an electron had wavelike properties.

Electron spin. Dirac was also responsible for proposing a relativistic equation for electrons—the Dirac equation that underlies the theory of atomic structure [Dir28; Bet57; Fes58; Gra06; Cow81]; see eqn (B.1-96). One of the consequences of the Dirac theory of electrons is that every electron has not only an intrinsic electric charge but also an intrinsic magnetic moment, associated with an intrinsic angular momentum or *spin* (and a consequent intrinsic magnetic moment). This degree of freedom for an electron is treatable as a *spinor* [Lap31; Hes67; Hes71; Car81; Lou01], a two-dimensional counterpart of the three-dimensional *vector* properties of a photon.

Electron-state specifiers: Quantum numbers. The discrete energies of the single electron of the hydrogen atom are associated with idealized stationary states of electron motion—quantum states—in which the spatial distribution of electron charge and electron currents remains constant with time unless the atom is acted upon by some colliding particle or by some externally-sourced electromagnetic field. As is noted in Appendix A.3.3 these quantum states are associated with complex-valued probability amplitudes (wavefunctions) that are solutions to the time-independent Schrödinger

---

[1]The seeming paradoxical possibility of the electron, as a "particle", moving across a wavefunction node of null electron density, is resolved by recognizing that an electron of known energy and angular momentum cannot be so minutely located: The electron defined by the 2p orbital cannot be more localized; it must be pictured as a wave.

equation. Their absolute squares are the spatial probability distributions that are depicted as the electron orbitals of quantum chemistry.

As I eventually understood, the letters *s, p, d, f,* ... labeling electron orbitals stood for integers $\ell = 0, 1, 2, 3, \ldots$ measuring magnitude of orbital angular momentum, in units of $\hbar$. For given ℓ there are $2\ell + 1$ possible distinct orientations of the angular-momentum vector. The complete specification of the quantum state of a single electron requires four quantum numbers: Three associated with spatial distribution of charge in three dimensions and one associated with the projection of spin along any selected direction. For a confined electron each of these four quantum numbers is discrete—an integer or, for spin projection, $\pm\frac{1}{2}$.

4.1.2   Discrete radiation-field modes

The counterpart of the electron wave equation that leads to pictures of bound-electron structure are the Maxwell equations for electric and magnetic fields, discussed in Appendix B. My undergraduate course in physics did not deal with differential equations *per se*, and did not discuss the Maxwell equations in their generality. But in later courses I learned how these equations, when applied to the sort of conducting-wall radiation enclosure that presumably modeled the Hohlraum of Planck, had only discrete solutions (*cavity modes*), distinguishable by the number of nodes (surfaces of null intensity) that cut the patterns of radiation intensity. With each discrete three-dimensional pattern came a discrete frequency of the oscillatory time dependence that accompanied the spatial pattern of intensity. It was these discrete frequencies of the allowed cavity modes, though based upon a classical description of radiation through Maxwell's equations, that gave connection to quantum theory through the identification of frequency (times Planck's constant) as a field-carried energy increment $\hbar\omega$, that is, to the notion of a single photon as a granular element of a quantum field.

Classes of field modes. The counterpart of the time-independent Schrödinger equation for electron wavefunctions is the *Helmholtz equation*, a second-order PDE for electromagnetic field modes [Lev16b]; see Appendix B.1.5. The *scalar Helmholtz equation* describes a three-dimensional spatial distribution of electric or magnetic field strength. The *vector Helmholtz equation* (VHE) incorporates directional properties, thereby adding another degree of freedom. Each of these equations, and the time-independent Schrödinger equation, involves the Laplacian second-order differential operator; see eqn (B.1-24). The equations have solutions expressible in terms of well-studied functions for several coordinate systems, most notably Cartesian (rectangular), cylindrical and spherical [Mor53]. Although the solutions to Helmholtz equations have application to defining photons, they are basically three-dimensional fields for use with classical electromagnetism and they appear for that purpose in standard references [Str41].

Standing and traveling waves. The spatial mode functions of the Helmholtz equations can be augmented with two classes of time variation, to produce either standing waves of stationary node patterns or traveling waves of nodal surfaces that move in free space with velocity c. Traveling waves may have, in addition to energy and linear momentum, orbital angular momentum around a propagation axis; see Appendix A.7.2.

Aside: Wave superpositions. A standing-wave pattern can be obtained by superposing two traveling waves, moving in opposite directions (counter-propagating), in accord with trigonometric identities such as

$$\cos(X)\cos(T) = \tfrac{1}{2}[\cos(X - T) + \cos(-X - T)]. \tag{4.1-1}$$

Figure 4.1 illustrates the effect of such a superposition.

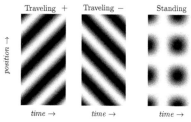

Fig. **4.1** *Traveling waves and their standing-wave sum*

Figure 4.1 shows examples of wave patterns from squared amplitudes. Peak intensities are dark, nodes are white. It shows two traveling waves, moving with increasing time toward positive and negative positions, and a standing wave formed by a coherent superposition of these. Superpositions of multiple waves leads to more elaborate pictures of node structure.

Cartesian coordinates. The Cartesian-coordinate modes (see Appendix B.2.2) are associated with plane-wave fields that have a definite propagation axis; the vectors of the electric and magnetic fields are, at any instant, constant over planes perpendicular to this axis—free-space radiation fields are transverse fields in which the propagation direction, the electric field direction, and the magnetic field direction are along three orthogonal (perpendicular) coordinate axes. These mode fields find most common application in describing plane-wave idealizations of radiation beams such as those found on optical benches threading a system of lenses and mirrors. They also have use in describing the standing waves formed in enclosed cavities that have conducting walls.

Cylindrical coordinates. Laser beams found in academic work have traditionally been idealized as having a circularly symmetric intensity distribution in the plane transverse to the propagation direction, a pattern that is most intense along the beam axis, falling exponentially with distance from that axis—so-called Gaussian beams; see Appendix B.2.5. Free-space modes that are defined in cylindrical coordinates offer a useful collection of beam characteristics in which nodal patterns occur. Whereas plane waves must have their electric and magnetic vectors perpendicular to the propagation direction (they are transverse fields), cylindrically defined modes have a small component of their field vectors in the propagation direction. This makes possible the existence of radiation beams that have well defined orbital angular momentum (OAM); see Appendix B.2.3.

Spherical coordinates. By contrast, free-space modes that are defined by spherical-coordinate solutions to the VHE describe waves that travel outward from a single point on spherical surfaces (see Appendix B.2.6). Near that point they are (vector) multipole fields, of which the three dipole fields are of greatest use in describing the fields emitted by or absorbed by an atom in free space [Sho90; Sho11]; see Appendix B.2.7. These fields carry angular momentum away from the source atom at the coordinate origin. When they are detected at a distant point they bring linear momentum to the detector—the source field is a coherent superposition of plane waves and the detector identifies one of these appropriate to the line of sight between source and detector.

Spherical modes occur not only as descriptors of radiation outgoing from an emitting atom but also descriptors of stationary fields found inside wavelength-sized dielectric spheres and other structures; see the discussion of whispering-gallery modes in Appendix C.3.4.

4.1.3   Superpositions and vortices

In practice, plane-wave field modes are used primarily as a basis for specifying more elaborate fields as coherent superpositions—spatial generalizations of the Fourier decomposition of time dependence, see Appendix B.5.1. For example, any elliptical polarization is specifiable as a coherent superposition of helicity states—states of right and left circular polarization.

Beams of radiation, formed by apertures and lenses, are specifiable as coherent superpositions of a distribution of momentum components in the plane transverse to the dominant propagation direction. The mathematics of all such superpositions adds field amplitudes rather than intensities, thereby offering interference patterns of intensity.

Although monochromatic plane-wave spatial modes fit well the description of fields in confined spaces, where the three momentum components become discrete by virtue of boundary conditions, they are not what is created by a single atom undergoing unhindered spontaneous emission. That process produces a field whose magnitude falls steadily with time, in keeping with the exponentially falling excitation probability. The wavepacket of this emission will have a spread in energies inversely proportional to the lifetime of the decaying atom. The photon from this process is a coherent superposition of monochromatic waves, of field-energy eigenstates that center around a carrier frequency equal to the Bohr transition frequency of the atom.

Phase singularities; Vortices.   Although simple laser beams with Gaussian intensity profiles in the transverse plane retain their long-standing interest, increasingly the ability to form arbitrary wavefronts has led to a veritable zoo of field patterns [Den09; Rub17]. Viewed over a plane there occur singular points where the field intensity vanishes and the phase becomes undefined. In three dimensions these points become singular lines known as optical vortices—lines of total darkness [Ber00; Sos01]. The helical phase fronts of beams that have orbital angular momentum (see Appendix B.2.3) have a phase singularity running along the beam center. Along a path that encircles a vortex the phase will change by an integer multiple of 2π. Multiple vortices can form elaborate patterns, including knots.

4.1.4   Knotted fields

Phase singularities, where field magnitudes vanish and phase becomes undefined, occur in many sorts of electromagnetic fields. Most simply these appear as a single node line (a "thread of darkness" [Pad11]), or vortex, along the center of a laser beam that has orbital angular momentum. More general field vortices, as three-dimensional curves in optics and in hydrodynamics, may form a variety of patterns, including linked loops and knots. Famously, Lord Kelvin thought [Tho67] that the many different ways in which curves (vortices of the aether, for him) can be stably linked or knotted offered an explanation for the variety of chemical elements, an approach to understanding atoms later replaced by the Bohr-Rutherford-atom model of electrons moving around a nucleus; see Section 3.2.

Tangles of field lines are not restricted to vortices but may be found for more general field lines. An electromagnetic field line is a curve in Euclidean space whose tangent defines the local direction of the field. Many of these can be defined as contour lines (level lines lines) of appropriate scalar functions of position at a fixed time [Ran90; Ked13; Arr17]. Field lines, as curves in three spatial dimensions, can form loops, linked loops, and knots, structures that can be classified by their topology (often stable), as parametrized by various integers [Ran92]; for example, by the number of holes in a volume and the nature of linkages of separate parts or connected parts. Papers by Antonio Rañada giving formulas for simple knotted fields [Ran90; Ran92; Ran95; Ran97] have led to a great variety of work on knotted electromagnetic fields, both stationary and traveling, theoretical and experimental [Ber01; Lea05; Irv08; Den10; Pad11; Dal12; Ked13; Hoy15; Arr17; Bia18; Cam18; Sug18; Cos19]. These solutions to the free-space Maxwell equations offer a variety of elaborate fields for constructing photons [Tem16].

4.2   Dirac's photons: Mode increments

Only when I began graduate studies and began to appreciate Quantum Theory did I meet the remarkable work of Dirac, who combined the world of atoms and the world of radiation into a unified treatment that provided a consistent quantum theory for both of these contributors to everyday physics.

4.2.1   Dirac quantization steps

The Dirac approach to defining photons proceeds through several stages, each of which can occupy several lectures in a course on quantum theory.

Canonical variables. The basic idea can seem straightforward in retrospect, once the quantum theory of particles is understood. To describe any classical system of particles one introduces a set of positions and, for each of these, an associated velocity or, more correctly, a momentum (mass times velocity). These two system attributes are termed *canonical variables* (see Appendix A.1). Equations of motion, such as those of Newton, provide expressions for potential energy as a function of position and for kinetic energy as a function of momentum. When the combined energy is expressed in terms of canonical variables it becomes the *Hamiltonian* (see Appendix A.1.6). All of this is a common academic formulation for problems of classical mechanics—for example the behavior of spinning tops and gyroscopes or the oscillation of a pendulum or a mass supported by a spring; see Appendix A.1.7.

Particle quantization. What turns the expression for conserved energy into a base for quantum mechanics is the replacement, in the Hamiltonian and the equations of motion, of paired canonical variables, coordinate and momentum, with symbols (operators) that do not commute (see Section 2.1.1 and Appendix A.4): The product of coordinate times momentum differs from the product of momentum times coordinate by an imaginary number having magnitude equal to $\hbar$ (Planck's constant divided by 2π). The equations, eqn (A.4-3), are actually much simpler than these words suggest.

Those are the essential principles underlying quantum theory, and they lead to partial differential equations (the Schrödinger equation). The details of working out their application, to a quantum gyroscope or a quantum harmonic oscillator or to the hydrogen atom "are left as an exercise for the reader" as textbooks and learned articles too often say.

Field quantization.   What Dirac did was to identify, in the equations of Maxwell, field descriptors that could be regarded as canonical variables. These then become quantum fields by requiring that, like the canonical variables associated with particles, these field variables should become noncommuting entities. Appendix B.3.2 discusses this step. Dirac said, in prefacing this work,

> Resolving the radiation into its Fourier components, we can consider the energy and phase of each of the components to be dynamical variables describing the radiation field.

The traditional quantum-optics approach to carrying out this program, sketched mathematically in Appendix B.1, is to start by expressing the free-space electric and magnetic field vectors, which are functions of coordinates and time, as superpositions of harmonic frequencies—basically the introduction of single-frequency Fourier components as is done in the treatment of heat flow or other behavior that involves both spatial and temporal derivatives. This powerful technique for separating space and time variables leads, for each frequency, to an ordinary differential equation (ODE) of second order that can be recognized as identical to the equation of motion for a classical harmonic oscillator—that is, to the equation of motion for a particle acted upon by a force proportional to the particle displacement from an equilibrium position (*Hooke's law* of force for the action of a springlike restoring force); see Appendix A.1.7.

The behavior of a simple harmonic oscillator is one of the standard soluble models of quantum theory (along with the particle in a box and the nonrelativistic hydrogen atom). Discussed in every elementary textbook on quantum theory, it is treated by replacing real-valued coordinate q and momentum p, and their noncommuting quantum versions, by an alternative complex-valued pair of variables a and a^* whose quantum-version commutators are unity; see Appendix A.7.1.

Associated with each of these oscillator frequencies, and with their associated time-dependent variables, there are second-order PDEs that govern associated spatial distributions of static electric and magnetic fields—a field mode [Lev16b; Mil19c]. The spatial equations are examples of the Helmholtz equations, see Appendix B.1. These are the electromagnetic-field counterparts of the time-independent Schrödinger equation for electrons. Their spatial characteristics depend upon the imposed boundary conditions that are needed to complete their mathematical definition for a specific system. With the usual simple plane-wave description of free-space radiation every frequency is associated with three possible propagation axes and, for each of these, two possible vector directions (polarizations).

4.2.2   Photons as field-mode increments

The solutions to the vector Helmholtz equation, eqn (B.1-28), provide distributions of complex-valued electric and magnetic field vectors (field modes) whose absolute squares give spatial distributions of electromagnetic field energy; see Appendix B.1.2. It is from the directional properties of the field modes that radiation beams acquire polarization character (linear, circular, or elliptical). From the spatial variation of the field modes the radiation acquires the nodes and antinodes that characterize the wavelike interference capability —the wave-defining continuum characteristics of both classical and quantum optics. If the chosen boundary conditions are those appropriate to an idealized metallic enclosure or other cavity then the squared field amplitudes give the spatial distribution of electromagnetic energy density within the cavity. When the differential equations are applied to fields constrained by a container the modes are discrete and can be ordered into a list denumerated by integers.

The combination of harmonic-oscillator behavior for the time dependence of a single-frequency field, and the spatial-mode structure of the VHE that is associated with every oscillator, leads to the following conventional interpretation of the quantum mechanics of electromagnetic radiation, as formulated with the aid of Fourier analysis and modes of the field. The possible quantum states of the harmonic oscillator associated with definite energy of a single frequency are *number states*, in which the monochromatic field is expressible as a discrete number of spatial field-mode increments. These discrete increments of electromagnetic field and energy are the photons—the discrete increments of the electromagnetic field. The energy distribution of the field can be regarded as a distribution of photon probability.

Individual field modes, or superpositions of modes, describe nodal structure that we attribute to waves. The stationary quantum states of monochromatic field modes are specifiable in terms of the number of photons in each mode—a so-called *Fock space* of integers ordered in accord with a listing of frequency and field modes.

This definition of a photon (as an increment of a field mode), following from the Dirac formulation of a quantum theory of electromagnetic fields, augments the two definitions mentioned earlier, of photons defined by single-atom emission or photons defined by single-atom absorption. Those two operationally defined sorts of photons presumably deal with creation and destruction of the single-mode monochromatic photons of the Dirac theory, i.e. of photons that are associated with particular solutions to the VHE supplemented by boundary conditions.

Photon spin.   It is desirable to realize that every electromagnetic field is a vector field, and that every physical vector has three components that express its direction. All of the vector properties of the field, as distinct from the node-bearing amplitude properties, are associated with an intrinsic angular momentum (spin), a property that is defined by how it is affected by rotations; see Appendices A.7.2 and A.18.5. For vector fields and photons this spin has three components and is an example of spin one. The spin of electrons has two components, an example of spin-half.

Traveling waves of free-space radiation, in the plane-wave idealization, have their field vectors aligned perpendicular to the propagation direction and so a specification of their vector properties (their spin orientation) requires values of the vector along two independent transverse directions (the third vector component is zero). This can be done by considering possible choices of linear polarization in two orthogonal directions, but it can also be accomplished by specifying components of right- and left-circular polarization. These latter vectors provide *helicity* states of the field; see Appendix B.1.6. Their photons have a well-defined spin component (either positive or negative) along the propagation direction and therefore they carry angular momentum of magnitude $\pm\hbar$ around the propagation axis.

Linearly polarized light can be regarded as a coherent superposition of right- and left-circular polarization, and the corresponding linearly polarized photons can be treated as a coherent superposition of two spin components. But linearly polarized light does not carry angular momentum. And the combination of right- and left-circular components is not an incoherent mixture, which would be unpolarized light.

Standing-wave fields do not carry either momentum or angular momentum. They can be treated as coherent superpositions of traveling waves, but in so doing it is essential that the coherence be retained and that the traveling waves always be treated in pairs.

Update: Spin.   Increasingly, as experimental techniques have enabled work with electromagnetic fields confined within tiny spaces whose dimensions are comparable to a wavelength or whose frequencies can be manipulated at will, it has been necessary to regard not just the simple beams whereby photons were first regarded, but to treat spatial distributions (densities) of energy and momentum that may become quite complicated [Pad03; Pad04; Pfe07; Bar10; Bar10b; Bar10c; Mil10; Gri12]. The notion of twisting fields associated with spin now appears through the separation of angular momentum density into orbital and spin parts [All92; Bar94; All99; Pad00; Lea02; Bar02b; Pad04; Cal06; Zha09; Yao11; Cer11; Wil13; All16; Bar16; Kre16; Pad17; Fra17; Man17]. Studies of radiation in bulk dielectrics bring further, often unexpected, contacts with notions of spin angular momentum—and opportunity for new views of photons. Appendices B and C offer details of the underlying mathematics and physics.

Photon localization; Photon wavefunctions.   As electromagnetism was being coordinated with the quantum theory of electrons it was recognized that the notion of a photon—a particle associated with the electromagnetic fields—brought particular mathematical problems because, being massless, it must always move at the speed of light. An electron, being a massive particle, could be brought to rest and so it was possible to consider the probability that it would be found within an arbitrarily small neighborhood of any specified point. It was that mathematical possibility that allowed the introduction of wavefunctions for electrons. But because the photon could not be localized in this way,[2] theoretical physicists sought ways of constructing a theory that would be compatible with the special relativity of Einstein and Lorentz and yet offer interpretation, in some way, in terms of particles. What was sought was a wavefunction for a single photon.

---

[2]The difficulty is in localizing a photon in *position space* [Bia09]. It is completely localizable in the *momentum space* of plane waves, wherein wavevectors provide the coordinates. Confinement of an electromagnetic field in any direction requires a superposition of wavevectors along that direction—the construction of a wavepacket.

A particularly useful approach has been to treat the electric and magnetic fields as real and imaginary parts of a complex-valued vector field, the Riemann-Silberstein (RS) vector (see Appendix B.1.11), and to regard the absolute square of this field (equal to the energy density) as offering interpretation as a photon-number density. In this approach one step involves recognizing that the Maxwell equations that describe the free-space electromagnetic field, when expressed in terms of the RS vector, have a structure very like the Dirac equation for free-space electrons, (see eqn (B.1.11)), but with spin-one rather than the spin-half of electrons [Bar14]. Just as solutions to the Dirac equation provide electron wavefunctions, so too does the RS vector, taken with Maxwell's equations, provide a photon wavefunction—a probability amplitude for electromagnetic energy.

It is worthwhile recognizing that the photons of Dirac (and the photons of Feynman discussed in Section 8.3) are idealizations of free-space fields, albeit fields that are often discretized by imposing boundary conditions. These mathematical constructs are typically used with the aid of perturbation theory to describe the interaction of radiation with matter. A different view, not limited to perturbation theory, was used by Chan, Law, and Eberly [Cha02] to define wavefunctions of photons that emerge from the spontaneous emission of a two-state quantum system. These are wavepackets of electromagnetic field which, because they originate with a single emission process, are single photons.

Static fields. The Fourier analysis that underlies the Dirac approach to defining photons expresses electromagnetic fields as superpositions of monochromatic constituents. It puts together constructions of assorted frequencies much as a cabinet maker puts together furniture from pieces of wood. Apart from possible boundary conditions, the frequency components have no limit—in free space their photons can have arbitrary values of ω. But to fit into the formalism of dynamics—to have equations of motion—the electric and magnetic fields must have *some* time dependence: ω cannot be exactly zero. Although static (zero frequency) fields can certainly cause observable changes to atoms, such changes are not customarily attributed to photons.

4.2.3 Specifying Dirac photons

The Dirac approach to defining photons uses a set of normal modes for spatial fields. That approach begins with Fourier analysis of time dependence, converting time variation into frequency variation. As discussed in the preceding paragraphs, a mathematically simple set of modes is obtainable by extending the single-dimension temporal Fourier analysis into three-dimensional spatial Fourier analysis. This is most commonly done with traveling plane waves whose Fourier components are specified by three Cartesian components k_x, k_y, k_z of the propagation vector $\mathbf{k}$; see Appendix B.3. To satisfy Maxwell's equations the angular frequency ω of such a wave in vacuum, and hence the energy $\hbar\omega$ of such a photon, is fixed by the magnitude of the propagation vector, $\omega = c|\mathbf{k}|$, so the totality of four-dimensional Fourier analysis involves only three independent parameters.

To the three linear-momentum quantum numbers k_x, k_y, k_z associated with the direction of plane-wave travel one must add some measure of the direction of the electric vector, which must be in the plane perpendicular to the vector $\mathbf{k}$. This is readily done by specifying the helicity, i.e. the projection $\pm\hbar$ of the intrinsic angular momentum (spin) onto the propagation axis. With this mode-enumeration scheme the quantum state of the photon is specified by four quantum numbers: The helicity and three momentum components. Alternatively one can specify the photon energy (or frequency), polarization, and two components of the momentum.

Superpositions and standing waves; Discretization. With whatever choice of parameters, the definition of a photon quantum state requires a minimum of four specifying parameters, as does the specification of a free-electron quantum state. Superpositions of such four-parameter states provide other choices for specifying modes and their photons. A simple example occurs with the replacement of traveling waves with superpositions of paired oppositely-directed waves, to produce standing waves; see eqn (4.1-1). This choice can be for just one coordinate, such as occurs between mirrors of an interferometer, for two coordinates, to provide an elaborate moving surface, or for three coordinates, thereby obtaining the sort of fields found in fully enclosed cavities. Each such boundary condition imposes the requirement that a discrete number of half wavelengths should fit into a finite coordinate interval. With Cartesian coordinates this means that one or more of the propagation constants k_x, k_y, k_z can take only discrete values. Discreteness of mode specifiers also occurs with angular variations in cylindrical and spherical coordinates, see Appendix B.2

Specifying beam photons. The full characterization of an electron or a Dirac photon requires a minimum of four parameters, four quantum numbers. When the radiation comprises a traveling beam, as discussed in Appendix B.2, two of the momentum-characterizing parameters are used in superpositions to confine the transverse spread of the field. There remains only the need to specify frequency and polarization. The two beam-defining parameters may either be continuous, from $\mathbb{R}$, or be discrete, from $\mathbb{Z}$. As discussed in Appendix B.2.1, the traveling wavefronts of beams may possess a twist associated with orbital angular momentum (OAM).

Specifying photon polarization. Polarization of a classical radiation beam can be specified by parametrizing the direction of a unit vector in the plane transverse to the beam axis; see Appendix B.1.6. In general the beams of interest are superpositions of two orthogonal polarizations. The specification of such states is quite commonly done with measured *Stokes parameters*, depictable as defining a point on a unit sphere, the *Poincaré sphere*; see Appendix B.1.7. This parametrization allows a quantitative description of unpolarized and partially polarized light, as well as the polarized light needed for Dirac photons.

Single-photon states. Whatever the choice of photon-defining parameters, whatever the set of four quantum numbers, a single-photon state is, by definition, one in which measurement will reveal a mean photon number of 1 and a variance around this number of 0. For every distinct mode of the field we can define a state that has a definite photon number—an infinite list of independent photon-number states known as a *Fock space*. In this regard, photons differ fundamentally from electrons: There is no limit to the number of photons (as bosons) that have identical quantum numbers, say the mode-occupation number, whereas no two electrons (as fermions) can have identical quantum numbers and cannot occupy the same mode.

4.3 Emission and absorption photons

With the Dirac approach every different set of boundary conditions for the VHE leads
to a different set of mode fields and hence to a different type of photon. The choice
of boundary conditions must fit particular experimental conditions, either those of
emission or those of absorption. When the mode fields used for defining emission differ
from those used in considering absorption it is necessary to convert from one type of
photon to another.

4.3.1 Emitted photon

A single atom in free space, brought to an excited state, will thereafter begin to
spontaneously emit its excitation energy as radiation—a photon. Quite often the radiation is in the optical portion of the spectrum and the emission process is that of
an electric transition-dipole moment, oriented along a specified quantization axis (one
of three possible orientations) of the atomic dipole-moment vector that is the source
of outgoing radiation. The quantum states of an atom in free space can be treated as
examples of states with well defined angular momentum, of specific magnitude J (an
integer or half-odd-integer) and of discrete orientation values as labeled by an integer
or half-integer magnetic quantum-number M (see Appendix A.7.2). Dipole transitions
between such states will change not only in energy but may change in angular momentum J by (at most) one unit and may change orientation, measured by M, by (at
most) one unit. Such changes, in energy, angular momentum, and orientation, must
be carried by the field, i.e. by the photon.

The field that results from this free-space time-varying moment—and the atomic
change in angular momentum—is a *multipole field* (see Appendix B.2.6), and so the
photon associated with this single-atom decay is a *multipole photon* , carrying angular
momentum. If, as is most often the case for optical transitions, the atomic moment is
that of an electric dipole, the photon is an electric-dipole photon; see Appendix B.2.7.
It is characterized by its carrier frequency (equal to the Bohr transition frequency
between two quantum states), by its dipole nature (it carries one unit of angular
momentum as contrasted with quadrupole and higher multipole field modes that carry
more units of angular momentum), and by its vector nature, which is selected by
the change of magnetic quantum number. The basic dipole-transition field may have
three orientations, identifiable by the change of magnetic quantum number 0, ±1, see
Appendix B.2.7. This multipole field is a classical electromagnetic field, a particular
solution to the vector Helmholtz equation. Its quantum nature comes, in part, from
the fact that its magnitude—the energy it carries—is $h\nu = \hbar\omega$, the energy of a single
photon, and in part from the fact that it originated in a single transition between
two quantum states. Free-space spontaneous emission produces a mixture of the three
basic multipoles.

Thinking about emission. Prior to the development of quantum optics, spontaneous emission was widely regarded as a rather mysterious quantum process, quite
beyond comprehension using classical equations for electromagnetism. In a charming
anecdote that typifies the mystery [Fey69], Feynman tells how, as a newly minted
physics PhD, he was asked by his father to explain spontaneous emission.

He said: "How do you ... think of a particle photon coming out [of the atom] without it having been there in the excited state?"
I thought a few minutes and I said: "I'm sorry. I don't know. I can't explain it to you."

Since then numerous textbooks oriented toward quantum optics offer a viewpoint, rooted in the Heisenberg equations that restate equations of motion for classical variables in a form that incorporates quantum restrictions; see eqn (A.9-3). With this Heisenberg picture [Dir65; Ack73; Sen73; Ack74; Mil76; Dal82; Sho90; Bar88; Sho93; Luk03; Har06; Lam06; Mil19], electron currents within an atom (a Lorentz electron, describable with the vector model of Feynman, Frank Vernon, Jr., and Robert Hellwarth [Fey57]) create a dipole moment that, like all oscillating charge dipoles, is a source of outgoing radiation [Scu72]. Although quantum theory imposes some restrictions (most notably the impossibility of radiating away the energy of a minimal-energy ground state), the picture is predominantly one that will be found in long-established textbooks on classical electricity and magnetism [Str41; Jac75].

Photons from nuclei. Just as the electrons bound to a nucleus can exist in a variety of energy states, corresponding to differing motions, so too can the protons and neutrons of an atomic nucleus have differing motions, individual and collective, with corresponding energy states. Nuclear excitations typically occur either as a result of alpha or beta decay (or fission) or as a result of particle bombardment. Like the bound-electron currents that serve as sources of visible light, so too can charge currents in nuclei produce radiation, but typically with photon energies of keV. The result, gamma or X-ray emission, is a form of spontaneous emission (as is alpha and beta decay), albeit associated with changes of nuclear structure rather than atomic structure. Whereas atomic sources are typically electric or magnetic dipoles, nuclear sources are often higher multipoles, with correspondingly longer lifetimes.

Constrained emission: Enclosed atoms. The model of a radiating atom in free space requires adjustment when the atom is confined, either by walls of a macroscopic cavity or by enclosure in a crystalline setting. When the excited atom is located in an enclosure, the field modes available to carry away the excitation energy are restricted. The continuum of field modes that are possible in free space become limited by the presence of matter, parametrized by complex-valued refractive indices, see Appendix C.3.1. By shaping the enclosure an experimenter can enhance particular directions, say along the axis of an optical fiber, and reduce emission into other directions, say transverse to the fiber axis. The available modes for linkage to atom excitation may favor photons that have a definite propagation axis and that are therefore well matched, beyond the confining region, to free-space beams. *Photonic crystals* are examples [Yab93; Rus03; Joa11].

An atom in such an environment does not have the rotational symmetry of free space, and its stationary quantum states (one of which is excited) are describable as coherent superpositions of angular momentum states, or as eigenstates suited to some nonspherical coordinate system. Because emission does not take place into free space, there is not a continuum of field modes to carry away the energy, and so the rate of spontaneous emission is not what is calculated with the traditional perturbation theory and Fermi's Golden Rule [see eqn (A.16-10) and eqn (C.5-11)]; Appendix A.16.2 notes the resulting *Purcell effect*. Nonetheless spontaneous emission will take place.

Stimulated emission. Although the optical field created by spontaneous emission is typically that of an outgoing electric multipole it combines with the field that prompted the emission, typically a traveling plane wave, to produce the field that will be observed further along the route of the incident wave. The two contributory fields have a definite phase relation with one another [Scu72]. Modeling of the coherent excitation of an atomic dipole moment and the radiation outgoing from this excitation [Cra82] shows that destructive interference occurs in all but the initial field direction, which exhibits the constructive interference needed to produce the expected stimulated emission addition to the incident wave.

A similar interference effect occurs with bulk matter. The effect of fields from atomic sources is destructive interference in all but the forward direction, where the resulting field undergoes a phase delay that is seen as a slowing of the propagation speed [Jam92; Fea96], see Appendix C.3.1.

4.3.2   Detected photon

Photon detectors are commonly based on planes of photosensitive material that respond to the arrival of an electromagnetic field by emitting an electron (the photoelectric effect); see Section 7.3. Various means are used to amplify that single electron into an avalanche of electrons that can be registered as an electric current. The basic concept here is that of a small surface oriented perpendicular to a straight-line path from the emitting atom. Thus what the detection apparatus detects is a photon that can be idealized as a traveling plane wave, a mode of the vector Helmholtz equation in Cartesian or cylindrical coordinates; see Appendix B.2. Such a photon is characterized by its energy (its wave frequency), by three components of its linear momentum (the direction between source atom and detector center), and by the direction of its electric-field vector (perpendicular to the propagation direction). That is, the photon that is detected is completely characterized by the labels of a traveling-plane-wave mode of the vector Helmholtz equation and by the time variation of its magnitude. This available information has little remnant of the multipole nature of the source (see eqn (B.2-47) of Appendix B.2.6). Its quantum nature, like that of the emitted photon, can only come from its energy content ($h\nu = \hbar\omega$) and from its time dependence as a single transition between quantum states.

4.3.3   Converting between modes

The picture I have presented above, of photon emission and photon absorption, may involve two very different electromagnetic field modes, two different classical fields that are solutions to the vector Helmholtz equation in different coordinate systems: We visualize an outgoing multipole field centered at the emitting atom and a traveling plane wave that passes by the detector. To evaluate the probability that a photon emitted by the atom produces a photoelectron at the detector we evaluate an overlap integral, involving the spherical harmonics of the multipole field and complex exponentials for the traveling wave; see eqn (B.2-47) of Appendix B.2.6. The square of this overlap integral gives the relative intensity of the radiation field in all directions around the emitter—it gives the angular distribution of the dipole radiation as well as the direction of the electric field at all locations around the emitter. It exhibits a very characteristic pattern in which no radiation is observed along the axis of the dipole and is circularly symmetric around this axis.

4.4   Comments on Dirac Photons

Planck and Bohr had considered radiation sources whose emission and absorption was discretized as photons. The relevant harmonic oscillators were associated with portions of ponderable matter; see Appendix A.14.4.

By contrast, the photos of Dirac, like those of Einstein and Compton, required no particular sources. The oscillators of Dirac were not material (nor did they pertain to an all-pervading ether): They were associated with massless fields in otherwise empty space; see Appendix B.3.

Both of these complementary viewpoints have merit and use; Appendix C describes their melding.

4.4.1   Next steps

My own recognition that the Dirac approach, described here and in Appendix B.3, provided a consistent definition of photons, eliminating the various historical wave-particle confusions, has not ended my interest in photon definitions. There remain questions concerning how free-field photons relate to radiation emitted from various sources. It has directed my more recent questions toward how it is possible to distinguish a quantum field from a classical field, a matter of contemporary concern for a variety of enterprises. And how can a field be manipulated without losing its quantum character? Chapter 7 discusses some possibilities. Appendix B.5 comments on some of the tools available for that purpose.

Although the Dirac notion of photons has had great appeal for theorists—persons who are at home in Hilbert space—it holds a somewhat less exalted position amongst experimentalists—persons who are inclined to think about what their tools can measure or produce. They tend to ask for operational definitions that rely on observable actions by photons. The following three chapters discuss various idealized procedures in which photon characteristics are to be seen and in which single photons are to be manipulated.

Update: Dirac and Quantum Electrodynamics.   The theory that brought a final unification of quantum theory with electromagnetism, now referred to as *quantum electrodynamics* or QED, came with papers by In-Itiro Tomonaga, Julian Schwinger and Richard Feynman that were awarded the 1965 Nobel Prize in Physics. These papers, and their underlying TSF formalism [Dys49], were a culmination of the work beginning with such physicists as Planck, Einstein and Bohr (see Section 3), as combined by Dirac with the nineteenth century description of electromagnetism in Maxwell's equations (see Appendix B.1). The formalisms offered by TSF overcame shortcomings of the Dirac work that had presented seeming calculational singularities; the TSF mathematical procedures, introducing renormalization to produce the QED of today, have subsequently allowed calculations of steady interactions of weak fields with electrons that agree with experiment to a remarkable number of decimal places, thereby establishing undisputed validity of QED in that regime—the idealization of eqn (5.1-1a), devoid of either the strong interactions of eqn (5.1-1b) or the decoherence of eqn (5.1-1c).

But this accuracy, and its guarantee of correctness, does not necessarily carry over to the description of multi-atom molecular energy levels nor to the raveling and unraveling of polymer strands, let alone to flawless design of computers, though TSF can be said to be the basic theory of electrodynamics that underlies such calculations. A sturdy foundation does not guarantee a watertight roof: Caveat homeowner.

The work of Dirac presented in this chapter, predating TSF, is quite adequate for understanding such concepts as photons and coherent excitation and Quantum Information; also for revealing the seemingly puzzling, even weird, behavior associated with wave-particle duality and with controlled entanglement between superpositions of paired degrees of freedom apparent in quantum theory; see Section 6.7.2. Much of what is today termed *Quantum Optics*, and all that this Memoir treats, requires none of the TSF formalism (and very little explicit special relativity). Its treatment of quantum states and superpositions would have been quite understandable in the 1930s, prior to the TSF works that completed the definition QED. That successor to classical physics is QED without TSF.

5

Photons as population changers

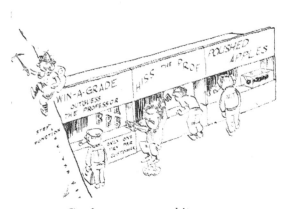

Grades can seem arbitrary.

Although it is now possible for an experimenter to trap a single atom or ion (or to construct and manage an artificial atom having chosen characteristics) and to subject it to controlled quantum-state manipulation with crafted pulses of laser radiation (see Section 5.7 and [Sho11]), earlier literature dealt primarily with situations in which atom samples contained a very large number of atoms, illuminated by a large number of photons. Quantum theory provided predictions for probabilities that individual atoms would be found in particular quantum states at any given time, given particular initial conditions and given a specific model of pulsed radiation.

This chapter opens with an overviewing sketch of the range of modeling approaches that are needed for various purposes, from the very simplest idealization of single, isolated atoms, to elaborate and complicated models that incorporate many details of actual experiments. The schematic diagrams, starting with eqn (5.1-1a), offer simple mnemonics for placing discussions and models into appropriate context.

It has proved convenient to refer to the time-dependent probabilities of finding atoms in particular states as *populations*;[1] these are the solutions to appropriate equations—either the rate equations applicable to incoherent radiation, see Appendix A.16, or the quantum-mechanical equations applicable to coherent radiation, see Appendix A.9. Bridging these two extremes is the density matrix discussed in Appendix A.14.

5.1   Interactions, decoherence, and ensembles

The simplified world of single, isolated atoms and free-space radiation is one that permits physicists to devise soluble equations for idealized behavior of Nature. As estab-

---

[1] The notion of "population transfer" used in discussions of coherent excitation has a very different, and often unsavory, meaning in descriptions of enterprises involving moving of human populations.

Our Changing Views of Photons: A Tutorial Memoir. Bruce W. Shore, Oxford University Press (2020). © Bruce W. Shore.
DOI: 10.1093/oso/9780198862857.003.0005

lished in the early part of the twentieth century, the time-*independent* Schrödinger equation of Appendix A.5 gave the observed discrete energy states of atoms and molecules and the Dirac photons provided a quantum theory of radiation. These dealt with versions of two classical systems, atoms and fields, each subject to quantum theory that dealt with particles and waves. Symbolically the view was:

$$\boxed{\text{atoms}} \quad + \quad \boxed{\text{fields}}. \qquad (5.1\text{-}1\text{a})$$

In the mathematics that attends this model (see Appendices A and B) each system (each box) has its own separate energy. When expressed in terms of parameters that define the particular system this is its *Hamiltonian*; see Appendix A.1.6. (With a quantum description this is an *operator*—an organized set of instructions for changes— in an abstract space that defines possible states of the system.) Within the atom system the sum of kinetic, potential, and chemical energy is conserved. In the picture that attends the rate-equation view of radiative action the energy increments are either in the atom (as excitation) or the field (as photons); see Appendix A.16.3.

Reversibility and entropy. The equations of motion associated with the names of Newton, Lagrange, and Hamilton, discussed in Appendix A.1, have in common the mathematical property that they apply equally well whether the independent variable of time t runs toward larger positive values (forward, to the future) or toward negative values (reverse, to the past). The equations as presented are commonly applied to motion of a single particle, or to a few degrees of freedom, and are *reversible*. However, systems of multiple independent degrees of freedom need not share the reversibility of their constituent equations of motion [Swe08; Nor17]. As time-evolving motions carry particles away from their initial positions the number of possible energy-conserving states increases. For any sizable collections of independent particles or many degrees of freedom, each specified by initial conditions, return to those original conditions requires that the starting state, defined by a collection of numerical values, be specified with increasing precision as the number of particles increases. The return of a gas of molecules to some initial confinement, even when the molecules are idealized as free of collisions, is so exceptional as to be regarded as negligible; their behavior is irreversible [Swe08; Nor17]. In statistical physics *entropy* is a quantitative measure of this number of options, a counting of degenerate states [Jay65; Sty00; Swe11]; its increase establishes an irreversible one-way "arrow of time" from past to future, for systems that have many degrees of freedom.

5.1.1 Energies and interactions

But radiation acts to heat matter, and the motions of electrically charged particles such as electrons are affected by electromagnetic forces. In turn, accelerating charges create electromagnetic radiation and hot matter radiates away its heat energy. Thus when my education proceeded beyond the most elementary recognition of atoms, electrons and photons as quantities for scientists to place into equations, I could understand that the needed formalism was in the form:

$$\boxed{\text{atoms}} \quad + \quad \boxed{\text{fields}} \quad + \quad \boxed{\text{interaction}}. \qquad (5.1\text{-}1\text{b})$$

Again, each boxed name is associated with an energy, and with a Hamiltonian operator in the quantum version; see Appendix C.5. With a quantum description of atoms and a classical description of radiation there is no overall conservation of energy: The radiation acts as a controlled but unconstrained source and sink of energy for atom excitation. The third term, *"interaction"*, deals with forces of fields on atom constituents (e.g. causers of transitions between discrete-energy states) and the modification of fields by atoms (e.g. the diffraction of light by obstacles and its absorption when passing through matter). It epitomizes the Maxwell equations with sources (see Appendix C.2) and the time-*dependent* Schrödinger equation (see Appendix A.9), and its appearance completes the basic theory of atoms, electrons, and photons.

With a fully quantum description of fields as well as atoms, energy is conserved. Energy passes between the several parts, not only between atoms and fields but also to an energy of interaction. The Jaynes-Cummings model of Section 5.8 provides a simple example of the three energies; see Appendix C.6.

The schematic of eqn (5.1-1b) offers, in principle, a means of describing the observable universe. For this purpose one includes all possible particles and fields and their interactions. The Hamiltonians have no explicit time dependence and there are no unincluded particles to act as the bath of Section 5.1.3. This approach underlies the Standard Model summarized in Appendix A.19.

In many situations the interaction has little effect on the atoms and fields that occur in its absence, and it can be regarded as introducing only a small perturbation of field-free behavior. This was how physicists viewed their world prior to the arrival of lasers and the opportunities of Quantum Optics. However, the actual effect of the interaction is to bring a blend of atom and field characteristics to the atom particles and the field waves. When these effects become appreciable it is necessary to treat the three portions of diagram (5.1-1b) as parts of a single system. The individual atoms and photons that occur with the picture of diagram (5.1-1a) then merge into collective entities (quasiparticles) that are neither atom nor photon. Appendix C.4.2 discusses an example known as a *polariton*. The dressed states [Coh77; Dal85; Aga90; Mil19] mentioned in Appendix C.5, obtained by diagonalizing a Hamiltonian matrix, are other examples. When the interaction is time dependent, the diagonalization produces *adiabatic states* [Sho08; Sho11; Sho13]; see Appendix C.5.

5.1.2   The quantum vacuum

When the interaction strength becomes sufficiently large, as it can with large laser systems or particle accelerators now in operation, it is necessary to revise the simple view of a vacuum as being the absence of the particles and waves that occur with the picture of diagram (5.1-1a). Instead it is common to define the vacuum as the lowest-energy state of the system of diagram (5.1-1b); see [Ait85; Mil94; Leu10; Wea16; Mil17; Mil19]. The observable elementary particles and fields occur as excitations of this starting state, and so they are "dressed" superpositions of the "bare" particles and fields from the boxes of diagram (5.1-1b) labeled "atoms" and "fields". Appendix C discusses models appropriate to this situation.

5.1.3   The environment; Decoherence

Despite the intellectual appeal of the world that diagram (5.1-1b) presents,[2] it cannot provide an adequate, practical description of many phenomena that the experiments of atomic physicists measure. What is missing are the inescapable and uncontrollable effects of the additional environment—the collisions of an isolated atom with environment particles, the noise in a laser beam. These originate with motions of particle aggregates that are beyond the control of an experimenter and cannot be specified with complete certainty. Their influence typically is treated phenomenologically, with the formalism schematically represented as

$$\boxed{\text{atoms}} \;+\; \boxed{\text{fields}} \;+\; \boxed{\text{interaction}} \;+\; \boxed{\text{bath}}\,. \qquad (5.1\text{-}1c)$$

The final term of this list, "*bath*", is the repository for effects of the vast world beyond the few parts that are explicitly included in the (controllable) degrees of freedom used in the several equations of motion. Its effects appear as friction on sliding sleds, converting kinetic energy into heat, or as photoionization transformation of an atom into an ion and an electron. Its effect on the field emerging from a laser is to randomize the phase associated with an otherwise monochromatic sinusoid, blurring the distinction between sine and cosine. And it is present, in the vacuum environment, as the electromagnetic field modes into which an atom is linked to produce spontaneous emission; see Appendix A.16.2. Like the other boxes of this diagram, there is a Hamiltonian for the bath and for its interaction with the controlled parts of the system.

The information that is used to predict behavior of a quantum system is encoded in complex numbers—as paired amplitudes and phases. A long as the phases remain constant, or change in a way that can be followed with certainty, the system retains *coherence*. The effect of the external-world bath on laser-induced changes to quantum states is to randomize the phases that provide the signature of quantum states. The result, collectively termed *decoherence*, is to require equations that can accommodate this randomization [Haa73; Bre02]; see eqn (A.14-24).

Each uncontrollable environmental influence has a characteristic time scale, a *decoherence time*, after which phase information is significantly lost. For time intervals much shorter than the decoherence time the environmental changes are sufficiently small to be ignored. The two-state time-dependent Schrödinger equation (TDSE) of Appendix A.10 applies only to time intervals that are much shorter than the time required for inevitable spontaneous emission to take place—times much shorter than the radiative lifetime of the excited state.

To treat longer intervals it is generally necessary[3] to deal with a *density matrix*, initialized as the bilinear product of probability amplitudes, see Appendix A.14. In the equation of motion governing changes to a density matrix the effects of an uncontrollable environment are incorporated as a few parameters, for example relaxation times or decay rates; see Appendix A.14.2. When these terms are negligible the predicted behavior has quantum mechanical characteristics and the equations are equivalent to the

---

[2]This picture underlies the Feynman vision of photons, see Section 8.3.

[3]A simplified description is possible when interest lies only in *loss* of population, not redistribution; see Appendix A.9.7.

TDSE. When the decoherence terms dominate, the equations become rate equations; see Appendix A.16.

The solutions to these density-matrix equations exhibit properties that, as time extends well beyond the decoherence time, typically undergo *relaxation* to a steady state. The decoherence that arises from phase interruptions is termed *homogeneous* because it affects every system equally; see Appendix A.14.2. (The following subsection, 5.1.4, discusses *inhomogeneous* relaxation, which occurs when observed properties involve an assortment of environments and the need to average their properties; see Appendix A.11.2.)

Vacuum-photon bath. The bath of eqn (5.1-1c) includes not only the influences of matter particles nudging the system but also the portions of electromagnetic fields that are not explicitly present in the equations of motion. Specifically, it includes the various modes of radiation that are available to gain energy by spontaneous emission from an atom in an excited state (but, as noted in Appendix A.16.2, there may be constraints). The modes act as an uncontrolled source and sink. When these can be regarded as a continuum (the set $\mathbb{R}$), as is traditional in treatments of spontaneous emission, the energy flow from atom to bath is irreversible.

Reversibility and friction. It is possible to fit onto the framework of reversible equations of motion, such as those in Appendix A.1, the existence of velocity-dependent frictional forces, with their consequent conversion of kinetic energy into heat—disorganized motions—that accompanies an irreversible increase of entropy. Such a system typically tends toward a static equilibrium that eliminates friction forces.

Reversibility and thermal bath.   The time-dependent Schrödinger equation for an N-state quantum system illuminated by fields of constant intensity is intrinsically reversible; its Rabi oscillations of excitation continue indefinitely, as long as the illumination remains constant; see Section 5.5. Although modifications of the TDSE to allow probability loss, or equations based on density matrices, do allow irreversible change, this happens only by considering the effect of a thermal bath. The two-state equation of eqn (A.10-1) has no loss of coherence, and the Jaynes-Cummings version of this equation (in Appendix C.6) describes strict energy conservation between atom, field, and interaction.

The notion of photon absorption—a permanent disappearance of a radiation increment—that is part of rate-equation descriptions of the radiation-matter interaction necessarily invokes memory loss, some mechanism that randomizes statevector phases. Such irreversible action has a place in density-matrix equations but not in the TDSE for coherent excitation.

Incoherent photons. The evident difference between coherent and incoherent excitation is a consequence of the difference in the fields—and photons—associated with the two limiting examples of excitation. A field of thermal radiation, and its packets of energy $\hbar\omega$ (interpreted as *thermal photons*), will not produce reversible coherent excitation. By contrast, a coherent radiation-field increment (say, a laser pulse) can be stored and recovered, but is not irreversibly absorbed; see Section 7.5. Although one may read of "incoherent photons" it is the <u>field</u>, rather than the individual increments, that carries the incoherence.

Partial coherence. The TDSE and the rate equations of Einstein represent limiting cases of complete coherence and complete decoherence, respectively. To model more realistic situations intermediate between these two extremes we require the density matrices of Appendix A.14. Appendix A.14.2 gives an example. The Stokes vector of Appendix B.1.7 provides experimental measures of partial coherence [Bor99].

Noise. A variety of random interactions with an environment can be regarded as small impulses. These may arise from brief collisions between the active system, say an atom, and various uncontrollable projectiles such as thermal atoms and molecules. Or they may occur as irregular shifts of the phase of the laser field responsible for the excitation—effects that are responsible for part of the laser bandwidth. Such effects underlie the diminution of system coherence and therefore affect the field (and the photons) radiated by the atoms. To evaluate steady, equilibrium behavior these random irregularities in the interaction Hamiltonian can be treated as a Monte Carlo process [Fox88; Mol93] in which a coherent propagator acts for a succession of irregular intervals, broken by revision of the statevector and the Hamiltonian to account for interruptions. Evaluation of light scattering, including spontaneous emission, has been done by considering the dipole correlation function responsible for emission [Ebe84; Wod84; Wod84b].

5.1.4 Ensembles

One further formalism modification must be included to bring basic quantum theory, such as appears with the two-state TDSE, into conformity with the real atoms and molecules of Nature: Quantum theory deals with single quantum states, but experiments often deal with systems that have quantum states whose energies are indistinguishable—the states are said to be *degenerate*—but that have other distinguishable characteristics, most notably discrete atomic orientations. Rather generally, observations average over results from an *ensemble* of systems—a collection that has been prepared in accord with some common rules of selection:

$$\text{ensemble of } \left\{ \boxed{\text{atoms}} \; + \; \boxed{\text{fields}} \; + \; \boxed{\text{interaction}} \right\} \; + \; \boxed{\text{bath}} . \quad (5.1\text{-}1\text{d})$$

Appendix A.11.2 discusses how to treat such situations: It is necessary to average over results for each individual degenerate quantum system—to average over individual orientations, for example. This averaging decreases some of the effects that distinguish quantum behavior from classical behavior—for example the Rabi oscillations of a two-state quantum system; see Section 5.5. The observed behavior is termed *inhomogeneous* relaxation, and it occurs whether or not there is significant homogeneous interaction with the external bath.

Quantum signatures. The overall result of ensemble averages and of observations that take place over times much longer than any decoherence time is to replace the two-state TDSE eqn (A.10-1) by two coupled rate equations, see Appendix A.16. (These are obtained from limiting conditions on density matrices.) If quantum signatures are to be sought, it is necessary to deal with times much shorter than any decoherence time and with single quantum systems rather than ensembles.

One must expect that with radiation any quantum properties are to be found with weak fields—fields in which the effects of a single photon are obvious. Laser fields are generally not of this nature, unless they have been strongly attenuated. Most often a laser beam contains vast numbers of photons and can therefore be regarded as an example of a classical field:

Classical: *A field in which one photon more, or one photon less,*   (5.1-2)
makes no discernible difference to measurable properties.

5.2   Einstein-equation populations; Equilibrium

The equations proposed by Einstein for treating the effects of incoherent radiation on ensembles of atoms, and presented in Appendix A.16, can be regarded as equations for probabilities that an atom will be found in a particular quantum state (or incoherent mixture of degenerate states) as time advances. The simplest model is that of an atom that has only two energy levels (a *two-level atom* [All87]) exposed to steady, monochromatic radiation whose angular frequency ω matches the Bohr transition frequency ω_{12} between a level 1 of energy E_1 and a level 2 with energy E_2, a condition known as *resonance*:

$$\hbar\omega = \hbar\omega_{12} \equiv |E_2 - E_1|. \qquad (5.2\text{-}1)$$

The population changes involve three radiative processes first identified by Einstein; see [Kle05] and Appendix A.16.1. With a beam of steady incoherent illumination the rates, per absorbing atom, are (here I assume $E_2 > E_1$):

- Absorption of radiation, at rate $B_{12} \times$ intensity (or $\bar{n}$);
- Stimulated emission of radiation, at rate $B_{21} \times$ intensity (or $\bar{n}$);
- Spontaneous emission of radiation, at rate A.

The so-called Einstein A and B coefficients appearing here[4] depend on the particular atomic transition and the radiation polarization but are independent of the radiation intensity. Appendix A.16.3 presents the differential equations governing population changes induced by these three processes, so-called radiative rate equations. (That presentation parametrizes the radiation effects of absorption and stimulated emission in terms of a mean photon number $\bar{n}$ rather than as the product of an intensity and an Einstein B coefficient.) The solutions to these equations steadily (and monotonically) approach equilibrium values appropriate to thermodynamic equilibrium. If the two levels have equal numbers of sublevels (equal degeneracies) and the radiation intensity is sufficiently high (so that stimulated emission and absorption dominates spontaneous emission) each level will equilibrate holding half the population.

Einstein changes are stochastic.   The three types of changes contributing to radiation alteration of internal energies of atoms are all examples of *stochastic processes*. That is, the moment when the change takes place is treated as a *stochastic variable*, determined only by a probability.

---

[4]The B coefficients defined by Einstein used radiation density, appropriate to an enclosure, rather than radiation intensity, appropriate to beams; see Appendix A.16.1.

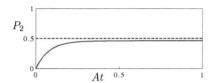

P_2

0.5

0

At 0.5

Fig. 5.1 *Rate-equation excitation*

Figure 5.1 shows an example of two-level rate-equation excitation probability $P_2(t)$, with equal degeneracies $\varpi_1 = \varpi_2$ and mean photon number $\bar{n} = 6$; see Appendix A.16.3. The abscissa here expresses time in units of the spontaneous-emission lifetime (the inverse Einstein A coefficient) and the steady-state excitation population $\bar{P}_2$ is 6/13; see eqn (A.16-18).

Photons as reagents. Einstein presented his model for radiation interacting with a discrete-state atom as a "heuristic" description of steady-state thermodynamic equilibrium [Ein05]. The use of Einstein A and B coefficients in actual equations for change came later. Rate equations—first-order ODEs for population changes—are commonly used in Astrophysics and in work dealing with Chemical Kinetics. When applied to photons as reagents (participants in reactions) [Hof12] they make no allowance for the radiation frequency: The field is assumed to be resonant with the Bohr transition, so that energy is conserved in photon absorption and emission—the energy increment resides either in the field or in the atom; see eqn (5.1-1a) below.[5] Treatment of non-resonant excitation or radiation that preserves the phase of oscillations requires other equations, based on the time-dependent Schrödinger equation (TDSE) of Appendix A.9.

Einstein-model intuition. The rate-equation description of radiation action proposed by Einstein (see Appendix A.16.3) gave a successful picture of how thermal light affects atoms, particularly in steady equilibrium. But this is not a reliable model for understanding how laser radiation acts—the topic of works on quantum optics. If we rely on the Einstein model for our intuition, we will find many examples of unexpected, counter-intuitive physics when we have a system driven by coherent radiation [Sho95]. Rabi oscillations, discussed in Section 5.5, are but one example. To develop intuition for such systems we require a different model, using the TDSE, though not necessarily requiring photons (see Appendix A.9).

5.3   Einstein-equation photons; Lasers

The changes to atomic populations described by the Einstein equations of Section 5.2 must accompany changes to the radiation that produces them—the fields give and receive energy from the atoms. Direct application of the Einstein equations leads to what we now know as lasers [Scu66; Sar74; Sie86; Mil88; Sal19]. Although laser construction and modeling has become appropriately complicated, the underlying principle remains fairly simple, relying only on the Einstein mechanisms.

Laser beam formation. The formation of a laser beam starts with a discrete-state quantum system (say, an atom at rest, perhaps bound in a solid) prepared, somehow, in an excited state that is able to emit dipole radiation; see Appendix B.2.7. The atom begins to spontaneously radiate its excitation energy, at a Bohr frequency. The

---

[5] With allowance for the environment, as in eqn (5.1-1c), energy exchanges also with uncontrolled atoms of a thermal bath.

outgoing wave from this atom passes a similar atom. If that atom is unexcited it will begin absorbing the radiation. But if it is in the same excitation state as the earlier atom then it is stimulated by the passing field to contribute to that field, thereby increasing the traveling-wave amplitude. If we follow a line of sight past a collection of similar atoms there will occur both absorptions and emissions—spontaneous and stimulated—each proportional to the populations available for these processes (see the next paragraph). When the emissions are more numerous than absorptions the field magnitude will increase: It will be amplified. The line of sight exhibits *gain*; see Appendix B.8.1. With growth the stimulated events will eventually exceed the spontaneous events. Each stimulated emission adds to the field that is already present and so with repeated enhancement the frequency becomes more narrowly defined, as does the direction of field travel; a laser beam forms.

The energy of the laser radiation has come from atom-excitation energy, supplied by some independent source (a *pump*, often incoherent) to atoms along a line of sight. The growth of the radiation can be enhanced by mirrors that pass the beam repeatedly past excited atoms and that select standing-wave frequencies.

Lasers without beams. The lasers found in everyday appliances and industrial applications typically produce beams of radiation. But when the gain is sufficiently large it is possible to observe such spontaneous-emission amplification into two or three dimensions; see [Hor12b].

Lasers without inversion. The basic concept presented here presumes that a (pump) mechanism places atoms into an excited state, and that emission from these atoms add to the passing radiation. What is necessary for amplification (*gain*) of a radiation beam is that the energy added (by emission) must exceed the energy removed (by absorption). For incoherent radiation traveling past two-level atoms that are in local thermodynamic equilibrium (LTE), as considered by Einstein, the relationships between the Einstein coefficients of eqn (A.16-3) require that, for energy levels that have the same degeneracy, there should be an excess of population in the excited state (a population inversion) – a condition that is clearly not thermodynamic equilibrium (that situation would require a negative temperature). Subsequent to the introduction of masers and lasers as routine tools, numerous scenarios have been shown to allow amplification without requiring population inversion. Known variously as amplification without inversion (AWI), gain without inversion (GWI), and lasing without inversion (LWI), these procedures rely on a variety of mechanisms for altering the restrictions of the Einstein relations of eqn (A.16-3). The simplest revisions, applicable to two-state systems, account for the momentum-induced Doppler shift that takes place between absorption and emission of a photon, or to the frequency structure of the spectrum affecting a weak field in the presence of a strong field of similar frequency [Koc92; Mom00]. Other procedures become possible when the system must be treated as having more than two discrete quantum states—having multiple states between which there are nonthermal coherence relationships. Autoionizing transitions, with their transparency window, offer one opportunity for having gain without inversion. Interference effects between path of three-state linkages offer others. Several reviews discuss the history of inversionless gain, the theory involved, and the experimental implementations [Koc92; Mom00; Svi13].

Lasers without photons. Although quantum conditions are not needed for the field arising in a laser (it is best described as a Glauber coherent state, see Appendix B.4.1), only the Einstein rates and discrete energy states of the atoms, the language of photons is often used: One says that the atoms add a succession of identical photons to those already present. However, the laser field does not have a definite number of photons; it is better treated as having a definite phase or by quadrature variables; see Appendix B.4.4.

5.4 Coherent population changes

Radiation-induced changes to quantum-state descriptions of atomic and molecular structure remain, as it was in the days of Einstein's first publications, the primary means by which radiation is characterized. The changes may involve photoionization, producing a current of electrons, or they may produce a more limited structural change that is interpretable as change of a discrete quantum state—a change in size and shape of a localized system. With the availability of lasers more subtle changes, involving coherent superpositions of quantum states, have become commonplace. The next sections discusses some of the possible changes, and their interpretation, with and without photons.

Semiclassical modeling. Many of the observations of quantum-state change caused by radiation require no quantization of the electromagnetic field—no photons. This is particularly true of changes induced by laser or maser radiation—fields that are coherent but not necessarily exhibiting quantum properties. Experiments with such coherent radiation typically deal with fields for which the mean photon number is very large, so that the removal of a single photon is hardly observable. The field can be regarded as completely under the control of an experimenter. It is then permissible to treat the change-inducing field as having a specified carrier frequency modulated by a controllable, specified envelope (a situation akin to a commercial radio signal). This acts on a discrete-state quantum system to produce changes in probability amplitudes. Such a model, of a quantum atom and a classical field, is termed a *semiclassical* model [Ebe68; Cri69; Str70; Sar74; Mil76; Ser78; Ser86; Bar94b; Hel18; Mil19], as distinct from a fully quantum model that deals with a quantized field and photon operators; see Appendix C.6.

    To model the effect of coherent radiation, such as that from a laser, upon an isolated atom, one must use an equation for probability amplitudes—the time-dependent Schrödinger equation (TDSE), a first-order ODE that links the probability amplitudes for states whose Bohr transition frequency matches (at least approximately) the carrier frequency of the field. Appendix A.10 presents the basic mathematics of this model.

5.5 Rabi oscillations

Unlike the excitation described by Einstein rate equations eqn (A.16-1) the solutions to the TDSE with steady illumination are oscillatory; see Appendix A.10 and Appendix C.6. The oscillation frequency of the probabilities, obtained as absolute squares of the probability amplitudes, is known as the *Rabi frequency*. Just as the carrier frequency of the radiation is a measure of single-photon energy, the Rabi frequency is a measure of the instantaneous energy of interaction between an atom (or other quantum system) and the electromagnetic field—an energy that does not require photons. (The accumulated Rabi frequency, its integral to a particular moment, is the *Rabi angle*.) For resonant excitation, in which the frequency of the light matches the Bohr frequency, the squared Rabi frequency is proportional to the instantaneous radiation intensity and, for most optical transitions, to the square of an electrical dipole-transition moment associated with the two states; see Appendix A.9.6.

Two-state Rabi oscillations. Oscillatory populations (Rabi oscillations) are a characteristic of steady, coherent illumination that begins suddenly (impulsively).

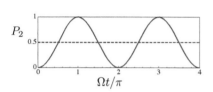

Fig. 5.2 *Resonant coherent excitation*

Figure 5.2 illustrates the oscillatory two-state excitation population $P_2(t)$ of resonant coherent excitation by steady illumination; see Appendix A.10.2. The abscissa expresses time in inverse Rabi frequencies scaled such that a value of 1 means a *pi pulse* (complete population inversion). The oscillations, and population returns, continue as long as the radiation remains unchanged.

Coherence requirement. Unlike the radiation treated by rate equations, the fields to be used with the basic TDSE must be fully coherent: They cannot suffer from random fluctuations in phase and amplitude (noise).[6] Such uncontrollable, irregular field behavior tends to destroy the phase coherence of wavefunctions and statevectors and thereby remove evidence of quantum behavior. Thus sunlight or other thermal sources of radiation, though filtered to select (very nearly) a single frequency, cannot produce Rabi oscillations.

Coherent radiation need not satisfy the resonance condition eqn (5.2-1) exactly to produce oscillatory quantum-state changes; periodic populations can occur with nonresonant radiation but the population transfer will be incomplete, $P_2 < 1$. For such situations the energy is to be found in three places: Atomic excitation, radiation (photons), and an interaction energy; see Section 5.1 and Appendix C.6.

5.6   Assured two-state excitation

There are two well-studied techniques for inducing complete population transfer—for producing coherent excitation with unit probability—of an initially unexcited two-state system. Each of these may, but need not, be associated with single photons.

Pi pulse. The most common procedure is to expose the system to a pulse whose carrier frequency equals the Bohr transition frequency—a resonant pulse. The pulse envelope must be so constructed that its time-integrated Rabi frequency (the *Rabi angle* or temporal pulse area) is exactly π (or an odd-integer multiple of π); see Appendix A.10.2. Not only must the pulse be carefully crafted to meet this requirement,

---

[6]However, various approaches to modeling noise make use of correlation functions constructed from the TDSE [Ebe84; Wod84; Wod84b].

but there can be no detunings, as caused by Doppler, Zeeman, or Stark shifts; see Appendix A.9.8.

Adiabatic passage. Various forms of pulsed radiation, most notably those in which the carrier frequency varies with time (for example, a steadily rising or falling frequency chirp), can produce population changes without oscillations. Such situations offer examples of *adiabatic passage*, discussed at length in various reviews [Sho08; Sho11; Sho13]; see Appendix A.14.5 and references [Cri73; Loy74; Hio83; Ore84; Kuk89] and [Sho08; Sho11; Sho13]. The induced change must proceed sufficiently slowly that the system remains in an adiabatic state (see Appendix C.5), but it must be completed before any decoherence can occur.[7] Adiabatic passage is *robust*: It does not fail when there are assorted but small Doppler, Zeeman, or Stark shifts.

With either of these techniques it is possible to produce, at least momentarily before spontaneous emission or decoherence occurs, complete population transfer into an excited state—complete removal by a single atom of a single energy increment from an electromagnetic field. The excitation will, if left unmolested, spontaneously emit its energy as radiation. Unless the system has been enclosed in surroundings that restrict the emission, the radiation will have the outgoing form of a multipole wave, whatever the form of the field that produced the excitation; see Appendix B.2.7.

Crafted pulses. A variety of pulse shapes and pulse detunings have found use in creating specified changes to multistate quantum systems. Often these pulses mimic the action of pi pulses or other simple change-inducing pulses, while satisfying various constraints upon pulse energy or peak intensity; see the discussion of composite pulses, page 249.

Population return. Coherent excitation processes can be reversed (though incoherent processes are not reversible): If the field is classical (i.e. comprising numerous photons) this same brief pulse will cause the system to return to the initial, unexcited state. Thus the pulse will induce exact population exchange, $1 \leftrightarrow 2$, of any initial population distribution. If the pulse is long-lasting the population return will be accompanied by spontaneous emission. When the field has a well-defined photon number [a Fock state, see eqn (B.4-14)], the reversal follows behavior discussed in the following paragraphs.

5.7   Single atoms, single boxed photons

It is not too difficult nowadays to follow the explanation of a graduate student whose work involves controlling the internal structure of a single atom. That story might begin with a tank of helium gas connected by tube to a small opening into a large vacuum chamber. As the atoms slowly leak into the evacuated surroundings a series of apertures in small metal plates stop all but a few atoms that move along a well-defined line. This is an *atomic beam*, in which the atoms can be so sparse that they make no collisions with one another: For experiments they are essentially freely moving individual atoms.

---

[7]The process is often referred to by the seeming oxymoron "rapid adiabatic".

With suitably arranged electric and magnetic fields the experimenter can select, or adjust, many of the properties of the beam atoms, even slowing them down and then holding them, without contact to any walls, by static fields. Such is the brief outline of the operation of the many research facilities that deal with single atoms, molecules, and ions in vacuum.

Rather than bring a single atom to rest, apparatus can (with much experimental care) pass it through a cavity whose walls impose boundary conditions upon possible electromagnetic fields that can be present: These can only occur with particular spatial structure (nodes must occur at the walls) that have correspondingly discrete frequencies. The earliest cavities were fully enclosed high-Q microwave resonators (boxes) [Rem91]. Subsequently the cavities have included one-dimensional optical resonators formed between mirrors [Tho92; Kim98; Kuh99]. When the cavity is cold there will be negligible thermal radiation inside at the frequency of interest, just a vacuum.

Into this empty cavity moves an atom that, somewhere along its path, has been given the energy of an excited state. This excitation energy, and the construction of the cavity, have been designed so that a Bohr transition frequency of the atom equals the frequency of one of the cavity-radiation modes. As the atom moves into the enclosure it will begin an exchange of its energy with the cavity field. It is important to recognize that the energy transfer described here does <u>not</u> rely on the incoherent radiative rates proposed by Einstein, of absorption, stimulated emission, and spontaneous emission. Instead, the effects are coherent Rabi oscillations that must be described by the time-dependent Schrödinger equation, as discussed in the preceding section; see Appendix A.10. The behavior is an example of the Jaynes-Cummings model of quantum electrodynamics discussed in the following section. When the velocity of the atom is just right (as adjusted by a velocity selector) it will leave the cavity at the moment when its excitation energy has been transferred to the cavity field. The cavity then contains just one single photon.

This very simplified description explains how an experimenter can obtain a single standing-wave photon. That field will ultimately suffer loss to the cavity walls, but with careful construction the photon will remain long enough to be of use in experiments—of *cavity quantum electrodynamics* (CQED) [Mes85; Wal88; Bra01; Wal06]. A standing-wave pattern can also be obtained from two counter-propagating monochromatic laser beams, see eqn (4.1-1), but this situation must be regarded as introducing a superposition of traveling waves that act always as pairs [Sho91].

5.8 The Jaynes-Cummings model; Evidence for photons

With the increasing availability of laser radiation in the 1960s and electronic-computer simulation in the 1970s there appeared a growing interest in the TDSE, primarily with idealization of laser radiation as a simple monochromatic classical field. Simulations used the semiclassical model in which the field was treated classically, typically as a controlled monochromatic (constant intensity) interaction.

In 1963 Ed Jaynes and his student Frederick Cummings showed that the two-state TDSE for monochromatic classical fields had direct application to the discrete-frequency cavity-photons of a quantum field, an exactly soluble, fully quantum-mechanical model of a field interacting with an atom that became known as the *Jaynes-*

Cummings model (JCM) [Jay63; Mil91; Sho93; Sho05; Lar07; Sho07; Gre13; Gro13; Mil19], see Appendix C.6. The simple exact solutions to the semiclassical equations, with their Rabi oscillations (see Section 5.4 and Appendix A.10), are also usable to describe a situation in which an excited atom, accompanied by n photons, undergoes cyclic transition to and from an unexcited atom plus $n + 1$ photons.

With the JCM the cavity photons need not be resonant (there can be a mismatch between the discrete cavity-mode frequencies and the Bohr transition frequency of the atom), and so energy conservation involves not only atom energy and field energy (photons) but also interaction energy; see eqn (5.1-1b). The model allows consideration of single photons, but it makes no use of the Einstein A and B coefficients: An excited atom in the photon vacuum will undergo a steadily evolving transition to a lower energy state, creating a photon, but this action is reversible; the changes are sinusoidal in time. This behavior is fully quantum mechanical—not only the atom but the field are fully described by quantum theory. It suggested a means for controlled creation of a single cavity photon, which might subsequently be observed when it leaked irreversibly out of the cavity.

Evidence for photons from the JCM. It was particularly noteworthy that the Jaynes-Cummings model could be used to describe superpositions of photon-number states. For fields having a statistical distributions of photon numbers the oscillations collapse to an expected steady value. In 1980 Joe Eberly and his colleagues discovered [Ebe80; Nar81] that with appropriate initial conditions (e.g. a near-classical "coherent-state" field of Appendix B.4.1), the Rabi oscillations would eventually revive, only to collapse and revive repeatedly in a complicated pattern. The existence of these revivals, present in the analytic solutions of the JCM, provided direct evidence for discreteness of field excitation (photons) and hence for the truly quantum nature of radiation; see Appendix C.6.

Coherent control; quantum control; optimal control. The intrinsically mono-chromatic light from a laser makes such radiation ideal for inducing selective quantum changes—for selecting only a single pair of quantum states that are resonantly linked by the light and which therefore undergo radiation-induced changes, leaving other states unaffected. The changes may result in *photoionization* (separation of a bound electron from its host) or in *photodissociation* (separation of an atom from a molecule), or they may merely induce internal changes of the quantum system—rearrangement of electronic structure or chemical bonds. Those internal changes may, in turn, affect chemical reactions. All of these processes give clues to what the activating radiation was like—to what an absorbed photon was.

The intended changes have become increasingly more complicated, involving ever more numerous quantum states. Such manipulations of quantum states with laser light quite commonly rely on maintenance of phases of the illuminating fields and of the affected quantum states. Researchers have used a variety of descriptors to distinguish such laser-based experimental techniques [Let07; Sho08; Sho11] from the much simpler radiative interactions that were customary prior to the widespread availability of lasers.

The term *coherent dynamics* emphasized the specialization of quantum changes to those induced by coherent radiation [Bia77; Ebe77; Coo79; Sho79] while the term *coherent population transfer* [Mar95; Sho95b; Mar96; Few97; Ber98] stressed the goal of altering populations. Later I regularly used the expression *coherent excitation* [Sho90], emphasizing the intent of producing population transfer by means of coherent radiation and *coherent manipulation* [Vit01; Sho08] to include coherent-superposition states in the objectives.

The expressions *coherent control* [War93; Gor99; Gru01; Sha03; Kra07; Ohm09; Koc12; Bau18] and *quantum control,* [Sha97; Wol05; Eis06; Sha06; Bri10; Don11; Ran18] emphasize the intended use of coherent fields to control quantum-state changes. To emphasize the optical wavelengths of the coherent radiation the term *optical control* has been used [Ric00; Liu10].

Much of the aforementioned work relied largely upon choosing a particular combination of pulses whose effects could readily be understood when acting on a relatively simple N-state linkage pattern. As the systems of interest became more complicated, with more numerous quantum states, it became useful to consider models that included feedback, so that desired final results and overall costs might be optimized by adjusting elements of the Hamiltonian. Considerable work follows this approach under the sobriquet of *optimal control* [Pie88; Dah90; Shi90; Rab00; Rab02; Bos02; Kum11]

In almost all of this work there is no need to invoke explicit quantum properties of the fields, i.e. photons: A semiclassical approach is quite adequate, with the time-dependent Schrödinger equation for the atoms and molecules being used with presumed classical electric and magnetic fields.

5.9 Coherent change; Interaction linkages

A common task for applications of quantum technology—and photons— is to manipulate the internal structure of an atom, molecule, or other localized system whose structures of interest can be described by N discrete quantum states.[8] An important class of quantum-state manipulations are those that take place *coherently*, meaning that during the specific time interval being modeled there are no irregular, random events that induce decoherence—no spontaneous emission, no collisions, no uncontrollable irregularities of the fields that induce change. Such randomization may have taken place earlier, or may take place later, but during a particular time interval that is being modeled their effect is assumed to be negligible.

Among other goals is the manipulation of the probabilities $P_n(t)$ for finding the system to be found in state n, having internal-excitation energy E_n, at time t. It is common to refer to $P_n(t)$ as the population of state n at time t, and to regard the time-varying set of internal-excitation probabilities as a flow of population [Ack76; Sho78; Ein79; Yat06]. For many purposes it is sufficient to consider just the quantum-state populations, but increasingly there is interest in dealing with coherent superpositions of quantum states, either as starting or concluding conditions for controlled manipulation.

Models of N-state quantum systems offer opportunities for $N(N-1)/2$ distinct pairs of two-state transitions. Not all of these will typically be active or of interest. Radiation-induced transitions between states of energies E_n and E_m are generally

---

[8]These physically identifiable states are known by various names, including basis states and bare states, to distinguish them from dressed states or adiabatic states discussed in Appendix A.13.2.

important only when a photon energy $\hbar\omega$ supplies the energy difference or, stated classically, when a radiation frequency ω matches a Bohr frequency ω_{nm}, a condition of *resonance* between states n and m. Various selection rules introduce further constraints (see Appendix A.11.1), but the resonance condition is usually required, at least momentarily.

A conceptually simple situation occurs when several resonant conditions hold simultaneously, say from steady monochromatic sources. The dynamics of the system can then be regarded as a flow of population through a set of energy states.

Linkage chains. As one considers more complicated quantum systems than the basic two-state atom [All87; Coo79; Sho79; Sho81; Sho90; Vit95], one of the models that offers exact analytic solutions for steady fields is the N-state excitation *chain* [Sho77; Bia77; Ebe77; Sho78; Coo79; Sho79; Sho90; Vit98b; Yat06; Sho11] in which successive states are linked by an interaction energy under experimental control to at most two other states. Overall the pattern of connections, with no branches or loops, is therefore that of a chain, from state 1 to state N with $N-1$ interactions, expressed schematically as

$$1 \longleftrightarrow 2 \longleftrightarrow \cdots \longleftrightarrow N. \tag{5.9-1}$$

Here each two-headed arrow indicates that population can flow in either direction, as determined by one or more fields. The energies E_n along the chain need not rise or fall monotonically: The connection $n \to n+1$ may be associated with either an increase or a decrease of system energy, i.e. with gain or loss of a photon.

With a chainlike pattern of linkages the populations only directly flow between nearest neighbors of the chain: Population can only increase in state n if there has been population in either state $n-1$ or state $n+1$. When the excitation interaction acts, population moves along this chain, becoming distributed amongst other states [Sho90]. In the present discussion the manipulation of interactions is assumed to be fully coherent (spontaneous emission and other uncontrollable influences are neglected), a constraint indicated by the double-headed arrows in the linkage pattern of eqn (5.9-1), and so the interaction generally produces a coherent superposition of the N quantum states. Particular instances include complete concentration of population in a single state, often assumed as a possible initial state or a desired final (target) state.

Three states and two fields. The three-state chain linked by two fields has three versions, depending on the ordering of the three energies E_1, E_2, E_3 [Sho11; Sho13; Sho17]. The simplest is that of a *ladder*, $E_1 < E_2 < E_3$. More common is the *Lambda linkage*, in which E_2 lies above the other energies (which may be degenerate). This occurs with so-called Raman transitions in which a controlled P field induces excitation from state 1 into state 2 and the S field is either spontaneously emitted or is stimulated by a second controlled field, so-called stimulated Raman scattering (SRS) [Hel63; She65; Blo67; Ray81; Sir17]. The third ordering of energies, in which E_2 lies lowest, is known as a Vee pattern. The initial state may be either the middle of the chain (state 2) or one of the ends.

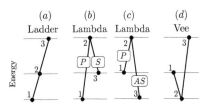

Fig. 5.3 *Three-state linkages*

Figure 5.3 shows the three distinct linkage patterns, Ladder, Lambda, and Vee, available with two fields and three states. The two Lambda frames differ only in the choice of ordering of the chain sequence: (b) is a *Raman* transition, wherein the P field produces a lower-frequency Stokes field S; (c) is an *anti-Raman* transition, in which the P field produces a higher-frequency *anti-Stokes* field, AS.

In such linkage patterns the dots, and their numerical labels 1, 2, 3, represent the quantum states, and the thick lines represent the radiative interactions (photons) that can move population between the end states of the line. (The horizontal position of the dot has no significance in this figure.) With each pair of dots in a linkage pattern there is associated a Bohr frequency—obtained from the difference between the two energies. With each thick line there is associated an interaction between fields and atoms, and a carrier frequency of the field associated with that interaction (the photon). Thus with each thick line there is also a *detuning*—the difference between a Bohr frequency and a carrier frequency (see Appendix A.10). All of these linkages have the same TDSE— the same set of three coupled ODEs; they differ only in the signs of the detunings that appear as diagonal elements of the Hamiltonian matrix. Although the schematic figure shows the three energies as distinct, they need not be: States 1 and 3 might be degenerate.

The four-state three-field chain adds a letter-N linkage, and the five-state four-field chain adds a letter-M and a letter-W linkage. In such patterns a particular field can occur in several linkages.

Loops and branches. As the number of essential states N increases there become available increasingly elaborate linkage patterns beyond the simple chain. Already with $N = 3$ and three fields there is the possibility of a closed loop [Buc86; Sho90; Una97; Kor02b; Mal04; Ran08]. With $N = 4$ it is possible to have branchings in which a single state has connection to more than two states (a *star*). A simple example is the *tripod* or *fan* [Una98; Una99; Vit00b], shown in Figure 5.4. As with simpler linkages, population can flow in either direction along any thick line, when the associated field is present. That field may either be imposed deliberately (say, a laser) or be present as the vacuum field available for spontaneous emission, from a higher to lower atomic energy E_n.

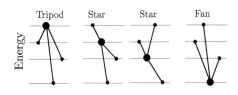

Fig. 5.4 *Four-state branched linkages*

Figure 5.4 shows the linkage patterns available with three fields and four states, one of which (shown by larger dot) has more than two partner states. The numbering of states is arbitrary; the associated ODEs differ only in detunings and the choice of state numbering, not shown here. Again, degeneracies may be present.

For discussion of more general multi-state branched linkages, with gating action on population flow, see [Sho84].

5.10   Morris-Shore photons

Fairly elaborate unlooped linkage patterns, of several states and many fields, can often be presented much more simply as a set of two-state linkages by introducing a change of coordinates—either in the three-dimensional Euclidean space used for field directions or in the larger Hilbert space of statevectors. The change, a *Morris-Shore (MS) transformation*, gives a picture in which excitation involves only two-state linkages. Figure 5.5 shows one of the simplest examples, for a four-state system.

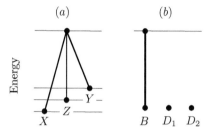

Frame (*a*) shows a tripod linkage involving low-lying stable states X, Y, and Z linked by resonant fields to a single excited state. Frame (*b*) shows the linkage pattern after an MS transformation that places all of the transition strength into a single link between the excited state and a bright superposition of the original X, Y, and Z states. In (*b*) the original energies for X, Y, Z are irrelevant and are not shown.

Fig. 5.5 *The MS transformation*

The Morris-Shore transformation [Mor83; Vit03; Kis04; Ran06; Vas07; Sho08; Sho13b] requires that the linked states form two disjoint sets: One of stable (or metastable) states that include the possible initial states, the other of excited states that can spontaneously radiate (fluoresce). Links occur only between, not amongst, the sets and must be resonant or must all have the same detuning. The resulting coordinate change gives a picture in which there are only pairs of linked ground and excited states (a pair of *bright states*) plus either unlinked ground states (*dark states*) or else unlinked excited states (*spectator states*).

In this simplest of examples the given system has a tripod linkage in which three low-energy states, distinguished by orientation, serve as possible initial states. These are linked to an excited state by as many as three fields, of arbitrary but constant intensity, distinguished by polarizations. (The three-state Lambda linkage is a special case in which one of the tripod links is absent.) The lower states may be degenerate, but every linkage is resonant (or has the same detuning).

The MS transformation illustrated here changes to coordinates in which all of the original interaction is concentrated within a single transition, between a state B (for bright) and the excited state. The remaining two states, labeled here D_1 and D_2 (for dark) remain completely unaffected by this particular field and therefore they do not produce fluorescence. (The Lambda linkage has only a single dark state.) In the language of photons, the system depicted in frame (*b*) is driven by single photons, each of which is a coherent superposition of at most three modes; see Appendix B.3.4. The bright quantum state B is a coherent superposition of the original three states X, Y, and Z, each of which will be affected by the superposition field. The system dynamics, as revealed by population changes, is the same with either picture, but the equations of motion, and the meaning of a photon, differ: With frame (*b*) a single photon can produce excitation in a superposition of several states, but it can never produce any excitation of the two dark states. To understand the relevant physics it is essential to distinguish between a (classical) <u>mixture</u> and a coherent <u>superposition</u>.

5.11 Pulsed excitation

Typically the control of the system change is accomplished by crafting a set of laser or microwave pulses whose effects on the system are expressed as an interaction energy— a Hamiltonian. The pulsed interaction can also be created by moving an atom or molecule across a laser beam, or moving it into a standing-wave field. Although resonant excitation has been a commonly imposed condition for steady illumination, a variety of alternatives using pulsed fields that comprise separate carrier frequencies, even fields whose carrier frequencies undergo controlled change during the course of a pulse, offer opportunities for a variety of population-transfer procedures that can be more useful than the use of steady illumination. There are also procedures that induce time-varying energy shifts of the basic energies E_n between which transitions are to be induced, thereby introducing resonance with fixed-frequency fields.

A continuing goal of quantum-state manipulation has been the moving of all population between states that have no direct radiative linkage, so that both the initial and final states can be stable. In one simple scenario pulses act separately and sequentially, to move population along a chain. Other pulse timings also prove useful, as will be discussed.

5.11.1 Pulse timings; Three states

The simplest example of a chain linkage is that of a three-state system [Sar76; Sho77; Sho90; Sho11], with radiation linkages $1 - 2 - 3$ in the pattern of Ladder, Lambda, or Vee. Typically the goal has been to move population from state 1 to state 3 but more recently interest has turned to the creation of various superpositions. In discussing three-state systems I will denote the field associated with the $1 - -2$ transition as the P field (for primary or pump) and the field of the $2 - -3$ transition as the S field (for secondary or Stokes). The two-photon transitions may occur simultaneously or sequentially, in any combination of absorptions and emissions, from two distinct fields or from a single field.

Simultaneous pulses. When the P and S fields are constant there will occur periodic three-state population cycling, generalizing the two-state Rabi oscillations; see Appendix A.15 and [Sho11]. The cyclic population changes depend upon the relative sizes of the two Rabi frequencies and upon the two detunings, particularly the presence or absence of one-photon resonance (either $\Delta_P = 0$ or $\Delta_S = 0$) and two-photon resonance (for individual detuning such that $\Delta_3 = 0$).

A commonly considered situation has fully resonant detuning ($\Delta_2 = \Delta_3 = 0$) and Rabi frequencies that maintain a constant ratio—simultaneous pulses. A change of independent variable from time t to either Rabi angle, gives coupled equations with constant coefficients. Their solutions involve sines and cosines and hence they exhibit oscillations. Population transfer can be complete, with full resonance, only if the two

Rabi frequencies are equal. When the Rabi frequencies differ appreciably the amount of population transfer induced by the weaker Rabi frequency diminishes significantly, and will approach zero as the imbalance increases. To restore the weak-field excitation its carrier frequency must be shifted by an amount equal to the stronger Rabi frequency—the *Autler-Townes splitting* produced by the strong field [Aut55; Gra78; Sho11]; see Appendix A.15.2. The Autler-Townes splitting appears, in steady state, as *electromagnetically induced transparency* (EIT) [Har90; Bol91; Har97; Kuh98; Hau99; Fle05] of a weak field in the presence of a strong field that completes a three-state linkage, when each is tuned to the relevant Bohr frequency.[9]

When the pulses are offset in time there are two possibilities for their ordering, discussed next.

Intuitively ordered pulses. When we consider pulsed fields we expect, intuitively, that a successful process might proceed through two steps, with a primary pulse $P(t)$ first moving population into state 2 and then a secondary pulse $S(t)$ moving that population on into state 3, each by either adiabatic passage or by a pi pulse. This *intuitive* pulse sequence is symbolized by the schematic diagrams

$$\text{start: } \boxed{1} \xleftrightarrow{P} 2 \longleftrightarrow 3, \quad \text{then: } 1 \longleftrightarrow \boxed{2} \longleftrightarrow 3, \quad \text{end: } 1 \longleftrightarrow 2 \xleftrightarrow{S} \boxed{3}, \quad (5.11\text{-}1)$$

where boxes enclose the states having population and double-headed arrows remind us that coherent excitation proceeds in both directions. This sequential-transfer technique for population change requires either a slow, adiabatically changing Hamiltonian or careful crafting of the integrals of pulse intensities (Rabi angles) of the two pulses. It further requires that population should temporarily reside in state 2, from which loss will occur by spontaneous emission.

Counterintuitive pulses; STIRAP. An alternative procedure for population transfer is possible, avoiding any state–2 population: A *counter-intuitive* pulse sequence in which the S pulse precedes (and then overlaps) the P pulse. Schematically, the coherent excitation sequence appears as

$$\text{start: } \boxed{1} \longleftrightarrow 2 \xleftrightarrow{S} 3, \quad \text{then: } \boxed{1} \xleftrightarrow{P} 2 \xleftrightarrow{S} \boxed{3}, \quad \text{end: } 1 \xleftrightarrow{P} 2 \longleftrightarrow \boxed{3}. \quad (5.11\text{-}2)$$

The mechanism that underlies the success of this pulse sequence, known as *stimulated-Raman adiabatic passage* (STIRAP) [Ber95; Sho95; Ber98; Vit01; Vit01b; Sho08; Sho13; Ber15; Vit17], requires relatively slowly changing pulses of fixed frequency that drive adiabatic change, but these must complete their action before decoherence acts. Appendix A.15.5 discusses the theory underlying this procedure and provides an explanation for the success of adiabatic processes. Although the language of photons is often used to describe pulsed effects, the discreteness associated with photons is usually not needed (but see Section 7.2).

---

[9]The theory of EIT relies on interference between two components of the dipole expectation value. This has an exact zero of transmission.

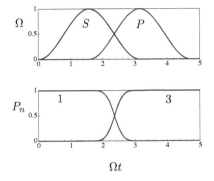

The top frame of Figure 5.6 presents an example of two pulsed Rabi frequencies that have equal peak value and equal temporal pulse shapes but are offset in time, with the S field acting first, the P field later. The two fields are each resonant with their Bohr frequencies.

The lower frame shows the populations of states 1 and 3 that occur when the STIRAP conditions of eqn (A.15-11) apply, so the population transfer is complete and negligible population appears in state 2 at any time.

Fig. 5.6 *Example of STIRAP*

5.11.2   Detuned adiabatic passage; Population return

The adiabatic passage of STIRAP proceeds satisfactorily, though more slowly, when there is single-photon detuning. This makes possible the use of STIRAP for ensembles in which there are Doppler shifts and counter-propagating laser beams. When this intermediate detuning becomes sufficiently large the three-state Hamiltonian can be replaced by an effective two-state Hamiltonian. Then the adiabatic passage produced by STIRAP can proceed equally well with either counterintuitive S–P or intuitive P–S ordering of the pulses. This means that the same pulse pair can move population $1 \rightarrow 3$ and then return it, $3 \rightarrow 1$. Such a pulse will induce population exchange (swapping), $1 \leftrightarrow 3$, of any initial population distribution.

The P–S ordering of this population return involves adiabatic following of a bright adiabatic state, rather than the dark-state following of the original STIRAP; see Appendix A.15.4. It has been called B-STIRAP [Vit97b; Kle08; Gri09; Bor10; Sca11; Sca11b; Sho11; Sho17].

5.12   Objectives of quantum-state manipulations; Superpositions

Preceding portions of this Section listed a variety of scenarios involving alteration of quantum systems by coherent radiation—by photons. With what purpose is such activity undertaken?

Many objectives motivate those who devise procedures for controlled alteration of quantum states [Sho11]. Historically the dominant motivator has been the desire to transfer population into excited states, and to place into these as much population as possible—to achieve *completeness* of transfer.

With availability of coherent radiation sources has come the opportunity to produce specified quantum states, coherent *superpositions* defined by predetermined amplitudes and phases, as needed for Quantum Information processing [Zol05; Bet05; Har06; Ved06; Pet08; Bar09; Kok10; Zol11; Kok16].

A superposition of two states that differ in energy will result in an oscillatory phase between them, a phase factor $\exp[\mathrm{i}(E_2 - E_1)t/\hbar]$ that persists after the creation of the superposition. Most often this phase is undesirable and so the intended target state will be a superposition of degenerate states whose relative phases remain constant and controllable [Vit99; Kis05; Yam08; Sha09; Vew10; Sha15].

The customary quantification of success is measured by the overlap between the target state and the produced state (the so-called *fidelity*).

Typical requirements. Even without detailed control of phases, it is common to desire the alteration to be *selective*—to choose between targets that have similar energies. When initiated by steady, near-monochromatic radiation this objective requires close matching of carrier and Bohr frequencies. As steady radiation becomes more intense, to produce a stronger interaction with atoms, the selectivity diminishes. A variety of techniques involving pulsed radiation give opportunity to improve this compromise and to use a given radiation source with high efficiency.

For many purposes it is desirable to avoid spontaneous emission and other forms of decoherence. This objective requires that, in addition to having initial and target states that are stable or metastable, any change to the quantum system should be completed within some minimal time. And always experimentally available radiation intensities are bounded, either by radiation-source limitations or by the need to avoid damage to optical elements. Thus designs of experiments must often include an optimization of costs and benefits.

Finally, it is often (though not always) desirable that the results be as insensitive as possible to unavoidable irregularities in the radiation pulses or in the environment of the altered quantum system—procedures that are *robust* to small changes in parameters have had particular value.

Creating superpositions. Quite commonly transitions into degenerate-state superpositions rely on Raman transitions in a Lambda linkage. When the intermediate state is significantly detuned from single-photon resonance, the behavior of the system is an example of the two-state excitation discussed above, between single sublevels. A superposition state occurs whenever a Rabi cycle is incomplete.

The creation of a specified superposition, like successful population transfer, can be accomplished in a variety of ways, offering opportunities for imposing a variety of secondary conditions upon pulse energy, peak intensity, pulse duration and sensitivity to error. Composite pulses offer opportunities for meeting simultaneous goals [Lev79; Lev86; Gen11; Tor11; Tor11b; Vit11; Gen14; Tor15; Tor18].

Sublevel superpositions. Any excitation involving Zeeman sublevels relies on a choice of quantization axis to define the sublevels being populated. Thus a procedure that is seen, with a particular choice of quantization axis, to produce complete transfer into a particular sublevel will, when viewed (and accessed) in another reference frame, be a well-defined superposition of Zeeman sublevels. The details of the superposition—amplitudes and phases—is completely determined by the relative orientations of the writing and reading lasers.

Manipulation verification. Once a superposition state has been constructed, various procedures are available for measuring the fidelity of the construction [Vit00; Vit00b; Vew03].

6

Photon messengers

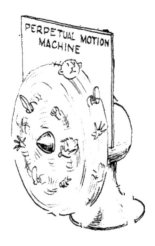

Sometimes the course load seems unending.

Photons of the traveling kind carry a history. Whether the photons are regarded as the monochromatic fields of Dirac or the wavepackets that accompany excited-state emission, they have an inevitable time dependence that accommodates their zero rest mass. Even a single-cycle pulse has a phase that can either add or subtract from a previous field. It is with that time dependence that photons become messaging agents.

The contemporary commercial activities clustered under the banner of "Photonics" (see page 6) rely, in part, on the sending of radiation through confining optical fibers [Qui06; Yar06; Mes07; Sal19]. The fields that are so confined to narrow tubes move at slower speed than in free space: They are altered by the atoms in the medium through which they travel; see Appendix C.2. A major economic driver of this technology is communication: Optical fibers, carrying radiation, are replacing copper wires carrying electrical currents. In turn, this replacement provides a new regime in which to define photons and contrast them with electrons. This chapter provides some background physics, rather different from that of the preceding chapters, that underlies the use of photons for communication. Appendix C.3.2 provides relevant mathematics.

To give a broad picture of photon messaging, with or without human initiation, it is useful to consider a particular historical setting, that of astronomy, where even now we are seeing scientific understanding being advanced on a galactic scale (and larger) much as it was being advanced on an atomic scale in the time of Planck, Einstein, and Bohr. And again, photons are the messengers.

6.1 Astronomical photons

The observation of light from the night sky has historically underlain our understanding of the universe as comprising relatively nearby planets and moons of our solar system along with much more distant "fixed stars", seen in apparent patterns on a celestial sphere. With the application of spectroscopy to telescopic observations it became possible to classify individual stars into a variety of types according to their

Our Changing Views of Photons: A Tutorial Memoir. Bruce W. Shore, Oxford University Press (2020). © Bruce W. Shore.
DOI: 10.1093/oso/9780198862857.003.0006

relative size and surface temperature (red giants and white dwarfs for example). From careful measurement of Doppler shifts of spectral-line wavelengths came values for the relative motions of stars and then of galaxies of stars. Details of the spectral-line intensities provided information about the chemical composition of stars [Ath72; Tuc75; Ryb85], and so courses in the quantum theory of atomic structure were part of the curriculum for astronomy students [Sho68].

When I came as a postdoc (with title of Research Assistant and Lecturer) to the Harvard College Observatory (HCO) in the early 1960s the understanding there of stars, nebulae, and galaxies seemed pretty complete. Much of the painstaking labor that led to this seemingly complete understanding had been done over many years by women at HCO, carefully examining images on glass photographic plates that were stored in stacks there.

But that picture of astronomy as a completed subject would alter dramatically during the next decade, when new observations brought evidence of entirely new classes of astronomical objects. As a member of the HCO community I attended the weekly colloquia where these discoveries were being presented by visitors to peers, and I listened to the speculations about their meaning. They were brought to HCO and my colleagues there as puzzles—since then largely (but not entirely) solved, as current textbooks and Wikipedia articles explain.

Quasars. The first of the strange new objects that was reported at HCO were known as "quasistellar objects" or QSOs. These were the sources of very strong signals first observed by radio telescopes and that came from what appeared to be bright point-sources of visible light, distinguishable from stars only by their unusual spectra and radio emission [Sch63]. Soon given the simpler name *quasar* (from *quasi* and *star*), they were found to lie at great distances from us and to have luminosities (rate of energy emission) that were a thousand times greater than our own Milky Way galaxy, yet they were evidently comparable in size to our solar system. These objects are now understood as being associated with radiation from the surrounding gas (an accretion disk) of a supermassive black hole in the center of a galaxy and they occur prolifically— hundreds of thousands have been cataloged [Sch10b]. Although such black holes are common in galactic centers (our own galaxy has one), not all these black holes become quasars [Sch63; Dav79b; Beg84; Elv00; Bar11].

Pulsars. The second class of unexpected astronomical object, termed a *pulsar*, came also from radio astronomy: In November of 1967 Jocelyn Bell, part of the Cambridge, England radio astronomy group, observed a point source that blinked with remarkable regularity, with pulses separated by 1.33 seconds [Hew68]. For this discovery Bell's advisor received a Nobel Prize in 1974. Other pulsing radio sources were soon found,[1] and members of the HCO community debated their origin. Pulsars are now considered to be highly magnetized, rapidly rotating, neutron stars that emit a lighthouse-like beam of electromagnetic radiation.

---

[1]The physics Nobel Prize of 1992, for the discovery of a binary pulsar, was awarded for work in the doctoral thesis of Russel Hulse, shared by his thesis advisor, Joseph Taylor. Incidentally, as an earlier doctoral student at Harvard, Joe took one of my lecture classes on scattering theory.

Cosmic background radiation (CBR). The third remarkable astronomical discovery of the 1960s was the microwave background radiation first reported in 1965 by Arno Penzias and Robert Wilson [Pen65]. Carrying out observations of discrete sources with a radio telescope they found a microwave background that seemed to come steadily and uniformly from all directions of the sky (their announcement called it "excess antenna temperature" of some 3 degrees K). The interpretation of this radiation, as being primordial photons arriving after the long journey from its origin shortly after the "Big Bang" with which our universe began [Sin05], led to Nobel Prizes for Penzias and Wilson in 1978.

Update: Cosmic microwave background photons. Subsequent studies have found that the spectral distribution of that radiation, now termed *cosmic microwave background* (CMB) [Sil68; Hu01; Hu02; Kin03; Sta18], follows remarkably closely that of the Planck blackbody-radiation formula for temperature $T = 2.725$ K (or 0.2348 meV). In the now well-established description of the early universe, the early plasma of electrons and protons (primarily) was in thermodynamic equilibrium with radiation. Expansion led to cooling and to the binding of electrons and protons to form hydrogen atoms which, being neutral, no longer coupled so strongly to the radiation. From the so-called *surface of last scattering* the photons of the radiation propagated freely thereafter, being Doppler shifted to lower energies and participating in the general expansion of the universe. These are nowadays being seen and studied as CMB photons. Small irregularities in the otherwise uniform distribution of the radiation give important information about the early universe and about its present large-scale structure. Contemporary fits to the data of CMB irregularities over a patch of sky require hundreds of spherical harmonics [Mil99; Sza01; DeB02; Kin03]. These studies also affect notions of the quantum vacuum and of particle physics. Wikipedia articles titled "Big Bang" and "Cosmic microwave background" provide a useful start to reading the growing literature on the subject.

X-ray astronomy. These three great observational astronomical discoveries of the 1960s, and many minor ones, all came first from radio telescopes, registering radiation that has no need for photons as explanation. Although follow-ons of these discoveries continue to define a changing view of astronomy and cosmology, with Nobel Prizes as rewards, the quantum nature of radiation has not been a part of that advance of understanding.[2]

The other portion of the electromagnetic spectrum that became open to study was that of very short wavelengths—X-rays and gamma rays. Our atmosphere extinguishes any X-rays from space, and so it was not until sounding rockets, with trajectories taking them briefly above the atmosphere, were supplied with X-ray detectors in the 1950s that astronomical sources could be observed. Soon after my graduate-student days earth satellites were bringing more regular reports of astronomical sources of X-rays, and these have now provided large catalogs of various types of source, most notably thermal emission from supernova remnants and gas infalling to neutron stars

---

[2]However, Roy Glauber has credited [Gla06] the astronomical intensity-interferometry work of R. Hanbury Brown and R. Q. Twiss [Han56] with prompting his own work that now underlies the theory of quantum coherence whereby photons are made evident [Gla63].

and black holes. The picture of the discrete objects of our universe, seen in X-rays, has been truly astonishing and quite unforeseen.

In addition, the space between galaxies in galaxy clusters is filled with a very hot, but very dilute gas at a temperature between 10 and 100 megakelvins (MK). The total amount of hot gas is said to be five to ten times the total mass in the visible galaxies.

Gamma-ray bursts (GRB). As earth-launched satellites became commonplace they offered platforms that remained steadily above the atmosphere, thereby permitting observation of X-rays and gamma rays. Detectors on Program-Vela satellites, launched to monitor compliance with the 1963 Partial Test Ban Treaty by the Soviet Union, gave the first indications that astronomical objects emitted short bursts of radiation, as announced in a 1973 publication [Kle73]. Most notable of these cosmic transients were termed *gamma-ray bursts* (GRB) [Mes06b; Nak07; Kum15], high-energy transient emissions of gamma rays and hard X-rays lasting as briefly as a few milliseconds or as long as hundreds of seconds. These are said to be the most energetic and luminous electromagnetic events since the Big Bang. They were found to originate in sources beyond our galaxy, impinging on the earth once or twice a day from apparently random directions in the sky. These photons from the edge of time remain a topic of much interest in cosmology [Mes06b; Nak07; Kum15].

Gamma-ray astronomy. Along with X-ray astronomy has come gamma-ray astronomy, based in part upon observations made from space but also from arrays of ground-based detectors. Such high-energy photons, on entering our atmosphere, create electron-positron pairs that produce extensive showers of secondary particles and radiations detectable at the ground or above. From measurements of the flux and arrival times of the shower constituents it is possible to determine whether these came from an incident photon or from a massive cosmic-ray particle and to determine the origin direction. The mean free path of gamma-ray photons decreases with energy and so PeV photons do not reach us from beyond our galaxy.

Regions of the electromagnetic spectrum are given identifying names largely in accord with the techniques used in measurement. With expanding technology it has been possible to extend observations of astronomical sources into the realm of gamma rays, and with that extension has come labels [Aha04] based on photon energies[3] for the segments of the gamma-ray spectrum now studied as *gamma-ray astronomy* [Wee89; Ong98; Cat99]:

Region:	Photon energy:
Low energy (LE):	below 30 MeV
High energy (HE):	30 MeV–30 GeV
Very high energy (VHE):	30 GeV–30 TeV
Ultra high energy (UHE):	30 TeV–30 PeV
Extremely high energy (EHE):	above 30 PeV

[3]See the powers-of-ten table on page 18 for the meaning of prefixes M$= 10^6$, G$= 10^9$, T $= 10^{12}$, and P $= 10^{15}$. A photon having an energy of 1 GeV has a wavelength of 1.2 fm, roughly the diameter of a nucleon.

Such energetic photons are thought to originate with accelerated electrons or colliding protons, rather than as thermal radiation. As of 2019, the highest-energy photons detected have come from a Crab Nebula source, bringing energy of 100 TeV [Ame19].

6.1.1 Missing photons; Dark matter

The laws of motion for masses moving under the influence of gravitational attraction have been known since the time of Newton, with revision to incorporate Einstein's general theory of relativity. As information increased about how stars and dust are moving in their galactic confines it became apparent that the apparent distribution of mass within galaxies was not in agreement with the apparent velocities of the visible matter. Increasingly there seemed to be "missing matter": The matter that was seen from its emission of radiation was not sufficient to account for the mass distribution that would be needed for the accepted laws of motion to produce the visibly evident velocities.

This discrepancy is now attributed to *"dark matter"* [Tri88], meaning gravitational mass that makes no contribution to visible light or other electromagnetic radiation. Explanations continue to be sought by theoreticians [Tri88; Bar01; Kha02; Ber05b; Clo06; McD11; Cha15c; Aru17], but it is no longer considered a subject unworthy of serious attention, as it was in my time at MIT and HCO. Although there is an inclination among astrophysicists to look toward modifications of general relativity for explanation, it seems likely that only when the general theory of relativity eventually becomes a long-sought unified theory of *quantum gravity* will there need to be consequent revisions of how photons are defined and regarded.

6.1.2 Photons in general relativity

General relativity and cosmology were regarded as esoteric and peripheral topics by the MIT faculty in my student days. Particle physics was widely seen as the route to Nobel Prizes. Nuclear physics, spectroscopy, chemical physics, even optics and acoustics, were seen as prosperous subjects. Definitive textbooks by MIT professors on my shelves today are reminders of the popular graduate-school topics of that time: acoustics [Mor36; Ber54], atomic structure [Sla60; Sla77], circuit theory [Gui49], electricity and magnetism [Str41], mathematical physics [Mor53], nuclear physics [Bla52; Eva55], plasma physics [Ros61], and statistical mechanics [Mor64]. But the study of large-scale gravity, with its mathematics, held little interest; it was outside the broad mainstream of physics. That was to change dramatically in the next decades, to become a trendy academic occupation with its folk heroes, though still not a part of my own expertise.

With general relativity, masses are regarded as producing a distortion of space itself, as well as time. Therefore the route of light from a point-source star to a point detector becomes curved as it passes a large mass. The observation of such an effect on light passing by the sun brought popularizing verification of Einstein's theory of general relativity [Isa07], and is observed nowadays as gravitational lensing.

The photons that comprise starlight are expected to follow geodesic paths through a space whose metric is established by masses. The photon itself has no mass and so it does not interact directly with the masses of stars and black holes. Its interactions are with charges, and these have mass. The possibility of a photon acquiring a mass, in a manner analogous to how it acquires a polarization field in passing through matter (see Appendix C.4.2), remains to be seen.

Photon weight. The general theory of relativity predicts that clocks are observed to run more rapidly in regions of stronger gravitational field, as will occur when the clock is closer to a large mass such as the earth center. The Bohr frequency of an atom, or the frequency of an electromagnetic wave (a photon) resonant with an atom, serve as forms of clock, and so this frequency undergoes a shift toward longer wavelengths (a gravitational red shift) as the radiation emitter or detector moves away from the earth. This effect, referred to as the weight of a photon, was first seen by Robert Pound and his student Glen Rebka [Pou59; Pou60].

Update: Gravitational bending of light rays. One of the experimental results that were considered proofs of the validity of Einstein's general theory of relativity (GR) was the deflection of light rays as they passed by the edge of the sun—an effect that had been foreseen by Johann Georg von Soldner in 1801 [Jak78; Wil88]. Gravitational redshifts of photons from a star and gravitational bending of a photon path grazing the solar disk (observable during an eclipse) can be derived by using only Newton's laws and the idea of a photon as a particle of mass E/c^2 and varying velocity. The acceleration of a test particle under the influence of a large mass is independent of the test mass, and this property can be used to evaluate the trajectory of a test particle past a large mass, such as the sun or a black hole. General Relativity predicts (and observations show) a deflection that is exactly twice as large as the simple nonrelativistic value. Several authors have discussed the reason for the GR doubling of the Newtonian result [OLe64; Had66; Jak78; Wil88; Ehl97; Bel02; Isa07; Per18].

Update: Distortion of the metric by light. Because radiation has energy density it will, according to the theory of general relativity, alter the metric of space. One might anticipate that an intense beam of radiation (a very powerful laser) could create such an alteration, equivalent to a gravitational field. Such a possibility has been confirmed with theoretical calculations by several physicists [Tol31; Scu79; Bro06], though the effects are too small to be observed.

Update: Astrophotonics. Much of the technology associated with observational astronomy is now part of what has been termed *Astrophotonics* [Bla09; Bla17]. The objects fitting this rubric include artificial guide stars created by lasers at the top of our atmosphere, large-scale interferometers, image formation techniques, light-detection interfaces, and optical waveguides with spatial filters, junctions, and couplers. These devices are allowing measurements that were hardly dreamed of when I was a postdoc.

Although the special and general theories of relativity are essential for understanding modern astronomy and cosmology, quantum theory has remained only partially included in those subjects. I have used here the expression "astronomical photons" primarily as a literary device—a synonym for electromagnetic radiation in an astronomical context— that does not imply the requirement of discreteness found with the other photons defined in this Memoir. It is primarily with considerations of the vacuum—the absence of photons— that quantum theory has begun to become important to cosmology [Pad03b; Mar12].

6.2 Scattered photons

Light passes readily through air, water, and glass, but ceramics and metals are quite opaque. Two mechanisms, termed absorption (removal of the light, for later appearance including heat) and scattering (redirection without loss), contribute to the opaqueness of matter; see Appendix B.8.1.

The language of photons is commonly used to describe these mechanisms, although the quantum nature of light is not involved: The passage of light through matter is regarded as akin to a stream of photons that are acted upon by discrete, random events along their way.

Absorption takes place when the light frequency matches, at least approximately, a Bohr frequency of some constituent of the matter, either a discrete frequency (as in a gas) or from a continuum (as in a metal). The light transfers energy, at least momentarily, to the matter, a process that will lead either to re-emission or to eventual randomization as thermal motion. Absorbed photons are essentially lost from a beam, although they can be replaced when excitation energy is present (and when lasing action is possible, see Section 5.3).

By contrast, scattering typically occurs without delay, and without significant change in frequency (other than what is required by momentum conservation with the scattering center); see Appendix B.8.2. In circumstances when the light frequency does not match a Bohr frequency (as can happen for infrared radiation passing into milk or biological tissue) it is scattering that attenuates a radiation beam. Scattered photons are essentially redirected. Because the scattering increases the path and hence the time of travel through a sample, it is possible to distinguish three classes of scattered photons [Yoo90; Wan91; Yod96; Dun03; Far07]:

Ballistic: Some small portion of the radiation goes directly along a straight line, with no scattering delay. This radiation retains the original coherence; its photons are known as *ballistic*.

Snake-like: Another portion undergoes only a few scatterings. When small angles dominate the scattering the track of the photons deviates only slightly from a straight line and the field acquires little random phase. These photons are known as *snake-like*.

Diffusive: After many scatterings the trail of a photon becomes very irregular, with loss of phase coherence, and the course of the radiation can be treated as *diffusion*; see eqn (A.1-43) in Appendix A.1.10.

Appropriate timing of the detection of a light pulse can separate these components and use the resulting signal to bring messages of (detect) embedded objects that would otherwise be hidden. More recently, techniques involving spatial light modulators (SLM) are able to adapt incident wavefronts to overcome the randomization produced by layers of random scattering centers or even irregular reflecting surfaces and be able to view objects that would otherwise be hidden [Vel07; Vel10; Kat12]. Such technology has given new meaning to the word "photon" and photon messaging.

Photon migration; The sun. In modeling the interior of a star it is sometimes useful to refer to the flow of radiation energy as diffusion of photons [Per04; Pie06;

Zab15], see eqn (A.1-42). These are pictured as following a succession of random paths between scatterings (a random walk), ultimately resulting in migration from the stellar center to the visible surface (the photosphere).

The release of nuclear-fusion energy in the central region (core) of our sun raises the local temperature to some 10^7K. This heat flows outward toward the solar surface, at first by a combination of conduction and radiation and then, in the outer portion, by convection (bulk fluid motion of convection cells visible as solar granulation). Throughout the solar interior the matter and radiation are expected to establish local thermodynamic equilibrium (LTE), and so the local distribution of radiation frequencies (or photon energies) is that of a blackbody at the local temperature. At the visible surface of the sun (the *photosphere*) the temperature has fallen to some 5×10^3 K.

From models of the temperature, density, and composition of the sun the photon journey has been estimated to require some 10^5 years [Mit92]. In this calculation the "photon" was defined only through the use of a mean free path—basically a spectrally averaged inverse absorption coefficient, see eqn (A.1-44) and eqn (B.8-3)—giving an average diffusion-step size of around a millimeter. The notion of "photon" used in such calculations has averaged over the frequency and polarization characteristics that underlie the usual definition of a photon, and therefore it shares only a name with the entities that appear elsewhere.

6.3   Electrical circuits

Atoms are regarded as the elementary units of matter, and photons are proposed as elementary units of radiation. There is yet another natural phenomena that has elementary units: The electron is the elementary unit of (negative) electricity.

In metals electrons are freed from their binding to individual atoms and can, given suitable impetus by applied voltage (electromotive potential), travel relatively freely. They leave behind stationary positive charges around atomic nuclei. Under appropriate circumstances the positive charges, treated as "*holes*" in the negative distribution of electrical charge, also move and form part of the electrical current.

My first memories of electricity are associated with that everyday object, the flashlight. I was intrigued that the two batteries could produce light from the little filament in the incorporated bulb, and I learned that Electricity was responsible. My subsequent experimental efforts came to naught until my father explained that electrical action required a complete *circuit*: Like any fluid, electricity had to flow from a source to a sink that, in turn, connected back to the source, thereby closing a circuit involving the two terminals of a battery. Wires soldered to resistors and capacitors and vacuum tubes, mail ordered from Radio Shack®, gave me some hands-on experience with electrical circuits, and a textbook helped me understand how currents, driven by voltages through resistance, originated with electrons, the elementary units of electrical charge.

Direct and alternating; Batteries and house current.   One of the things I soon learned about electricity was that it came in two sorts. What I experienced with my flashlight was an example of Direct Current (DC), whereby the electrons flowed steadily in one direction through the light-producing filament of a lamp. The other type, Alternating Current (AC) came from a wall outlet into lamps. Its electrons flowed back and forth, reversing direction some sixty times each second. Not until very much later did I learn how Ohm's law of DC electricity,

$$V = I R, \tag{6.3-1}$$

relating voltage V, current I, and resistance R, had to be modified to deal with the complications introduced by AC through replacement of real-valued R by a more general complex-valued *impedance Z* in the world of Electrical Engineering. There, just as in quantum physics, phase is important.

Analog and digital; The telephone and telegraph. No one who grew up with sighted eyes in the twentieth century could ignore the wire-supporting telephone poles to be seen in neighborhood alleyways and along highways. In my childhood these wires brought to our household telephone voices of neighbors and distant relatives. But my reading brought knowledge of the earlier use of such wires, for telegraph signals. Those earlier messages, made audible as a succession of clicks from electromagnets, were encoded in patterns of intervals known as dots and dashes, the Morse code that many of us boys memorized either to gain Cub Scout advancement or to bypass classroom prohibitions against messaging. Morse code had the great advantage that it could be used over distances not previously linked by wires: Short and long pulses of light from a flashlight or from a shipboard military beacon would suffice. Every letter of the alphabet had its own unique pattern of dots and dashes, from the briefest letter E (just a single dot, "dit") and T (just a dash, "dah") to the more complicated combination of letter V ("dit dit dit dah") that became famous during World War II as the signal for hoped-for Victory of the Allied forces over those of the Axis powers.

The distinction between telephone and telegraph had to do with how they carried information. The older, telegraphic method used a sequence of pulses, a succession of on and off electrical intervals, a procedure subsequently termed *digital*. By contrast, the telephone, and various other devices such as the phonograph, dealt not with simple on and off states, but with a continuum of values for voltages and currents, vibrating at frequencies sensed by the human ear. These were termed *analog* signals, as contrasted with the digital signals of the telegraph. Most measurement instruments available prior to the late years of the twentieth century were analog devices, with displays that could be read as an apparent continuum of values: Compasses, voltmeters and ammeters, tachometers, settings of radio-station dials, all provided numbers from the set of reals $\mathbb{R}$. I learned how analog electrical signals could be amplified using triode vacuum tubes, technology that later converted to transistors and eventually grew into the economically powerful music business.

Quantum circuits. The models of atoms and fields used in quantum optics introduce collective coordinates that incorporate attributes of fixed material structures that are subjected to classical time-varying forces and for which it is possible to find canonical variables that can be taken as obeying appropriate quantum rules. The behavior of a superconducting circuit is another example of a quantum system that has no immediate need for elementary particles of the sort Feynman imagined: Instead of using Kirchoff's circuit laws (essentially expressions of charge conservation) to analyze relationships of voltages and currents, it is possible to construct a Lagrangian from which the needed nondissipative equations will be obtained. This leads to a Hamiltonian involving macroscopic variables such as charges and magnetic fluxes, and a quantum mechanics of their circuit [Dev04; Gir11; Wen17; Voo17]. Charges and magnetic fluxes are quantized, but there are no obvious photons of the sort imagined by Planck and Einstein.

6.4 Information

The content of the various signals that passed along copper wires to our telephone, or over the airwaves to our radio receiver, brought primarily amusement and entertainment—then as now. But I was aware that in businesses and government offices decisions were made from what was to be learned from these sources as well as from various printed communications. This economically or militarily valuable sort of communication underlay the mathematical theory of communication described in 1948 by Claude Shannon [Sha48], work that founded what we now know as Information Theory and which eventually replaced Norbert Wiener's analog world of Cybernetics [Wie48; Dys06].

6.4.1 Classical information: Bits and bytes

The basic idea of Shannon was that communications could be considered the passage of information from source to receiver. Information, for him, meant what was required to make an informed choice between various alternatives. The simplest such choices, for which one needed only the most elementary amount of information, was the choice between two equally probable alternatives, termed variously yes and no, or true and false, or on and off, or up and down. The amount of information needed for such a basic decision, the elementary unit of information, was termed the *bit*—a binary unit of information.

In my days as a graduate student electronic computers based upon digital signals (digital computers) were just beginning to be of use—they were large, room-occupying machines fed by punched cards. We newcomers to science, particularly those with an inclination towards theory, could recognize the importance of instructing such machines to carry out arithmetic operations that were either tedious or beyond reasonable hope of accomplishment. These instructions, known then as computer programs, dealt with digital logic and conditional changes to a stored binary unit of information. The designers of early digital computers found it useful to define a unit of information that was sufficient to specify an alphanumeric character. This became the *byte*, nowadays taken to mean 8 bits.

As a basis for creating a useful mathematical model of digital signals (as contrasted with analog signals) the work of Shannon, based on binary choices, holds a place comparable to that of the founders of quantum theory two decades earlier. Nevertheless I was prepared to consider alternatives to the simple two-choice concept. Those have now arrived, along with a new view of photons, as Quantum Information, noted below. A word of preparation is useful.

The on-off position of a traditional household light switch is still a common example of a binary operations, one that I readily understood as a child. But nowadays room-lighting control has become more analog. Many of the wall switches in my present home offer a continuum of shades of dimness.

The concept of yes-no choices underlay the game with which my siblings and I passed the time on long automobile journeys—the game of *Twenty Questions* whereby, given only the hint of animal, vegetable, or mineral, we were to guess the object, real or imaginary, being considered by our adversary. It is astonishing what tangible-object identities can be encoded in twenty yes-no answers to imaginative questions. (And how difficult it can sometimes be to give a yes-no answer to an apparently simple question –"Have you stopped beating your wife?" being the classic courtroom example.)

This *binary logic* is obviously a simplification of the real world. I learned, as I suspect all parented children do, that the "No" of a parent does not inevitably lead to rejection but is subject to negotiation and modification. That world of requested permissions was not rigidly restricted to just two options. Communications between humans requires such options as "maybe" and "possibly" and the conditional "if you are good". (This leads to notions of *"fuzzy logic"*.) Communication with pets requires inaudible body language that is not readily amenable to recording on paper. And commodities such as Goodness, Beauty, and Integrity are notoriously difficulty to quantify.[4] Nonetheless, the notion of the bit remains the foundation for commercialization of Information.

6.4.2   Quantum information: Qubits

The notion of a system (say a light switch) with two possible positions has an obvious application to the two-state atom considered by Einstein and then by the multitudes of practitioners of quantum theory [All87]. But the quantum description deals with probabilities, say the probability that an atom drawn from a population of identically prepared systems will be found to be in the excited state rather than the ground state. This probability is the square of a probability amplitude. So quantum information deals not with just two numbers, 0 and 1, but with a bounded continuum of numbers. The basic unit of quantum information, a *qubit*, is defined as the information needed to describe fully a two-state quantum system [Pre98; Nie00; Jam01; Ved06; Bar09]. This may be taken as two complex-valued probability-amplitudes whose absolute squares sum to unity. This is therefore the information that can be taken away from (or brought to) such a system by a photon. In theory the numbers involved with quantum information are from the continuum set $\mathbb{R}^2$, usefully organized as complex, $\mathbb{C}$. However, in practice impositions of finite resolution act to discretize the numbers.

6.4.3   Stored and moving information

Practical usage of information falls into two categories: Data that is being stored, for eventual use, and data that is being moved—retrieved from storage—for use in decision making, re-storing, or simply amusement.

The stored, tangible records used during millennia of recorded history have relied on a variety of physical and chemical alterations of objects for that purpose—clay, papyrus, and paper, or transiently on bottles, boxes, and birthday cakes. When we deal

---

[4]This was the practical difficulty for Jeremy Bentham with his Utilitarian principle of "The greatest good for the greatest number"; see [Sho00].

with transient quantum information it is necessary to maintain individual quantum states (or superpositions of states). A variety of atoms or other quantum systems offer opportunities for storing the phases and amplitudes of quantum information. The electrons of individual atoms (or artificial atoms) provide commonly used examples of storage mechanisms. Each active electron provides a large number of potentially usable quantum states. The number is limited, in part, by the range of available excitation frequencies, by the confinement volume of the manipulated atom (and consequent abbreviation of Rydberg series), and by the resolution being used for distinguishing quantum-state characteristics. But all of that information can be encoded in the spatial region occupied by the electron. It is only necessary to distinguish, experimentally, the various electron sites.

This notion of a photon as a carrier of quantum information is an alternative to the earlier notions of photons espoused by Einstein, Bohr, Dirac, and others. It underlies the contemporary notion of Quantum Information [Nie00; Har06; Kni06; Lam06; Ved06; Bar09; Kok10; Kok16; Str16] and its elementary carrier, the *information photon*. It offers a *"flying qubit"* [Ben00; Nor14] that can not only carry information but that can maintain a correlating link (an entanglement) with its source; see Section 6.7.2 and Appendix C.8.5.

Photon polarization as a qubit. A quantum bit (or qubit) of information can be encoded in any degree of freedom. Traveling-wave electromagnetic fields, and their photons, offer several useful degrees of freedom for encoding information, including spatial modes and discretized time history. Beams of radiation can be polarized, meaning that they can maintain a definite direction of the electric vector in the plane transverse to the propagation direction; see Appendix B.1.6. Because photon polarization is a property that can be manipulated accurately, rapidly, and almost losslessly, it has held particular interest for carrying information. It is such polarized light, rather than incoherent unpolarized light, that is commonly used for quantum communication.

The electric vector is often described either by two independent options of linear polarization, say horizontal (H) and vertical (V), or as a superposition of two circular polarizations, right (R) and left (L). Figure B.1 in Appendix B.1.7 shows how these particular choices, and more general elliptical polarization, map onto a point on the surface of a sphere, the *Poincaré sphere*; see Appendix B.1.7.

Once the orthogonal pair of unit polarization vectors are fixed (say H and V or R and L or any two polarizations represented by opposing points along a great circle of the Poincaré sphere) the electric vector of a traveling radiation field, with its magnitude and direction, can be specified by either a single complex number or by two real numbers. Within the restrictions imposed by a given intensity, the components of the superposition can take a continuum of values. This situation is analogous to the parameters needed to specify (apart from an overall phase) the quantum state of a two-state system. Thus the polarization state of a photon (specified in terms of a point on a Poincaré sphere) provides an example of a qubit. For free-space propagation this travels at the speed of light with the photon. To store the photon qubit it is necessary that the photon storage procedure should allow for arbitrary polarization.

Time-binned photons. The creation of a single traveling-wave photon can be done with some control of the time variation of the electric field. This field can be regarded as a succession of controllable amplitude values, each of which provides an example of a time-binned qubit [Vas09; Vas10; Nis11].

6.5   Photons as information carriers

The Einstein-Bohr notion of photons discussed in Section 3.4 gives us a picture of atom emission and absorption in the schematic form

$$\boxed{\text{atom emits}} \longrightarrow \boxed{\text{photon}} \longrightarrow \boxed{\text{atom absorbs}}. \tag{6.5-1}$$

What is the nature of the electromagnetic field—the photon— that carries the energy between the emitting atom and the absorbing atom? The discreteness of the internal-energy states of two atoms implies that these atoms are quantum systems, i.e. the internal structures are governed by quantum mechanics. It was Dirac who most famously proposed that because the sources of the radiation were quantum systems, then the radiation—the electromagnetic fields that carry the energy between atoms—should be treated as a quantum system, one in which the electric and magnetic fields become replaced by noncommuting operators; see Chapter 4 and Appendix B.3. That is, the electromagnetic field is a quantum system, a quantum field. It is natural to ask: How is this quantum nature observable? Such a question rephrases my original question concerning the identity of the elementary quantum increment of radiation, the photon.

Although we can relate the properties of some fields to attributes of the field source, for example the spontaneous-emission field from a two-state atom or the field from a thermal source such as frequency-filtered sunlight, we can also measure quantum (or classical) characteristics of fields independent of sources. Generally we deal with a continuum of possibilities, ranging from completely quantum, with no decoherence, to completely classical with no quantum (non-classical) properties.

The emission and absorption processes bring about a connection, an interaction, between the emitting and absorbing atoms. If fields are to be used for that connection, it is natural to ask: What must their properties say about their sources, in addition to incorporating measures of energy (discrete or continuous)? Are there measurable attributes of a radiation field that will reveal the quantum nature of the field source? Appendix B.5 takes up this question.

6.5.1   Photon time dependence

The picture of an isolated atom, spontaneously emitting a dipole photon of specified carrier frequency after it has been suddenly excited (see Appendix B.2.7), leads to considering the time dependence of the outgoing field and its spectrum—the Fourier transform of this time dependence; see Appendix B.5.1. The atom, once excited, begins to radiate and as it does so the probability of being in the excited state falls exponentially. The outgoing field magnitude carries this same exponential decline. The resulting spectrum of the radiation has a Lorentz profile centered around the carrier frequency (i.e. around the Bohr transition frequency), see eqn (B.8-8) on page 363. This spectrum originates with the time dependence of the emitted photon, and it will be this time dependence that is available at the detector. It is from this time variation that any quantum nature of the field (any evidence of a photon) is to be found.

Monochromatic fields. The simplest, idealized time dependence of an electromagnetic field is that of a monochromatic field. Such a field oscillates sinusoidally (or cosinusoidally) and extends indefinitely in time—it has no beginning and no end. The mathematical expression for such time variation is $\sin(\omega t + \varphi)$ for standing waves or $\exp(i\omega t + i\varphi)$ for traveling waves, where ω is the (carrier) angular frequency of the oscillation and φ is a constant phase. Such time dependence is often taken as a conveniently simple prescription when one models the response of a quantum system to radiation. It is an extreme example of a <u>coherent</u> field. Radiation from a steadily operating laser is often idealized as a simple sinusoid, albeit one that has a randomly fluctuating phase.

Pulsed fields. A simple means of modeling a pulse of radiation that is localized in time, or has a definite duration, is to multiply the infinite-duration sinusoid by an envelope function that becomes very small, or zero, outside some prescribed interval. Often the center of this envelope is taken to be time $t = 0$, so that the time interval of interest begins at negative times and continues to positive times. Alternatively, the time $t = 0$ is taken to be the start of a field that, for prior times, was strictly zero.

Fourier analysis presents any temporal disturbance as a superposition of monochromatic components—pure sine waves or oscillatory exponentials; see Appendix B.5.1. Pulses of radiation that are localized in time, or have a definite duration, have Fourier transforms that have components within a finite frequency interval, by contrast with a monochromatic wave which has a transform of infinitesimal frequency width (and infinite duration). The longer the pulse duration, the smaller the frequency distribution; the broader the frequency distribution, the shorter the pulse.

In practice radiation pulses obtained from various devices (e.g. sunlight and lasers to name two extreme examples) often undergo uncontrollable fluctuations of their phases φ. This random phase fluctuation adds to the breadth of the underlying Fourier-transform spectral width of the pulse. It diminishes the coherence of the radiation. The exponentially decaying field that emerges from an atom as it undergoes spontaneous emission is an example of a pulsed, coherent field.

Thermal sources. Light from the sun can be imagined as a superposition of light from an enormous number of atoms along a line of sight. Each atom becomes excited, either from collisions or from absorbed radiation, and then it generates an outgoing radiation field. There is no particular phase relationship amongst these atoms, and so the resultant field, a superposition of a vast assortment of sinusoids having different phases, is not coherent. Sunlight can be filtered to select a very narrow range of frequencies, but it retains the incoherence of its origins. In this regard it differs significantly from the coherent field that emerges from the radiative decay of a single atom and which is often imagined as the prototype of a single photon.

6.5.2 Characterizing radiation by its coherence

Consider, from a classical viewpoint, the fields we expect to emerge from two sorts of samples. We understand, at the outset, that the radiation of interest here is a traveling electromagnetic wave. That is, at any position (say that of a detector) and any instant, the radiation comprises two orthogonal vector fields—an electric vector and a magnetic vector—whose vectors are perpendicular (transverse) to the propagation direction, i.e. to the line of sight between detector and source atoms. Two examples of radiation sources raise the notion of coherence of the radiation, i.e. the degree to which a monochromatic wave maintains a constant phase.

Spontaneous-emission fields. The first example is the field that results, after an excitation event, from the spontaneous emission of a single atom. Following whatever the field may be during the initiation event, the field that results from spontaneous emission must exhibit a temporal decay as the stored energy of the atom diminishes. The atom must be re-excited before it can again begin to radiate. In the language of photons, a second photon cannot be emitted immediately after a first photon, because the atom must first be re-excited. This requirement will affect the autocorrelation function of the electromagnetic field; see Appendix B.5.3. It is from that quantity that we are able to assess the quantum-theoretical nature of a given field.

Thermal fields. The second example is the field emerging from a frequency-filtered sample of sunlight, or from a pinhole in a Hohlraum. The latter radiation is basically an example of blackbody radiation, in which radiation within an enclosure is in thermodynamic equilibrium with atoms within the enclosure. The sunlight arriving at the earth originates in a similar source, although the radiation has undergone various absorption and emission events that imprint a structure upon the spectrum. But a filtered sample, of narrow spectral bandwidth, will be getting a superposition of radiation fields from a large number of atoms in the line of sight. Each atom has a definite but random orientation of its radiating dipole moment and of the phase of its emitted electric field, and these combine to be treatable as unpolarized, incoherent light. That means that the electric-field vector, which at any single instant must point along some axis transverse to the propagation direction, undergoes a random change during any short observation time. The field phase undergoes a similar random change. Various mathematical models exist for treating such *stochastic* processes [Fox72; Van76; Wod79; Gar85]. The consequence of the randomizing nature of the source atoms—either atoms in the sun or atoms in the Hohlraum—is that the electric field of the radiation differs from that of a single spontaneously emitting atom. A packet of radiation from either source has experimentally observable characteristics, based upon autocorrelation measurements, that will distinguish it; see Appendix B.5.4.

6.6 The no-cloning theorem

Making copies of macroscopic objects requires both skill and appropriate raw materials, but given those there is no fundamental limitation on what can be copied. Things are different at the quantum level.

The quantum theory of radiation builds in a requirement that copying a photon requires use of a photon-creation operator. Each such event reproduces existing photons (as does the Einstein stimulated-emission B coefficient) but it also includes spontaneous emission of a single photon (as does the Einstein A coefficient of spontaneous emission). The photons need not have any particular frequency association with atoms, it is only necessary that there be a mechanism for activating photon creation. Because of this unavoidable property of quantized radiation it is not possible to duplicate any given field without introducing some additional contribution from spontaneous emission [Mil82]. This effect can go unnoticed when the field is strong, as in a laser, but it becomes quite apparent when one deals with a single photon. As Milonni and Hardies state [Mil82]:

> One cannot make an exact copy of (or clone) a single photon of unknown polarization.

This prohibition against exactly duplicating unidentified photons is a special case of a more general characteristic of quantum systems. With quantum theory one can construct a linear superposition of two quantum states. This can, in principle, be accomplished regardless of the quantum states. It is an example of the linearity of quantum theory. That very simple characteristic of quantum theory rules out the possibility of making an exact copy of an arbitrary quantum state—a state that is completely unknown. This prohibition, the "no-cloning theorem" [Woo82; Woo09; Ort18; Mil19] can be stated as [Sca05]:

> No quantum operation exists that can duplicate perfectly an arbitrary quantum state.

However, if one is willing to accept some imperfections (or if one has *a priori* knowledge), then effective copying can be accomplished [Buz96; Buz98; Sim00; Sca05; Kok16].

Secure messages. The impossibility of making exact copies of photons underlies one of the major contemporary interests in them: A message constructed with single photons cannot be intercepted without that action being apparent to the intended recipient of the message. This property of photons provides the incentive for financial support of the newly created field of quantum communication [Eke91; Kuh02; Zol05; Kim08; Dua10].

6.6.1 Quantum teleportation

The oft-mentioned procedure of *"quantum teleportation"* [Ben93; Bou97; Nol13; Kok16; Zei20] uses the quantum nature of photons to transport quantum-state information between two locations. It does not transport *matter*, as do the fictional teleportation machines of entertainment media, it only transports *information* about how a particular quantum system is constructed from its given parts. The parts must necessarily already be at hand.

6.7 Correlation and entanglement

In this next section I will set aside aversion to equations in order to discuss one of the physics notions that has become central to practical applications of quantum theory and photons, a subject that Appendix C.8.4 presents in more mathematical detail. It is a source of much contemporary discussion, a modern counterpart to the wave-particle duality that drew lively disputes during the origin of quantum theory (recall Chapter 3) and now concerns notions of objective reality, see page 12.

6.7.1   Correlation

In any system, quantum or classical, that has more than one part or more than one distinguishable attribute or *degree of freedom* (DoF), and hence more than one measurable property about which information (attribute value) is available, it may occur that information about one of the attributes implies some information about other attributes. The attributes linked in this way are said to be *correlated*. The amount of correlation is quantifiable by a *conditional probability* $P(A|X)$ of finding attribute value A given a preparation procedure (or prior information) X:

$$P(A|X): \quad \text{measure } A \quad \leftarrow \quad \text{given condition } X. \tag{6.7-1}$$

Appendix C.8.1 discusses the mathematics used for quantifying this characteristic.

A simple classical example occurs in the idealized model of red and blue marbles (attribute color, values red and blue), paired in a closed sample envelope. If the preparation process, conducted by an assistant, is known to pair only marbles of different colors then an observation of the color of just one of the two envelope marbles will immediately inform the observer of the certain color of the unseen partner. If we take the preparation process X to include viewing one of the marbles, then we deal with the chance of finding the second marble to be red through the conditional probabilities

$$P(\text{red}|\text{red}) = 0, \qquad P(\text{red}|\text{blue}) = 1. \tag{6.7-2}$$

That is, no matter where the second marble may be, still hidden in its envelope, we can be certain of its color because we know the color of its mate.

This simple classical situation deals with just two possibilities for the color value, and with the assumption that the preparation step, placing two marbles into an envelope and later viewing one of them, brought certainty of a relationship of the two colors: Whatever the color of the first marble, the second one very definitely differed.

Quantum correlation.   Quantum theory does not deal with the actual physical objects, the marbles in the present example, but with *information* about them—with probabilities about their attributes. The seemingly simple procedure of selecting a marble of different color from a first one requires some measurement of color and some information—a step that passes without notice in the formulation of classical models but which is crucial in dealing with quantum systems.

Furthermore, quantum theory deals not only with limited certainty, quantified by probabilities, but it constructs the probabilities from *probability amplitudes*, complex-valued numbers whose absolute squares provide the desired probability. Information about the system is emplaced in the amplitudes, which are additive, not directly in the probabilities. The mathematics of quantum probabilities is therefore akin to what must be used for treating classical radiation: What is there observed is intensity, and this is expressible by the square of electric-field amplitudes.

Consider a quantum system (a single degree of freedom) that has just two mutually-exclusive attribute values, either A or B. To present the quantum theory it is useful to employ the bracket notation devised by Dirac, the *bra* $\langle \cdots |$ and the *ket* $| \cdots \rangle$ (see eqn (A.3-5) in Appendix A.3.2). Amplitude products then give the formulas

$$|\langle A|A\rangle|^2 = P(A), \qquad |\langle B|B\rangle|^2 = P(B), \qquad \langle A|B\rangle = \langle B|A\rangle = 0. \qquad (6.7\text{-}3)$$

With that notation, consider the particular probability amplitude $|\Psi\rangle$ associated with a coherent superposition of two probability amplitudes, in the form

$$|\Psi\rangle = \mathcal{N}\Big(|A\rangle + |B\rangle\Big), \qquad (6.7\text{-}4)$$

where $\mathcal{N}$ is a normalizing constant, to be chosen so that $|\Psi\rangle$ has unit probability.[5] Then, given the requirement that the attribute value can only be A or B and the certainty that the system has been prepared as eqn (6.7-4), the individual conditional probabilities (of finding A or B given the superposition condition Ψ) will sum to unity,

$$P(A|\Psi) + P(B|\Psi) = P(\Psi|\Psi) = 1. \qquad (6.7\text{-}5)$$

Next consider two independent degrees of freedom, 1 and 2, say two atoms or two electrons or two characteristics of a single radiation beam, each of which has just two mutually-exclusive attribute values, either A or B. Let us denote by $|A\rangle_i$ the probability amplitude for associating attribute value A with DoF i. We denote by $|A\rangle_1|B\rangle_2$ the probability amplitude appropriate to describing the joint attributes of the two DoFs – A with DoF 1 and B with DoF 2. In this model the initial condition for a conditional probability can include both requiring a particular superposition as well as specifying the result of a measurement of one of the DoFs. Then consider the particular probability amplitude[6]

$$|\Psi\rangle = \mathcal{N}\Big(|A\rangle_1|B\rangle_2 + |B\rangle_1|A\rangle_2\Big). \qquad (6.7\text{-}6)$$

In this construct the two DoFs are known to have different attributes. Thus, if we are certain that the paired probability amplitude is $|\Psi\rangle$ and we find that the DoF 1 has attribute value A then we can be certain that DoF 2 has attribute value B: The conditional probabilities for attribute values of DoF 1, given specific information about DoF 2 (and the certainty of superposition $|\Psi\rangle$), are

$$P_1(A|\Psi; A) = 0, \qquad P_1(A|\Psi; B) = 1. \qquad (6.7\text{-}7)$$

Here $P_1(A|\Psi; X)$ is the conditional probability of finding A for DoF 1 given the prior knowledge of X for DoF 2 and the specific superposition Ψ. Crucial to this model of activity is the requirement that we can be certain of somehow creating the particular superposition eqn (6.7-6), and that we can be certain of our subsequent measurement on one part of this prepared system.

---

[5]The assumption that attribute values are exclusive (and that $A \neq B$) gives for the normalizing constant the formula $\mathcal{N} = 1/\sqrt{2P(A)P(B)}$.

[6]This $|\Psi\rangle$ and the $|\Phi\rangle$ of eqn (6.7-8) are examples of the Bell basis, eqn (C.8-16).

An alternative superposition, expressing certainty that the two attributes are the same,

$$|\Phi\rangle = \mathcal{N}\Big(|A\rangle_1|A\rangle_2 + |B\rangle_1|B\rangle_2\Big), \tag{6.7-8}$$

leads to the conditional probabilities

$$P_1(A|\Phi; A) = 1, \qquad P_1(A|\Phi; B) = 0. \tag{6.7-9}$$

These several conditional probabilities are extremes found with the simple models used here. More generally the values range between 0 and 1; see Appendix C.8.

Alternative bases. The choice of attribute measurement is often arbitrary. For example, with the two values A and B taken to be directions on a table top we might choose north and south for the axes, but we might also choose east and west by a rotation of the table. Or we might deal with vertical directions of up and down (as with spin direction), but with alternatives of tilted axes. The attributes of radiation beams offer options of linear (H,V) and circular (R,L) polarization, and the specification of computer-screen colors offer several standard options. We can introduce an alternative mathematical description of the two attributes (an alteration of the underlying abstract vector space associated with attributes, see Appendix A.3), say

$$|a\rangle - |b\rangle = \sqrt{2}|A\rangle, \qquad |a\rangle + |b\rangle = \sqrt{2}|B\rangle. \tag{6.7-10}$$

The probability amplitude $|\Psi\rangle$ retains the correlation because

$$|A\rangle|B\rangle \pm |B\rangle|A\rangle = |a\rangle|b\rangle \pm |b\rangle|a\rangle. \tag{6.7-11}$$

In all cases, when we know that the initial conditions define a particular superposition then the instant we become informed about one DoF we immediately have information about the correlated DoF.

6.7.2 Entanglement

These situations, the conditional probabilities of eqn (6.7-7) and eqn (6.7-9), are examples of *entanglement* discussed in Appendices C.8.1 and C.8.5, in which there occurs correlation of separate attributes (degrees of freedom), expressed by probability amplitudes. We gain knowledge of an as-yet-unobserved attribute (DoF 1) immediately upon learning about a second attribute (and the certainty of having a particular superposition). The entanglement expressed by relationships eqn (6.7-7) or eqn (6.7-9) is a consequence of the inherent probability-amplitude structure, of eqn (6.7-6) or eqn (6.7-8), known to exist with certainty (because of how we prepared the system) between two independent degrees of freedom. (The mathematics involves rays in two independent abstract-vector spaces.) Were we to replace either of the symbols $|\cdots\rangle_1$ or $|\cdots\rangle_2$ by plain numbers we would be dealing, as in eqn (6.7-4), with a single degree of freedom. There would then be no attribute to entangle.

Entanglement can be created with any number of DoFs greater than one. The correlated (entangled) degrees of freedom may be associated with subsystems that can be physically separated, such as two electrons, an atom and an electron, or two radiation beams. Alternatively, they may be associated with independent attributes that cannot become spatially separated, for example, the color and polarization of a single, classical radiation beam [Spr98; Lee04; Lui09; Qia11; Gho14; Kar15; Ebe16; Qua17; Sch17].

An increasingly important class of quantum systems, particularly for applications in quantum information, comprises those in which two independent degrees of freedom, with their two distinct Hilbert spaces, are associated with particles (or photons) that can be spatially separated. For example, the electron and proton of a hydrogen atom, or the two electrons of a helium atom, can each be prepared as two-particle bound states that have definite total angular momentum, a state in which individual spins are correlated. Photoionization can then produce quantum states of free particles that can each subsequently be measured at distances far from the original coordinate origin and well separated from each other. With suitable procedures the two particles can retain their collective correlation. Such physically separated particles (or photons) are said to be *entangled* [Ben96; Hil97; Ved97; Kni98; Nie00; Rai01; Bar03; Min05; Mes06; Har06; Van07; Ami08; Hor09; Ebe15; Mil19], a term that is attributed to Schrödinger [Sch35; Sch35b; Sch36; Tri80], having consequences discussed in a 1935 paper of Einstein with his postdoctoral research associates Boris Podolsky and Nathan Rosen (known as the EPR paper) [Ein35] that was much later brought to wider attention through experimental tests proposed by John Bell [Bel66; Eke91]. It was this form of correlation, allowing a measurement at one location to give immediate knowledge about a distant object, that drew the attention of Einstein as indicative of the incompleteness of quantum theory—a concern now known to be unwarranted.

Spooky weirdness. This retention of entanglement with alteration of measurement choice for spatially separated objects has led some physicists to regard entanglement as a "spooky action at a distance", as "quintessential quantum attribute", and as *the* essential "weirdness" trait, of quantum theory [Kni98; Qia18]; see Appendix C.8.5. Application of such entanglement underlies much of the contemporary activities of quantum technology and quantum engineering; see their literature.

7
Manipulating photons

Midnight oil is for burning before exams.

If electromagnetic fields can be treated as comprising photons, and if single photons are to be descriptors of the weakest fields, then it is natural to ask whether single photons can be routinely created and detected. And whether they can be manipulated—whether their temporal and spatial characteristics can be altered in controlled and useful ways.[1]

It is such questions as these that I have seen researchers answering affirmatively [Ray12] using technology that in earlier years remained in the fantasy world of thought experiment. And it is here that my own established view of photons – an inheritance from Planck, Einstein, Bohr, and Compton described in Chapter 3—has been undergoing most notable revision.

Studies of single photons are no longer just idle academic exercises, to be discussed in senior commons rooms of ancient universities or on pages of esoteric peer-reviewed journals. They are now seen to have significant commercial use, as evident from media accounts of quantum computing, quantum cryptography [Eke91; Eke94; Pho00; Gis02], and, more generally, quantum-information processing (QIP) [Zol05; Bet05; Har06; Ved06; Pet08; Bar09; Kok10; Zol11; Kok16]. There is evidently money to be made from single photons as well as from the crowds of photons used in surgery and for industrial cutting and welding.

7.1 Particle conservation

There is an important distinction between particles that have rest mass—atoms and electrons—and the quanta of radiation, which are massless. To a good approximation

---

[1] Feynman held the view, discussed in Section 8.3, that photons could not be altered in flight: They were created by an interaction with a particle, propagated freely in vacuum, and were annihilated with another interaction. Such a view fits the needs of those who deal with constant Hamiltonians and fundamental particles, as in Appendix A.19, but it does not provide a practical approach to dealing with systems in which collective coordinates are to be quantum variables, as with light traveling through matter (see Appendix C.4), or when the Hamiltonian is being adjusted to impose pulsed forces that are most easily described classically, as in Section 5.4.

Our Changing Views of Photons: A Tutorial Memoir. Bruce W. Shore, Oxford University Press (2020). © Bruce W. Shore.
DOI: 10.1093/oso/9780198862857.003.0007

the total mass of chemical reagents remains unchanged as they undergo chemical reactions. The count of protons and neutrons (regarded collectively as *nucleons*) remains unchanged when nuclear fission or nuclear fusion takes place. Electrons may be removed and recombined with binding centers, all the while maintaining a fixed total number. But there is no such limitation on the creation and absorption of photons; see the Darwin quote on page 60. Like the heat generated when boring a cannon on a lathe, there is no limit to how much energy can be converted to light.

The heat and light we see from a candle flame [Far05], or from burning fuel in a fireplace, comes from rearrangement of the atoms that are bound together as molecules in the fuel. The energy given off comes from a chemical reaction that creates, from large molecules of fuel, smaller gaseous molecules of carbon dioxide, carbon monoxide and water. The energy released by the restructuring of chemical bonds maintains the burning process. The Einstein formula $E = mc^2$, rewritten as $m = E/c^2$, allows one to express the chemical energy change, E, as a change in mass, m. But because c is so large this mass change by the atoms is extremely small.

The light produced when electrons flow through a flashlight-bulb filament is the result of chemical energy stored in the battery. Again, it is rearrangement of atomic structures within a solid environment (the battery) that drives the electrons through the circuit and produces radiation. There is no change in the number of atoms during the lifetime of a battery. The number of electrons, though very large, remains fixed; the electrons break free from their bindings in the battery, flow through the circuit, and return to the battery.[2] In so doing they are responsible for producing visible light, but the photons so created cannot be considered as stored in the battery.

In all these various processes the electrons, protons, and neutrons maintain unchanging rest mass. They acquire, by virtue of their chemical structuring into molecules, potential energy (of binding) that can subsequently be released as heat and light, but their intrinsic properties remain forever fixed constants of nature. In brief,

> *Particles that have rest mass are conserved in chemical and nuclear reactions, whereas photons, being massless, have no such constraint.*

There are significant apparent exceptions to particle conservation. A gamma-ray photon traveling through matter can, if its frequency is sufficiently high, create an electron-positron pair, a process known as *pair creation*. The positive electrical charge on the positron exactly balances the negative charge of the electron, thereby maintaining charge neutrality as electromagnetic energy converts to two masses. The positron will soon meet an electron (not necessarily the same one, they are indistinguishable) and the two masses will revert to pure energy. At much higher frequencies, photons can create matching pairs of more massive particles. The assortment of particles, and their conservation rules, that appear in experiments at high-energy accelerator facilities do offer ongoing employment to Particle Physicists, but they have limited relevance to the more mundane world of everyday photons that I am discussing in the present Memoir (but see Appendix A.19).

---

[2]Because all electrons are identical (indistinguishable), there is no way to follow the course of a particular electron and to know that it travels the circuit from end to end.

7.2 Creating single photons

Numerous procedures are now being used for creating single photons [Kuh99; Kur00; Kir04; Mau04; Moe04; Sch07; Hij07; Wil07; Bar09c; Kuh10; Vas10; Muc13; Far16], and numerous review articles discuss these [Oxb05; Lou05; Bar09b; Sch09; Bul10; Kuh10; Vas10; Eis11; Buc12; Sin19]. Amongst the simplest for creating optical photons are the following.[3]

Weak laser pulse. A particularly simple idea is to use a very weak (i.e. strongly attenuated) laser beam, idealized as a coherent state with very small mean photon number; see Appendix B.4.1. In such a beam the probability of finding n photons is given by Poisson statistics (a Poisson distribution of photon numbers, eqn (B.4-11)), and so although the probability for having more than one photon can be made arbitrarily small, the probability for no photon at all becomes very large. The detector must be poised to register a photon during all of these dark intervals, and so inevitably spurious dark-current pulses intrude. Furthermore, the probability of two photons being present simultaneously, though small, is never zero.

Spontaneous emission. Conceptually the simplest procedure for creating a single photon remains based on spontaneous emission from a single two-state quantum systems—an atom, ion, molecule, or any of the various discrete-state constructs that have become available. An experimenter first exposes the system to a brief excitation-producing pulse, typically a resonantly tuned pulse whose Rabi frequency and duration produce complete population inversion—a *pi pulse* (see Appendix A.10.2). Then the experimenter waits as the system undergoes inevitable spontaneous emission, thereby converting the single excitation into a single photon.

This procedure has been demonstrated not only for emission from suitably accessed individual atoms, molecules, and ions, but also for a variety of discrete-state quantum systems in which electrons are physically constrained in one dimension (quantum wells), two dimensions (quantum wires), and three dimensions (quantum dots).

Although the free-space emission typically produces an outgoing dipole-radiation field (see Appendix B.2.6), the surroundings can be modified so as to produce a more directed beam, for example a field that is confined to an optical fiber [Dom02; Sto09; Sto10b].

In the atom-plus-field-plus-interaction model of Section 5.1.1 that underlies the exactly soluble Jaynes-Cummings description of a single mode of confined quantum radiation interacting with a two-state quantum system, the field is always available for re-exciting the atom in an ongoing periodic sequence; see Section 5.8. By contrast, the effects of spontaneous emission into free space, not into mode-defining cavities, have traditionally been treated with perturbation theory, a standard topic of quantum theory textbooks. With this approach, of a quantum atom interacting weakly with a continuum of radiation modes [see eqn (A.16-11)], the excitation energy of the atom is predicted to flow steadily away from the atom. The radiation begins to flow at the instant the atom becomes excited, and during the subsequent exponential decline of excitation the outgoing field will correspondingly decline. The result is a Lorentzian

---

[3] In addition to the mechanisms listed, pairs of gamma-ray photons can be created from annihilation of positrons by electrons.

distribution of field frequencies that create a photon wavepacket. However, an interesting work by Chan, Law, and Eberly [Cha02], treating the dynamics of an instantly excited two-state atom interacting with an unconstrained quantum field (no cavities, waveguides, or fibers) derives a set of entangled Schmidt-decomposition modes (see Appendix C.8.5) for joint atom-photon wavefunctions that show remarkable modulated temporal and spatial structure.

Cavity STIRAP. In the basic process of stimulated Raman adiabatic passage (STIRAP) described in Appendix A.15.5, a single atom, in state 1, is subjected to a pair of pulses, P and S, in counter-intuitive ordering, S-preceding-P, to move population adiabatically from state 1 to state 3 in the linkage chain $1 - 2 - 3$, see eqn (5.11-2). The STIRAP process, in adiabatically transferring population between two quantum states that are linked by a stimulated-Raman interaction, also transfers a photon from the P field to the S field. The energy of the P photon passes to the S photon together with Raman excitation of the atom. Each excitation transfer accompanies the creation of a single S photon. There is no spontaneous emission and therefore the timing of photon emission is under experimental control [Kuh99; Kuh02; Hen03; Kuh10; Vas10].

This action can be used to control the placement of a single S photon into a cavity. With cavity STIRAP the S field is initially the vacuum state of the cavity and the P field is a laser pulse. The stored S-field photon subsequently leaks out of the cavity, through a slightly transparent bounding mirror, to become a traveling-wave photon. Because the process involves a single-atom excitation it produces a single photon in the cavity. The field that subsequently emerges from the cavity retains this single-photon character and its association with a single de-excitation.

With cavity STIRAP the experimenter can craft the P pulse to place into the cavity a field which, upon leaking from the cavity and becoming a traveling wave, has a predetermined time dependence. It is a traveling photon having a predetermined temporal envelope.

Spontaneous parametric downconversion; Heralded photons. One of the processes treated in discussions of nonlinear optics is the ability of various crystals to spontaneously convert a portion of a single pump-field beam into pairs of beams, termed signal and *idler* with frequencies and directions set by energy and momentum conservation. Known as spontaneous parametric downconversion (SPDC) [Man86; Har96; Bur02; Tho04; Gal05; Mos08; Pea10; Zha12] the polarization states of the created fields are correlated in a manner characteristic of the particular crystal: In type I conversion the signal and idler have parallel polarizations. In type II conversion the two polarizations are orthogonal. The photons in each pair are created with negligible time separation and are emitted on opposite sides of two cones, whose axis is the original beam. They produce a set of rainbow-colored rings [Har96]. In one cone the polarization is *"ordinary"* and in the other *"extraordinary"*. Along the intersection lines of the cones the photons are polarization entangled but in other directions the emission direction determines the polarization.

Although the occurrence of the conversion is probabilistic (as is spontaneous emission of a single photon), the detection of one of the two fields is a guarantee that the other field must be present. One says that the detection of an idler photon *heralds the presence* of a signal photon [Pir10; Eis11; Rie14; Rie16]. And because the frequency, momentum, and polarization of the signal and idler were linked at their formation, measurements on the idler photon can provide complete specification of the signal photon without measuring it. For further discussion see Appendix C.8.5.

7.3 Detecting photons

The notion of a photon as an absorbed increment of radiation energy raises the question: How does one detect such a photon? A great variety of devices have been used to make electromagnetic waves perceptible to human senses. Several of these were part of my childhood introduction to technology. Others have become available only in recent years.

The human eye. The patterns of star locations in the night sky have long held human attention, and amateur astronomers continue to be delighted in pointing out dim objects that are barely visible to the unaided eye. The human eye is a remarkably efficient system for converting optical images into brain-stored information through an elaborate series of stages; its complexity has long been a source of wonder [Daw86]. An eye that is properly dark adapted can perceive very weak sources, can detect optical signals that contain only a few photons, perhaps only a single photon [Hub95; Rie98; Bru09; Art14; Nel17]. So efficient are the combined elements of the eye for photon detection that as late as 1998 it was argued that the performance of the human eye "equals or exceeds that of man-made detectors in several ways" [Rie98]. More recently there have been interesting suggestions for use of human vision in verification of quantum effects such as efficient photon cloning [Sek09].

Photography. Permanent records of visible light were familiar to me, and to my parents and grandparents, as photographs—events and objects that had been imaged in cameras onto rolls of photosensitive film that had then undergone chemical treatment to make the record permanent. In this process a brief exposure of a flat surface of photosensitive molecules (silver nitrate or silver chloride in the earliest examples) created a *latent image*, formed by single grains responding to light by altering their structure. The subsequent chemical processing acted on these grains to make a stable and visually evident molecule where the initial radiation change had occurred. When the exposure was weak the graininess of the final photograph gave evidence of the individual absorption events, a demonstration of photon action.

However, the granularity of traditional photographic images originates with exposure of microcrystals to sufficient energy in sufficiently short time to produce a latent image—to the arrival of several photons [Gar08]. They are not single-photon detectors. Not until the invention of electron-multiplier technology was there a detector of single photons.

Photoelectric detectors. The predominant procedure for detecting the arrival of a photon-associated field is by means of the photoelectric effect. The arriving electric field produces an electron-hole pair in the detector. A variety of devices nowadays can convert single photoelectrons into detectable events. Typically these devices apply voltages that accelerate the unbound photoelectron. Collisions of the moving electron with bound electrons generate secondary free electrons. These in turn are accelerated and produce tertiary electrons. The swarm of moving electrons, in an exponentially growing avalanche, are detectable as an electric current in a recording circuit. This mechanism underlies the widely used avalanche photodiode (APD) and the older photomultiplier tube (PMT) that was an essential tool in my graduate-school days.

Digital cameras. Contemporary digital cameras use two applications of the photoelectric effect to create digitizable electrical signals. In them the photoelectric process takes place in a two-dimensional array of thousands or millions of pixels, independent light-sensitive semiconductor surfaces that convert light intensity into freely movable conduction electrons.

The first digital cameras used charged coupling devices (CCDs) that created the separate pixels by means of an overlying electrode structure. Following the exposure to light, and the creation of photoelectrons, the timed electrode structure brings the photoelectron charges to a corner of the sensor by means of sequential row transfer and column transfer. There an analog-to-digital converter converts the time-varying charge into digital values, to be treated as a digitized raster scan of the image. These devices are capable of converting as much as 80% of the photons into electrons, compared with an efficiency of 20% for a human eye and around 10% for photographic film.

The later complementary metal-oxide semiconductor (CMOS) chips have an array of independent areas of semiconductor as the pixels. With each photosensitive pixel there is a transistor that amplifies the original photoelectrons into larger signals, led by wires to a central processor where they arrive simultaneously rather than as a raster scan. The manufacturing procedure is cheaper for CMOS chips than for CCD sensors and they require less power to operate so they are the basis for common cameras found in smart phones, although they are less efficient in converting photons to signal electrons.

Recently developed intensified charge coupled devices (ICCDs) are fast and sensitive enough to image in real-time the effect of the measurement of one photon on its entangled partner [Fic13].

Bolometers. A photoelectron, without further amplification, can increases the local temperature and in turn alter the resistance of a superconductor, with consequent measurable change in an electric circuit. Such devices, used for astronomy as early as 1878, are said to be capable of registering the arrival of single photons [Bul10].

Fluoroscopy. X-rays, with their ability to look through shoe leather and display foot bones, were a common adjunct (as fluoroscopes) to the retailing of child-fitting shoes during my childhood. The X-rays, in stopping, generate photoelectrons that, in coming to rest, create atom excitation and subsequent fluorescence that, in turn, can be made visible on a display screen. Children were greatly amused to stand on the fluoroscope and watch their wiggling toes as X-rays illuminated everything. No longer is such exposure tolerated. Its potential damage to living cells and genetic instructions therein, already understood by my own parents, has subsequently prompted regulation of all uses of X-rays.

Scintillators. Gamma rays passing through crystals deposit their energy as kinetic energy of photoelectrons. Those particles, in turn, create fluorescing atoms whose light, in turn, can be used to start a cascade of photoelectrons moving from cathode to anode in a photomultiplier tube, producing at the end a current in an electronic circuit. The individual gamma-ray photons thereby become apparent as discrete, recordable events such as I watched during long hours of patient graduate-school data taking.

Radio receivers. Radio receivers use rather different principles. In my youth I strung a long wire atop our house roof and brought it down through a window into my bedroom where a tiny galena crystal and a "cat's whisker" of copper wire together formed a "crystal-set radio" that could turn radio signals into faint sounds on earphones. I learned that radio waves, seemingly surrounding everything, induced weak response of the conduction electrons in the antenna wire. These could be brought into the house where a tuned circuit of coil and variable condenser (or inductance and capacitance as I would much later learn to call them) would select the frequency being broadcast by a particular station. (Then, as now, frequencies were allocated by government overseers in discrete sets to commercial applicants.) The signals of interest—music or spoken words—were encoded as audio-frequency modulation of the much higher-frequency radio carrier frequency. Various electrical circuits in the radio receiver acted to extract this signal and amplify it.

Radio waves were not stored, nor were the audio signals that came to us over the radio. Recording devices for consumers relied on either creation of spiraled groves of mechanical surface modulation (the vinyl platters known as "records") or in variations of the magnetization of wires (wire recorders) and, later, magnetic tapes (tape recorders). It was a time of analog, not digital recordings.

Nowhere in this entire procedure of radio reception was there to be found evidence for photons. Radio photons have so little energy that their effects are to be observed primarily when they act collectively. Only when I learned about various forms of magnetic resonance phenomena, in which atomic nuclei with magnetic moments are induced by radiation to flip their orientations in a magnetic field, did I find the photon concept of use (see Appendix A.14.7). And even there, as with the theoretical description of masers and lasers, only the nuclear behavior, not the field, required quantum theory.

Aside: A bit of trig. One of the routine tasks assigned in my college trigonometry class was to prove various relationships amongst sines and cosines—called trigonometric identities. One of these identities states that the product of two sines is equal to half the difference between two cosines whose arguments are the sum and difference of the two sine arguments z_1 and z_2:

$$\sin(z_1)\sin(z_2) = \tfrac{1}{2}\left[\cos(z_1 - z_2) - \cos(z_1 + z_2)\right]. \qquad (7.3\text{-}1)$$

This, and similar algebraic relationships with function arguments $z_i = \omega_i t$, had important application in radio receivers. It says, for example, that when a radio-frequency signal at frequency ω_R is modulated by (multiplied by) an audio-frequency signal at the much smaller frequency ω_A the result is two sidebands of the original radio frequency signal, offset by the much lower audio frequency, $\omega_R \pm \omega_A$. Identity eqn (7.3-1) underlies the heterodyne protocol for radio receivers and procedures used in nonlinear optics, to combine (multiply) two sinusoidally varying waves and produce waves at sum– and difference-frequencies. And it expresses the effect of coherent population changes (Rabi oscillations) on the steady monochromatic radiation that produces them. Little did I anticipate in college the ongoing encounters with this simple formula.

Nondemolition detection. Traditional methods for detecting photons rely on converting the field energy into electron excitation in some medium – turning the photon energy into either a chemical change (as occurs in making a photograph) or into a measurable electric current (storable, in turn, in electromagnetic memories). The photon is thereby lost: Such devices irreversibly destroy the original photon, replacing its quantum state by an incoherent, mixed state and thereby preventing any continued coherent manipulation of the original quantum state.

Various techniques exist for detecting the presence of a field, even one as weak as a single photon, without irretrievably losing the field. Many of these techniques rely on nonlinear optics (see Appendix C.3.6), whereby the presence of the test field alters the phase of a reference field. This phase change is observable with the aid of interferometric procedures.

A variety of other procedures, known as *quantum nondemolition* (QND) techniques, have been identified for measuring quantum systems without inducing irreversable changes to them [Cav80; Bra96; Kok02; Pry04; Ral06; Gre09; Eis11; Rei13; Sat14]. These procedures rely on the fact that physical observables are associated with quantum operators and that any such operator has a companion (a conjugate variable) such that the paired operators cannot be simultaneously measured with complete accuracy. The position and momentum of a particle, or the electric and magnetic components of the electromagnetic field, are common examples of conjugate variables. In a QND measurement of an observable (say one associated with operator $\hat{A}$), the inevitable disturbance generated by the measurement is restricted to the observable which is canonically conjugate to $\hat{A}$. The subsequent evolution of $\hat{A}$ is therefore left unaffected. For the photon number observable the conjugate variable is the phase of the electromagnetic field [Bar86; Peg88; Peg89; Hra90; Lyn95; Peg97]. One QND method for determining photon numbers relies on the existence of a phase (or energy) shift which depends on photon number.

The electric field of a photon can, in principle, participate in nonlinear optical effects that produce a measurable change to the phase of an ancillary field, a change that is measurable using interferometry. Such effects leave the original photon available for further manipulation [Cav80; Bra96; Kok02; Pry04; Ral06; Gre09; Eis11; Rei13; Sat14].

7.4   Altering photons

As will be noted in Section 7.5, it is possible to store the quantum-state-defining information carried by a photon and recover it. Several of the methods used to create single photons in a cavity allow an experimenter to craft the temporal shape of the photon. The technique of cavity STIRAP discussed on page 155 is one such procedure.

Numerous optical techniques are available to alter characteristics of beams of radiation after they have been launched. These offer opportunities to alter characteristics of photons after they have been created.

Redirection. The most obvious optical devices are flat mirrors that alter a beam direction. Versions of partially reflecting mirrors (beam splitters, see Appendix C.3.5) allow the controlled superposition of two fields. Lenses provide redirection of wavefronts, into and out of a focus. Such optical elements are part of the conventional toolbox with which experimenters manipulate laser beams on optical tables. They have application not only to weak optical beams but to single-photon states. (But note the view of Feynman regarding photons and mirrors, in Section 8.3. We are not dealing here with Feynman photons—nor with Einstein photons.)

Polarization change. Transparent material that has suitable structure can rotate the direction of the electric vector in controllable ways. In many materials the polarization-rotating structure can be altered by application of static electric or magnetic fields. As discussed in Appendix B.1.7, a common way of characterizing beam polarization is by Stokes parameters. With conventional linear-optics elements it is possible to convert between any two sets of Stokes parameters.

Carrier-frequency change. Photons that are created by transitions between discrete quantum states acquire a carrier frequency equal to the Bohr transition frequency. The creation procedure alters the envelope of the field but not its carrier frequency. However, various techniques of nonlinear optics can be used to shift the carrier frequency [Ray12]. Appendix C.3.6 discusses some possibilities for accomplishing that change, after explaining the underlying physics of nonlinear optics.

Mode conversion. Waves, of whatever sort, obey equations that permit traveling waves that are classifiable into distinct modes. When the wave propagates through some medium that undergoes a change of properties with distance (a change in wave velocity) the nature of the modes change and the ongoing wave may differ structurally [Bat05; Sal10; Wan16b]. An example is the conversion of transverse-electric (TE) waves into transverse-magnetic (TM) waves [Alf80] or the conversion of transversely structured Gaussian beams from rectangular (Hermite) into circular (Laguerre) nodal description [Pad00]; see Appendix B.2.5. More recently there are devices that convert orbital angular momentum of a beam into helicity (spin angular momentum) [Kar09; Slu11]. Although all of these devices rely on concepts of classical optics, their spatial mode manipulations offer opportunities for altering the spatial structure of a single photon.

7.5 Storing and restoring photons

Photographic images store a record of radiation intensity, localized on an image plane of some detector, either film or an electronic device. Underlying this image record is some form of photoelectric effect, and this response to radiation is insensitive to the phase of the radiation—its response is to photon number. Such a record does not include sufficient information to restore the original electromagnetic field, typically described either by time-varying amplitude and phase or complex-valued components of vectors.

Several procedures exist for capturing more information about a pulsed beam of known-polarization radiation and storing this in a collective quantum state of excitable atoms; see Appendix C.4.2.

Polariton storage. A popular technique for coherent storage of photons relies on dynamic alteration of electromagnetically-induced transparency (EIT) [Har90; Bol91; Har97; Kuh98; Hau99; Fle05]. In this nonlinear process an experimenter-adjustable field applied to an optically thick medium controls the group velocity of a much weaker signal pulse as it travels into the medium. Acting on the excitable atoms of the medium the two fields form the two legs of a Lambda-linkage stimulated *Raman process* shown at the left in eqn (5.11-2): The relatively weak signal field (the Raman pump field, P, to be stored), links a populated ground state 1 to an excited state 2 while the adjustable relatively strong control field (the Raman Stokes field S) forms a linkage between state 2 and an unpopulated metastable state 3. The storage procedure is a variant of stimulated-Raman adiabatic passage (STIRAP) (see page 129) in which the signal field and the collective quantum state of the atoms form a coherent superposition, a dark-state polariton (DSP); see Appendix C.4.2. The experimenter adjusts the control field to first slow the signal field in the medium and then bring it to rest. In so doing, the original traveling-wave field becomes a stored collective excitation, termed a *spin wave*, involving a distribution of atom coherence of states 3 and 1. Such a technique has been used for planar images that generalize traditional photographs [Hei10; Hei13].

The original signal field can be regenerated by reversing the dynamics of the control field: As its strength grows the stored polariton is given a traveling-wave electromagnetic component, and will subsequently emerge from the storage medium. This storage procedure maps the original time dependence of the signal field onto the spatial variation of a spin wave. The experimenter, by crafting the control field, has the ability to adjust this spatial variation and, in turn, the time variation of the restored field.

Storing a single photon. A variant of EIT–based procedures, known as the DLCZ scheme after the initials of its originators [Dua01], relies on unstimulated, spontaneous Raman scattering to signal single-photon storage. Into an optically thick medium send a weak P field (the *write* pulse). The medium through which this field passes is regarded as an ensemble of N identical three-state atoms in state 1. The P field is tuned near, but not exactly equal to, the Bohr frequency for a transition to excited state 2. Some of the P photons undergo Raman scattering to produce (spontaneously) a Stokes photon and an excitation of metastable state 3, a sequence $1 \rightarrow 2 \rightarrow 3$. The detection of a Stokes-field photon heralds the loss of a P-field photon and the creation of an atom in state 3. The procedure does not identify which of the many atoms has become excited, and so the P-field photon is written onto a collective superposition state, one that expresses an equal likelihood of the excitation being in any single atom along the optical path. Later, the application of a strong *read* pulse pumps the single excitation back to its original state, thereby generating a single anti-Stokes photon.

7.6   Verifying photons

To establish that detection events signal the arrival of single electromagnetic-field quantum states (single photons) one can, in principle, measure the autocorrelation function of the field, $g^{(2)}(\tau)$. Measurements of the field can establish whether it is expressible as a single quantum state or is a mixed state (see page 261). Various quasi-probability functions derived from the field density matrix, such as the Wigner function [Ste78; Ber87; Roy89; Asp01; Sil07; Cas08; Lvo09; Alo11], can reveal whether the field has quantum-mechanical characteristics, as contrasted with a demonstrably classical field of many photons.

One might wonder whether a detected field, demonstrably that of a single photon, is the same field (the same photon) as was sent through various optical elements to the detector. Feynman argued against such a possibility (see Section 8.3), but the discussions of Appendix C suggest otherwise—that useful definitions of photons do allow their manipulation. It is certainly possible to ask whether all the measurable attributes of a quantum field are identical after the field has passed through a variety of devices. This is what is required for quantum-information processing. Such a determination requires some means of measuring correlation between the detected photon and some reference field. The paired photons that are generated in a spontaneous parametric down-conversion process [Man86; Har96; Bur02; Tho04; Gal05; Mos08; Pea10; Zha12] offer an opportunity for such measurements.

8

Overview; Ways of regarding photons

The major exam is the big one.

The historical record of how radiation quanta and photons have been regarded by physicists (and who should know better?) that I have presented in Chapters 3 and 4 marks a route taken not only by myself but by many colleagues. The story of my own ongoing education in photon matters is one that many of them would echo.

Whereas definitions of atoms and electrons as elementary units of matter and charge leave little room for disagreement, the notion of photon offers several differing operational and theoretical definitions, several of which appear in preceding chapters. It is common for physicists (and others) to use the term "photon" to describe a variety of radiative interactions with matter. When a beam of light passes through a portion of matter one may speak, as did Feynman, of its individual constituent photons as encountering a succession of many atoms. The more numerous the atoms the more likely that the photon will be absorbed. When a single atom is placed into a reflecting-wall cavity one speaks of the atom as encountering many passes of a photon that bounces between the walls. The more numerous the passes (and the higher the quality factor of the cavity) the greater the likelihood of atom excitation. In more specialized discussions, atomic changes are said to occur by one-photon transitions, but some are regarded as two-photon transitions, requiring the simultaneous presence at the atom of two single photons. Multiphoton transitions similarly require simultaneous presence of multiple photons. These transitions take place by means of induced dipole moments, an example of nonlinear optics, see Appendix C.3.6.

This penultimate chapter summarizes some notions of photons that appeared in preceding chapters, and comments on some further photon concepts that find use in contemporary physics. It concludes with remarks on some of the alternative approaches to treating radiation without photons, and notes some of the many photon applications not mentioned elsewhere in this Memoir.

Our Changing Views of Photons: A Tutorial Memoir. Bruce W. Shore, Oxford University Press (2020). © Bruce W. Shore.
DOI: 10.1093/oso/9780198862857.003.0008

8.1 Historical photons

Historically, the first definition of radiation quanta, discussed in Chapter 3, dealt only with radiation-energy increments to steady radiation—the cavity-enclosed photons of Planck discussed in Section 3.1. To this was added the notion of traveling photons, carrying momentum—the photons of Einstein described in Section 3.4. As noted in Sections 3.8 and 8.6, the fields responsible for these manifestations of absorption and emission need not have quantum properties. They were examples of the photon crowds discussed below, in Section 8.4.

Experiments with various forms of cavities led to definitions involving stationary field modes, of photons that lacked both momentum and angular momentum. The theoretical formulation of Dirac (Chapter 4) involving field oscillators and associated photon creation and annihilation operators encompasses all of these sorts of photons; see Appendix B.3.2. The spatial modes associated with these oscillators can incorporate, through boundary conditions, basic interference (wave) effects.

These observations, gathered over time, have led me to conclude that, unlike atoms and electrons, there is no unique entity to be called a photon. The photon is a concept that finds use in many alternative ways, some of them seemingly contradictory. In any context the word "photon" must be given meaning—an operational definition—from its connection with experiment.

The historical view of photons being irreversibly absorbed, awaiting subsequent emission, has given way to recognition of reversible changes in which a coherent field induces Rabi oscillations between atom and field energy, as embodied in the Jaynes-Cummings model of Appendix C.6, or in which adiabatic changes induced by crafted field pulses produce predetermined quantum-state alterations. Such reversibility leads to notions of photon storage and recovery that differ fundamentally from what happens with the incoherent fields available to Planck, Einstein and Bohr; see Section 7.5.

What is important nowadays in many optics laboratories can be regarded as treating particular superpositions of the Fourier components of a field—*wavepackets*. These, when suitably normalized to carry energy $\hbar\omega$, are commonly termed "photons" by experimentalists, although these entities are not the simple monochromatic photons in free space that were presented by Dirac and discussed by many other theoretically-inclined authors over the years; see Section 8.3.3.

8.2 Pulsed photons

It seems advantageous these days to think of defining a single photon not as an increment of some spatial mode (defined through solutions to the vector Helmholtz equation with suitable boundary conditions, see Appendix B.1.5) but as a field that can completely excite exactly one atom [Sto10] or which emerges ultimately from spontaneous emission of a single quantum-system excitation. Chapter 5 discussed these operational definitions. Such a photon is certainly not monochromatic; it has a pulse envelope that varies with time. The pulse envelope can have an amplitude and a phase that an experimenter can control, either during the storage step when the field is mapped onto a spin wave of coherent matter excitation (see Appendix C.4.2), or during the retrieval step, when the distributed coherent excitation is brought into free space again. Chapter 7 discussed such manipulations.

Such fields, sourced in quantum changes, are experimentally distinguishable from weak classical fields by their coherence properties, i.e. from various correlation functions; see Appendix B.5.3. Single photons from thermal sources tend to arrive in bunches, whereas photons from demonstrably quantum sources (such as single-atom emission) tend to be *antibunched* [Kim77; Pau82; Zou90; Bia91; Wod92; Kol99; Hen05]. Much contemporary research aims to find procedures that will identify, and put to use, such quantum-field increments—what one might call *contemporary photons*, as distinct from Einstein photons or the Feynman photons discussed next. In much of that work, and the discussion in the remainder of this chapter, there is an implicit assumption that the fields are steady, rather than being pulsed. This fits well the notion of a monochromatic photon but is less convenient for treating pulsed excitation and the photons of Chapter 7.

8.3 Steady, Feynman photons

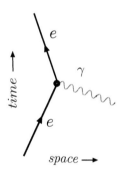

Fig. 8.1 *Basic electron-photon interaction according to Feynman*

Traditional quantum electrodynamics (QED) deals with properties of steady-state systems. A calculation tool introduced by Feynman [Fey48; Dys49; Kai05] for obtaining numerical values from relativistic perturbation-theory treatment of steady-state scattering systems presented a notion of photons that has become widespread throughout theoretical physics and chemistry. In this view, displayed in simple patterns of path lines (*propagators*) meeting at nodes (instantaneous interactions), the static Coulomb force between two charges can be regarded as originating with exchanges of "virtual" monochromatic photons (see page 167): A repulsive force between two like-charged electrons, an attractive force between an electron and a proton.[1] As seen in Figure 8.1, the straight lines represent schematic space-time charged-pointlike-particle paths (for fermions), meeting wiggly photon lines (bosons) at interaction nodes. Each portion of a diagram brings a recipe for calculating, by means of an integral, as in eqn (A.9-35c) of Appendix A.9.9, one of an infinite number of contributions to a perturbation series for a scattering amplitude that is applicable to stationary states.

Aside: Propagators. As explained by Freeman Dyson [Dys49], the Feynman propagators, with their exponential arguments, are integral expressions for solutions to the differential equations used by Julian Schwinger in his formulation of

---

[1] You might naively think that the exchange of momentum-carrying photons would produce repulsion between the exchanges, rather like tossing a ball back and forth between ice skaters, but instead it produces attraction between oppositely charged particles.

quantum electrodynamics [Sch48; Sch49; Sch49b]. As used in this Memoir, propagators are operators (or matrices) that act to turn an initial statevector into a later one, as dictated by the time-dependent Schrödinger equation. A *Green's function* serves this purpose for other PDEs.

Such calculations, with their diagrammatic foundation,[2] have led to remarkably accurate Nobel-Prize-winning values for such (stationary) experimental quantities as the magnetic moment of the electron and the energies of the excited states of hydrogen, avoiding various mathematical divergences that had attended earlier attempts to use the Dirac theory of electrons with the Dirac theory of photons.[3]

The monochromatic photons appearing in common Feynman-inspired diagrams occur often as intermediaries of interactions, not as observable field increments—not as the free-space photons discussed in Section 4 and Appendix B.3. They appear as theoretical constructs in many-body perturbation-theory (MBPT) calculations that deal with an infinity of possible spacetime trajectories [Mat67; Pal73; Sap06]. They have been of less use in the world of quantum optics, where Rabi oscillations are to be found and perturbation theory is of limited use. Figure 8.2 shows two diagrams that contribute to calculation of free-photon effects, either Compton scattering from an electron or resonance scattering from an atom. In the latter interpretation the photon absorption *a* (with energy gain by the atom) may occur either before or after the photon emission *b*, and the corresponding formulas involve what are often termed resonance and anti-resonance energy denominators [Pow64].

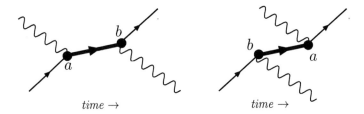

Fig. 8.2 *Feynman diagrams for photon scattering by a charged particle or by an atom.*

Feynman developed, in general lectures (his 1979 Sir Douglas Robb lectures at Aukland University in New Zealand), what became his book *QED—The Strange Theory of Light and Matter* [Fey85]. In his lectures Feynman made the following comment, quoted in [Dud96]:

> A most wonderful fact is that light never does anything, really, when you get down to it. Except go, in a vacuum, from one place to the other. It's emitted by

---

[2] Quite commonly QED is regarded as meaning the combination of Feynman diagrams and perturbation theory, but Schwinger was famously opposed to using those diagrams, and work on Cavity QED or Quantum Optics has little need for perturbation series and Feynman diagrams, only the quantum discreteness of photons.

[3] In the resulting quantum electrodynamics a single electron can be regarded as traveling with a cloud of virtual Feynman photons. In turn, a photon travels with a cloud of virtual electron-positron pairs.

one atom or particle and absorbed by another, and it never goes and gets slowed down or gets reflected. ...Reflected light is really not the same photon coming back as went in. Photons from the source went into the glass, and from the glass comes out a new photon.

With an alternative, traditional wave description of optics, as in Section 8.6, we would say that, in being reflected by a mirror an electromagnetic field has created, in an atom that it is passing, a transition dipole moment. This dipole is set in motion, with phase fixed by the incident field, and it then creates an outgoing electromagnetic wave that, along with others from the many atoms of the mirror surface, travels back toward the original source, properly phased to reproduce the initial field but in a different direction. Appendix C.2.1 discusses how this sequence of events is quite simply incorporated into a refractive index, whose surface discontinuities produce reflection— a basic beam splitter, see Appendix C.3.5.

The photons that Feynman spoke and wrote about are imagined as occurring in a a steady state in which all matter is made from elementary units whose every motion can be expressed mathematically. The Feynman photons are part of the ongoing enterprise of scientists who exploit unifying theories of the most elementary of particles, indivisible units from which nuclei can be constructed, as is discussed in Appendix A.19.

However, these are not the photons that are imagined by scientists who daily experiment with storage and recovery of quantum fields, as discussed in Chapter 7. The photon wave equations for fields in an inhomogeneous medium is describing propagation of some collective excitations of the whole system—atoms and fields together—and not just of free photons; see Appendix C.2.2. The electromagnetic fields appearing in these models for quantization (and thereby for interpretation as photons) incorporate the equilibrium behavior of bulk matter as susceptibility and a refractive index.

Virtual photons; Virtual states. The notion of photons presented by Feynman, of field increments that carry away a sharply defined energy and momentum from interaction with a particle, creates an opportunity to describe situations in which the field is possibly present only very briefly between its encounters between energy-momentum measuring interactions. Such photons, having only weak bounds upon their energy and momentum, have been termed *"virtual photons"* [Ait85; Luk03; Hol04; And04; Yan09; And14; Scu18]—as contrasted with the "real" photons that are controllable and detectable, and that satisfy energy-momentum conservation constraints with matter. Both types appear in various Feynman diagrams, such as the energy-altering ongoing emission and reabsorption of an electromagnetic field by a single electron. When they do not conserve energy these diagrams describe "virtual transitions" into a "virtual level" [Sho79b; Lam06].

The Feynman approach uses integrals over all possible intermediate states—all possible routes between two observable states—even those for which the energy-momentum relationships of free-space photons do not hold. Such photons cannot be observed as free-space fields but they affect such perturbation calculations as scattering amplitudes and energies, wherein there are innumerable successions of intermediate states.

8.3.1 Special relativity math

Any quantitative discussion of reference frames that are moving steadily but with different speeds must take into account the finite speed of light when comparing timings and lengths—and masses. The needed mathematics—the special theory of relativity—deals with connections between space and time that occur when one describes motions in different steadily-moving coordinate frames. Special relativity treats four-dimensional (4D) vectors in a *Minkowski space* of space-time coordinates (events) for which the *norm* (space-time interval between events),

$$\ell^2 = (ct)^2 - x^2 - y^2 - z^2, \tag{8.3-1}$$

is invariant—it is unchanged by introducing an alternative, steadily-moving coordinate system

$$\{t, x, y, z\} \to \{t', x', y', z'\}. \tag{8.3-2}$$

The Minkowski norm ℓ shown in eqn (8.3-1) is positive for *timelike* vectors, negative for *spacelike* vectors, zero for *lightlike* vectors. (Other conventions, with opposite signs, are often used.)

The *Minkowski norm*, which must be used to compare measurable lengths and times between moving reference frames, extends the familiar norm for lengths in Euclidean geometry,

$$|\mathbf{r}|^2 = x^2 + y^2 + z^2. \tag{8.3-3}$$

Rotations and translations of coordinate systems leave this length unchanged: They move points in a four-dimensional $\mathbb{R}^4$ over a three-dimensional sphere $\mathbb{S}^3$.

An important vector in Minkowski space, the *energy-momentum vector* for a particle, has as its elements the three components of momentum $\mathbf{p}$ and the energy E. Its invariant defines the particle mass m,

$$(E)^2 - (\mathbf{p}c)^2 = (mc^2)^2. \tag{8.3-4}$$

This equation applies both to photons, for which $m = 0$, and to electrons or other particles, for which m becomes the rest mass when $\mathbf{p} = 0$. A *Lorentz transformation* expresses the change of any four-vector from an inertial reference frame having zero momentum to one moving with constant relative velocity $-\mathbf{v}$. For the energy-momentum vector this reads [Bai07]

$$\{E_0, \mathbf{0}\} \to \{E_0\gamma, E_0\gamma\mathbf{v}/c\}, \qquad \gamma \equiv 1/\sqrt{1 - (v/c)^2}. \tag{8.3-5}$$

Here E_0 is the rest energy and γ is the Lorentz factor of eqn (A.1-10), quantifying time dilation and length contractions.

Group theory, summarized in Appendix A.18.2, provides a powerful tool for treating symmetries—for associating some physical symmetry, such as that of a sphere or cylinder, with mathematical functions that are consistent with the symmetry—as being *representations* of a symmetry group. The relationships above are examples of *Orthogonal transformations* of the three Cartesian coordinates, linear relationships, representations of the *Orthogonal group*, the Lie group O(3) that leaves the Euclidean norm, eqn (8.3-3), unaltered; see Appendix A.18.3.

Lorentz transformations (of the 4D coordinates) are linear relationships, representations of the *Lorentz group*, the Lie group O(1,3) that leaves the Minkowski norm, eqn (8.3-1), unaltered; see Appendix A.18.3. A more general set of Minkowski-norm symmetries, the *Poincaré group*, consists of three spatial rotations, three space translations, one time translation, and three transformations to a moving frame (boosts). This group incorporates the defining symmetries of special relativity. Reflections in space and time of the 4D Minkowski vector also leave the norm unchanged.

The following paragraphs describe connections of the abstract mathematics of group theory, Appendix A.18, and algebras, Appendix A.17, with three notions of photon definition.

8.3.2   The quaternion photon

The algebra of quaternions described in Appendix A.17, an extension of the algebra of Euclidean vectors, incorporates the requirement of special relativity for maintaining the Minkowski norm, eqn (8.3-1). Quaternions are therefore well suited to representing the components of a free-space electromagnetic field associated with each of the two possible helicities [Maj76; Edm78]. These are *quaternion photons*. To treat more general free-space electromagnetic fields it is necessary to superpose both helicities, a doubling of the quaternions that leads to biquaternions (*octonions*) [Dix94; Bae01; Con03], mathematical objects that continue to interest a few of the physicists concerned with electrodynamics [Gun73; Gog06; Cha10; Tan11; Cha14b; Cha15b; Cha15c].

A reviewer has, justifiably, pointed out that quaternions have long been regarded as outmoded predecessors to tensors. I mention them in this Memoir because not long ago I was asked by PhD friends to comment on a recent press report about octonions as novel approaches to contemporary physics:

https://www.quantamagazine.org/the-octonion-math-that-could-underpin-physics-20180720/.
As indicated by the citations above, I found that the subject still retains the interest of some researchers.

8.3.3   The Wigner photon

Eugene Wigner [Wig39] described in detail the representations of the Lie groups appropriate to preserving symmetries required by special relativity of four-dimensional space and time coordinates, expressed as preservation of the Minkowski norm, eqn (8.3-1)—that is, of functions that could describe traveling waves of electromagnetic fields. Subsequently authors have suggested that one could regard a photon, rather esoterically, as an "irreducible representation of the inhomogeneous Lorentz group" [Pow64] or of "the (proper) *Poincaré group*" [Bia06c; Bia17; Bia18]. This approach, emphasizing the relationship of abstract mathematical groups and special relativity to solutions of the Maxwell equations (photons), leads to the *Riemann-Silberstein vector* as a photon wavefunction; see Appendix B.1.11. It has had less appeal to workers in quantum optics than to particle physicists and cosmologists, whose notion of photons involves a contributor to the Standard Model of Particle Physics outlined in Appendix A.19.

The Wigner photon is notable because it is regarded by some physicists as "the one <u>true</u> photon", in keeping with the requirements of special relativity; all other notions of photons then require superpositions of the Wigner photon. In this Memoir

I am suggesting that, just as it is often useful to treat an atomic nucleus as a point particle, ignoring any internal structure, and to treat fluid behavior without regard for its molecular constituents, so too is it useful to accept a variety of definitions of "photon" to fit particular conditions.

8.3.4   Gauge photons

The three fundamental non-gravitational interactions of physics deal with fields generated by, and acting on, a number of fundamental (i.r. indivisible) spin-half particles. The Standard Theory of Particle Physics outlined in Appendix A.19 organizes these particles into several sets—of low-mass leptons and of heavier quarks. The three non-gravitation interaction-force fields, when treated with quantum theory, are associated with a set of quantum fields that are identified with particular space-time symmetries and allowable alterations of reference frames, bearing labels from their governing Lie group: $U(1)$, $SU(2)$, and $SU(3)$; see Appendix A.18.3. These are termed *gauge fields*[4] because they are associated with Lie-group symmetries in which conservation of a current leads to the need for an associated field. In this context the photon originates with a gauge field associated with $U(1)$, the symmetry of a phase change in field amplitudes, and conservation of electric charge.

The effects of these interactions are quantified by evaluating various Feynman diagrams in which the gauge field (the photon, for example) travels freely in vacuum until it meets a point particle (an electron or other charge), whereupon a quantitative interaction takes place in accord with a definite prediction of strength. Upon summing over all possible histories of such events (a *Feynman integral*) one obtains a value for a stationary physical observable such as a magnetic moment or a scattering cross section.

By contrast, the photons that make their appearance in quantum optics occur in a nonrelativistic time-dependent Schrödinger equation that often presents its results (Rabi oscillations, for example, see Appendix A.10.2) in closed analytic form—formulas involving sines and cosines.

Treatments of very intense fields (near or beyond the Schwinger limit, see page 173) assume that the basic free-space photon, when interacting with hadrons (uncharged neutrons as well as charged protons) must be treated as superposition of field and hadron states, in much the same way as ordinary optical radiation traveling through matter must be regarded as a superposition of electromagnetic field and matter states to form polaritons (quasiparticles); see Appendix C.4.2.

8.3.5   Photon uncertainty

The traditional relativistic appearance of a photon is within momentum space; no proper position operator exists for such a photon. There have been various suggestions for alleviating this difficulty, particularly for wavepacket photons, and to obtain photon counterparts to the Heisenberg uncertainty relations for particles. [Bia13b; Bia12; Bia12b] A useful approach, accompanying the interpretation of the RS vector as the traveling-photon wavefunction, starts by defining a center-of-energy location **R**

---

[4]The term "gauge" also occurs to describe the flexibility available in choosing vector potentials, as discussed in Appendix B.1.9.

(first moment of the energy distribution) in configuration space as a position vector. Components of this vector with those of the vector of total momentum $\mathbf{P}$ obey the canonical position-momentum commutation relations. For fields in three dimensions there is an uncertainty relation [Bia12b]

$$\sqrt{(\Delta \mathbf{R})^2}\sqrt{(\Delta \mathbf{P})^2} \geq \frac{3}{2}\hbar \times 1.27\ldots . \qquad (8.3\text{-}6)$$

The requirement from the Maxwell equations that free-space electromagnetic fields must be transverse implies that the three spatial degrees of freedom are not independent and so the sharpest lower bound to the uncertainty is slightly larger than three times the one-dimensional value $\hbar/2$.

8.4   Crowds and singles

Human mobs are notorious for behaving very differently from single individuals. An organized work crew can create structures that unco-operating individuals could not. The discipline of Sociology, though dealing with human activity, encounters social constructs that have no counterpart with individual humans—the subject for study by Physiologists. Similarly, in physics a spoonful of water must be treated by a different formalism—different variables, different equations—than we would use when describing only half a dozen diatomic molecules. We should not be surprised that large crowds of photons have uniquely collective properties and that single photons might exhibit properties associated with discreteness. Just as studies of collective human behavior justify separate university departments from studies of human bodies, so too might we expect that there would be differences in the equations used to describe single photons and crowds of photons. It is with the weak radiation fields of single photons that we may require quantum theory. It is with more intense fields, comprising a vast number of photons, that quantum effects are not evident and the formalism of nineteenth century electromagnetic theory usually proves fully satisfactory.[5]

There is an unlimited variety of possible photons, differing in frequency, momentum, and angular momentum for example. But whereas experience tells us that every human is unique—even "identical twins" have physiological differences that distinguish them—it is quite possible to have large numbers of fully identical photons. (The no-cloning theorem of Section 6.6 tells us that a single, completely unknown photon cannot be exactly replicated, but crowds of known identical photons can be created without relying on replication.)

8.4.1   Multiple-photon and multiphoton

The photon-mediated model of radiation effects draws a distinction between what can be accomplished when photons arrive well separated in time (a *multiple-photon* process) that proceeds by distinctly separate energy-conserving steps, and the so-called nonlinear effects that become possible when several photons are present simultaneously (a *multiphoton* process [Gol65; Bon65; Van77; Sho84; Hio84; Del06; Sto19]). In the

---

[5]Classical electromagnetic theory fails when energy density is so great that the weak and strong interactions must be considered along with electromagnetic interactions; see Appendix A.19.

latter situations, often involving photons of different frequencies, the various fields need not be separately resonant with specific transitions. The requirement is that the summed photon energies must exchange with energies of the material quantum system. In place of the simple two-state Bohr condition of eqn (5.2-1) we have the N-photon requirement for adding and subtracting energy increments,

$$|E_n - E_m| = \hbar \, |\omega_1 \pm \omega_2 \pm \cdots \pm \omega_N|. \tag{8.4-1}$$

Whereas each excitation rate of a multiple-photon process is linearly proportional to an intensity, the N-photon rate is proportional to N such intensity factors (and $2N$ amplitude factors); see Appendix A.12. As with the two-state Bohr condition, this multiphoton energy-conservation equation is a guideline for producing maximum effect, not a strict equality requirement.

Although I have succumbed to the convenience of referring to these radiative processes by using photon terminology, the behavior of the atoms is almost always what can be described by a *semiclassical* approach in which the atoms are governed by quantum theory but the fields are all classical fields, experimentally controllable.

8.4.2   Creating multiphoton fields

Just as single photons can produce, or be produced by, an energy change in a single atom, so too can the simultaneous presence of multiple photons cause, or be caused by, more elaborate energy changes in single atoms. These are well-studied multiphoton transitions that are attributable to field-induced transition moments; see Appendix A.12.

Many articles have noted that an atom, moving through a resonant cavity, can add or remove a single photon, depending on how many Rabi cycles it undergoes during its passage. With such techniques it is possible to create a single-mode cavity field that has a specified number of photons: A *Fock state* [Var00; Bra01; Var04; Wal06]. More generally, it is possible to create an arbitrary coherent superposition of single-mode cavity Fock states, as proposed by C. K. Law and Joe Eberly [Law96].

The spontaneous parametric downconversion of a single photon into two correlated photons, mentioned in Section 7.2, produces an example of a two-photon traveling-wave state.

8.4.3   Ultra-high intensity

Finally, it should be noted that theoretical studies of laser radiation having intensities far above what is of interest in atomic physics and quantum optics predict that the vacuum can no longer be considered empty space [Ait85; Mil94; Leu10; Wea16; Mil17; Mil19]. Under those extreme conditions electromagnetic fields create electron-positron pairs that alter the properties of field increments identifiable as photons. Relativistic quantum theory is needed to describe the physics of such high intensities. This is a regime in which photons, however defined, are present in vast numbers.

8.4.4   Mixed-state fields

Any randomizing influence upon a quantum system will tend to remove the coherence that allows it to be described as a single quantum state (or as a coherent superposition of quantum states). This situation requires description by a density matrix rather than by a wavefunction or a statevector: The system can only be regarded as an incoherent mixture of pure states (see Appendix A.14). The same required alteration of description occurs with radiation when it acquires a random nature. As with other quantum systems, incoherent radiation requires a density-matrix description [Lou73b]; it is an unphased mixture of the differing pure-state photons that occupy primary attention is this Memoir. The radiation produced by spontaneous emission is an example; see Appendix B.2.7.

8.5   Interacting photons

The entities familiarly known as (free-space) photons have no rest mass, and so they are unlike such particles as electrons that can be brought stably to rest. What is commonly termed Particle Physics (see Appendix A.19) deals with a variety of entities that, although possessing rest mass, are unstable and therefore cannot be brought to rest for very long. Positrons, muons, pions and numerous other particles are of this sort. Like photons, they can be created, formed into beams, and destroyed. In particular, such particles scatter from each other. Their scattering properties, and those of stable particles, have long been used to detect them and study their interactions.

8.5.1   Free-space photons

In free space the wavelike solutions of Maxwell's equations describe fields that propagate freely at the vacuum speed of light. Unlike beams of massive particles, radiation beams in free space readily pass through each other undisturbed: Commonly observed radiation beams in vacuum do not affect each other when they intersect. Unlike electrons or other massive particles, everyday photons, regarded as modes of free fields, do not collide or scatter from each other: The observed interference patterns are superpositions of independent fields. However, this lack of photon-photon interaction no longer applies to light beams passing through matter; see Appendix C.2. And intense fields induce polarization of the vacuum that allows photon-photon scattering [Ait85; Mil94; Leu10; Wea16; Mil17; Mil19].

> **Aside: The Schwinger limit** When radiation fields become sufficiently intense it is no longer possible to avoid considering the creation of particle-antiparticle pairs. The weakest of these nonlinearities, the spontaneous generation of electron-positron pairs and consequent *vacuum polarization*, becomes significant at the *Schwinger limit* of electric field magnitude
>
> $$\mathcal{E}^S = \frac{m_e c^2}{e \lambdabar_c} = \frac{\mathcal{E}^{at}}{\alpha^3} \approx 1.32 \times 10^{18} \text{ V/m}, \tag{8.5-1}$$
>
> Regarded as a plane-wave amplitude this correspond to an intensity of $I^S \approx 10^{33}$ W/m^2. This predicted nonlinearity would allow photon-photon scattering.

8.5.2   Photons in matter

Fields that propagate through matter are affected by material response to the radiation; see Appendix C. Amongst other effects, the matter alters the propagation speed, typically slowing (and distorting) pulses. Group velocities of "slow light", little faster than a bicycler or vigorous jogger, have been demonstrated [Hau99; McD00; Fle05; Khu10].

Lenses and other common optical devices owe their usefulness to the effect of matter on fields, typically parametrized by a wavelength-dependent refractive index; see Appendix C.2.1. In such situations, regarded as Linear Optics as presented in basic texts [Hec87; Bor99; Sal19], light rays may follow curved paths and undergo frequency-dependent refraction that separates colors. Light that encounters abrupt discontinuities of refractive index may undergo frequency-dependent reflection and diffraction into multiple beams.

There are two general ways of incorporating quantum properties of radiation into treatments of field propagation through matter.

Scattering photons. A conceptually simple approach is to regard the atoms of a vapor, liquid, or solid as a distribution of discrete scattering centers between which free-space photons travel; see Appendix B.8.3. This is a view in keeping with that of Feynman, see Section 8.3. At each scattering center there will occur elastic or inelastic scattering of photons arriving from previous scattering events. A photon emerging from the material will have undergone a succession of scatterings, with appropriate probabilities for 1, 2, ... distinct scatterings. Although it is convenient to use the word "photon" in describing multiple scatterings, the description of individual scatterings requires only classical fields [Lax51; Gol64; New66; Sho67; Fre18; Kra19].

Bulk-matter photons. Alternatively, it is possible to incorporate a refractive index into quantization of the electromagnetic field [Dru90; Gla91; Hut92; Gar04]. The photons defined in this way move at a frequency-dependent speed. They are not purely disturbances of the electric and magnetic fields; they incorporate a small amount of atomic excitation; see Appendix C.2.1.

As the radiation-induced atomic changes become larger, and interest shifts to short pulses, the field-atom coupling endows the traveling disturbance (in some sense a "photon") with a large component of atomic excitation (described as a continuously distributed polarization field) and a consequently slower group velocity. This is the regime of the quasiparticle *polariton*, see Appendix C.4.2.

Field increments that exhibit such behavior are hardly the free-space photons envisaged historically, but the name "photon" continues to be applied to whatever disturbance travels into matter as a continuation of a free-space quantum field. If it is a photon in free space then it is regarded as a photon when it continues into matter. Accepting that definition, one finds situations in which "photons" have mass, can interact with one another, and can form liquids and even crystals. One reads of "quantum fluids of light" [Car13] and "crystallization of strongly interacting photons ..." [Cha08]. Such organized quantum structures, though discussed in scholarly articles under the rubric of "photon", are very far from the photons of a century ago. They owe their properties to the strong coupling of electromagnetic modes with matter modes as treated in quantum nonlinear optics [Hil09; Pey12; Fir13; Rei13; Dru14; Cha14; Fir16; Mur17].

Photons and nonlinear optics. The topic of Nonlinear Optics [Blo82; Boy08;

Gae06; New11; Dru14] deals with the many situations in which one beam of radiation, through its influence on atoms, affects either itself or another beam that encounters those atoms; see Appendix C.3.6. As the interaction of the fields with matter becomes stronger, either from increased radiation intensity or by increased atomic dipole-moment response in the medium that transmits the radiation, nonlinear optics becomes important.

For example, a field of one frequency can alter the propagation of another frequency or can create new frequencies. The effect of the medium can then be regarded as supporting "photons" that can scatter from each other and that acquire mass. Such entities are therefore quite different from the free-space photons whose corpuscular attributes have been discussed in previous chapters—the photons of Einstein and Dirac and Feynman. They represent a superposition of electromagnetic fields and collective coherent alterations of microscopic matter structure—polaritons that superpose light and matter degrees of freedom.

In another effect of nonlinear optics, electromagnetically-induced transparency (EIT) [Har90; Bol91; Har97; Kuh98; Hau99; Fle05], a strong control field causes a thick absorber to become transparent to a probe field. The needed large transient dipole transition moments are available when one of the two fields is linked to an atomic *Rydberg state* in which a single electron moves at considerable distance from its binding center [Cub05; Gal06; Gal08; Saf10; Pet11; Dud12; Fir16].

The literature on strongly interacting "photons" and related topics of nonlinear quantum optics is already very large and is growing rapidly. I will not attempt to offer even a modest introduction here.

8.6   Doing without photons

In several areas of physics the notion of photons is helpful, but is not essential—it is "sufficient but not necessary", as my college math instructor used to say. When one does introduce a notion of photons, it is advisable not to argue from some preconceived notion of a photon, valid in one regime, to predict expected experimental results in another regime. Better to deduce the particle aspects of radiation from an appropriate application of underlying quantum electrodynamics.

Lasers without photons. Following the discovery of lasers, just at the time I was concluding my graduate studies, the deployment of lasers to physics laboratories revolutionized the experimental possibilities in many branches of physics and chemistry. The underlying principle of laser-induced change is quite simple, though not without challenge to implement in practice. It can be understood from the line of reasoning used by Einstein in identifying stimulated and spontaneous emission as the two ways in which equilibrium populations of discrete-energy quantum states convert their energy of excitation into radiation energy. The basic concept can be presented without the need of photons, although reference to photons simplifies the narrative somewhat: A single atom or molecule that has excitation energy and a transition dipole moment can be induced by an appropriate arriving electromagnetic field (so says the semiclassical theory of quantum behavior) to behave as a miniature antenna acting in phase with the existing radiation. The result will be additional electromagnetic energy in the radiation field, whatever that field may be. The requirements are that the field must have a frequency that satisfies the Bohr resonance condition and that the vector direction of the field must match a possible dipole-transition moment of the atom. When the original field is a traveling wave it can acquire energy, and thereby become amplified, as it passes each emission-capable atom. The resulting radiation is an enlarged copy of the original field. Its characteristics, and the output from a laser, depend on the properties of the individual atoms from which it gains energy, from the distribution of these gain sources in space, and from such optical elements as may confine and direct the field.

With a photon description one says that accompanying atomic de-excitation there will be created a new photon that is identical with those that are already present. They form a coherent multiphoton assemblage. However, such a view, invoking quantization of the field, is not required in order to explain the amplification that becomes a laser.

8.6.1   Semiclassical, photonless fields

As was discussed in Section 3.8, several of the effects that were at first regarded as strong evidence for granularity of electromagnetic fields—for photons—have subsequently been shown to be explainable by treating atoms with quantum mechanics and leaving the electromagnetic field without explicit quantum properties. These models, termed *semiclassical*, are able to explain the observed characteristics of the photoelectric effect and are able, with the mathematics of time-dependent perturbation theory, to provide correct values for the Einstein A and B coefficients that parametrize the interaction of atoms with incoherent light [Ebe68; Cri69; Str70; Sar74; Mil76; Ser78; Ser86; Bar94b; Hel18; Mil19].

Floquet theory.   When one uses the semiclassical approach to the interaction of monochromatic radiation with matter, the fields are treated as periodic forces acting on a multistate quantum system. The resulting time-dependent Schrödinger equation that describes changes to the N-state quantum system comprises N coupled first-order linear ODEs; see Appendix A.9. These can be rewritten as a single ODE of order N, i.e. an equation involving time derivatives as high as order N. Because the fields are taken to be monochromatic, these equations are periodic—the Hamiltonian involves sums of frequency components that appear as coefficients of the time derivatives in the ODE. Such equations with periodic coefficients were studied in the nineteenth century by Gaston Floquet, who proposed a simple change of independent variable that would convert the periodic ODE coefficients into constants, and the N coupled equations with periodic coefficients into an infinite set of linear equations with constant coefficients. This mathematical procedure, producing from the periodic Hamiltonian a *Floquet Hamiltonian*,[6] is now known as *Floquet theory*. The relevance of this theory was first brought to the attention of the atomic, molecular, and optics community in 1965 by Jon Shirley and subsequently used by many others [Shi65; Sal74; Bar77; Ho83; Gue97; Gue03]. In the simplest application, to a single periodic frequency ω, successive blocks of the Floquet Hamiltonian originate with successively higher harmonics $n\omega$. For all but the smallest values of n the Floquet Hamiltonian for classical frequencies of

---

[6]The eigenvalues of the Floquet Hamiltonian are often called *quasienergies*.

atomic response is just what one gets using a fully quantum-mechanical Hamiltonian for the field and number states for the photons, as in Appendix C.6.

Neoclassical theory. During the 1960s and 70s Ed Jaynes and his colleagues examined what became known as *neoclassical theory* [Jay63; Ebe68; Cri69; Str70; Jay73; Gib73; Van74; Cho75; Mil76; Cri96], a semiclassical approach to radiation-matter interactions in which atoms (and matter more generally) obey the Schrödinger equations (time-dependent and time-independent) but radiation is described fully by the classical Maxwell equations. The neoclassical theory showed that some of the effects of radiation-matter interaction that had been thought to require photons could be obtained with classical fields, but in detail the theory failed to fit results of experiments: A correct detailed description requires fields that are described by operators.

8.6.2 Heisenberg equations

Well after I concluded my graduate studies, as I began my career in what became known as quantum optics, I learned that there were two approaches to quantum dynamics, two pictures that dealt with exactly the same underlying physics of quantum-state change.

In the traditional *Schrödinger picture* one deals with such concepts as transitions between quantum states and quantum transitions—*quantum jumps*, an abrupt change in the *information* about a quantum system. One works with probabilities and wave-functions and statevectors, in a manner that brings out notable differences from the world of classical particles, and leads to ongoing ontological disputes.

The alternative *Heisenberg picture* [Dir65; Ack73; Sen73; Ack74; Mil76; Dal82; Sho90; Bar88; Sho93; Luk03; Har06; Lam06; Mil19] uses equations of motion for operators (see Appendix A.9.2), and so it rests upon the same quantum theory, but the equations and their expectation values appear closer to the world of classical mechanics and its dynamical variables. Transitions and quantum jumps are not evident. The fields can, if one chose, be quantized, but the discreteness of photons (through number states) is not clearly evident. The Heisenberg equations, pioneered in quantum optics by Jay Ackerhalt [Ack73; Ack84], provide an alternative (but fully quantum-mechanical) view of radiative phenomena, with the spontaneous emission as radiative reaction [Ack73; Sen73; Ack74; Mil75; Dal82; Ser86; DiP12; Bur14; Mil19]. In this view [Ack73],

> ...it is not the presence of vacuum fields [and photon fluctuations] but of the dipole's own radiation field, the source field, that modifies the atom's characteristics in such a way as to produce a finite decay rate [spontaneous emission] and a [Lamb] shift of the transition frequency.

8.6.3 Denial of photons

In his discussion of Einstein and photons, Pais points out [Pai79] that

> ...the physics community at large received the light-quantum hypothesis [of Einstein] with disbelief and with skepticism bordering on derision. ... The hypothesis seemed paradoxical: Light was known to consist of waves, hence it could not consist of particles.

The presumed particulate nature of radiation, or at least radiation that was in thermo-dynamic equilibrium with enclosure walls, renewed a very old dispute between Newton and Thomas Young and Christian Huygens over the nature of light: Young and Huygens had argued from evidence of interference and nodes that it was a wave, whereas Newton had advocated a particulate view, based on his observation of straight-line rays and a notion of colored corpuscles.

The use of the term "photon" to describe radiation is not without controversy even in recent years. Nobel Prize winner Willis Lamb famously raised objections, stating in a 1995 article titled "Anti-photon" [Lam95] that, although a radiation field must, admittedly, be treated according to the laws of quantum mechanics,

> ...there is no such thing as a photon. Only a comedy of errors and historical accidents led to its popularity among physicists and optical scientists. ...Photons cannot be localized in any meaningful manner, and they do not behave at all like particles. ...Radiation does not consist of particles, and the classical, i.e., non-quantum, limit of [the quantum theory of radiation] is described by Maxwell's equations for the electromagnetic fields, which do not involve particles.

Despite these well-reasoned objections, I have found the word "photon" to be a useful one that shows no signs of disappearing from accounts of research activities or corporate investment. This Memoir is my attempt to provide some justification for the continued value of concepts associated with the term "photon" and to explain my own reasoning, shared by many others. Section 8.8 summarizes contemporary evidence for photons.

8.6.4 Macroscopic quanta, without photons

From an education stressing historical foundations of physics we think of electrons within atoms, or perhaps quarks and their milieu, as being the natural environment where quanta—our photons—are to be found and where quantum theory governs all behavior (apart from gravity). That is the viewpoint presented by particle physicists, who attribute all fundamental interactions (except gravity) to boson fields that obey quantum theory. So it can come as a surprise to find that there are macroscopic systems in which quantum theory provides the needed description, and does so without photons. One such regime is the physics of superconductors: Systems of very cold semiconductors wherein resistanceless electric currents and voltages are the basic dynamical variables whose commutation relations install the character of quantum theory to produce qubits of *quantum circuits* [Gir09; Wen17]. But the field quanta that occur in this context are not elementary increments of free-space electromagnetism, they are collective effects attributable to organized assemblies of electrons and their guiding lattice. As with the electromagnetic field, the imposition of commutators on macroscopic canonical variables of a Lagrangian (see Appendix A.1.4) brings possibilities for discrete number-state excitations and behavior, not limited to either photons or elementary particles.

8.7   Alternatives to photons

Although the concept of photons has demonstrably provided a useful way of under-standing various observations, there has remained interest in determining whether photons are not only *sufficient* for explaining measurements but whether they are also *necessary* [Ste73; Gar08; Buo16]. With Heisenberg equations there need be no reliance on photon numbers. Is it possible to find alternative models that will reproduce the established results without requiring that radiation be intrinsically granular? In such a search it is well to keep in mind the advice from Sherlock Holmes:[7]

One should always look for a possible alternative, and provide against it.

Altering the physics.   Some past non-photon theories required refinement of basic electromagnetic theory, or of the nature of the light-matter interaction. Very early in the discussions of quantum theory a proposal by Bohr, Hendrik Anton Kramers, and John Clark Slater [Boh24; Pai79; Bia06] went so far as to propose abandoning the principles of conservation of energy and momentum, except on average. Subsequent experimental evidence ruled out this proposal. They also considered spontaneous emission to be induced by a virtual field that contains all possible frequencies.

A model known as *random* or *stochastic electrodynamics* developed by Tim Boyer [Boy69; Boy75; Boy11; Boy18; Boy19] briefly offered a modification of the general radiation laws.

In those cases where the alternatives to conventional quantum theory and quantum electrodynamics have been tested the alternatives have not prevailed. And without definitive testable predictions there is little incentive for following a new course.

8.8   Contemporary evidence for photons

The customary argument for the correctness of the quantum version of electromagnetic radiation is the remarkable agreement, to many decimal places, between QED calculations and measurements of the bound-state energies of electrons. These are typically interpreted with the use of Feynman photons, see Section 8.3.

Several lines of experiment, not available until lasers became laboratory staples, produce results that are widely regarded as demonstrating quantum characteristics of electromagnetic radiation and therefore requiring discrete radiation quanta—photons of some sort [Ray90].

In one, presented in discussions of the Jaynes-Cummings model (see Appendix C.6), patterns of temporal modulation are attributed to superpositions of discrete frequencies that are associated with discrete photon numbers.

In another, various experiments that make use of beamsplitters (see Appendix C.3.5) demonstrate that a photon cannot be further divided (a good indication of an elementary particle) and that identical photons exhibit correlations not found with classical radiation (see Appendix B.5.3).

The Hamiltonian for n photons differs insignificantly from that of $n \pm 1$ photons when n is 10^3 or more (as in the Floquet Hamiltonian of Section 8.6.1), whereas there are immediately measurable differences when $n = 1$. Notably, it is not possible to remove a photon from the vacuum state, $n = 0$.

---

[7] As recorded by his biographer John Watson in *The adventure of Black Peter.*

These effects go beyond the notion of radiation quanta that were being considered in the early decades of the twentieth century, beyond the simple notions of photons that were associated with the names of Planck, Einstein, and Bohr in Chapter 3 and which were part of my own limited physics education prior to graduate school. Nowadays the indisputable experimental evidence for photons is accessible for undergraduates at well-equipped physics departments, as noted next.

Photons for undergraduates. The questions of contemporary quantum optics have shifted away from whether photons exist—whether radiation must be regarded as comprising discrete elementary units. Instead, the questions center on particular radiation sources and whether they demonstrate distinct quantum properties. These issues, key to the advancement of quantum technology and quantum engineering, involve use of various optical and electronic tools, most notably beamsplitters and correlation measurements.

Although the original discoveries and proofs of principle that have defined quantum optics have traditionally taken place at major research institutions, the remarkable advances in commercialization of optics technology, particularly of sources and detectors as well as signal-processing software, have made formerly difficult experiments now accessible to college physics-laboratory facilities. Pages of *The American Journal of Physics* regularly present detailed descriptions of physics-lab experiments in which undergraduates can carry out their own demonstrations of single-photon creation and manipulation, dealing not just with idealized notions and mathematical abstractions such as I am presenting but with all the challenges of real apparatus [Tho04; Gal05; Dim08; Pea10; Gal14; Sch18]. I experience envy as I read what has replaced the simple inclined planes and rolling balls or lens-and-mirror arrays that remain my memory of my own weekly undergraduate physics-lab experiences.

8.9   Overlooked photons

Contemporary media use the label "photonics" to appear *au courant* with ongoing changes of technology, without requiring that there be any underlying quantification of the radiation (to the annoyance of anti-photon Willis Lamb, see page 178). Thus the word appears in reference to health care, crop management, home security and many other subjects of practical interest. The Memoir format of this book has allowed me to be somewhat selective in choosing topics to discuss. Amongst the topics omitted are the following.

Photons in industry.   Nowadays few readers of this book will be unaware of how ubiquitous are laser photons in manufacturing and construction. They drill holes and weld joints, of objects large or miniscule. Engineers of all disciplines routinely measure distances and altitudes with the aid of photon crowds, to accuracies that were impossible before lasers became tools. Every possible attribute of photons—their energy content, their directed momentum, their colors, their pulses—all find application in commerce.

I have primarily observed that world of photons as a consumer and a spectator. My more direct involvement has been through academic enquiries into the basic physics and mathematics of electromagnetic radiation, and it is that background that frames this Memoir. For presentation of industrial and engineering uses, and details of device construction, readers must look elsewhere, e.g. under the title of Photonics [Sal19].

Photons in biology.   Notably missing from this Memoir has been any mention of biological systems—of organic molecules and cells and living things. The omission is not because these are not important but simply because they have not been part of my own studies and research. To the extent that the interesting aspects of biological systems cannot be regarded as mere atoms and photons, thereby fitting into my narrative, it is necessary to bring entirely new views of photons: How they are generated in bioluminescence (fireflies, for example), their physiological effects (vision, for example [Hub95; Rie98; Bru09; Art14; Nel17]), and the great difficulty of reliable measurements and sampling (a theory to itself). Ongoing advancement of device technology brings opportunities marketed as healthcare photonics.

Although this may well be the aspect of photons that is most relevant to all of us, lack of space (and familiarity) forces me to rely on merely referencing a portion of the literature of biophotonics [Pra03b; Bel07; Pav08b; Kei16].

9

Finale

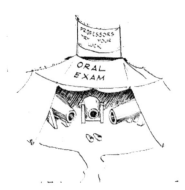

The oral exam can be stressful.

Having devoted scores of pages to offering explanations for what a photon has been regarded, it is appropriate for me to address a natural follow-on that has come from friends who have heard portions of this story: So what? What does it matter what a photon might be?

It is notoriously difficult to answer questions of this sort, calling for justification of research into topics that seem to offer little economic return, little improvement of social conditions. The mathematician Godfrey H. Hardy acknowledged satisfaction that his lengthy study of the distribution of prime numbers[1] had no practical use whatever [Har92]. But in recent years prime numbers have been a vital part of commercial cryptography: The difficulty of finding two prime factors of a huge number has offered some assurance of security for encryption methods that are routinely used.

This sort of encryption has led to interest in constructing ever more powerful computers for use in decrypting intercepted encoded messages. Basic principles of quantum theory, long regarded as an esoteric activity for philosophers to ponder, have become regarded as a foundation for quantum computers that could far exceed the digital computers of today in solving certain sorts of mathematical challenges that are very difficult, very time consuming, when addressed with contemporary computers. Computers using qubits (and their continuum of defining parameters $\mathbb{R}$) rather than traditional *bits* (with discrete values $\mathbb{Q}$) offer possibilities for implementing powerful data-base searching and code-breaking prime-number factoring that have interest for corporations and governments. Photons (or at least quantum fields), with their ability to transport quantum *information* (but not any amount of matter), play an important part in designs for quantum computing [Ste98; DiV00; Haf08; Liu10; Bar15].

The quantum properties of photons also offer possibilities for sending secure messages. Because photons cannot be copied without error (the no-cloning theorem), any eavesdropper who attempts to intercept a message that is encoded in single photons

---

[1]The distribution of prime numbers is associated with the Riemann zeta function and one of the great unsolved problems of mathematics [Der04].

Our Changing Views of Photons: A Tutorial Memoir. Bruce W. Shore, Oxford University Press (2020). © Bruce W. Shore.
DOI: 10.1093/oso/9780198862857.003.0009

must inevitably reveal the interception. The interception, though not entirely preventable, can become known to the message sender, a first step toward making a secure encrypted communication [Eke91; Kuh02; Zol05; Kim08; Dua10].

It takes little imagination to anticipate that the ongoing development of experimental techniques for manipulating the radiation-induced changes of quantum systems will continue to produce devices that have commercial and societal uses that can presently only be guessed. The crystal ball of anticipation remains a cloudy glass into which we see only imperfectly.

9.1   A concluding thought

Around us are many individuals and communities—scientists, engineers, inventors, entrepreneurs, and many others—for whom the word "photon" has meaning in their activities. They calculate, they measure, they devise and use. It should hardly be surprising that the commonly used word photon has, like many other nouns used by diverse associations, a variety of interpretations. In this Memoir I have noted many of these meanings and uses. Each finds reasoned support. Throughout this exposition I have suggested that this diversity of views offers useful opportunities—for personal intellectual satisfaction and as a basis for technological advancement.

Acceptance of such diversity can be, as a Rotary Club principle says,[2] *"beneficial to all concerned"*.

9.2   Basic references

For discussions of the basic physics discussed here, with careful explanations of the origin of equations and models that were part of my academic training, the published lectures of Feynman are an excellent, and unique, resource [Fey63].

The literature on the quantum theory of radiation, and on photons, is too vast to be presented here in any detail. In addition to what is written in the many textbooks on Quantum Theory there are books and monographs that deal specifically with the quantum theory of radiation [Hei54; Bia75; Ber82; Coh97; Lou00]. Numerous texts on Quantum Optics treat the coupled effects of field-induced atom excitation and atom-induced field changes [Ste73; Ebe87; Kni89; Scu97; Yam99; Pur01; Ved01; Sch01; Bar02; Ger04; Pau04; Lam06; Vog06; Gar08; Wal07; Mey07; Ber11; Ken11; Har13b; Mil19]. Much of this work takes a semiclassical approach: It treats the atoms with quantum theory and the field as classical. Nevertheless, the realm of coherent excitation has relevance to photon creation and detection. The work of Dirac [Dir27; Dir58] and Fermi [Fer32] remains a foundation for descriptions of quantized radiation.

The nature of photons.   The article by Pais presents a detailed history of the early introduction of radiation quanta and photons [Pai79]. Numerous journal articles offer insights into the nature of photons [Scu72; Lam95; Mar96; Coh98; Lig02; Mut03; Bia06c; Fri09] including discussions of photon wavefunctions [Coo82; Sip95; Bia96; Kel05; Mut05; Bia06c; Smi07; Haw07; Sal11; Bar12; Bia13; Bia17].

---

[2]This is the final item of The Four-Way Test created by Herbert J. Taylor in 1932 and now used as an ethical guide for Rotarians internationally; see the web page http://thefourwaytest.com/history-of-the-four-way-test/.

Cavity photons. The physics of atoms and photons in an enclosure, known as cavity quantum electrodynamics (CQED) is particularly relevant to notions of monochromatic photons [San83; Wal88; Har89; Kim98; Vah03; Mil05; Mes06; Har06; Har13].

Traveling field modes. Whereas the literature of the twentieth century dealt primarily with simple beams of radiation, more recent experimental work, supported by theoretical advances, present a panorama of traveling electromagnetic field constructions that have elaborate node structure [Ber09; Yao11; Aie15; Lev16; Rub17], even tangled knots of field that require topological characterization [Ber01; Lea05; Irv08; Den10; Pad11; Dal12; Ked13; Hoy15; Arr17; Bia18; Cam18; Sug18; Cos19].

Single photons. The creation of single photons is mentioned by many authors. Reviews include [Oxb05; Lou05; Bar09b; Sch09; Bul10; Kuh10; Vas10; Eis11; Buc12; Sin19].

Atomic and molecular structure. The properties of the atomic and molecular quantum states between which radiative transitions occur are discussed in numerous texts on atomic structure [Con53; Bet57; Sla60; Hin67; Sob72; Cow81; Sob92; Bra03; Fis06; Dra06; Gra06; Dem10] and molecules [Her50; Her50b; Jud75; Bra03; Dem10].

Atomic changes. The Einstein rate equations of Appendix A.16 are commonly used in astrophysics [All62; Tuc75; Ryb85; Sal19]. Rabi oscillations as the behavior of a two-state quantum system were early recognized, although not with that name [Maj32; Rab36; Rab37; Sch37]. A definitive reference for two-state excitation by coherent radiation remains the monograph of Allen and Eberly [All87]. The model of a two-state atom interacting with a single-mode quantum field, whose analytic solutions were found by Jaynes and Cummings [Jay63], has been extensively discussed; reviews include [Sho93; Sho05; Lar07; Sho07; Gre13; Gro13]. The widely used Bloch equation of Appendix A.14.2 is discussed in [Fey57; All87; Sho90; Sho11].

Reviews of coherent excitation. I have discussed the underlying physics of the coherent excitation of quantum systems in a number of books and review articles [Sho90; Sho95; Sho08; Sho11; Sho13; Sho13b; Sho17].

9.3 Acknowledgments

The insights presented in this book have come from years of enlightening discussions (pun intended) with colleagues, encouragement of friends, and support from various organizations and individuals.

My father, Walter Shore, gave me my first understanding of scientific enquiry, and my mother Madeline Shore steadily encouraged my drawing as well as my academic pursuits.

From Paul Sunderland at East Bakersfield High School I first encountered the fascinating world of atoms and their electronic orbitals.

From Emerson Cobb at College of the Pacific I learned what chemistry was really about, and was encouraged to pursue a graduate degree.

From my thesis advisor Jack Irvine at MIT I learned the importance of careful technical writing.

I was cheered by Avery Ashdown, retired chemistry professor, whose benevolent oversight of the MIT Graduate House made it truly "A Gentlemen's Club and a Home away from Home", as he often said.

I was particularly inspired by the lecture courses of Victor Weisskopf at MIT and Julian Schwinger at Harvard, two very different styles (neither of them knew me).

From Donald Menzel at the Harvard College Observatory I had my start toward an academic career and learned the importance of avoiding the passive voice in the technical writing of my first book.

From Reg Garton at Imperial College, London and Carl Moser at CECAM in Paris I learned to appreciate European life and scientific endeavor.

It was my good fortune in the Laser Program of what was then the Lawrence Livermore Laboratory to be encouraged by program leaders Ben Snavely and Jim Davis to pursue applications of the time-dependent Schrödinger equation beyond the elementary (two-level semiclassical rate-equation) models that were being employed as lasers became routine research tools.

From collaborations with Joe Eberly at the University of Rochester I learned much about the connection between the time dependent Schrödinger equation and the behavior of model atomic systems. His insightful advice and lively debate on the present Memoir has been much appreciated.

From Peter Knight at Imperial College, London, I learned still more about the world of quantum optics (and English life) and was encouraged to undertake my second, two-volume, too massive, book.

With Klaas Bergmann at the Technical University of Kaiserslautern I had many years of collaborative exploration of his experimental technique of stimulated-Raman adiabatic passage.

From the Alexander von Humboldt Foundation I had support for many trips to participate in German research on coherent atomic excitation.

With Herbert Walther at the Max Planck Institute for Quantum Optics I have enjoyed light-hearted discussions (pun intended).

With his remarkably powerful DOS-based app "PS" Murray Sargent III introduced me to the joys (and frustrations) of computer typesetting.

With Nikolaly Vitanov at Sofia University I have enjoyed co-authoring technical articles and have gained an appreciation for Bulgarian life and history.

Most recently Thomas Halfmann has regularly brought me into active participation in his research group at the Technical University of Darmstadt. He very kindly edited a Festschrift collection of articles honoring my 70th birthday.

Markus and Tina Roth have graciously hosted some of those visits, and Markus has set an example of broad scientific interest.

I have greatly enjoyed working with many others to create peer-reviewed journal articles relevant to the present book: Mike Johnson, Jim Morris, and Dick Cook at the Lawrence Livermore National Laboratory; Jay Ackerhalt at Los Alamos National Laboratory; Iwo Bialynicki-Birula of Warsaw University; Wolfgang Schleich at University of Ulm; Leonid Yatsenko of the Institute of Physics, Ukrainian Academy of Sciences; and colleagues at the Technical University of Kaiserslautern: Razmik Unanyan and Michael Fleischhauer.

The Macintosh operating systems of Apple have long underlain my writing activities, beginning with the famous little Mac-plus box that first brought black-on-white display of recognizable fonts and symbols to this pastime, now extended with the TeX typesetting language of Donald Knuth.

I am grateful to Sonke Adlung of Oxford University Press for seeing merit in my draft of this Memoir. In its writing I have appreciated the encouragement and helpful corrections of Stephen Barnett, Peter Milonni, and Uli Gaubatz.

I am beholden to son Tim for numerous wide-ranging insightful discussions, to daughter Hilary for her enthusiastic reading of the start of this st Memoir, to sister Betty for encouraging me to write technical essays, and to brother Dick for demonstrating with his own writing how lively even technical topics can be.

Finally, I appreciate the patience of my wife Randi in encouraging me to spend time with my writing, and for listening.

The appendices

The following three appendices provide mathematical underpinnings for the various concepts that the main text presents primarily with words. In them I explain the origins of needed equations and discuss their use, aiming the presentation to be accessible to students of atomic, optical, and quantum physics as well as to working physicists who are turning to a new subject. Proofs are left for other resources. Although these appendices present many equations, the underlying mathematics is of limited variety— as mathematics today goes. What is required for understanding photons, and quantum physics more generally, includes the familiar version of arithmetic as well as functions and their derivatives—linear differential equations of first and second order—and basic linear algebra, all of which was understood in the nineteenth century. Still more— Hilbert space and matrices—became part of mathematical physics a century ago.

Appendix A presents the basics of quantum theory as it is applied to bound particles—to electrons in atoms and atoms in molecules, as well as in structures such as artificial atoms and any other system whose behavior involves discrete quantum states. In so doing it relies on a Hilbert-space description, presented after an elementary introduction to abstract vector spaces.

Appendix B presents the free-space Maxwell equations of electromagnetism and the spatial Helmholtz equations for modes It introduces the quantum theory of radiation for the photons of Dirac—free-space field modes that become photons.

Appendix C includes bulk matter in the Maxwell equations. Here we find the refractive index that underlies classical optical elements of mirrors and lenses. In the end the atoms and the radiation are treated as a single, coupled system, the radiation dynamically altering the atom structure and the atoms affecting the field.

Symbol typography. I generally follow the default typesetting rule of TEX that Latin letters appearing in equations, and therefore denoting mathematical variables, are italicized: $a, b, c \ldots i, j, k \ldots x, y, z$. There are no hard and fast rules about what symbols one should choose for the variables one uses, and so different authors, or even single authors at different times, choose different symbols for particular quantities[3]. The letter c is commonly used for the speed of light in vacuum. The letter E is commonly used for energy, but it also appears as a measure of electric-field strength. I use t for time and x, y, z for Cartesian coordinates, but these latter symbols also appear with other meanings. In these appendices, where equations abound, I use typographic options available in the LATEX typesetting software, such as boldface (for vectors, $\mathbf{F}, \mathbf{\Psi}$), sans serif (for matrices, H, W), and calligraphic ($\mathcal{E}, \mathcal{H}$) to assign different interpretations for individual Latin and Greek letters. Subscripts ($E_1, E_2, F_x, F_y, \boldsymbol{\psi}_n, W_{nm}$) identify particular elements of a set of numbers (organized as vectors and matrices); see Appendix A.3. Superscripts ($E^{\mathrm{kin}}, E^{\mathrm{pot}}, \mathsf{H}^{E1}, \mathsf{H}^{M1}$) generally indicate classes of objects. I have used an unconventional typography m for masses, to differ from the m used for magnetic quantum numbers. The Dirac notation $|A\rangle$ often appears in place of $\mathbf{A}$ for a vector in an abstract space; see eqn (A.3-5).

---

[3] Unfortunately it is not possible to avoid conflicting definitions of α ans γ.

Appendix A
Atoms and their mathematics

*Experimental apparatus can be mastered
...I was told.*

In preparation for discussing equations that describe controlled changes in behavior of electrons in atoms, and other quantum system, this first Appendix begins, in A.1, with a summary of three traditional formalisms for creating equations that describe temporal changes to systems of particles—*equations of motion*—having names associated with Newton, Lagrange, and Hamilton. These versions of classical mechanics are the foundation for quantum mechanics and statistical mechanics.

Quantum theory, and its photons, is generally thought of as a description of very small objects (but see Section 6.3). Appendix A.2 defines several of the length parameters commonly used to give size to atoms, electrons, and photons.

Appendix A.3 extends the concepts of vector found in Section 2.5.2 and Appendix A.1.1 to introduce notions of abstract vector spaces, and matrix operators acting within them, and discusses their use for describing discrete-state quantum systems (as contrasted with quantum scattering theory that deals with continua of energy values [Gol64; New66; Rod67; Sho67; Tay72; Wat07; Mis10; Fre18]). This vector-space formalism provides a very simple and direct approach to the mathematics of quantum theory and the interpretation of its results.

Appendix A.4 presents the conventional method for converting classical mechanics into quantum mechanics by way of noncommuting operators whose representation leads either to wave mechanics and wavefunctions (Appendix A.5) or to matrix mechanics (Appendix A.7) and statevectors (Appendix A.8). Building upon this foundation, Appendix A.9 presents the equation of motion for statevectors—the time-dependent Schrödinger equation that underlies all nonrelativistic quantum-state changes. For that use it presents the definition of the Hamiltonian commonly used in quantum optics. Appendix A.10 applies this formalism to the basic model of a two-state quantum system. For steady illumination the solutions oscillate indefinitely (Rabi oscillations, Appendix A.10.2), a marked contrast with the exponential approach to steady state that occurs with rate equations that embody the physics (and photons) of Einstein that were used prior to the availability of laser light sources, as discussed in Appendix A.16. Other parts of this Appendix discuss various foundational theoretical tools of quantum optics, including density matrices, adiabatic change, linkage patterns and pulsed manipulation of three-state systems. Although the atoms are here treated quantum mechanically, radiation is treated (classically) as specifiable fields. This is the *semiclassical* theory of atomic excitation.

This Appendix closes, in A.18, with another dose of mathematics: A summary of group theory, giving particular attention to the Lie groups (and their algebras) that are used to label representations in the Standard Model of particle physics (summarized in Appendix A.19), aspects of gauge fields used in defining photons in the context of high-energy physics. Appendix A.18.5 notes Noether's theorem that connects Lie-group symmetries with conservation laws for such constants of motion as energy, momentum, and angular momentum.

A.1   Classical equations of particle motion

Descriptions of classical (i.e. non-quantum) behavior of particles or mechanical objects typically take the form of ordinary differential equations (ODEs) whose solutions provide values for positions, orientations and velocities of the various parts of a mechanical system. These equations, part of the study of Dynamics (motions) or Mechanics (equilibria), appear in various forms. In their simplest, as the equations attributed to Newton, they deal with relationships of three-dimensional vectors of force, position, and (nonrelativistic) velocity.

A.1.1   Physical, Euclidean vectors

The vectors $\mathbf{V}$ that we high-school physics students dealt with—forces and positions—could be quantified by their components V_x, V_y, V_z along three perpendicular Cartesian coordinates. Those were real-valued numbers whose squared components gave the magnitude of the vector as

$$|\mathbf{V}| = \sqrt{V_x^2 + V_y^2 + V_z^2}. \tag{A.1-1}$$

Each of the the three vector components V_i is a positive or negative real-valued number, a *scalar*. When one of the components of $\mathbf{V}$ is absent, or set to zero, the vector lies in a two-dimensional plane perpendicular to the absent component.

Such objects are often termed vectors in *Euclidean space* to distinguish them from the four-dimensional *Minkowski space* of special relativity or the curvilinear Einstein space-time coordinates of general relativity. All such vectors, available for describing positions and velocities in physical space, differ from the most general examples of abstract vectors to be discussed in Appendix A.3, by having lengths—that is, by having a definite relationship amongst the several vector components, as in eqn (A.1-1) for Euclidean vectors: They are examples of vectors in *metric spaces* (see Appendix A.3.2). The length of the Cartesian vector $\mathbf{V}$ above does not depend on whether the coordinates are fixed as north and south, east and west, up and down, or as somehow aligned with instantaneous values of wind direction—as suits an airplane pilot or sailor.

Vectors have not only magnitude (length) but direction. The definition of direction necessarily involves notions of angles, and trigonometry, and is dealt with most simply for two-dimensional vectors, where a single angle θ suffices to define the direction. Each additional dimension requires an additional angle, a periodic real variable. Three-dimensional objects of finite, nonspherical size require additional angles to specify their orientation.

Essential trigonometry.   Almost the only portion of a required semester-long college course in trigonometry that has proven of continuing use to me has been the definition of the trigonometric functions for a right triangle of sides x, y, and hypotenuse r:

$$\sin\theta = \frac{y}{r}, \qquad \cos\theta = \frac{x}{r}, \qquad \tan\theta = \frac{y}{x} = \frac{\sin\theta}{\cos\theta}. \tag{A.1-2a}$$

Applied to line drawings on a page or map these provide the two Cartesian components of a radius vector,

$$\begin{aligned} \text{horizontal:} & \quad x = r\cos\theta, & \text{vertical:} & \quad y = r\sin\theta. \\ \text{(east)} & & \text{(north)} \end{aligned} \tag{A.1-2b}$$

Although these defining formulas pertained to the angle θ of a triangle, they appear with application to steady rotations and angles $\theta = \omega t$ with the use of Euler's formula for relating complex-argument exponentials to trigonometric functions, eqn (A.3-2).

Unit vectors.   It is often useful to distinguish the magnitude $|\mathbf{V}|$ of an Euclidean vector $\mathbf{V}$ from its direction. I shall sometimes use the notation $\check{\mathbf{V}}$ for this direction measure,[1] an example of a three-dimensional *unit vector*—a vector whose length is unity:

$$\check{\mathbf{V}} \equiv \frac{\mathbf{V}}{|\mathbf{V}|} \qquad \text{so} \qquad |\check{\mathbf{V}}| = 1. \tag{A.1-3}$$

Vector pairs and vector multiplication.   Only later did I learn the useful typographic shorthands for describing geometric properties of pairs of such physical vectors as force and position. Two abbreviations, part of the vector calculus developed by J. Willard Gibbs, are commonly used for combining pairs $\mathbf{A}$ and $\mathbf{B}$ of three-dimensional real-valued vectors: The *dot* (or scalar or inner) product that produces a scalar, $\mathbf{A} \cdot \mathbf{B} = C$, and the *cross* (or vector or outer) product, $\mathbf{A} \times \mathbf{B} = \mathbf{C}$, that produces a vector. Expressed in Cartesian coordinates x, y, x, labeled by $1, 2, 3$, the products are defined by the constructions for the scalar

$$C = A_1 B_1 + A_2 B_2 + A_3 B_3, \tag{A.1-4a}$$

and the vector components

$$C_1 = A_2 B_3 - A_3 B_2, \quad C_2 = A_3 B_1 - A_1 B_3, \quad C_3 = A_1 B_2 - A_2 B_1. \tag{A.1-4b}$$

The *magnitude* of a vector $\mathbf{V}$ is the square root of the dot product of the vector with itself,

---

[1] I avoid the common hat notation $\hat{\mathbf{V}}$ to prevent confusion with the notation for operators.

$$|\mathbf{V}| \equiv V = \sqrt{\mathbf{V} \cdot \mathbf{V}}. \tag{A.1-5}$$

The dot product quantifies the projection of one vector on another: The angle θ between two vectors is obtainable from the formula

$$\mathbf{A} \cdot \mathbf{B} = AB \cos\theta. \tag{A.1-6a}$$

The vectors are *parallel* if $\theta = 0$ and are *antiparallel* if $\theta = \pi$. Two vectors are *orthogonal* (perpendicular to each other) if their scalar product vanishes, as occurs if $\theta = \pi/2$.

The *cross product* of two vectors, such as occurs in the definition of angular momentum or torque, is a vector perpendicular to each of them. Two vectors $\mathbf{A}$ and $\mathbf{B}$ that are not parallel (or antiparallel) define a two-dimensional plane. In turn, the orientation of this plane can be defined by a direction—a vector perpendicular to the plane. This vector is the cross product $\mathbf{A} \times \mathbf{B}$. The magnitude of the cross product can be expressed in terms of individual magnitudes and an angle,

$$\mathbf{A} \times \mathbf{B} = -\mathbf{B} \times \mathbf{A}, \qquad |\mathbf{A} \times \mathbf{B}| = AB \sin\theta. \tag{A.1-6b}$$

The direction of the cross product is determined by the right-hand rule, the arrangement of two fingers (for $\mathbf{A}$ and $\mathbf{B}$) and thumb (for $\mathbf{C}$) of a right hand. Two vectors that are parallel (or antiparallel) have a vanishing cross product,

$$\mathbf{A} \times \mathbf{B} = 0 \quad \text{when} \quad |\mathbf{A} \cdot \mathbf{B}| = AB. \tag{A.1-7}$$

Euclidean vectors such as position and velocity are known as *polar vectors*. Vectors which are defined by a vector product of polar vectors, as are torques and angular momentum, are known as *axial vectors*. (The electric field, with its origins in charges and charge moments, is a polar vector. The magnetic field, with its origins in circulating currents, is an axial vector.) Axial vectors are also known as *pseudo-vectors* because under coordinate reflection they change sign.

These basic relationships, taken with the spatial derivative ∇ as a possible vector [see eqn (B.1-3)], formed the essentials of the mathematics (vector calculus) that I needed for dealing with the equations of physics at the start of my professional training. They are the starting point for more elaborate examples of algebras discussed in Appendix A.17.

Essential calculus. Only two concepts from calculus underly the basic equations of particles and fields to be discussed here. One is the *derivative* of a function, say, $(d/dx)F(x)$ or $dF(x)/dx$ for function $F(x)$ of variable x. This quantifies the slope of the curve $F(x)$ at point x. The rate of change in a slope is quantified by a second derivative, $d^2 F(x)/dx^2$. When there are several independent variables, say $F(x,y)$, the variation with one of them is a *partial derivative*, $\partial F(x,y)/\partial x$.

The second needed calculus concept is the *integral*, $\int dx\, F(x)$. This sum of increments $dx F(x)$ quantifies the area under the curve $F(x)$.

A.1.2   Newton's equations

It is the motion of charged particles that creates the electromagnetic fields of radiation, and hence discussions of photons soon deal with the dynamics of their source particles.

A purpose of classical dynamics is to predict how forces change systems of particles and mechanical parts. The predictions come through examining results of equations of motion for system parts. The simplest examples are for systems of separate point-like particles having mass but no shape.

Trajectory. The three-dimensional time-varying particle position—the particle *trajectory* or *orbit* (a curve in $\mathbb{R}^3$)—is obtained, for given initial values at time $t = 0$, by integration of the particle velocity $\mathbf{v}(t)$—the time derivative of the $\mathbb{R}^3$ position vector $\mathbf{r}(t)$,

$$\frac{d}{dt}\mathbf{r}(t) = \mathbf{v}(t), \qquad \mathbf{r}(t) = \int_0^t dt'\mathbf{v}(t'), \qquad \text{(A.1-8a)}$$

which in turn is obtained by integration of the acceleration $\mathbf{a}(t)$,

$$\frac{d}{dt}\mathbf{v}(t) = \mathbf{a}(t), \qquad \mathbf{v}(t) = \int_0^t dt'\,\mathbf{a}(t'). \qquad \text{(A.1-8b)}$$

The equations of motion typically appear as (ordinary) *differential equations* (on the left, here of first order), rather than as integrals, but both forms have their uses.

Force. Equations of motion relate observable trajectories to applied forces. The basic equation of motion attributed to Newton relates the changing momentum vector $\mathbf{p}$, or the acceleration vector $\mathbf{a}$ of a (slowly moving) particle, to the force vector $\mathbf{F}$,

$$\mathbf{F} = \frac{d}{dt}\,m\mathbf{v} \quad \text{or, if } v \ll c, \quad \mathbf{F} = m\,\mathbf{a}. \qquad \text{(A.1-9)}$$

The proportionality of $\mathbf{F}$ to $\mathbf{a}$ defines the particle mass m. The acceleration produced by a given force is smaller for larger masses: They have larger *inertia*.

Relativistic mass. According to the special theory of relativity, a particle that has mass m_0 when seen at rest will appear, when seen moving steadily with speed v, to have an inertia that can be attributed to [Adl87; Oku89; Bai07] a (transverse) *relativistic mass* γm_0 perpendicular to $\mathbf{v}$ and a longitudinal relativistic mass $\gamma^3 m_0$ along $\mathbf{v}$, where γ is the *Lorentz factor*

$$\gamma = \frac{1}{\sqrt{1 - (v/c)^2}}. \qquad \text{(A.1-10)}$$

As the particle speed approaches the speed of light c the force required to accelerate it further (its apparent inertia or mass) becomes larger without bound.

Lorentz force. The interaction between radiation and matter builds upon the *Lorentz force* [Jac75] between a moving particle, of charge e localized at $\mathbf{r}$, and given electric and magnetic fields, $\mathbf{E}(\mathbf{r},t)$ and $\mathbf{B}(\mathbf{r},t)$ respectively:

$$\mathbf{F}(\mathbf{r},t) = e\big[\mathbf{E}(\mathbf{r},t) + \mathbf{v}\times\mathbf{B}(\mathbf{r},t)\big]. \qquad \text{(A.1-11)}$$

The vector multiplication associated with the symbol $\times$ is defined by Cartesian components, as in eqn (A.1-4b); The Lorentz force at a given position and time has the three vector components

$$F_x = e(E_x + v_y B_z - v_z B_y), \qquad \text{(A.1-12a)}$$
$$F_y = e(E_y + v_z B_x - v_x B_z), \qquad \text{(A.1-12b)}$$
$$F_z = e(E_z + v_x B_y - v_y B_x). \qquad \text{(A.1-12c)}$$

Whereas the electric field can act to accelerate the particle along its direction of motion, thereby adding kinetic energy, the magnetic field always acts to turn the trajectory without altering the kinetic energy.

In traditional formulations of electromagnetic theory bulk matter is typically regarded as an averaged assemblage of individual point charges, each acted upon by a local Lorentz force. The fields, in turn, arise from distributed charges and currents, in accord with Maxwell's equations, see Appendix B.1.1.

SI Units. With the now standard SI units (acronymed as SI from the French *Système international d'unités*) that have replaced the cgs and MKSA metric systems I learned, energy is expressed in joules (J), force in newtons (N), electric charge in coulombs (C), electric current in amperes ($A = C/s$), the E field in volts per meter (V/m), and the B field is in teslas ($T = 10^4$ gauss). The following relationships hold:

$$1\,J = 1\,N\,m = 1\,W\,s = 1\,C\,V = 1\,C\,T\,m/s. \qquad \text{(A.1-13)}$$

Degrees of freedom. The notion of a single point particle subject to a force provides the simplest example of a situation that allows straightforward construction of a mathematical trajectory. As we consider inclusion of additional point particles into the model we add additional variables – additional *degrees of freedom* (DoF)—whose behavior is to be predicted by equations of motion. In the world of classical dynamics each point particle requires specification of three position descriptors and, for each of these, a three-dimensional velocity. Fields, and their photons, also have degrees of freedom; see Appendix B.

An essential difference between a classical system and a quantum system is that the information about the latter, as embodied in the quantum observables, is subject to an intrinsic constraint expressed by the Heisenberg *uncertainty principle*: It is not possible to obtain arbitrarily accurate simultaneous values for positions and velocities of a particle; see Appendix A.4.2. This means that whereas a classical system of N unconstrained point particles requires $3N$ Cartesian coordinates and $3N$ corresponding velocities for complete characterization, an unconstrained quantum system of N particles requires only $3N$ variables. For structured particles these are most commonly chosen as center-of-mass positions and spatial orientations.

Initial conditions; Reversibility. Equations of motion provide only *possibilities* for behavior. Particular cases obtain from specifying *initial conditions*: Values of all the positions and velocities at some initial time, after which the behavior is to be followed. In some situations the requirement is that a trajectory pass through specified initial and final values—a boundary-value problem. It is only the combination of equations and conditions that lead to particular examples that we can compare with observation: Particular apples on trees, particular planetary systems, particular distributions of molecules into structures.

It is often noted that, apart from the impracticality of knowing positions and velocities for a vast number of particles, the Heisenberg uncertainty principle prevents any possibility of having exact simultaneous values of paired position and velocity variables. Nonetheless, it can be useful to overlook this limitation and examine the properties of the equations of motion.

In the absence of friction, the equations of Newton, and those of Lagrange and Hamilton that are derived from these below, have the property of being *reversible*: Any solutions that are taken from a final description of positions and velocities will, when evaluated for negatively-running time, reproduce the original initial conditions. This property of the equations originates in the nature of the forces: As written in the following, these force equations do not include the effects of friction or other irreversible processes that inevitably affect any attempt to apply Newton's equations to actual real-world situations. The forces are assumed to be *conservative*, such that energy is either kinetic or potential, and heat energy is not considered.

The equations of motion for collective variables, equations such as those of fluid motion, typically have solutions that are extremely sensitive to small changes of initial conditions. This is why predictions of the weather are notoriously unreliable.

A.1.3 Collective coordinates

When we treat aggregates of single particles it often proves useful to introduce a few collective coordinates that characterize the collection as a whole. A simple collective coordinate is the position $\mathbf{R}$ of the *center of mass* for a collection of point particles at positions $\mathbf{r}_i$, the average weighted by individual masses m_i totaling mass M,

$$\mathbf{R} = \frac{1}{M} \sum_i m_i \, \mathbf{r}_i \quad \text{with} \quad M = \sum_i m_i. \tag{A.1-14}$$

Angular momentum. Another important collective coordinate is the sum of motional (orbital) angular momenta, expressible with a dimensionless vector $\mathbf{L}$ as[2]

$$\hbar \mathbf{L} = \sum_i m_i \, \mathbf{r}_i \times \mathbf{v}_i. \tag{A.1-15}$$

Electrons and nuclei have also intrinsic (spin) angular momenta $\hbar \mathbf{s}_i$ that combine (vectorially) to give total spin $\mathbf{S}$,

$$\mathbf{S} = \sum_i \mathbf{s}_i. \tag{A.1-16}$$

The sum of orbital and spin angular momentum is a vector, the total angular momentum,

$$\mathbf{J} = \mathbf{L} + \mathbf{S}. \tag{A.1-17}$$

These expressions refer to collections of discrete particles. Fields, being spatially distributed quantities, have angular momentum densities. These too allow separation of orbital and spin contributions.

---

[2]The rationalized Planck constant $\hbar$ is the atomic unit of angular momentum.

Multipole moments. In treating the energy of a collection of bound particles (electrons in atoms, for example) in electric and magnetic fields it is useful to introduce electric and magnetic moments. The most important of these are the electric-dipole moment **d**, built from electric charges e_i

$$\mathbf{d} = \sum_i e_i \mathbf{r}_i, \tag{A.1-18}$$

and the magnetic-dipole moment **m**, built from circulating electric currents expressible in terms of collective angular momenta. For electrons collective magnetic moment is proportional to the *Bohr magneton* μ_B and can be written

$$\mathbf{m} = \mu_B(\mathbf{L} + 2\mathbf{S}), \qquad \mu_B = \frac{e\hbar}{2m_e} \approx 9.27 \times 10^{-24}\,\mathrm{J/T}, \tag{A.1-19a}$$

where m_e is the electron rest mass. The magnetic moment of a nucleus that has spin **I** is proportional to the *nuclear magneton* μ_N, and hence inversely proportional to the proton mass m_p,

$$\mathbf{m} = g_I\,\mu_N\mathbf{I}, \qquad \mu_N = \frac{e\hbar}{2m_p} \approx 5.05 \times 10^{-27}\,\mathrm{J/T}. \tag{A.1-19b}$$

Here the dimensionless *g-factor* g_I incorporates nuclear structure. The relationship between magnetic moment and an angular momentum $\hbar\mathbf{J}$ is often expressed by a *gyromagnetic ratio* γ (not to mistaken for the Lorentz factor γ) in the form

$$\mathbf{m} = \gamma\,\hbar\mathbf{J}. \tag{A.1-20}$$

For charged particles this is proportional to the charge-to-mass ratio.

The electric dipole moment is but the first of a series of moments of electric charge. The next moment, the quadrupole tensor Q of order 2, is definable as (the factor 3 is by convention)

$$Q = 3 \sum_{ij} e_i \mathbf{r}_i\, e_j \mathbf{r}_j. \tag{A.1-21}$$

When treating solids it is desirable to introduce collective patterns of small displacements, mode patterns that correspond to sound waves and whose quantized behavior is regarded as *phonons*—acoustic counterparts of photons [Bor54; Zim60]. Their randomization is heat, the end result of absorbed radiation.

A.1.4 Lagrangian dynamics

You cannot get very far in quantum theory without encountering a Hamiltonian—a prescription for energy. The Hamiltonian gets its full definition through momenta defined from a Lagrangian. In turn, Lagrangians rely on arbitrary definitions of variables. The following sections present an overview of these steps needed to obtain a Hamiltonian for a given system.

Part of the challenge I faced with my fellow students of mechanics was the proper identification of force components. When there are constraints, such as a tabletop that constrains motion to two directions, or linkages between parts of a mechanism, this task can take long evenings with homework assignments. In graduate school I learned that such work could be made much easier by considering energies rather than forces. One begins by devising some set of independent generalized coordinates q_i and associated velocities $\dot{q}_i$, one for each degree of freedom (DoF). The generalized coordinates can be distances or angles or areas or any other convenient measurables—even currents and voltages. The kinetic energy associated with the set of coordinates q_i (denoted collectively as $\mathbf{q}$) is the sum of individual motions. For nonrelativistic behavior these are generally expressible as proportional to squares of generalized velocities $\dot{q}_i$,

$$E^{\text{kin}}(\dot{\mathbf{q}}) = \sum_i \tfrac{1}{2} m_i (\dot{q}_i)^2 \quad \text{where} \quad \dot{q}_i \equiv \frac{d}{dt} q_i. \tag{A.1-22}$$

Here the inertial factor m_i serves as a generalized mass, having units appropriate to the coordinate choice for its DoF. For forces that are derivable from a potential energy E^{pot} as derivatives along generalized coordinates (*conservative* forces),

$$F_i = -\frac{\partial E^{\text{pot}}}{\partial q_i}, \tag{A.1-23}$$

we form the *Lagrangian* $\mathcal{L}$ as the difference between kinetic and potential energies

$$\mathcal{L}(\mathbf{q}, \dot{\mathbf{q}}) = E^{\text{kin}}(\dot{\mathbf{q}}) - E^{\text{pot}}(\mathbf{q}). \tag{A.1-24}$$

The equations of motion, expressed in Lagrangian form, are then

$$\frac{d}{dt}\left(\frac{\partial \mathcal{L}}{\partial \dot{q}_i}\right) = \frac{\partial \mathcal{L}}{\partial q_i}. \tag{A.1-25}$$

These are to be completed by specifying initial conditions on all of the generalized coordinates q_i and their derivatives $\dot{q}_i$.

The Lagrangian approach, with its emphasis on potential energy rather than on forces, lends itself well to treatments of constrained motions. As a graduate student I was delighted to learn how it could simplify the tedious analysis of forces acting on objects that had to move on curved surfaces or that were linked together in some way.

It should be noted that the assumption of conservative forces, eqn (A.1-23), that underlies the Lagrangian description of dynamics presented here, rules out incorporation of friction and thermal energy into the dynamics. These irreversible effects must be treated separately, whether for a classical system or for one governed by quantum mechanics; see Section 5.1.

Lagrangians have been used as the basis for relativistic quantum electrodynamics and for quantum field theory, but quantum optics and related subjects rely more on an approach discussed next, the Hamiltonian approach to mechanics. Neither the Lagrangian nor the Hamiltonian are unique for a given system: A variety of choices for the underlying variables are always possible—alternatives that produce the same equations of motion and hence the same description of system dynamics but with differing definitions of coordinates and momenta.

A.1.5 The principle of least action

By attending the Harvard lectures of Julian Schwinger I was introduced as a graduate student to the remarkably elegant way of incorporating a great deal of physics—equations of motion, for example—into variational principles. These are formulations of behavior in which all activity can be predicted as consequence of minimizing or maximizing some quantity constructed from what are taken to be the basic system variables – generalized coordinates and their time derivatives (generalized velocities); see texts on the calculus of variations [For60; Fox87; Cas13].

To derive Lagrange's equations, for example, one defines the *action* S to be the time integral of the Lagrangian between two times, $t1$ and $t2$,

$$S = \int_{t1}^{t2} dt \, \mathcal{L}(q, \dot{q}). \tag{A.1-26}$$

The values of the dynamical variables that will actually occur during that time are those for which the action is smallest, a conclusion known as Hamilton's principle or the *principle of least action*. At the minimum (or maximum) of a function its first derivative vanishes. For the action integral for a single degree of freedom this minimization property, with respect to a coordinate q and its velocity $\dot{q}$, leads to the variational equation,

$$\delta S = \int_{t1}^{t2} dt \left[\frac{\partial \mathcal{L}}{\partial q} \delta q + \frac{\partial \mathcal{L}}{\partial \dot{q}} \delta \dot{q} \right] = 0. \tag{A.1-27}$$

expressing the change in action, δS, as a time integral over changes that originate with the coordinate variable. By integrating the bracketed integrand by parts and setting the result to zero, one obtains the Euler-Lagrange equations of motion [For60; Fox87; Cas13].

Such variational formulations of physics behavior as examples of mathematical extrema have a long history, including Fermat's principle that light rays take the shortest path between two points. In the hands of Schwinger and Feynman it was used in formulating quantum electrodynamics. In lighter vein it has inspired the principle of maximum inconvenience [Sho00].

A.1.6 Hamiltonian dynamics

One further reorganization of the equations of motion is needed for converting classical mechanics into quantum mechanics. From each generalized coordinate of the Lagrangian we define a "canonically conjugate" momentum,

$$p_i = \frac{\partial \mathcal{L}}{\partial \dot{q}_i} \qquad \text{or} \qquad p_i = \frac{\partial E^{\text{kin}}}{\partial \dot{q}_i}. \tag{A.1-28}$$

In simple cases the result is just mass times velocity, $p_i = m_i \dot{q}_i$. The sum of potential and kinetic energies, expressed in terms of generalized coordinates and canonical momentum, is the *Hamiltonian*

$$\mathcal{H}(\mathbf{q}, \mathbf{p}) = E^{\text{kin}}(\mathbf{q}, \mathbf{p}) + E^{\text{pot}}(\mathbf{q}, \mathbf{p}). \tag{A.1-29}$$

where, as above, $\mathbf{q}, \mathbf{p}$ denotes the set of canonical variables, however numerous. *Hamilton's equations* of motion are paired first-order ODEs,

$$\frac{d}{dt}q_i = \frac{\partial \mathcal{H}}{\partial p_i}, \qquad \frac{d}{dt}p_i = -\frac{\partial \mathcal{H}}{\partial q_i}, \qquad \text{(A.1-30)}$$

to be solved subject to initial conditions on coordinates and momenta. As the definition of eqn (A.1-29) shows, the Hamiltonian is the total energy, expressed in terms of coordinates and momenta.

A.1.7   Example: The harmonic oscillator

A common model of classical mechanics is motion in one dimension of an idealized point particle of mass m in which the force, that of Hooke's law, is proportional to the separation distance x away from an equilibrium position, which is conveniently taken to be the coordinate origin $x = 0$. Newton's equation for changing values of $x(t)$ reads

$$F(x) = -\kappa\, x(t) = m\frac{d^2}{dt^2}x(t). \qquad \text{(A.1-31)}$$

Assuming the particle is released from rest at position $x_{\max}$ at time $t = 0$, its subsequent motion, a solution to eqn (A.1-31) that fits the initial conditions, is purely oscillatory—an example of *simple harmonic motion* at angular frequency ω,

$$x(t) = x_{\max}\, \cos(\omega t), \qquad \omega = \sqrt{\frac{\kappa}{m}}. \qquad \text{(A.1-32)}$$

In the absence of any other force these oscillations persist indefinitely. In reality, there are always friction forces, proportional to the velocity $\dot{x}$, that will eventually bring the particle to rest at an equilibrium position, here $x = 0$. Under these circumstances the motion is a damped oscillation.

The oscillator Lagrangian.   The Lagrangian is the difference between kinetic and potential energies, regarded as functions of positions and velocities. For the frictionless one-dimensional harmonic oscillator it is

$$\mathcal{L}(x, \dot{x}) = E^{\text{kin}}(\dot{x}) - E^{\text{pot}}(x), \qquad E^{\text{kin}}(\dot{x}) = \tfrac{1}{2}m\dot{x}^2, \qquad E^{\text{pot}}(x) = \tfrac{1}{2}\kappa x^2. \quad \text{(A.1-33)}$$

The resulting Lagrange's equation of one-dimensional motion, eqn (A.1-33), are recognizable as Newton's equation (A.1-9) for the harmonic oscillator,

$$\frac{d}{dt}\left(\frac{\partial \mathcal{L}}{\partial \dot{x}}\right) = \frac{\partial \mathcal{L}}{\partial x} \quad \Rightarrow \quad \frac{d}{dt}m\dot{x} = -\kappa x. \qquad \text{(A.1-34)}$$

By using the definition of the oscillation frequency in eqn (A.1-32) we can write the oscillator equation as

$$\frac{d^2}{dt^2}x(t) = -\omega^2 x(t), \qquad \text{(A.1-35)}$$

a form that is often useful for recognizing simple harmonic motion; see Appendix B.3.2. Specific solutions are obtained by fixing two initial conditions, say $x(0)$ and $\dot{x}(0)$ for time $t = 0$.

The oscillator Hamiltonian. From the Lagrangian for the one-dimensional harmonic oscillator we evaluate the momentum to be

$$p \equiv \frac{\partial \mathcal{L}}{\partial \dot{x}} = m\dot{x}. \tag{A.1-36}$$

The Hamiltonian for the one-dimensional harmonic oscillator therefore reads

$$\mathcal{H}(x, p) = \frac{1}{2m}p^2 + \frac{\kappa}{2}x^2, \qquad \kappa = m\omega^2, \tag{A.1-37}$$

and Hamilton's two equations of motion become the pair of coupled first-order ODEs,

$$\frac{d}{dt}x = \frac{\partial \mathcal{H}}{\partial p} = \frac{p}{m}, \qquad \frac{d}{dt}p = -\frac{\partial \mathcal{H}}{\partial x} = -\kappa x, \tag{A.1-38}$$

recognizable as a definition of momentum and a rewriting of Newton's equation of motion. The Hamiltonian formulation shown here has replaced the single second-order ODE of Lagrange with two first-order ODEs, a commonly-used procedure that is advantageous for obtaining numerical solutions to second-order ODEs.

All of these equations are obtained from the same idealized model, a classical, conservative system that that has no friction, maintains a constant energy (the Hamiltonian) and which undergoes simple harmonic oscillations indefinitely. When friction forces are present they convert the original kinetic and potential energies into heat as the particle settles to its stationary equilibrium position, here $x = 0$, but that behavior cannot be deduced from the Lagrangian of eqn (A.1-33) nor the Hamiltonian of eqn (A.1-37).

A.1.8 System points and phase space

Equations of motion, whether from Newton, Lagrange, or Hamilton, provide a general framework for describing possible changes of a particular system. We pick particular cases by specifying initial conditions of the basic dynamical variables, say a position and momentum for each degree of freedom. Thereafter the properties of the system are completely defined by listing, for each time t, the collection of position and momentum pairs. It often proves useful to regard this information as a *system point* in a space of dynamical variables, a so-called *phase space* that has coordinates q_i and p_i for each degree of freedom i. As time proceeds we then follow the path of the system point to view the system behavior. Phase-space pictures of system behavior have many uses, including the presentation of quantum mechanical dynamics [Sch01].

Phase space is but one of several useful ways of organizing the description of system behavior. The incorporation of quantum mechanics into descriptions of system dynamics is commonly done by organizing the system information into elements of an abstract mathematical space, as described in Appendix A.3 below.

A.1.9 Summary of particle dynamics

Any problem that is amenable to treatment with a Hamiltonian or Lagrangian, as are many descriptions of photons and their actions, can proceed toward resolution through several steps, possibly overlapping:

System: Identify the system and its parts. An auto on a road? An atom in a box? What kind of auto or atom?

Variables: Choose suitable dynamical variables as general coordinates. This choice fixes the Lagrangian, the momenta, and the Hamiltonian.

Initial conditions: Specify initial values for every relevant coordinate and, if appropriate, every velocity or momentum. This selects a particular case from the infinity of possible motions.

Solve: Set up and solve the relevant differential equations of motion, either for the variables, for their averages, or for properties of experimental interest.

Share: Organize and present the results—what do they show?—for discussion and publication.

A.1.10 Distributed mass: Fluid equations

Photons are viewed both as particles and as waves. The preceding sections presented a variety of equations pertinent to particles, and it is with such mathematics that we should expect to describe the discrete, particulate nature of radiation. To find comparable equations applicable to waves it is necessary to consider fluids and classical fields—charge and mass and other properties that are regarded as being distributed continuously through space, rather than being discretely localized.

Some aspects of liquid and vapor behavior can be understood by modeling the matter as a collection of distinct, independent molecules that move, apart from brief collisions, along Newtonian trajectories. For treating gases the molecules are regarded as having mass but negligible size. However, behavior of condensed matter—liquids and readily deformable portions of matter—are better treated as *fluids*, defined by continuous distributions of mass and velocity.[3] The relevant equations of motion are those of hydrodynamics, expressing conservation of mass, energy, momentum, and angular momentum [Lam45; Mil60; Lan87; Fal11]. The mass and size of individual molecules sets the value of the fluid density. The forces between molecules appear as distributed viscosity in these equations.

An important form of fluid equations expresses, through partial differentials, the rate of increase of some material density $\rho(\mathbf{r}, t)$ in an infinitesimal volume element (say, mass per unit volume) to the *flux* of that quantity into that volume, $\mathbf{j}(\mathbf{r}, t)$ (say, momentum per unit area). In the absence of sources or sinks this relationship is a *continuity equation*

$$\frac{\partial}{\partial t}\rho(\mathbf{r}, t) + \nabla \cdot \mathbf{j}(\mathbf{r}, t) = 0. \tag{A.1-39}$$

Continuity equations. A general continuity equation describes the flow of a fluid or field by expressing a distributed form of a conservation law. For a quantity X, say mass or charge (or energy or angular momentum or velocity), the continuity equation expressing conservation of X has the dimensional structure,

$$\frac{1}{T}\left(\frac{X}{L^3}\right) + \frac{1}{L}\left(\frac{X}{L^2 T}\right) = 0, \tag{A.1-40}$$

---

[3] The two sorts of equations are akin to the mathematician's sets of discrete ($\mathbb{Z}$ or $\mathbb{Q}$) and continuum ($\mathbb{R}$ or $\mathbb{C}$) numbers.

with length L and time T. The quantity X/L^3 is a *density* of X, while $X/(L^2T)$ is a *flux* of X. A flux of X is often expressible as a density of X times a velocity (L/T).

Diffusion equations. Gases and fluids are readily seen to flow from regions of high mass density to regions of low mass concentration. So too does heat (thermal energy density) flow, from hot to cold regions. Such flows are governed by *Fick's law*, that a flux is proportional to the spatial gradient of a local density,

$$\text{flux} = \text{diffusion coefficient} \times \text{density gradient.} \tag{A.1-41}$$

Expressed with flux $\mathbf{j}(\mathbf{r}, t)$ and local density $\rho(\mathbf{r}, t)$ the relationship reads

$$\mathbf{j}(\mathbf{r}, t) = -D\,\nabla\rho(\mathbf{r}, t). \tag{A.1-42}$$

The diffusion coefficient D has the dimensions of area per unit time (or velocity times length).

This diffusive flow, originally applied to distributions of chemical concentration, is commonly applied to heat energy density. Taken with a continuity equation this leads to the heat equation

$$\frac{\partial}{\partial t}\rho(\mathbf{r}, t) = D\,\nabla^2\rho(\mathbf{r}, t). \tag{A.1-43}$$

This equation does not have oscillatory, wavelike solutions—it is missing the imaginary unit i of the TDSE that brings oscillations. Instead, solutions approach a steady equilibrium density as time increases. In application to heat (or energy) flow there are three recognized mechanisms, each with its distinct diffusion coefficient:

Conduction: The irregular thermal motions of atoms and molecules (or freely moving electrons) is imparted to adjacent matter.

Convection: Aggregates of matter, typically globs of fluid, move together from hot to cold regions.

Radiation: Electromagnetic energy travels from emission in hot regions to absorption in cold regions.

Photon mean free path. The diffusion coefficient for radiation is often expressed as $D = cL_{\mathrm{mfp}}/3$ where L_{mfp} is the mean free path between scattering or absorption events [Pie06]. This length is the inverse of the attenuation coefficient α of eqn (B.8-3), defined for monochromatic light from the product of number density of scatterers, $\mathcal{N}$, and an attenuation cross section, σ (see Appendix B.8.1) as

$$L_{\mathrm{mfp}} = \frac{1}{\alpha} = \frac{1}{\sigma\mathcal{N}}. \tag{A.1-44}$$

When the radiation is regarded as a crowd of photons (see Appendix B.8), L_{mfp} is said to be the *photon mean free path*. The *optical depth* associated with a propagation distance z is the dimensionless ratio z/L_{mfp}.

Opacity. Astrophysicists parametrize the hindrance of radiation flow through a mass density ρ by the *opacity* κ, defined such that the absorption coefficient for monochromatic light is factored as [Fle05]

$$\alpha = \kappa\rho = \sigma\mathcal{N}. \tag{A.1-45}$$

What is required for a photon mean free path in a star is the *Rosseland mean opacity* $\bar{\kappa}$ [Lan80; Bad10], defined for an isotropic medium as[4]

$$1/\bar{\kappa} = \int d\omega \, \frac{u(\omega, T)}{\kappa(\omega)} \, / \int d\omega \, u(\omega, T), \tag{A.1-46}$$

where $u(\omega, T)$ is the Planck distribution function for thermal radiation; see eqn (B.7-3). Because the absorption coefficient varies from atom to atom, the mean opacity depends upon the local composition.

The Navier-Stokes equation. The conventional equation for describing flow of an incompressible classical fluid is the nineteenth century *Navier-Stokes equation* [Lam45; Mil60; Lan87; Fal11] for a velocity field $\mathbf{u} \equiv \mathbf{u}(\mathbf{r}, t)$,

$$\frac{\partial}{\partial t}\mathbf{u} + (\mathbf{u} \cdot \nabla)\mathbf{u} = \frac{1}{\rho}\nabla P + \nu\nabla^2\mathbf{u}, \tag{A.1-47}$$

where ρ is the mass density, P is the pressure and ν is the kinematic viscosity. From the ratio of inertial forces to viscosity forces one defines a measure of possible turbulence around an object through the dimensionless *Reynolds number*,

$$\mathrm{Re} = uL/\nu, \tag{A.1-48}$$

where u is the velocity along the object surface and L is a characteristic length appropriate to the geometry of the object. The Reynolds number provides an indicator of whether the fluid motion will be *laminar* (low Re) or *turbulent* (large Re).

A.2 Measurement; Sizes

Various granular samples are routinely sorted by size through the use of sieves—surfaces containing holes whose size establish a maximum granule-dimension for passage. For oblong granules a round hole only fixes the minimum dimension of the granule, which may be a poor measure of size: A long stick passes readily through any small opening that is larger than the stick diameter [Dom15; Sho15].

Determining the location of a single, sizable but movable object is relatively easy using the sense of touch or sight or, if you are a bat, sound. As the object becomes smaller, the task becomes more difficult. Locating the last rice grain on a supper plate without eyesight poses a straightforward challenge: The fork or spoon will move the rice. Minuscule objects are particularly hard to locate by touch, and easily displaced, even with delicate tools. Eyesight too has problems, because the light (the photons) we see reflected from an object will impart a small, but measurable, momentum and thereby alter the object position. Inevitably we realize that the measurement process itself can alter whatever we wish to measure.

Such reaction effects are not confined to measurements of physical objects—grains of food or atoms—but are encountered by sociologists who hope to measure public

---

[4]For consistency I here use angular frequency ω where astrophysical custom uses frequency ν.

opinion: The very questions that are intended to elicit information inevitably initiate thought processes that may change opinions. Moreover, the order in which questions are asked may make a significant difference in the answers. This is an interesting observation, but hardly anything notably strange.

The concept of measurement has a large literature, to which physicists and philosophers and general academics have contributed at length. Quantum theory, with its particular constraints, has provided a notable share of this literature, and a person could easily devote many evenings (even several lifetimes) to examining the assorted issues associated with the notion of measurement that arise within that topic. I shall not go there. I shall assume that, although a workable definition for measurement, as a concept, is just part of the understanding we bring to thinking about our world, nonetheless it is important to organize some basic aspects of measurement into the presentation of physics.

A.2.1 Electron size

In treatments of particles as waves the de Broglie wavelength of eqn (2.7-1) provides the basic length scale. For an electron this wavelength is

$$\lambda_e = \frac{h}{m_e v} \sqrt{1 - (v/c)^2} \quad \text{so} \quad \lambda_e \approx \frac{h}{m_e v} = \frac{h}{p}. \tag{A.2-1}$$

At nonrelativistic speed $v \ll c$ this length varies inversely with velocity v or momentum p and so it increases without limit as the electron slows. More pointlike localization becomes possible as the speed increases.

Although calculations treat the electron as a point particle, having no physical extent, several derived length scales are useful in dealing with the distribution of charge within atoms. The fundamental unit of charge e and the electron rest mass m_e combine with two other fundamental constants, the vacuum light speed c, and the reduced Planck constant $\hbar$, to define these lengths.

The smallest useful electronic scale of length, the *classical radius of the electron*,

$$r_e = \frac{1}{4\pi\epsilon_0} \frac{e^2}{m_e c^2} \approx 2.8 \times 10^{-15} \text{ m}, \tag{A.2-2}$$

is defined as the distance at which the electron-proton Coulomb energy $e^2/(4\pi\epsilon_0 r_e)$ of an idealized hydrogen atom equals the rest energy $m_e c^2$ for the electron.

Two larger distances are expressible in terms of r_e and the dimensionless Sommerfeld fine-structure constant [Kra03]

$$\alpha = \frac{e^2}{(4\pi\epsilon_0)\hbar c^2} \approx \frac{1}{137}. \tag{A.2-3}$$

The *reduced Compton wavelength* of the electron,

$$\lambda_c = \frac{r_e}{\alpha} \approx 3.86 \times 10^{-13} \text{ m}, \tag{A.2-4}$$

occurs as a characteristic length in the scattering of high-energy photons by electrons; see Appendix B.8.3.

A.2.2 Atom size

Larger by $1/\alpha^2$ than r_e, the Bohr radius of eqn (2.2-3),

$$a_0 = \frac{r_e}{\alpha^2} \approx 5.29 \times 10^{-11} \text{ m,} \qquad (\text{A.2-5})$$

(roughly half an Ångström or 5 nm) serves as the atomic unit of length: It is the radius of the smallest Bohr orbit in hydrogen, and so it is a characteristic size for atoms.[5] The internal structure of atoms involves complicated distributions of electrostatic fields, so that no single definition of atomic (or electron) size fits all requirements. Near a nucleus of Z protons the length scale becomes a_0/Z.

A.2.3 Photon size: Resolution and wavelength

Microscopes and telescopes, with their lenses that focus rays of light into small spots in a focal plane, traditionally provide the means of viewing objects and measuring their shapes. The wavelength of the illumination limits the size of objects that we can distinguish by means of focusing optics. The resolution of distinct adjacent objects by an optical system acting on unstructured light is set not only by the wavelength of the light but also by the range of angles that are combined into the detector image: In accord with the mathematics of Fourier analysis, the wider the cone of light the smaller is the resolvable distance between details at the focus. Although precise specifications of resolution depend on details of the optical system, including the detector, a long-used formula for the transverse distance r between two points that can be resolved by monochromatic light of wavelength λ is the *Rayleigh criterion*, often written as [Dri78]

$$r \approx \frac{0.6\,\lambda}{\text{NA}}. \qquad (\text{A.2-6})$$

Here NA is the *numerical aperture*, a measure of the range of angles that contribute to the image:

$$\text{NA} = \text{n}\sin\theta_{\text{max}}, \qquad (\text{A.2-7})$$

where n is the refractive index and θ_{max} is the largest angle of beam deflection being collected. The square of its NA measures the light-gathering ability of an optical system.

The *Abbé diffraction limit* replaces 0.6 by 0.5 in eqn (A.2-6). It is this diffraction limit, applicable to conventional laser beams used for scanning, that underlies the common notion that the resolution limit (with NA = 1) is half a wavelength. That estimate originates with simplifications of the field distant from a focus by many wavelengths (the far field). The use of nonlinear optics, such as *stimulated-emission depletion* (STED) microscopy [Hel99; Kla01], and near-field optics (where evanescent non-propagating fields become important) make possible resolutions appreciably smaller than the diffraction limit. The use of radiation beams that have transverse nodal structure, as discussed in Appendix B.2, offers further opportunity for improving resolution beyond the classical limit of Rayleigh and Abbé [Lin12].

---

[5]There is no sharp boundary to the electron charge distribution of an atom: It falls exponentially with distance from the nucleus. For an unexcited hydrogen atom the Bohr radius a_0 is the scale of the exponential fall.

Subwavelength resolution. The traditional derivation of a resolution limit rests on the assumption that the wave phase across the aperture (or focusing lens) is constant or slowly varying. It is now possible to piece together spatial light modulators (SLM) that can adjust the focusing field to have a variable wavefront, and to use feedback techniques to adapt this wavefront to give maximum contrast variation for a given illuminated object. The effective resolution for that object can be a fraction of a wavelength, set in part by the number of phase elements [Vel07; Vel10; Kat12].

Locating a single emitter. The Abbé limit applies to attempts to distinguish the light originating with two distinct emitters. Locating the position of a single small emitter does not have this restriction. For such measurement the lens system collects the light from a beam and brings it to constructive interference around a point. The result is an intensity pattern spread as a focal spot. The full width at half maximum of the spot in the (transverse) focal plane is Δr and longitudinally, along the optical axis, it is Δz, where [Hel07]

$$\Delta r \approx \frac{\lambda}{2\mathrm{n} \sin \theta_{\max}}, \qquad \Delta z \approx \frac{\lambda}{\mathrm{n} \sin^2 \theta_{\max}}. \qquad (\text{A.2-8})$$

Although the diameter of the spot in the focal plane is governed by the wavelength of light and the action of optical elements, the location of the spot center (and the inferred position of the object) is restricted only by noise and optical imperfections, and can be obtained with precision much less than the wavelength [Yil05; Arn05].

However, structure of the emission source can introduce appreciable error in the apparent location of an emitter whose dimensions are much smaller than a wavelength, as are single-atom sources of resonant scattering and spontaneous emission. Radiation emitted by a localized source appears, many wavelengths away (in the *far field*), to travel along straight lines. These rays appear to emerge from a point, the apparent location of the emitter. The measured quantity at any position is the intensity—the time-averaged momentum density, the Poynting vector. In the far field region the field lines of the Poynting vector asymptotically approach straight lines, but in the subwavelength region around the source (the *near field*), the lines of the Poynting vector are generally curved, and may have a complicated structure that affects the far field.

Closed-form expressions for the Poynting vector of radiation emitted by a multipole of arbitrary order are available [Arn05]. The simplest example is the field produced by an electric dipole moment that oscillates, with frequency ω, along the z axis, such as occurs with a transition between Zeeman sublevels for $\Delta m = 0$. The field lines for the Poynting vector of this multipole are straight, and in the far field they are seen as linear polarization. The field produced with $\Delta m = \pm 1$ is that from a rotating dipole moment. The near field Poynting vector has lines that swirl around the z axis before departing [Arn04; Arn05; Shu08; Li12; Arn16]. Asymptotically the lines approach straight rays, but their apparent center is not necessarily that of the atom center of mass [Hel07; Li08; Shu08; Li10; Li12].

A.3   Abstract vector spaces

The definition of a vector as an ordered list of elements (*components*), as proposed in Section 2.5.2, offers many opportunities for practical application. In my first foray into the world of electronic-computer programming, with the FORTRAN language of instructions, I recognized the power of this definition, a way of regarding the apportioning of various electronic words into machine storage locations that takes place when a computer program is run. In such usage the elements of a vector need not be numbers, they can be words of a sentence or computer instructions. Whatever may be the possible vector elements—numbers, words, or even just symbols—the totality of possibilities, and the rules for manipulating them, form an (abstract) *vector space*—an organization of information.

A.3.1   Property spaces

A simple example of an abstract vector that has elements of different sorts occurs in the computer-generated instructions sent to my printer by the operating system of my computer. The printer uses five ink cartridges, containing the colors black (B), cyan (C), yellow (Y), magenta (M), and a second black (Bk). An arrangement of controlled hydraulic pathways allows fluid from each cartridge to produce a small colored dot on a piece of paper. The instructions to the printer must specify the location of the center of a cluster of colored dots (two Cartesian coordinates x and y, each of which is bounded by the paper size) and amounts of each of the five colored fluids that will construct what my eye sees as part of a continuum of color. The two Cartesian coordinates define vectors of a two-dimensional subspace that, together with a five-dimensional cartridge subspace (four distinct colors, but with doubling of black), form a seven-dimensional abstract space.

Within my computer a similar multidimensional abstract space organizes information about the display I see on the screen. There is again a two-dimensional vector referencing position on the screen, supplemented with a component that specifies which of several simultaneously-running apps have generated this particular image. With my drawing and painting apps I typically have the choice of specifying color in a four dimensional subspace CMYK or a three-dimensional subspace RGB of red, green, blue. If I am creating a moving display, to be stored as a video, then the stored information must include, for each frame of the video (i.e. at each discretized time), all of this spatial and color information. All of this organization takes place out of sight, as part of the operating system of the computer.

These abstract spaces, incorporating a subspace of Euclidean-space coordinates, are of the sort that Wilczek refers to as a *property space*, for use with quantum chromodynamics [Wil15]. In my printer-based examples of such spaces only two of the several dimensions have any association with everyday Euclidean space.

The occurrence of discrete properties of atoms, and of radiation, makes possible their description by other forms of abstract-vector spaces, as the sections below will discuss.

A.3.2 Normed and metric spaces

An abstract vector space in which lengths are defined for vectors is known to mathematicians as a *metric space* or a *Banach space*. When the metric is obtained from some form of scalar product between pairs of vectors, as it does for Euclidean vectors, the abstract space is termed a *Hilbert space*. These are the specialized abstract spaces, allowing notions not only of length but angles between vectors, underlying the matrix mechanics that became quantum theory. Apart from the possibility of more than three dimensions (and complex numbers), their mathematics has much in common with the Euclidean space of high-school geometry and house construction.

Coordinates. To find our way through everyday journeys it was once common to employ maps showing available roads or trails. Even a cursory glance at a street map of a small town will often reveal a remarkable set of choices for street orientations. Portions of an older town commonly follow the Roman tradition of aligning streets in accord with the north star, while other portions often follow terrain markers of canals and hills and railroads—although it is not uncommon for street layouts to ignore terrain entirely, as they do in San Francisco. Quite clearly the choice of journey coordinates—along compass directions or terrain directions—is arbitrary. So too are our choices of coordinates in a metric space, although measures of distance and relative angles are independent of the coordinate system. Moreover, we may either make our coordinate choice as a permanent decision, or we may alter it at any time. The notion of household vertical may require change as better measurements become available (or as the hillside foundations shift). But if we wish to compare details of a journey with an acquaintance, then it is important for both of us to know the underlying reference frames in use by the two of us, whatever those may be.

Complex numbers: Magnitude and phase. A simple example of an abstract space with norm is the two-dimensional plane in which complex numbers are defined: Each number is an example of a vector (with magnitude and direction). Complex numbers are used routinely in quantum theory as well as in electrical engineering. The magnitude of a complex number $z = x + iy$ is the positive real number

$$|z| \equiv |x + iy| = \sqrt{x^2 + y^2}. \tag{A.3-1}$$

Complex numbers are known to electrical engineers as *phasors*, entities expressible with exponentials by means of *Euler's formula* relating exponentials to trig functions[6]

$$z = |z|\, e^{i\phi}, \qquad e^{i\phi} \equiv \exp(i\phi) = \cos\phi + i\sin\phi. \tag{A.3-2}$$

The *phase* ϕ of a complex number is obtainable by inverting any of the formulas

$$\sin\phi = \frac{y}{\sqrt{x^2 + y^2}}, \qquad \cos\phi = \frac{x}{\sqrt{x^2 + y^2}}, \qquad \tan\phi = \frac{y}{x}. \tag{A.3-3}$$

---

[6]The functions $\sinh(z)$ (hyperbolic sine) and $\cosh(z)$ (hyperbolic cosine) are defined as

$$\sinh(z) = \tfrac{1}{2}\left[e^z - e^{-z}\right], \qquad \cosh(z) = \tfrac{1}{2}\left[e^z + e^{-z}\right].$$

Addition of two phasors z_1 and z_2 proceeds by adding the components of their independent real and imaginary parts:

$$z_1 + z_2 = (x_1 + x_2) + \mathrm{i}(y_1 + y_2). \tag{A.3-4a}$$

Multiplication of two phasors leads to addition of their phases:

$$z_1 \times z_2 = |z_1| \times |z_2| \, \mathrm{e}^{\mathrm{i}(\phi_1 + \phi_2)}. \tag{A.3-4b}$$

The phase of a complex number specifies the direction in the two-dimensional complex plane of a point having Cartesian coordinates x and y. Multiplication of phasors amounts to a rotation and an expansion or contraction.

Scalar products.   When we define a vector **A** by a list of N components A_n it is useful to liken these to a single vertical *column* of a spreadsheet table, using the Dirac symbol $|A\rangle$ to denote the set. In order to introduce notions of length in this vector space we define, for each vector $|A\rangle$ an *adjoint* $\mathbf{A}^\dagger \equiv \langle A|$ that can be likened to a horizontal row from a spreadsheet table. When the vector components are complex numbers the elements of the adjoint are taken to be complex conjugates of the elements of a single row,

$$\mathbf{A} = |A\rangle = \begin{bmatrix} A_1 \\ A_2 \\ \vdots \\ A_N \end{bmatrix}, \qquad \mathbf{A}^\dagger = \langle A| = \{A_1^*, A_2^*, \cdots, A_N^*\}. \tag{A.3-5}$$

With such vectors and adjoints we can form numbers that serve as lengths and angles, thereby defining a metric space. This procedure involves defining a relationship, the *scalar product*, between pairs of vectors that maps all their elements onto a single number. For any pair of vectors $|A\rangle$ and $|B\rangle$ that have the same number of elements (a requirement) we define the scalar product, denoted in the fashion of Dirac[7] as $\langle B|A\rangle$, as the sum of paired products from **A** (column elements) and the adjoint of **B** (row elements),

$$\mathbf{B}^\dagger \cdot \mathbf{A} = \langle B|A\rangle = \sum_n B_n^* A_n. \tag{A.3-6}$$

The scalar product $\langle B|A\rangle$ can be regarded as the projection of a *bra vector* $\langle B|$ onto a *ket vector* $|A\rangle$ (or vice versa). The length $|A|$ of any vector **A** is the square root of scalar product of the vector with its adjoint,

$$|A| = \sqrt{\mathbf{A}^\dagger \cdot \mathbf{A}} = \sqrt{\langle A|A\rangle} = \sqrt{|A_1|^2 + |A_2|^2 + \cdots |A_N|^2}. \tag{A.3-7}$$

This definition of length generalizes to N dimensions the two-dimensional Pythagorean theorem.

Orthogonality.   Two vectors that have the same dimension and whose scalar product vanishes, $\langle A|B\rangle = 0$, are termed *orthogonal*, a generalization of perpendicular lines in

---

[7] Dirac referred to the scalar product $\langle B|A\rangle$ as a *"bracket"*, and referred to its constituents as a "bra" $\langle B|$ and a "ket" $|A\rangle$. Mathematicians often use the notation (B, A) for scalar products.

two dimensions. Such vectors provide independent coordinates in their vector-space setting. An N-dimensional space is definable by N independent vectors, but these need not be orthogonal—they only need to not be parallel or anti-parallel.

The Cauchy-Schwarz inequality. The magnitude of the scalar product of any two vectors from a Hilbert space is bounded by their individual lengths through the *Cauchy-Schwarz inequality* [Ste04],

$$|\langle A|B\rangle|^2 \le \langle A|A\rangle \langle B|B\rangle. \tag{A.3-8}$$

Unit vectors. A vector $\mathbf{A}$ of any dimension that has unit value for its length, $|\mathbf{A}|$ is a *unit vector*. The set of unit vectors in N dimensions are representable by N column vectors whose elements are all zero except for a single unit element,

$$\mathbf{e}_1 = |e_1\rangle = \begin{bmatrix} 1 \\ 0 \\ 0 \\ \vdots \\ 0 \end{bmatrix}, \quad \mathbf{e}_2 = |e_2\rangle = \begin{bmatrix} 0 \\ 1 \\ 0 \\ \vdots \\ 0 \end{bmatrix}, \quad \cdots, \quad \mathbf{e}_N = |e_N\rangle = \begin{bmatrix} 0 \\ 0 \\ 0 \\ \vdots \\ 1 \end{bmatrix}. \tag{A.3-9}$$

These all have unit length, and each one is orthogonal to all the others, as embodied in the orthonormality condition expressed by the *Kronecker delta*,

$$\langle e_n|e_m\rangle = \delta_{nm} \equiv \begin{cases} 1 & \text{if } n = m \\ 1 & \text{if } n \neq m. \end{cases} \tag{A.3-10}$$

Vector components. Unit vectors provide the basic framework for expressing directions in an abstract space and relationships amongst abstract vectors. They allow us to write any vector as a sum,

$$\mathbf{A} = \sum_n A_n \, \mathbf{e}_n, \tag{A.3-11}$$

and so they serve as a *basis* for defining any vector $\mathbf{A}$ in terms of components A_n. But much about them is for users to choose: The choice of a first direction and the assignment of labels $1, 2, \ldots N$. It is not necessary that the unit vectors be orthogonal, as in eqn (A.3-10), only that the set be complete: There must be N independent unit vectors for an N-dimensional space.[8] Often a particular choice seems preferable, but this may refer to a moving reference frame, such as we deal with as a passenger in an automobile or airplane. Any superposition of an independent set of unit vectors will do equally well for a basis. This is true for the statevectors of quantum theory and other metric spaces, but also for property spaces, such as the RGB and CMYK color definitions use in computers.

Coordinate changes. Euclidean vectors differ in an essential way from the vectors of abstract spaces that have no metric—property spaces, for example. The N components of vectors in a metric space are linked by a length-defining relationship—in

---

[8] A choice of nonorthogonal unit vectors defines a *metric*: $g_{nm} = \langle e_n|e_m\rangle$.

two dimensions the Pythagorean theorem—and so it is not possible to change just one coordinate value without changing a length. Given a metric space, the numbers that are listed as components of a vector will be changed in value by an alteration of the coordinate system—a redefinition of the unit vectors.

However, there are many possible collaborative changes of coordinates that do not alter lengths. The hypotenuse that defines the screen size of an iPad is unaltered whether we place it flat or hold it up. The length of a vector defined by two Euclidean points does not depend on where we locate the coordinate origin or whether it is moving. Translations, rotations, and reflections all leave lengths unchanged, although the N coordinate values will be very different. Relationships between vector components that are embodied in such *coordinate transformations*, discussed in Appendix A.18.3, are an essential characteristic of Euclidean vectors – of the space in which we live and find objects, including photons.

Dimensions; Fractals.   When we define a vector as an ordered set of components we assign it a dimension, namely the number of components. Dimensions also occur in a variety of other contexts, most notably in geometry, where we encounter points (dimension 0), lines (dimension 1), planes (dimensions 2) and the every-day Euclidean space of 3 dimensions. But measures of size for irregular curves and rough surfaces depend upon the measurement resolution. For such objects, termed *fractals* by Benoit Mandelbrot [Man77], mathematicians have been able to assign fractional-integer dimension. A commonly-cited example of the scale-free self-similar nature of these objects is the coastline of Britain, with dimension 1.25 [Man67]. For a possible connection with photons, see Appendix B.8.3.

Function space.   Hilbert spaces allow not only the N-dimensional column vectors discussed above but also vectors whose components form a continuum of values (they are *function spaces*). In place of the sum over n in eqn (A.3-6) the scalar product of two complex-valued one-dimensional functions takes the form of an integral,

$$\langle B|A \rangle = \int dx \, f(x) \, B(x)^* A(x), \qquad \text{(A.3-12)}$$

where $f(x)$ is a factor chosen so that the scalar product satisfies some desirable condition, such as the normalization $\langle A|A \rangle = 1$ for bounded functions. This sort of scalar product is used with sets of electron wavefunctions or other functions that are defined by differential equations: In one dimension the definition is

$$\langle \psi_n|\psi_m \rangle = \int dx \, f(x) \, \psi_n(x)^* \, \psi_m(x), \qquad \text{(A.3-13)}$$

where the weighting factor $f(x)$ depends on the particular coordinates and the desired normalization of lengths. A less common form for a scalar product, used for some versions of photon wavefunctions, [Haw07; Smi07] is a nonlocal integral

$$\langle B|A \rangle = \int dx' \int dx \, f(x', x) \, B(x')^* A(x). \qquad \text{(A.3-14)}$$

A.3.3   Quantum Hilbert space

The mathematics of abstract vectors provides a natural framework for the description of quantum phenomena. Just as with the property spaces discussed above, it provides a way of organizing information. The vectors of quantum physics store information about quantum states.

Discrete quantum states.   The notion of photons, and the history of their use in physics, is linked to idealizations of single atoms and their interaction with radiation, primarily their gain and loss of structural energy in discrete increments. That energy change, responsible for creating or removing single photons, is attributed to the motion of electrons bound within atoms and to atoms bound within molecules—to internal energies of composite groups of elementary charges—rather than to unrestricted translational motion and unbounded acceleration or scattering. For a system whose motion is constrained, such as an electron bound within an atom or a rotating, vibrating molecule, the possible energies form a discrete set—the energies are *quantized* (as wave mechanics predicts; see Appendix A.5). Quantization of atomic structures (energies and motions) leads quite naturally to quantization of radiation energy increments and to notions of photons as these discrete increments.

   Such stationary states of internal motions—denumerable as distinguishable discrete *quantum states*—form a set that can, in principle, be ordered into a list and given natural-number labels $1, 2, \ldots$. Most commonly the chosen ordering is by increasing energy so that, if there are no duplicate values, $E_1 < E_2 < \cdots$, but other choices, and other labelings, prove useful when dealing with chains of linked energy states, as in eqn (5.9-1).

Essential states.   For many idealized models of discrete-state quantum systems (for example, the free-space hydrogen atom and the harmonic oscillator) the number of quantum states is mathematically unbounded, but for practical purposes it generally suffices to consider only a small number of states, say, two or three. The description of such a quantum system requires information about the N important quantum states— the *essential states*—that may participate perceptibly in its behavior. These discrete states, and their defining information, can, in principle, be ordered into a list. When one recalls the definition of a vector as just an ordered list, it is not far fetched to consider an abstract-vector space in which each coordinate is associated with a distinguishable, unique quantum state. Subsequent sections examine that possibility, after first discussing some of the tools available for manipulating abstract vectors.

Quantum probabilities.   The description of a quantum system requires information about the N possible quantum states $1, 2, \ldots, N$ that may participate perceptibly in its behavior.[9] The mathematics associated with abstract vector spaces offers a useful framework for describing these quantum states and their changes. The set of N probabilities P_n for finding the system in state n could be treated as coordinates in an N-dimensional abstract vector space, but it turns out to be more useful to express these probabilities as absolute squares of complex-valued *probability amplitudes* $C_n(t)$,

---

[9]As with other mathematical models, the accuracy generally improves as one takes more states into consideration.

$$P_n(t) = |C_n(t)|^2, \tag{A.3-15}$$

and to use these amplitudes as the coordinates in an abstract vector space—as ordered elements that are collectively denoted $\mathbf{C}(t)$. The vector space employed for this purpose is one in which there are lengths and angles defined by scalar products, making it a *Hilbert space*, as discussed in the next section.

Like the postulated commutator relationships of conjugate variables, eqn (A.4-3), the formula of eqn (A.3-15) is part of the defining mathematical structure of quantum theory, contributed by Max Born [Ber05].

Degeneracy. Each distinct stationary state of motion is distinguishable from all other possible quantum states by labels (quantum numbers) that identify such measurable attributes as energy and angular momentum. It usually happens that a variety of bound motions share a common energy: Internal energy does not depend on whether a symmetry axis of an atom or molecule in free space points up or down or in some arbitrary direction. Quantum states that share a common energy are termed *degenerate* and their number is the *statistical weight* ϖ_n of energy level E_n—a count of the number of quantum states that share energy E_n. It should be noted that the connection between probability and probability amplitude in eqn (A.3-15) refers specifically to *nondegenerate* states, not to the more general probabilities $\overline{P}_n$ of eqn (A.3-16). Appendix A.11.2 explains how to deal with degeneracy.

Populations. Historically, the most common notion of an observable is the number of atoms in a fixed given volume that share internal energy E_n, quantified as the *number density* $\mathcal{N}_n$, with units of particles per unit volume. Rather than treat number densities it proves convenient to introduce dimensionless fractions, *populations*,

$$\overline{P}_n = \frac{\mathcal{N}_n}{\mathcal{N}}, \qquad \mathcal{N} = \mathcal{N}_1 + \mathcal{N}_2 + \cdots, \tag{A.3-16}$$

where the summation for total number density $\mathcal{N}$ includes all energies that are being considered. Here the overbar on the symbol $\overline{P}_n$ for population in energy level n is a reminder that the number densities deal with a sum of degenerate discrete states.

A.3.4   Coherent superpositions

Quantum theory deals with probabilities, but unlike such examples as weather forecasts, these probabilities are expressed as absolute squares of probability amplitudes. The notion of probability amplitudes fits well into the formalism of a normed abstract-vector space. Suppose, as in Section 6.7, we have a simple system that has a single degree of freedom, and that this attribute can take only two values, such as yes and no (or up and down or blue and red). Let us designate these mutually-exclusive attribute values as A and B. We associate the corresponding probability amplitudes with abstract vectors $|A\rangle$ and $|B\rangle$ such that their lengths provide probabilities,

$$P(A) = \langle A|A\rangle, \qquad P(B) = \langle B|B\rangle. \tag{A.3-17}$$

The symbol $\langle A|B\rangle$ is associated with the probability of simultaneous occurrence of values A and B. This is zero if the attribute values are exclusive (either-or).

A key property of normed vector spaces is that we can consider superpositions of the vectors. For the example of a two-dimensional space of probability amplitudes we can construct the two probability amplitudes

$$|\Psi_+\rangle = |A\rangle + |B\rangle, \qquad |\Psi_-\rangle = |A\rangle - |B\rangle. \tag{A.3-18}$$

To avoid a null superposition $|\Psi_-\rangle$ the assumption $A \neq B$ is needed. We shall here assume that the two component amplitudes provide a complete set of exclusive attribute values, so the probability associated with the superposition amplitude $|\Psi_\pm\rangle$, with either sign, must be unity,

$$\langle\Psi_\pm|\Psi_\pm\rangle = \left(\langle A|\pm\langle B|\right)\left(|A\rangle\pm|B\rangle\right)$$
$$= \langle A|A\rangle + \langle B|B\rangle \pm \left(\langle A|B\rangle + \langle B|A\rangle\right) = 1. \tag{A.3-19}$$

That is, the vector $|\Psi_\pm\rangle$ must have unit length. Let us require that the attribute values be disjoint, being either A or B. Then

$$\langle A|B\rangle = \langle B|A\rangle = 0, \tag{A.3-20}$$

and the individual probabilities of eqn (A.3-17) sum to unity,

$$P(A) + P(B) = 1. \tag{A.3-21}$$

The observable probabilities appearing here are unaffected if we multiply the vectors by complex-valued phase factors,

$$|A\rangle \to e^{i\phi}|A\rangle, \qquad |B\rangle \to e^{i\phi'}|B\rangle, \tag{A.3-22}$$

and so it is often possible to choose phase factors for their mathematical convenience; see Appendix A.8.4. In treating quantum systems it is traditional to introduce a set of vectors having unit length that are orthogonal to each other; see Appendix A.8.4. For the present discussion these could be any pair $|e_A\rangle$ and $|e_B\rangle$ having the properties

$$\langle e_A|e_A\rangle = \langle e_B|e_B\rangle = 1, \quad \langle e_A|e_B\rangle = \langle e_B|e_A\rangle = 0. \tag{A.3-23}$$

Using such unit vectors we write the superpositions of eqn (A.3-18) as

$$|\Psi_\pm\rangle = C_A|e_A\rangle \pm C_B|e_B\rangle, \quad \text{or} \quad \mathbf{\Psi}_\pm = C_A\mathbf{e}_A \pm C_B\mathbf{e}_B, \tag{A.3-24}$$

so that the probabilities are absolute squares of the complex-valued coefficients C_i,

$$P(A) = |C_A|^2, \qquad P(B) = |C_B|^2. \tag{A.3-25}$$

The vectors $|A\rangle$ and $|B\rangle$, with their adjustable lengths and phases, form a *ray space*. The superposition $|\Psi_\pm\rangle$ is a unit vector in this ray space.

A.3.5 Matrices and operators; Eigenvectors

Matrices. Just as a vector can be defined as a single, ordered list, a *matrix* can be defined as a two-dimensional table, of horizontal rows and vertical colulmns. Whereas a vector **V** require a single subscript index, V_i, to identify its components, matrices require two indices, M_{ij}, for the elements of matrix M. When the matrix is displayed as a table, the first index identifies the row, the second index identifies the column:

$$\mathsf{M} = \begin{bmatrix} M_{11} & M_{12} & \cdots \\ M_{21} & M_{22} & \cdots \\ \vdots & \vdots & \ddots \end{bmatrix}. \tag{A.3-26}$$

The elements of a matrix may be symbols, complex numbers, or smaller matrices. A *square matrix* is one that has the same number of rows and columns. The matrices used in quantum mechanics are primarily square matrices.

Determinants. For any square $N \times N$ matrix M the *determinant*, denoted $|\mathsf{M}|$ or DetM, is defined as the sum of all possible products of N matrix elements M_{ij} from differing rows and columns, taken with a sign that counts the permutation of the subscripts ij from the ordering 1,2,3, ... The number of terms in the sum is $N!$ (factorial N), a function defined as the product of all positive integers less than or equal to N (factorial zero is defined as 1),

$$N! \equiv N \times N - 1 \times N - 2 \times \cdots \times 1. \tag{A.3-27}$$

For a 2×2 matrix the determinant is

$$|\mathsf{M}| = M_{11}M_{22} - M_{12}M_{21}. \tag{A.3-28}$$

For a 3×3 the determinant is

$$\begin{aligned} |\mathsf{M}| = \; & M_{11}M_{22}M_{33} - M_{11}M_{23}M_{32} \\ & -M_{12}M_{21}M_{33} + M_{12}M_{23}M_{31} \\ & +M_{13}M_{21}M_{32} - M_{13}M_{22}M_{31}. \end{aligned} \tag{A.3-29}$$

Interchanging any two adjacent columns (or rows) of the array M reverses the sign of the determinant. If two columns (or two rows) are numerical multiples of each other the determinant is zero.

Matrices as vector mappings. Simple changes to N-dimensional Hilbert-space vectors, being ordered lists, can be regarded as the result of operators, represented by arrays of number pairs collected as matrices. The elements of a matrix provide rules for changes of vectors. The changes (mappings between two vectors) are expressed as multiplication of a vector V by a matrix **M** to produce the vector V', as embodied in the rule

$$V_i' = \sum_j M_{ij} V_j = M_{i1} V_1 + M_{j2} V_2 + \cdots . \tag{A.3-30}$$

For a two-component vector the definition of matrix multiplication can be presented as

$$\mathbf{M}\mathbf{V} \equiv \begin{bmatrix} M_{11} & M_{12} \\ M_{21} & M_{22} \end{bmatrix} \begin{bmatrix} V_1 \\ V_2 \end{bmatrix} = \begin{bmatrix} M_{11}V_1 + M_{12}V_2 \\ M_{21}V_1 + M_{22}V_2 \end{bmatrix}. \tag{A.3-31}$$

Vectors $|A\rangle$ can be regarded as a special case of single-column matrices; their adjoints $\langle A|$ are single-row matrices. The mathematical properties of vectors and matrices are typically taught in courses on linear algebra and found in specialized monographs [Gan60; Gan05]. The following paragraphs summarize the properties of matrices that are most used in quantum mechanics.

Matrix algebra: Multiplication and addition. The multiplication of two matrices obeys the rule

$$\mathbf{M} = \mathbf{A}\mathbf{B}, \quad M_{ij} = \sum_k A_{ik}B_{kj}. \tag{A.3-32}$$

For two dimensions this algorithm reads

$$\begin{bmatrix} M_{11} & M_{12} \\ M_{21} & M_{22} \end{bmatrix} = \begin{bmatrix} A_{11}B_{11} + A_{12}B_{21} & A_{11}B_{12} + A_{12}B_{22} \\ A_{21}B_{11} + A_{22}B_{21} & A_{21}B_{12} + A_{22}B_{22} \end{bmatrix}. \tag{A.3-33}$$

Matrices that have the same dimensions can be added, with the rule

$$\mathbf{M} = \mathbf{A} + \mathbf{B}, \quad M_{ij} = A_{ij} + B_{ij}. \tag{A.3-34}$$

The multiplication rule of eqn (A.3-32) only requires that the number of rows in $\mathbf{A}$ must match the number of coluns in $\mathbf{B}$. Only if the matrices A and B are square and commute will the order of multiplication not matter, giving $\mathbf{M} = \mathbf{M}'$ where

$$\mathbf{M}' = \mathbf{B}\mathbf{A}, \quad M'_{ij} = \sum_k B_{ik}A_{kj}. \tag{A.3-35}$$

Some matrix classes. A *diagonal* matrix is a square matrix that has elements M_{ij} only along the principal diagonal (where $i = j$), as

$$\mathbf{M} = \begin{bmatrix} M_{11} & 0 & \cdots \\ 0 & M_{22} & \cdots \\ \vdots & \vdots & \ddots \end{bmatrix}. \tag{A.3-36}$$

A *unit matrix* of dimension N, denoted $\mathbf{1}_N$, is a square diagonal matrix that has unity as all its nonzero elements, $M_{ii} = 1$. The *trace* of any square matrix $\mathbf{M}$ is the sum of its diagonal elements,

$$\text{Tr } \mathbf{M} = \sum_n M_{nn}. \tag{A.3-37}$$

Relationships amongst elements of a matrix have important consequences. From a matrix $\mathbf{M}$ having elements M_{ij} we form the following matrices

Transpose	$\mathbf{M}^T$	elements	M_{ji}
Complex conjugate	$\mathbf{M}^*$	elements	M_{ij}^*
Hermitian adjoint	$\mathbf{M}^\dagger$	elements	M_{ji}^*

A *symmetric matrix* is one that is equal to its transpose, $M = M^T$; it has reflection symmetry across the principal diagonal, with $M_{ij} = M_{ji}$. An *Hermitian matrix* is equal to its Hermitian adjoint $M = M^\dagger$. Reflection across the principal diagonal produces the complex-conjugate element, $M_{ij} = M_{ji}^*$.

The *inverse* M^{-1} of an N-dimensional matrix M, acts to produce the unit matrix,

$$M^{-1}M = \mathbf{1}_N. \tag{A.3-38}$$

A *unitary* matrix U is one that preserves lengths of complex-element vectors, implying that its inverse is its Hermitian adjoint,

$$|U\mathbf{V}|^2 = |\mathbf{V}|^2, \quad \text{so} \quad U^{-1} = U^\dagger. \tag{A.3-39}$$

The determinant of a unitary matrix has unit magnitude.

Eigenvectors. If the action of M is to multiply the vector upon $\mathbf{V}$ by a number (possibly complex) ε,

$$M\mathbf{V} = \varepsilon\mathbf{V}, \tag{A.3-40}$$

the vector $\mathbf{V}$ is said to be an *eigenvector* (or *eigenstate*) of the operator M. The scalar ε is the *eigenvalue* of M. All the eigenvalues of any Hermitian matrix are real numbers and so the association of measurable quantities is with Hermitian matrices. The Hamiltonian is chief amongst these: When it is represented by an Hermitian matrix the observable energies are real valued.

Diagonalization. The product $M'\mathbf{V}'$ of a square matrix M' and a vector $\mathbf{V}'$ can often be expressed as the action of a transformed matrix $M = UM'U^{-1}$ upon a vector $\mathbf{V} = U\mathbf{V}'$ in an altered coordinate system:

$$\text{if } \mathbf{V} = U\mathbf{V}' \text{ and } M = UM'U^{-1} \quad \text{then } M'\mathbf{V}' = M\mathbf{V}. \tag{A.3-41}$$

If $\mathbf{V}$ is an eigenvector of M, as in eqn (A.3-40), then the transforming matrix U has brought M' to diagonal form—it has *diagonalized* M'. The diagonal elements of M are the eigenvalues of M and M'. Numerous procedures are available, both analytically and numerically, for finding the transformation matrix U. When M is Hermitian then U is unitary and the transformation is a length-preserving rotation (in a metric space).

Matrix functions. The definitions of matrix multiplication and addition allow the definition of matrix polynomials involving matrix powers,

$$M^2 = MM, \quad M^3 = MMM, \quad \text{etc.,} \tag{A.3-42}$$

and, from these, more general functions of square matrices. Formulas that are used to define power-series expressions for functions of complex variables allow direct application to functions of matrices. The simplest of these is the geometric series formula, applicable for $|M| < 1$,

$$[1 - M]^{-1} = \sum_{n=0}^{\infty} M^n = 1 + M + M^2 + M^3 + \cdots. \tag{A.3-43}$$

An important example is the definition of the exponential of a matrix,

$$\exp(\mathsf{M}) = \sum_{n=0}^{\infty} \frac{\mathsf{M}^n}{n!} = 1 + \frac{\mathsf{M}}{1!} + \frac{\mathsf{M}^2}{2!} + \frac{\mathsf{M}^3}{3!} + \cdots . \qquad \text{(A.3-44)}$$

Integrals involving matrices are obtainable in a similar way, for example:

$$\int_{-\infty}^{\infty} ds \, \exp(s\mathsf{M}) = \mathsf{M}^{-1}. \qquad \text{(A.3-45)}$$

From the exponential of an Hermitian matrix H one obtains a unitary matrix U,

$$\exp(-\mathrm{i}\mathsf{H}t) = \mathsf{U}(t). \qquad \text{(A.3-46)}$$

Matrices have a place in the mathematics of Linear Algebra, but it is from their use as tools for quantum physics that matrices find direct application to the world of photons, a formalism that will be introduced (as Matrix Mechanics) in Appendix A.7.

A.4 Quantization

The notion of abstract vector spaces described in Appendix A.3, particularly Hilbert spaces, is well suited to storing the information associated with states of motion, particularly the discrete-energy states that occur with quantum theory. But how are these information-storing abstract vectors to be used for particular quantum systems—say, for the description of photons? And what mathematics is to describe their alteration to accommodate experimental observations that will change the information they embody? Somehow the rules of system behavior expressed by equations of motion—those of Newton, Lagrange, and Hamilton discussed in Appendix A.1 – must be associated with the regulation of Hilbert-space changes with time. Paragraphs below outline the underpinning essentials of that mathematics—quantum theory leading to the time-dependent Schrödinger equation, the fundamental equation of motion for quantum systems in the absence of decoherence.

A.4.1 Quantum measurement math; Commutators

One of the notable properties of quantum systems is that simultaneous measurements of different observables—different dynamical variables—may only be possible with limited accuracy, as stated by the uncertainty principle. This limitation is particularly relevant to measurements of the paired canonical coordinate and momentum variables defined by the Lagrangian and appearing in the Hamiltonian. The measuring A of one such variable will unavoidably alter the system and affect the subsequent measuring B of a canonically-linked second variable: The two measurement actions do not commute,[10] as symbolized by the expression $AB \neq BA$.

Quantum operators. The world of classical mechanics is most simply and directly converted into quantum mechanics by treating canonical coordinate and momentum

---

[10]Noncommuting actions in classical mechanics occur in group theory, where they underlie non-Abelian groups; see Appendix A.18. An example is a succession of rotations about different axes: The resulting rotation depends upon the ordering of the two constituent rotations.

variables as *operators* acting in a Hilbert space. These operators are assumed to induce noncommuting action on Hilbert-space vectors. In this action the commutators are postulated to be expressed as complex numbers, imaginary multiples of Planck's constant. In one dimension, a single degree of freedom, the most common representation of quantum theory is invoked by replacing a coordinate x and its canonical momentum p by *operators* $\hat{x}$ and $\hat{p}$,

$$x(t) \Rightarrow \hat{x}, \qquad p(t) \Rightarrow \hat{p}, \tag{A.4-1}$$

that have the commutation relations

$$[\hat{x}, \hat{p}] \equiv \hat{x}\hat{p} - \hat{p}\hat{x} = i\hbar. \tag{A.4-2}$$

More generally, for collections of independent variables, the proposed change from classical to quantum theory is made by replacing pairs of classical canonical variables by paired operators that obey the commutation relations

$$[\hat{q}_i, \hat{p}_j] \equiv \hat{q}_i\hat{p}_j - \hat{p}_j\hat{q}_i = \delta_{ij}\, i\hbar. \tag{A.4-3}$$

Such relationships of paired canonical variables can be imposed for any degree of freedom, whatever may be the generalized coordinate—not just single particles but collective motions of any size (see Section 8.6.4). With this algebra of variables (see Appendix A.17) quantum theory becomes installed in the dynamics; this is where the quantum rabbit enters the seemingly magic hat of quantum dynamics and it becomes *quantized.*

Much literature deals with possible justification for these commutator expressions. Here I accept them as falsifiable postulates that have consistently provided demonstrably correct predictions—the foundation upon which quantum theory is to be constructed, as described in Appendix A.4.3.

Quantum observables as eigenvalues. Quantum theory requires that any observable—anything that is measurable, say a generalized coordinate—be associated with an operator, and that the measured values of the observable come from the set of eigenvalues of the operator. Discrete eigenvalues means discreteness in measured values of some dynamical variable.

When it exists[11], a statevector provides a probability distribution of those measured values for a specific situation—a link between relevant experiment and theory. Each degree of freedom has its paired canonical operators and its probabilities. The statevector, being a complete repository of information, can also provide conditional probabilities associated with multiple degrees of freedom.

A.4.2 Uncertainty; Variance

A traditional measure of uncertainty in measurements of some observable for either a quantum system or a classical stochastic system, say $\hat{X}$, is the spread of values that occur in repeated measurements, the *standard deviation* or *dispersion*

---

[11]As a minimum, a statevector requires coherence in the described dynamics.

$$\Delta X = \sqrt{\langle \hat{X}^2 \rangle - \langle \hat{X} \rangle^2}. \tag{A.4-4}$$

The square $(\Delta X)^2$ is the *variance* of $\hat{X}$; see eqn (2.11-3). The size of ΔX depends upon both the particular variable and upon the particular system. For a quantum system it depends on the particular quantum state used for the measurements. If the state is an eigenstate of operator $\hat{X}$ then there is no uncertainty in measuring $\hat{X}$.

The uncertainty product of two operators $\hat{X}$ and $\hat{Y}$ must exceed half the absolute value of the expectation value of their commutator [Car68],

$$\Delta X \, \Delta Y \geq \tfrac{1}{2} \langle \hat{C} \rangle, \quad \text{where} \quad \hat{C} = [\hat{X}, \hat{Y}] \equiv \hat{X}\hat{Y} - \hat{Y}\hat{X}. \tag{A.4-5}$$

For quantum systems this limit gives the traditional position-momentum uncertainty relation

$$\Delta p \, \Delta q \geq \tfrac{1}{2} \hbar. \tag{A.4-6}$$

Two operators can be measured simultaneously with unlimited precision only if they commute.

This brief description of variance and uncertainty gives a simple and direct procedure for converting the description of a mechanical system into one that satisfies all the requirements of quantum theory—of quantizing its system variables, whatever they may be. The primary restriction is that the mechanical system be conservative—maintaining constant energy, unaffected by friction or other energy losses.

A.4.3 Quantization procedure

The steps to convert a conservative classical system to a quantum one are:

First: Define generalized coordinates.

Second: Construct a Lagrangian.

Third: Determine the canonical momenta.

Finally: Require that the variables have the commutators eqn (A.4-3).

This procedure produces a physics in which expectation values of variables obey uncertainty relations, a defining characteristic of quantum mechanics. It is with eqn (A.4-3) that quantum behavior is installed in any dynamics. This is "where the rabbit goes into the hat" (as one says of the stage magician). This is the origin of the difference between classical and quantum mechanics and, as discussed in Appendix B, the difference between classical and quantum electromagnetic fields.

The procedure need not be limited to electrons in their atomic and molecular setting, but can be applied to macroscopic-scale resistanceless electrical circuits of superconductors in which voltages and currents are basic variables [Yur84; Dev04; You05; Wen05; Gir11; Xia13; Wen17; Gu17; Voo17].

Not only do the coordinates and canonical momenta that enter the Hamiltonian find use in establishing quantum mechanics, through commutators, but the Hamiltonian itself is critical: The Hamiltonian operator governs the time evolution of a (nonrelativistic) quantum system as embodied in a statevector. In many situations the essential properties of interest derive directly from the Hamiltonian, and its energy values, requiring no concern with behavior of the underlying canonical variables.

Operators: Wave mechanics or matrix mechanics. Operators, as the term is being used here, must operate on (act on) something. Two possibilities occur in quantum physics: (a) functions (wavefunctions) and (b) abstract vectors (statevectors). The two choices gave rise to two complementary approaches to quantum mechanics, known historically as (a) wave mechanics (see Appendix A.5) and (b) matrix mechanics (see Appendix A.7).

Operator ordering. Once the basic canonical variables of a system have been converted into noncommuting operators the classical-physics expression for the Hamiltonian follows. However, because the Hamiltonian involves products of operators this step requires some choices when those operators originate with complex quantities A or sums of positive and negative frequency parts, as they do for harmonic oscillators and field operators:

$$A \to \hat{A}, \qquad A^* \to \hat{A}^\dagger. \qquad (A.4\text{-}7)$$

Two choices for ordering the operators are common [Mil73; Mil94]:

$$\text{normal ordering: } |A|^2 \Rightarrow \hat{A}^\dagger \hat{A}, \qquad (A.4\text{-}8a)$$

$$\text{symmetric ordering: } |A|^2 \Rightarrow \tfrac{1}{2}[\hat{A}^\dagger \hat{A} + \hat{A}\hat{A}^\dagger] = \hat{A}^\dagger \hat{A} + \tfrac{1}{2}[\hat{A}, \hat{A}^\dagger]. \qquad (A.4\text{-}8b)$$

When applied to quantization of harmonic oscillators the symmetric ordering leads to a zero-point energy $\hbar\omega/2$ for each mode, an energy that is absent with normal ordering. With symmetric ordering the Lamb shift and spontaneous emission are attributed to vacuum fluctuations of photons [Gla06], whereas with normal ordering these effects are attributed to radiation reaction of a dipole moment [Mil73; Mil94].

A.5 Wave mechanics and wavefunctions

The form of quantum mechanics most commonly taught to chemists, known as wave mechanics from its use of spatial wave fields, introduces differential operators, acting on functions of position, to meet the commutator requirements. For coordinate x the operator $\hat{x}$ becomes just the number x while the conjugate momentum operator $\hat{p}$ becomes an imaginary multiple of the partial derivative,

$$\hat{x} \Rightarrow x, \qquad \hat{p} \Rightarrow -i\hbar \frac{\partial}{\partial x}. \qquad (A.5\text{-}1)$$

This replacement[12] makes the kinetic energy of the Hamiltonian a differential operator that is second order in spatial variation. For motion of a single structureless particle of mass m in one dimension the Hamiltonian becomes a differential operator involving x and its second derivative

$$\mathcal{H}(\hat{x}, \hat{p}) = \frac{\hat{p}^2}{2m} + V(\hat{x}) \Rightarrow \hat{\mathcal{H}}(x) = -\frac{\hbar^2}{2m}\frac{\partial^2}{\partial x^2} + V(x), \qquad (A.5\text{-}2)$$

where $V(x)$ is the potential energy. The complex-valued functions of position this Hamiltonian acts on are known as *wavefunctions*—so named because they show the

---

[12]The choice in eqn (A.5-1) defines the *position representation* of quantum theory. The alternative *momentum representation* introduces a derivative for the position operator. The associated momentum wavefunction, eqn (A.6-1), provides a distribution of velocities rather than positions.

nodes and interferences characteristic of standing waves.[13] They provide, via the *time-independent* Schrödinger equation,

$$\mathcal{H}(\hat{x}, \hat{p}) \, \psi(x) = E \, \psi(x), \tag{A.5-3}$$

wavefunctions $\psi(x)$ whose absolute squares are interpreted as the stationary, steady-state probability $P(x)$ of finding the one-dimensional particle to be within an arbitrarily small neighborhood of the point x,

$$P(x) = |\psi(x)|^2. \tag{A.5-4}$$

For one-dimensional motion of a structureless particle having mass m the time-independent Schrödinger equation becomes the second-order ODE:

$$\left[\frac{\hbar^2}{2m} \frac{d^2}{dx^2} - V(x) + E \right] \psi(x) = 0. \tag{A.5-5}$$

Solutions to this time-independent Schrödinger equation represents *stationary states*. The energy E appearing in the Schrödinger equation eqn (A.5-3) is an example of an *eigenvalue*, a number that is fixed by the Hamiltonian operator and its boundary conditions.

Expectation values from wavefunctions. The wavefunction provides, in addition to a probability amplitude, a weighting function for evaluating the stationary expectation value of any function of the quantum variables, through the integral

$$\langle F \rangle = \int dx \, \psi(x)^* F(x, \hat{p}) \psi(x). \tag{A.5-6}$$

The wavefunction. What are we to make of the wavefunction that occurs with the time-independent Schrödinger equation? How are we to think of it? This question has drawn answers throughout the decades of its use, for it is often seen as embodying the essence of quantum theory. An article by Raymer [Ray97] offers a particularly useful basis for understanding wavefunctions: The wavefunction is not a physical field, such as an electric or gravitational field. It is an *"information structure"* that incorporates all the details, consistent with such restrictions as the uncertainty principle, that can be used to describe the state of a particular quantum system—to quantify the details that define alternative possible distinguishable motions of the system constituents. This information can be determined, using techniques such as quantum tomography [Ber87; Leo95; Ray97; Zav04; Lvo09], from measurements on multiple systems prepared under identical conditions.

Boundary conditions. A wavefunction $\psi(x)$ is *defined* as the solution to a PDE, and is *interpreted* as a probability-amplitude distribution. The full definition of the solution to this PDE requires some specification of boundary conditions. These are of two sorts: Bound states and scattering states.

---

[13]Electromagnetic wave equations are second order in time derivatives and have real-valued solutions. The time-dependent Schrödinger wave equation is first order in time and deals with complex-valued solutions.

Bound states. The wavefunction is spatially confined (possibly by imposing box-like artificial boundaries). In order that the possible probabilities of being somewhere add to unity it is necessary that the wavefunctions be *square integrable*, meaning its norm $|\psi(x)|$ is everywhere finite and can be adjusted (normalized) to give unit integral,

$$\int dx \, |\psi(x)|^2 = 1. \tag{A.5-7}$$

It is this requirement of square integrability, when applied to attractive potentials, that leads to discrete solutions to the time-independent Schrödinger equation and to the prediction of discrete-energy quantum states for confined systems. In application to atomic and molecular physics there are commonly an infinite set of discrete energy eigenvalues for a given Hamiltonian. Initial conditions pick out the particular combination of interest for each specific situation.

Scattering states. Alternatively, the wavefunction may be unconfined (a scattering state that extends indefinitely in space). Such wavefunctions have energy values taken from a continuum of eigenvalues. The scattering states describe situations such as photoionization into an electron and an ion, and molecular dissociation into two atomic fragments. Scattering-state wavefunctions are traveling waves, whereas bound-state wavefunctions are stationary, standing waves. It is with the discrete bound states that I shall primarily be concerned because it is with their discrete-energy states that discrete packets of radiation energy (photons) appear and disappear.

Quantum degrees of freedom. We generalize the Schrödinger equation eqn (A.5-3) and its wavefunction to systems involving multiple variables by interpreting the argument x as denoting an appropriate set of coordinates. Each of the spatial variables of the resulting wavefunction represents one *degree of freedom* (DoF) for the particle. When the particle is constrained to move in only one direction, then its wavefunction has just one DoF. A system with two DoFs, from two coordinates or two distinct one-dimensional particles (or two separable sets), is termed *bipartite*.

A group of N spinless quantum particles bound together has $3N$ DoFs. Three of these can be taken as coordinates (or momentum) of the center of mass for the group. When dealing with molecules the motion of the molecular framework of N nuclei is treated by separating the center of mass motion (3 DoFs) and the overall orientation (3 angles in general, 2 angles for a linear molecule because rotation about the figure axis is not detectable). This leaves $3N - 6$ DoFs for the framework of a nonlinear molecule, to be treated as various stretching and bending motions of the chemical bonds that hold the molecule together.

A collection of N spin-half particles (fermions) has additional degrees of freedom, expressible as as N projections, $\pm\frac{1}{2}$, of their spins along any arbitrary quantization axis (see Section 2.2.8). Most often these spins are paired in stable structures, leaving no overall collective spin when N is an even integer.

Bound-state wavefunctions. The possible energy eigenvalues appearing in the time-independent Schrödinger equation eqn (A.5-3) and its multidimensional generalizations typically form an infinite set, unbounded above. For simple systems it is often possible to label the wavefunctions by a set of quantum numbers (not necessarily integers), one for each degree of freedom. More generally, we can simply arrange the set of solutions according to some convenient algorithm (typically including energy) so that the Schrödinger equation can be written

$$\mathcal{H}^0(x, \hat{p})\psi_n(x) = E_n\psi_n(x), \tag{A.5-8}$$

where x denotes a complete set of coordinates for all the degrees of freedom being considered and the label n is an identifier of the multi-coordinate wavefunction. Here $\mathcal{H}^0$ is shown as a function of x, but this dependence includes derivatives that originate with momenta, indicated by argument $\hat{p}$. The superscript on the Hamiltonian operator $\mathcal{H}^0$ anticipates the introduction of interaction Hamiltonians. The wavefunctions $\psi_n(x)$ are termed *eigenfunctions* of the given time-independent Hamiltonian $\mathcal{H}^0(x)$. If it is known with certainty that the system is in a stationary state whose wavefunction is $\psi_n(x)$ then the probability that the degrees of freedom associated with coordinate x will be found within an arbitrarily small neighborhood of the point x is

$$P_n(x) = |\psi_n(x)|^2. \tag{A.5-9}$$

The probability that the system will be in the state n associated with energy E_n is the integral over all possible positions associated with that state,

$$P_n = \int dx |\psi_n(x)|^2. \tag{A.5-10}$$

Summation of these probabilities over a complete set of states must give unit probability,

$$\sum_n P_n = 1. \tag{A.5-11}$$

Separation of variables; Product functions. Often it is possible to express a function $F(x_1, x_2, \cdots)$ of several independent variables (one for each DoF), here $x_1, x_2, \cdots$, in terms of products of functions,

$$F(x_1, x_2, \cdots) = \sum_{a,b,\cdots} C_{ab\cdots} f_a(x_1) g_b(x_2) \cdots . \tag{A.5-12}$$

Such a separation of variables is always possible for the wavefunctions of a single electron bound in a central potential (i.e. a force directed toward a single point). Omitting spin, the needed variables can then be taken as three spherical coordinates r, ϑ, φ and the separable functional relationship is that of a single-electron spatial orbital,

$$\psi_{n\ell m}(r, \vartheta, \varphi) = R_{n\ell}(r)\, \Theta_{\ell m}(\vartheta)\, \Phi_m(\varphi) = R_{n\ell}(r)\, Y_{\ell m}(\check{r}), \tag{A.5-13}$$

in which the directional properties appear as a spherical harmonic $Y_{\ell m}(\check{r})$, see eqn (A.7-22). Product functions occur in treatments of the wavefunction of two independent particles bound together; the variables are the sum of two sets of individual variables.

Other product-function constructions occur with separable solutions to the Helmholtz equation for classical electromagnetic fields discussed in Appendix B.2 and eqn (A.5-16) below.

In all cases it is necessary that the set of individual functions form a *complete set*, as do, for example, spherical harmonics as a basis for describing behavior involving two angles. The numerical coefficients (possibly complex valued) $C_{ab...}$ of eqn (A.5-12) must be chosen to fit specific subsidiary conditions, thereby picking out a particular case from all possibilities.

Spin. Electrons (and photons) have not only dynamical degrees of freedom associated with motion, such as momentum and angular momentum, they also have an *intrinsic* angular momentum, termed *spin*.

With inclusion of electron spin and the use of spherical coordinates a stationary-state electron wavefunction is expressible as

$$\psi(\check{s}, \mathbf{r}) = \sum_{\mu n \ell m} C_{\mu n \ell m} \, \chi_\mu(\check{s}) \, \psi_{n \ell m}(r, \vartheta, \varphi). \tag{A.5-14}$$

The unit *spinors* $\chi_{mu}(\check{s})$ appearing here are implementations of spin-half states in an abstract space independent of the spatial wavefunction. The argument $\check{s}$ is a unit vector that defines the direction of the quantization axis. Spinors are counterparts for electrons (or other fermions) of the unit vectors used for electromagnetic fields. For any established quantization direction for spin, $\check{s}$, they have two possible projections ($\mu = +\frac{1}{2}$ or "up" and $\mu = -\frac{1}{2}$ or "down"),

$$\chi_+ = |+\rangle = |\uparrow\rangle, \qquad \chi_- = |-\rangle = |\downarrow\rangle. \tag{A.5-15}$$

These two spinors are mathematical counterparts of the three unit vectors $\mathbf{e}_i$ used for electric fields. As will be discussed in Appendix B.1.3, an electric field is expressible in product form as amplitudes $\mathcal{E}_i(\mathbf{r}, t)$ of three independent, orthogonal unit vectors $\mathbf{e}_i(\mathbf{r})$,

$$\mathbf{E}(\mathbf{r}, t) = \sum_i \mathbf{e}_i(\mathbf{r}) \, \mathcal{E}_i(\mathbf{r}, t). \tag{A.5-16}$$

The unit vectors $\mathbf{e}_i(\mathbf{r})$, discussed in Appendix B.1.6, are implementations of spin-one states in a space independent of function values. Commonly they are presented in Cartesian coordinates, with index i denoting x, y, z components. Their argument, an angle-defining unit vector $\mathbf{r}$, is a reminder that they depend upon the chosen coordinate orientation.

Although the spin, like the electric charge, is an intrinsic, unchangeable characteristic of a particle, it is like any angular momentum: Its projection onto a specified direction (the *quantization axis*) can take different values and must be reckoned as one additional degree of freedom to the three spatial DoFs for every particle. Thus a single electron (or a photon) requires four numerical parameters—four quantum numbers—to characterize its motion. Those individual attributes can be combined into collective states in various ways. Most notably the individual spins can combine into a total collective spin, part of a total collective angular momentum [Bri68; Zar88; Sho90; Lou06].

Collective excitations; Quasiparticles.   Electrons and nuclei that are bound collectively as a molecule are best regarded not as individual point charges but as a framework of (heavy) nuclei surrounded by an adapting cloud of electron charge. The needed degrees of freedom for the molecular framework are most conveniently taken to be from translational motion of the center of mass, from overall rotation of the molecule (its angular momentum), and from various normal modes of vibration. Placing energy into molecular structure means adding energy to collective modes of nuclear motion.

Other aggregates of electrons and nuclei also require treatment by treating motions collectively. Solids, particularly crystals, involve degrees of freedom too numerous to allow individual-atom consideration; they pose a problem in *many-body physics* that is best treated by introducing *normal modes* of small displacements whose patterns correspond to various wave patterns—a concept from classical physics that leads to quantization by way of harmonic oscillators for each mode. These collective vibrations are observable as sound waves. Like the disturbances of the electromagnetic field that are observed as radiation waves and regarded quantum-mechanically as photons, quantized collective wavelike disturbances (acoustic waves) in the rigid framework of solids are given a name suggestive of particle-wave duality, *phonons*—acoustic counterparts of photons [Bor54; Zim60], a "quantum of lattice vibrations" [Joh02]. Their randomization is heat, the end result of absorbed radiation in bulk matter.

When the lattice-framework distortion of a dielectric solid produces small charge separations, and thereby an alteration of the polarization field that occurs in Maxwell's equations, the quantized charge-displacement modes are known as *polarons*. (The polaron, a fermionic quasiparticle, should not be confused with the *polariton*, a bosonic quasiparticle that combines a photon and an optical phonon.) Collective magnetic-spin oscillation modes are called *magnons*.

A.6 Phase space

The momentum distribution.   For the simplified model of eqn (A.5-5) above, of a system that has one coordinate x, the wavefunction $\psi(x)$ is associated with a conditional probability: Its absolute square, evaluated for position x, is the probability that the system coordinate will have the value x given that the system was prepared in accordance with some set of reproducible conditions.

Associated with every coordinate x used in describing a quantum state there is a canonical momentum p_x. The Fourier transform of the spatial wavefunction $\psi(x)$ provides a *momentum wavefunction*

$$\tilde{\psi}(p_x) = \frac{1}{\sqrt{2\pi\hbar}} \int dx \exp(-ixp_x/\hbar)\,\psi(x), \tag{A.6-1}$$

whose absolute square gives the (conditional) probability of finding momentum value within an infinitesimal neighborhood of the value p_x,

$$P(p_x) = |\tilde{\psi}(p_x)|^2. \tag{A.6-2}$$

From the mathematical nature of Fourier transforms it follows that a sharply localized spatial wavefunction requires a broad distribution of momentum values. Conversely, to create a state with sharply defined momentum, as we have with radiation beams (and photons), we require a broad distribution of position values.

The Wigner distribution. Descriptions of classical dynamics often make use of simultaneous values of position and velocity (or momentum). The resulting abstract space, of position and momentum values for each degree of freedom, is termed *phase space* [Ray97; Sch01; Mac03; Lee05]. The statistical description of a point in this space, for a single coordinate x, involves the joint probability $P(x, p_x)$ of finding specified values of coordinate and momentum. The uncertainty relations of quantum theory, expressed through commutation relations between operator versions of these canonical variables, is incompatible with such a joint probability for a quantum system. However, a variety of alternative functions of position and momentum have been found useful. These differ from proper probability distributions by allowing negative values—they are known as *quasiprobability* distributions.

The most commonly encountered example is the Wigner function, definable as the integral of a product of two spatially-offset wavefunctions [Ste78; Ber87; Roy89; Asp01; Sil07; Cas08; Lvo09; Alo11]. For the simple example of a single degree of freedom, with coordinate x, the definition reads

$$W(x, p_x) = \frac{1}{2\pi\hbar} \int dx' \, \exp(-ix'p_x/\hbar) \, \psi^* \left(x - \tfrac{1}{2}x'\right) \psi \left(x + \tfrac{1}{2}x'\right). \qquad \text{(A.6-3)}$$

This is a Fourier transform, but in addition to having dependence upon p_x it retains dependence upon x. The subsequent integration over either x or p_x produces a probability distribution of the remaining variable. The (conditional) probability of finding the particle has coordinate x given that it is in quantum state ψ is ,

$$\int dp_x \, W(x, p_x) = |\psi(x)|^2 = P(x|\psi). \qquad \text{(A.6-4)}$$

Similarly, the probability of finding momentum p_x is obtained from the integral over coordinate x,

$$\int dx \, W(x, p_x) = |\tilde{\psi}(p_x)|^2 = P(p_x|\psi). \qquad \text{(A.6-5)}$$

The occurrence of negative values of the Wigner function is evidence that the system has quantum character; see eqn (B.4-36) of Appendix B.4.

Wavefunction drawbacks. The generalization of the time-independent Schrödinger equation from one dimension to many dimensions, as is needed for a multi-electron atom or other elaborate system, brings a complicated multi-variable partial differential equation that poses a computational challenge for all but the simplest idealizations. Typically it is necessary to introduce a variety of approximations, such as expressing the wavefunction as a superposition of products of factors for each degree of freedom. The result of such calculations, when they are possible, provide vastly more information that is usually required for treating gains and losses of energy. The alternative approach of matrix mechanics, based on much more limited information, is better suited to the needs of quantum optics and its studies of photons.

A.7 Matrix mechanics and operators

The alternative to wave mechanics, and to the second-order PDEs that were obtained by Schrödinger, is based upon representing noncommuting operators as finite-dimensional square matrices (acting on vectors in a Hilbert space), a formulation devised by Heisenberg, Max Born, and Pascual Jordan [Bor25; Bor26; Bow80; Ber05] and known for a time as matrix mechanics. This formalism is well suited to descriptions of changes in discrete-energy states induced by radiation. The connection between wavefunctions and statevectors is as follows.

Wavefunctions as Hilbert-space vectors. Given a discrete set of wavefunctions we obtain a Hilbert space by identifying each wavefunction with a *basis vector*, combining the quantum numbers as a vector label:

$$\psi_n(x) \Rightarrow \boldsymbol{\psi}_n \quad \text{or} \quad |\psi_n\rangle \quad \text{or} \quad |n\rangle. \tag{A.7-1}$$

Each of these vectors represents a possible stationary state of motion, a possible *statevector* that fully describes what can be known about the system when undisturbed. Wavefunction arguments (x) do not appear in the abstract-vector picture.

We define a scalar product between such abstract vectors in terms of multiple integrals over the several coordinates, symbolized as

$$\langle \psi_n | \psi_m \rangle = \int dx\, \psi_n(x)^*\, \psi_m(x). \tag{A.7-2}$$

The wavefunctions can be chosen to be orthogonal, meaning that the quantum states are disjoint—the system can be determined to be unambiguously in one specific state, and to be normalized to unity, as is needed for a probability interpretation of eqn (A.5-10). The mathematical expression for these requirements, expressed in terms of scalar products, is, for discrete states, a Kronecker delta [see eqn (A.3-10)]

$$\langle \psi_n | \psi_m \rangle = \delta_{nm}. \tag{A.7-3}$$

The statement that the set of wavefunctions is complete, meaning that the system must be in one of the states, reads

$$1 = \sum_n |\psi_n\rangle\langle\psi_n|. \tag{A.7-4}$$

The sum must include an integral over any continuum wavefunctions.

Integrals of the Hamiltonian between two wavefunctions can be organized as a square matrix H^0, with matrix elements[14]

$$H^0_{nm} = \int dx\, \psi_n(x)^*\, \mathcal{H}^0(\hat{x})\, \psi_m(x) = E_n\, \delta_{nm}. \tag{A.7-5}$$

Because, by definition, the wavefunctions satisfy the Schrödinger equation eqn (A.5-8) the matrix H^0 is diagonal, with entries E_n attributed to the stationary states of the system and the degrees of freedom that are being subject to quantum theory.

---

[14]The Hamiltonian $\mathcal{H}(\hat{x})$ can include derivatives with respect to x.

More generally, any function $F(\hat{x}, \hat{p})$ of the basic quantum variables can be represented as a square-matrix F having elements evaluated from a given set of basis functions $\psi_n(x)$,

$$F_{nm} = \int dx \, \psi_n(x)^* \, F(\hat{x}, \hat{p}) \, \psi_m(x). \tag{A.7-6}$$

Most models of atoms and molecules involve an infinite number of discrete quantum states for each degree of freedom. Typically only a few of these states are needed for satisfactory descriptions of coherent manipulation of quantum states (the *essential states approximation*). These are states which, during the course of an experiment, are occupied with noticeable probability. The resulting formalism requires only a few matrix elements, quite independent of any details of the wavefunctions that may, in principle, underlie their definition. The details of wavefunctions (e.g. the difference between hydrogen and oxygen or their combination as water) enter the formalism only indirectly, through matrix elements such as the integral of eqn (A.7-5).

Matrix mechanics and Lie algebra. All of the quantitative properties of quantum systems—the energy values and the transition rates—are embodied in the numerical values of matrix elements. Like other students of my generation I was taught that the mastery of differential equations (of wavefunctions) was the key to understanding the quantum theory of atoms and molecules. It therefore came as a pleasing surprise to discover that the needed numbers could be obtained more directly through simple algebra, and that the properties of operators acting on Hilbert-space vectors can be determined completely from commutator relationships (I subsequently learned that the required mathematics was an example of Lie algebra; see Appendix A.18.4). Two examples are particularly important: The harmonic oscillator and angular momentum.

A.7.1 The harmonic oscillator; Number operator

The harmonic oscillator is one of the models of a system that permits exact closed-form expressions for behavior in classical mechanics and quantum mechanics, and so it is invariably discussed at some length in all introductory textbooks of these two subjects. For one-dimensional motion of a particle of mass m about the equilibrium position $x = 0$, subject to a restoring force that varies linearly with displacement, the Hamiltonian can be written as

$$\mathcal{H}(\hat{x}, \hat{p}) = \frac{\hat{p}^2}{2m} + \frac{m\omega^2}{2}\hat{x}^2, \tag{A.7-7}$$

where the fundamental frequency ω replaces the force constant κ as the characterizing parameter. The properties of primary interest in the present context (of photons) are not pictures of the oscillator motion, such as would be revealed by a wavefunction, but energies. Therefore rather than represent the momentum by a derivative, and thereby obtain the time-independent Schrödinger equation for the oscillator wavefunction, it is more useful to work directly with an operator equation. This is most simply done by expressing the conjugate position and momentum operators with alternative operators $\hat{a}$ and $\hat{a}^\dagger$:

$$\hat{x} = \sqrt{\frac{\hbar}{2m\omega}} \, [\hat{a} + \hat{a}^\dagger], \qquad \hat{p} = -i\sqrt{\frac{m\hbar\omega}{2}} \, [\hat{a} - \hat{a}^\dagger]. \tag{A.7-8}$$

From the commutation properties of $\hat{x}$ and $\hat{p}$ it follows that these new operators have the commutator

$$[\hat{a}, \hat{a}^\dagger] = 1. \tag{A.7-9}$$

When used with the number operator $\hat{N}$ and its eigenvectors (eigenstates) $|n\rangle$

$$\hat{N} \equiv \hat{a}^\dagger \hat{a}, \qquad \hat{N}|n\rangle = n|n\rangle, \qquad n = 0, 1, 2, \ldots, \tag{A.7-10}$$

they act as raising and lowering operators of *occupation numbers*, the non-negative integers n that are the eigenvalues of $\hat{N}$,

$$\hat{a}^\dagger|n\rangle = \sqrt{n+1}\,|n+1\rangle, \qquad \hat{a}|n\rangle = \sqrt{n}\,|n-1\rangle. \tag{A.7-11}$$

The hierarchy of energy states begins with the ground state $|0\rangle$,

$$\hat{a}|0\rangle = 0, \qquad \hat{N}|0\rangle = 0. \tag{A.7-12}$$

From direct substitution of the operator definitions eqn (A.7-8) into the operator Hamiltonian eqn (A.7-7) it follows that this Hamiltonian is expressible, with attention to the ordering of noncommuting operators, as

$$\hat{\mathcal{H}} = \frac{\hbar\omega}{2}[\hat{a}\hat{a}^\dagger + \hat{a}^\dagger\hat{a}]. \tag{A.7-13}$$

It is at this point that the choice of operator ordering can be seen: As written here the choice is symmetric ordering but, as remarked in Appendix A.4.3 with eqn (A.4-8), other orderings have their use. By using the commutator relation eqn (A.7-9) and the definition of the number operator, eqn (A.7-10), we write the oscillator Hamiltonian for symmetric ordering as

$$\hat{\mathcal{H}} = \hbar\omega(\hat{N} + \tfrac{1}{2}). \tag{A.7-14}$$

Here the number operator $\hat{N}$ measures the number of excitation increments that are present in the system. The eigenvalue $n = 0$ and its eigenstate $|0\rangle$ represent no excitation; this is the state of minimum allowable energy, the ground state of motion.

Zero-point energy. From eqn (A.7-14) for the Hamiltonian we see that the allowed energies of an harmonic oscillator whose basic frequency is ω are spaced equidistantly, separated by the excitation increment $\hbar\omega$. Courses in quantum theory, and in quantum chemistry, point out that the lowest allowable energy of an harmonic oscillator, the energy when there is no excitation present, is not zero. There is a ground-state energy of $\hbar\omega/2$, meaning half an excitation quantum. The energies occurring for the harmonic oscillators of chemistry (for particles of nonzero rest mass) are therefore

$$E_n = (n + \tfrac{1}{2})\hbar\omega, \qquad n = 0, 1, 2, \ldots. \tag{A.7-15}$$

The factor of $\tfrac{1}{2}$ represents the zero-point of the oscillator energy. It is responsible for preventing coordinate and momentum from each having precisely known values (zero) when the oscillator is in its ground state; the oscillator, like other quantum systems, is never permanently at rest: It undergoes zero-point motion.

A.7.2 Angular-momentum operators; Coupling and spin

In addition to the excitation-number states that are appropriate for an harmonic oscillator, states of angular momentum offer opportunity for using operator algebra. From the conventional definition $\hbar\mathbf{J} = \mathbf{r} \times \mathbf{p}$ of a point-particle motional angular-momentum vector (in units of $\hbar$) there follow commutation properties for three dimensionless angular-momentum operators $\hat{J}_i$ with $i = 1, 2, 3$ or $i = x, y, z$,

$$[\hat{J}_1, \hat{J}_2] = i\hat{J}_3, \quad [\hat{J}_2, \hat{J}_3] = i\hat{J}_1, \quad [\hat{J}_3, \hat{J}_1] = i\hat{J}_2. \qquad (A.7\text{-}16)$$

Angular momentum eigenstates $|j, m\rangle$ of the two commuting operators

$$\hat{J}_3, \qquad \mathbf{\hat{J}}^2 = (\hat{J}_1)^2 + (\hat{J}_2)^2 + (\hat{J}_3)^2, \qquad (A.7\text{-}17)$$

involve two quantum-number labels, known as the total angular momentum j and the magnetic quantum number m,

$$\mathbf{\hat{J}}^2|j, m\rangle = j(j+1)|j, m\rangle, \quad \hat{J}_3|j, m\rangle = m|j, m\rangle, \qquad (A.7\text{-}18)$$

where $2j$ is a non-negative integer and $2m$ is an integer bounded by $-j \le m \le j$.

As with other operators, it is possible to represent the needed angular-momentum properties with either differential operators acting on functions or with matrices acting on vectors. The paragraphs below note these two possibilities.

Coupled angular momentum. When any system, quantum or classical, moves without any orientation-dependent forces or energies, it angular momentum is conserved. For a quantum system whose Hamiltonian is independent of spatial orientation it is convenient (but not necessary) to introduce angular momentum states for its description.

Angular momentum of a single point particle originates with the (orbital) motion of its center of mass around any given point. When the particle has structure then it can also have angular momentum from rotation about its center of mass (spin). A collection of particles may have both kinds of angular momentum for each constituent. Both particles and fields (and their photons) may have angular momentum, identifiable as spin or orbital in origin. What is conserved, in the absence of orientation-dependence of a Hamiltonian, is the total angular momentum of all the parts.

When we look toward defining single photons in terms of spatial modes—either beams or multipoles—we have the opportunity to deal with angular momentum eigenstates, either of photon spin or of spin and orbital angular momentum, as part of the photon definition. For this purpose it is useful to have at hand the mathematics of angular momentum theory that deals with combining (coupling) angular momentum.

From two independent angular momenta vectors $\mathbf{\hat{J}}(1)$ and $\mathbf{\hat{J}}(2)$, of whatever origin, we can define a total angular momentum by summing components,

$$\hat{J}_i = \hat{J}_i(1) + \hat{J}_i(2). \qquad (A.7\text{-}19)$$

The eigenstates of total angular momentum are expressible as a superposition of products of the states $|j_1 m_1\rangle$ and $|j_2 m_2\rangle$ of the constituent angular momentum,

$$|j_1 j_2 JM\rangle = \sum_{m_1 m_2} (j_1 m_1, j_2 m_2 | JM) \; |j_1 m_1\rangle \, |j_2 m_2\rangle, \quad \text{with } m_1 + m_2 = M. \quad \text{(A.7-20)}$$

The coefficients of the expansion are Clebsch-Gordan coefficients [Edm57; Bie81; Zar88; Lou06]. This coupling procedure, to produce eigenstates of total angular momentum, can involve any independent angular momentum operators.[15]

Orbital angular momentum. A special case of angular momentum is the orbital angular momentum (OAM) operator,

$$\hat{\mathbf{L}} = -i\mathbf{r} \times \nabla, \quad \text{(A.7-21)}$$

that acts on function arguments. Its eigenfunctions are spherical harmonics $Y_{\ell m}(\theta, \phi)$, typically identified by the non-negative integer ℓ used for expressing the eigenvalue of $\hat{J}^2$ and the integer m (the magnetic quantum number) as the eigenvalue of angular momentum component $\hat{L}_z$ along a quantization axis:

$$\hat{\mathbf{L}}^2 Y_{\ell m}(\theta, \phi) = \ell(\ell+1) Y_{\ell m}(\theta, \phi), \qquad \hat{L}_z Y_{\ell m}(\theta, \phi) = m Y_{\ell m}(\theta, \phi). \quad \text{(A.7-22)}$$

For spinless particles in a central field wavefunctions can be obtained that are eigenfunctions of $\hat{\mathbf{L}}^2$.

Spin angular momentum. Spin angular momentum is associated not with altering function values by rotations but with altering component values; see Appendix A.18.5. It is an intrinsic, unchangable property of a particle or field that distinguishes, for example, spinors (two components), Euclidean vectors (three components), and Euclidean tensors (nine components).

The three Pauli matrices σ_j and the two-dimensional unit matrix $\mathbf{1}_2$,

$$\sigma_1 = \begin{bmatrix} 0 & 1 \\ 1 & 0 \end{bmatrix}, \quad \sigma_2 = \begin{bmatrix} 0 & -i \\ i & 0 \end{bmatrix}, \quad \sigma_3 = \begin{bmatrix} 1 & 0 \\ 0 & -1 \end{bmatrix}, \quad \mathbf{1}_2 = \begin{bmatrix} 1 & 0 \\ 0 & 1 \end{bmatrix}, \quad \text{(A.7-23)}$$

provide a complete set of four matrices with which to express any arbitrary two-dimensional matrix. They form the basis of a *Pauli algebra* [Bay92; Sug04], a particular example of a Clifford algebra or geometric algebra; see Appendix A.17. These matrices are often given labels of Cartesian axes, $\{1, 2, 3\} \rightarrow \{x, y, z\}$. The Pauli matrices are multiples of the spin-half matrices, which can be taken as[16]

$$\mathsf{s}_j = \tfrac{1}{2}\sigma_j. \quad \text{(A.7-24)}$$

Two-component eigenvectors of these spin matrices are used to describe the orientation of an electron, which has two possible values with respect to any arbitrary direction,

---

[15]The inverse to coupling—the decomposition of the total angular momentum of a photon field into orbital and spin contributions—has been subject to some debate [Lea14; Bar16].

[16]Alternate definitions are possible with sign changes of two of the matrices.

traditionally taken as the z axis. The components are often termed spin up, with state $|\uparrow\rangle$ and spin down, with state $|\downarrow\rangle$:

$$s_z|\uparrow\rangle = +\tfrac{1}{2}|\uparrow\rangle, \qquad s_z|\downarrow\rangle = -\tfrac{1}{2}|\downarrow\rangle. \tag{A.7-25}$$

For three-dimensional Euclidean vectors the relevant spin matrices, those of $j = 1$ (spin one), can be taken as

$$s_x = \begin{bmatrix} 0 & 0 & 0 \\ 0 & 0 & -i \\ 0 & i & 0 \end{bmatrix}, \qquad s_y = \begin{bmatrix} 0 & 0 & i \\ 0 & 0 & 0 \\ -i & 0 & 0 \end{bmatrix}, \qquad s_z = \begin{bmatrix} 0 & -i & 0 \\ i & 0 & 0 \\ 0 & 0 & 0 \end{bmatrix}. \tag{A.7-26}$$

These matrices allow the vector product of two three-dimensional vectors to be written

$$\mathbf{a} \times \mathbf{b} = -i(\mathbf{s} \cdot \mathbf{a})\mathbf{b}. \tag{A.7-27}$$

The three spin-one matrices do not form a complete set of basis vectors for describing arbitrary three-dimensional matrices: A total of nine matrices is needed for that purpose. The additional matrices can be chosen in a variety of ways; often the choice can fit particular symmetries of the Hamiltonian [Hio87; Hio88].

Total angular momentum. The two forms of angular momentum combine to give representations of total angular momentum

$$\hat{\mathbf{J}} = \hat{\mathbf{L}} + \hat{\mathbf{S}}, \tag{A.7-28}$$

in which $\hat{\mathbf{L}}$ acts on function arguments and $\hat{\mathbf{S}}$ acts on components. The construction of eigenstates of total angular momentum from any independent eigenstates of orbital and spin angular momentum involves a superposition with Clebsch-Gordan coefficients $(S\mu, \ell m | JM)$, a *coupled state*

$$|S\ell JM\rangle = \sum_{m\mu}(S\mu, \ell m | JM)\,|S\mu\rangle\,|\ell m\rangle, \qquad \text{with } m + \mu = M. \tag{A.7-29}$$

These coupled states are labeled by eigenvalues of $\hat{S}^2$ and $\hat{L}^2$ and of

$$\hat{\mathbf{J}}^2 = (\hat{\mathbf{L}} + \hat{\mathbf{S}})^2, \quad \text{and} \quad \hat{J}_3 = \hat{L}_3 + \hat{S}_3. \tag{A.7-30}$$

They provide a complete set of orthonormal states for describing angular motion and spin,

$$\langle S\ell JM | S\ell' J'M'\rangle = \delta_{\ell\ell'}\,\delta_{JJ'}\,\delta_{MM'}. \tag{A.7-31}$$

The uncoupling into states that have well-defined individual magnetic quantum numbers reads, with $M = m + \mu$,

$$|S\mu\rangle|\ell m\rangle = \sum_{J}(S\mu, \ell m | JM)\,|S\ell JM\rangle, \quad \text{with } J \geq 0, \quad |L-S| \leq J \leq L+S. \tag{A.7-32}$$

Such constructions can be applied not only to massive particles such as electrons (in atoms) and atoms (in molecules) but to field modes (photons); see the discussion of vector harmonics, eqn (B.2-33).

A.8   The statevector

A statevector is an element of a multidimensional metric space, an abstract vector whose components are interpreted as probability amplitudes for finding a quantum system in one of a set of possible quantum states. It is an embodiment of information about a particular system at a particular time.

A finite-dimensional Hilbert space offers an infinite collection of vectors from which to make selections. The choices, for any statevector, are mathematically constrained by the requirement that the absolute squares of its components, being probabilities, must sum to unity. So, allowable statevectors must have unit length, but their phase is arbitrary. The individual probability amplitudes that contribute to the statevector might be thought to also have arbitrary phases, but it turns out that various observable quantities depend on products of probability amplitudes, and so these phases are not without observable consequence. As with Euclidean vectors, the values of statevector components depend on the choice of the coordinate system.

The preceding sections have given examples of several operators whose eigenvectors provide possible basis vectors for Hilbert spaces. The operator eigenvalues provide possible labels for these vectors, but for simplicity the vectors will be assumed to have a well-defined ordering and be labeled by integers $1, 2, \ldots, N$.

A.8.1   Abstract-space unit vectors

The fixed, time-independent unit-length coordinate vectors in the statevector Hilbert space, termed *basis vectors* and denoted variously with such notation as

$$\boldsymbol{\psi}_n, \quad |\psi_n\rangle, \quad |n\rangle, \tag{A.8-1}$$

are associated with the possible quantum states of a given system (one for each degree of freedom). The independence of the quantum states is expressed by the orthogonality of the unit vectors, symbolized as

$$\langle \psi_n | \psi_m \rangle = \delta_{nm}. \tag{A.8-2}$$

These orthogonality conditions mean that the set of quantum states is *disjoint*: Measurements of properties such as energy and angular momentum that identify the quantum state with unit vector $|n\rangle$ exclude the possibility that the state can be, in part, in any other state. When the measurements only distinguish energies, and some states share a common energy (are *degenerate*) then the needed distinguishing quantum-state characteristics can be chosen, arbitrarily, for mathematical convenience. The set of quantum states must be complete, as expressed by eqn (A.7-4).

A.8.2   Degrees of freedom; Multiple particles

In the remainder of this appendix I shall be considering quantum states of a single atom, or other isolated quantum system. Most often I shall consider just a single degree of freedom—for example a particular vibrational mode of a molecule or an electronic excitation of an atom. But this is an appropriate place to mention how quantum theory deals with multiple degrees of freedom, or systems of multiple particles, by means of products.

The wavefunction for multiple independent degrees of freedom, involving canonical coordinates q_i, can always be expressed as superposed products of functions of separate variables in the form

$$\Psi(q_1, q_2, \cdots) = \sum_{ab\cdots} C_{ab\ldots}\psi_a(q_1)\psi_b(q_2)\cdots, \tag{A.8-3a}$$

where $C_{ab\ldots}$ is the probability amplitude for finding the system to have the set of attributes $a, b \ldots$. Typically these amplitudes are obtained by requiring that the wavefunction satisfy a time-independent Schrödinger equation for a Hamiltonian that involves all of the degrees of freedom, $\mathcal{H}^0(q_1, q_2, \cdots)$. The construction of coupled angular-momentum states is an example; see eqn (A.7-29).

This superposition formula generalizes, for N distinguishable spinless particles, to a wavefunction construction of the form

$$\Psi(\mathbf{r}_1, \mathbf{r}_2, \cdots) = \sum_{ab\cdots} C_{ab\ldots}\psi_a(\mathbf{r}_1)\psi_b(\mathbf{r}_2)\cdots. \tag{A.8-3b}$$

or, with allowance for individual spins, to

$$\Psi(\mathbf{r}_1\mathsf{s}_1, \mathbf{r}_2\mathsf{s}_2, \cdots) = \sum_{ab\cdots} C_{ab\ldots}\psi_a(\mathbf{r}_1\mathsf{s}_1)\psi_b(\mathbf{r}_2\mathsf{s}_2)\cdots. \tag{A.8-3c}$$

Expressed more succinctly in terms of Hilbert-space vectors this construction has the form

$$|\Psi\rangle = \sum_{ab\cdots} C_{ab\ldots}|a\rangle_1|b\rangle_2\cdots. \tag{A.8-3d}$$

where the unit vector $|a\rangle_i$ is associated with the single-particle state having attributes a and being occupied by distinguishable particle or degree of freedom i.

A.8.3 Multiparticle states and symmetry; Spin and statistics

When we deal with groups of identical particles their spins (not their spin projections) have an important effect on their overall quantum state. Quantum states of several identical particles are found to be of only two types: Those that are *symmetric* (undergoing no sign change) with respect to interchange of particle labels and those that are *antisymmetric* (undergoing a sign change) under interchange. With the notation $|a\rangle_i$ for the state having attributes a being occupied by particle i, the possibilities are

symmetric: $|a\rangle_1|b\rangle_2 = |b\rangle_1|a\rangle_2$, $\quad$ antisymmetric: $|a\rangle_1|b\rangle_2 = -|b\rangle_1|a\rangle_2$. (A.8-4)

The *spin-statistics theorem* [Duc98; Cur12] says that when dealing with quantum states (or wavefunctions) of multiple identical particles those for *bosons* (particles having spin $0, 1, 2, \ldots$, including photons) are symmetric with respect to particle interchange (*Bose-Einstein statistics*). For *fermions* (particles having spin $\frac{1}{2}, \frac{3}{2}, \ldots$) the states are antisymmetric (*Fermi-Dirac statistics*). These constraints allow multiple photons to have identical descriptors (four quantum numbers) but it prevents any two electrons from occupying the same single-electron quantum state (the *Pauli exclusion principle*):

> No two fermions can have exactly the same values
> for all their identifying quantum numbers.

In constructing a many-electron atom, no two electron orbitals (including spin projection) can be the same—no two electrons can occupy the same space. It is this constraint that necessitates the variety of electron orbitals found with multi-electron atoms and which prevents their condensation into a single orbital—and which maintains the size of solid objects.

Symmetrization or antisymmetrization is readily incorporated into the formalism of matrix elements by means of operators $\hat{S}$ and $\hat{A}$ that include normalization, for example

$$\hat{S}\,|a\rangle_1|b\rangle_2 \equiv \frac{1}{\sqrt{2}}\Big[|a\rangle_1|b\rangle_2 + |b\rangle_1|a\rangle_2\Big], \qquad \hat{A}\,|a\rangle_1|b\rangle_2 \equiv \frac{1}{\sqrt{2}}\Big[|a\rangle_1|b\rangle_2 - |b\rangle_1|a\rangle_2\Big]. \tag{A.8-5}$$

Anyons. The particle interchange that leads to symmetric or antisymmetric wavefunctions accompanies a sign change of ± 1 in eqn (A.8-4). These two values are special cases of a more general change of phase by $e^{i\delta}$, in which $\delta = \pi$ for fermions and $\delta = 2\pi$ for bosons. This restriction of phase δ holds rigorously for particle interchanges in three-dimensional Euclidean space, but it need not apply when particle motion is restricted to two dimensions, as happens with the fractional quantum Hall effect [Ait91; Sch98b]. With such quantum systems δ can take any value. The particle-like excitations for such systems, intermediate between fermions and bosons, are known as *anyons* [Wil82; Ait91; Ste08; Nay08].

Slater determinants. To the generation of students who learned their atomic physics from the lectures and textbooks of John Slater [Sla60], the antisymmetrization required for multi-electron wavefunctions came from the mathematical properties of determinants; see Appendix A.3.5. The needed antisymmetrized products of individual electron orbitals were regarded as *determinantal wavefunctions*, defined more succinctly in statevector form,

$$\begin{vmatrix} |a\rangle_1 & |b\rangle_1 & \cdots \\ |a\rangle_2 & |b\rangle_2 & \cdots \\ \vdots & \vdots & \ddots \end{vmatrix} = |a\rangle_1|b\rangle_2 \cdots - |b\rangle_1|a\rangle_2 \cdots + \cdots . \tag{A.8-6}$$

The sum has $N!$ terms if there are N electrons. For the actual task of constructing matrix elements the explicit use of an antisymmetrization operator $\hat{A}$ has advantages, particularly when angular momentum must be considered [Sho65b].

Second quantization. In dealing with multiparticle states of identical particles (say, electrons) it often proves convenient to introduce creation and annihilation operators analogous to those of photons, i.e. analogous to excitation increments of harmonic oscillators, but with quantum numbers ν that identify a mode (or an electron orbital) and which have undisturbed energies E_ν. Whereas the creation operators $\hat{a}_\nu^\dagger$ and annihilation operators $\hat{a}_\nu$ for photons (or other bosons) are related by commutator relations, those for fermions have anticommutators:

$$\text{bosons: } \hat{a}_\nu \hat{a}_{\nu'}{}^\dagger - \hat{a}_{\nu'}{}^\dagger \hat{a}_\nu = \delta_{\nu,\nu'}, \qquad \text{fermions: } \hat{a}_\nu \hat{a}_{\nu'}^\dagger + \hat{a}_{\nu'}^\dagger \hat{a}_\nu = \delta_{\nu,\nu'}. \tag{A.8-7}$$

This formalism is commonly used in developing descriptions of multi-electron atoms or other multi-electron systems as antisymmetric products of single-electron states. It takes the name "second quantization" to acknowledge a "first quantization" that produces the single-particle wavefunctions used as Hilbert-space unit vectors.

A.8.4 Statevector time dependence; Superpositions

It is a basic premise of quantum theory that all of the possible information about the quantum state of the system at any time t can be incorporated into the coordinates of a single time-varying *statevector* $\mathbf{\Psi}(t)$, in a Hilbert space whose multiple coordinate axes eqn (A.8-1) provide a complete set of possible quantum states. Such a statevector is expressible as coherent superposition of basis vectors that will satisfy given initial conditions. One has considerable leeway in choosing these, as is evident when one writes the general form of a statevector as

$$\mathbf{\Psi}(t) = \sum_{n=1}^{N} C_n(t) \, e^{-i\zeta_n(t)} \, \boldsymbol{\psi}_n. \tag{A.8-8}$$

Here the $\boldsymbol{\psi}_n$ are constant unit vectors that provide a complete Hilbert-space coordinate system, as is required for describing a particular quantum system. (They will be chosen, as noted in Appendix A.9.3, as eigenvectors of an appropriate undisturbed Hamiltonian.) The complex-valued numbers $C_n(t)$ that appear as projections of the statevector along these fixed coordinates, denoted with the symbols

$$\langle \psi_n | \mathbf{\Psi}(t) \rangle = C_n(t) \, e^{-i\zeta_n(t)}, \tag{A.8-9}$$

are *probability amplitudes*. These are to be determined from initial conditions and the assumed Hamiltonian: The absolute square of $C_n(t)$ gives the probability $P_n(t)$ that the system will be found in state n at time t, as exhibited in the fundamental formula of eqn (A.3-15):

$$P_n(t) = |C_n(t)|^2. \tag{A.8-10}$$

Because probabilities must sum to unity, the sum of squared probability amplitudes must equal unity: The statevector has unit length:

$$|\mathbf{\Psi}(t)|^2 = \sum_{n=1}^{N} |C_n(t)|^2 = 1. \tag{A.8-11}$$

Note that the sum here and in eqn (A.8-8) is not just over discrete <u>energies</u>, as one finds with thermal systems, but over discrete <u>quantum states</u>, an important distinction when the energies are degenerate.

Phases and pictures. The factor $e^{-i\zeta_n(t)}$ of eqn (A.8-8) has no effect on the probability of eqn (A.8-10); it fixes a (possibly time varying) phase $\zeta_n(t)$ of each probability amplitude.[17] The phase choice does affect the appearance of the equations governing the time evolution of a statevector, the time-dependent Schrödinger equation of

---

[17]This is an example of a simple *gauge transformation*, that of the U(1) Lie-group symmetry discussed in Appendix A.18.3.

Section A.9.4, and when treating effects of laser radiation it is common to choose the phases so as to simplify these equations and their solutions by omitting rapidly varying terms—the the *rotating-wave approximation* (RWA) of Appendix A.9.6. That choice of phases, discussed in some detail in the following sections, is known as the *rotating-wave picture*. A common alternative choice, the *Dirac* or *interaction* picture, uses phases $\zeta_n(t) = E_n t/\hbar$ and thereby places zeros into diagonal elements of the coefficient matrix W(t) of eqn (A.9-7). The *Schrödinger picture*, useful for quasi-static interactions, simply sets all phases to zero, $\zeta_n(t) = 0$.

Expectation values from statevector. Given a statevector constructed as in eqn (A.8-8), the expectation value of any operator $\hat{F}$ is evaluated from its matrix representation eqn (A.7-6) as

$$\langle F(t) \rangle = \langle \Psi(t) \hat{F} \Psi(t) \rangle = \sum_{mn} C_m(t)^* \, C_n(t) \, e^{+i\zeta_m(t) - i\zeta_n(t)} \, F_{mn}. \tag{A.8-12}$$

This expression exhibits the phases $\zeta_n(t)$ that accompany the choice of coordinates.

Phase coherence. The superposition construct of eqn (A.8-8) brings a very important constraint on what sort of system it describes. It proposes definite phase relationships between the complex numbers that embody system information. These phases are affected by any alteration of the environment, say a neighboring atom or heat radiation. When these phase alterations are random and unpredictable it is no longer possible to describe the system by a statevector. Pure quantum mechanics then requires supplementation by more elaborate descriptions, such as that offered by a density matrix, and the system attributes will exhibit non-quantum characteristics.

Rotating coordinates. To simplify equations, and their interpretation, it is useful to incorporate the phases of eqn (A.8-8) into a coordinate system of rotating unit vectors $\psi'_n(t)$:

$$\Psi(t) = \sum_{n=1}^{N} C_n(t) \, \psi'_n(t), \qquad \psi'_n(t) \equiv e^{-i\zeta_n(t)} \, \psi_n. \tag{A.8-13}$$

It is these time-dependent unit vectors, each rotating in a two-dimensional complex plane, that I (and other authors of works on quantum optics) most often use in presenting pictures of statevector motion as well as Bloch-vector motion. They allow a viewer to appreciate the Rabi oscillations unhampered by considerations of carrier frequencies. This simplification is akin to stepping aboard a merry-go-round to view the up-and-down motion of the horse-mounted riders. The choice of phases ζ_n (a simple example of a gauge choice) is quite arbitrary, but not without mathematical consequences. An example can be seen in the ways two-state excitation dynamics is pictured with different phase choices in Figure A.2 .

A.9 The time-dependent Schrödinger equation

The preceding paragraphs present a formalism for treating probability amplitudes as coordinates of a vector $\Psi(t)$ that moves with passing time in an abstract vector space. When used as a descriptor of coherent excitation this statevector retains unit length, a mathematical property associated with the assumption that the N Hilbert-space coordinates provide a complete set of quantum states—that there is no probability lost to states that are not part of the N-state set. This means that any changes to the statevector can be regarded as a generalized rotation in an N-dimensional Hilbert space, expressible as the action of a unitary matrix operator that is a representation of the Lie group SU(N); see Appendix A.18. The operator that generates incremental time changes is the Hamiltonian.

A.9.1 The statevector equation of motion

Time dependence in nonrelativistic quantum theory is governed by the time-dependent Schrödinger equation (TDSE). In elementary introductions this is typically first presented through a time-varying single-particle wavefunction $\psi(\mathbf{r}, t)$ or, more generally, for a multidimensional many-particle wavefunction, a function of time and multiple position variables, $\psi(x, t)$,

$$i\hbar \frac{\partial}{\partial t} \Psi(x, t) = \mathcal{H}(x, \hat{p}, t) \Psi(x, t). \tag{A.9-1}$$

Here x denotes the set of canonical positions and $\hat{p}$ denotes the corresponding momentum operators. The equation involves the partial derivative with respect to time, while the Hamiltonian operator $\mathcal{H}(x, \hat{p}, t)$ involves spatial derivatives through dependence on momentum.

When we approach quantum theory through matrix mechanics rather than wave mechanics we deal with a statevector rather than a wavefunction. The resulting linear equation describing temporal change of a statevector $\Psi(t)$ is TDSE expressed as an ODE, without reference to spatial coordinates but using Hilbert-space coordinates:

$$i\hbar \frac{d}{dt} \Psi(t) = \mathsf{H}(t) \Psi(t). \tag{A.9-2}$$

The rotation-producing matrix $\mathsf{H}(t)$ appearing here is the matrix representation of a Hamiltonian energy operator $\mathcal{H}(\hat{x}, \hat{p}, t)$, the sum of parts that reflect the idealization used in describing the system and its interactions with the external world; see Appendix A.1.6. Appendix A.10 presents the commonly-encountered form for the TDSE for a simple two-state quantum systeem.

Initial conditions. The TDSE of eqn (A.9-2) is a first-order ODE and so it requires, for complete specification, initial conditions upon the statevector. Typically this takes the form of a set of N initial values for statevector components. When the system is known at time $t = 0$ this means specifying $\Psi(0)$.

Reversibility. As are the classical equations of motion for a conservative system, the TDSE is reversible when the Hamiltonian matrix is Hermitian. Given the initial conditions $\Psi(0)$ it is possible to determine earlier states that led to this state. Only with the introduction of a non-Hermitian Hamiltonian (as when modeling probability loss by photoionization or spontaneous decay) or by incorporation of decoherence into a density-matrix description does the behavior become irreversible.

A.9.2 Heisenberg equations

The TDSE provides a picture of quantum behavior by means of a moving statevector $\Psi(t)$ in a fixed Hilbert-space coordinate system—coordinate unit-vectors ψ_n associated with stationary quantum states of an undisturbed system. In this approach, termed the *Schrödinger picture*, the various possible operators associated with observables such as canonical position and momentum variables or dipole moments and electric fields are represented by static matrices. An alternative approach, known as the *Heisenberg picture*, maintains a static statevector Ψ but places all time dependence into the operators. These satisfy Heisenberg equations in which the commutator with the Hamiltonian drives the time dependence. For any operator $\hat{O}$ the Heisenberg equation reads [Dir65]

$$i\hbar \frac{d}{dt}\hat{O}(t) = [\hat{O}(t), \mathsf{H}(t)] \equiv \hat{O}(t)\mathsf{H}(t) - \mathsf{H}(t)\hat{O}(t). \qquad \text{(A.9-3)}$$

The measurable quantities are associated with expectation values. In evaluating these with quantum theory it is important to recognize that expectation values of a product of two operators is not necessarily equal to the product of two expectation values,

$$\langle \mathsf{H}(t)\hat{O}(t)\rangle \neq \langle \mathsf{H}(t)\rangle \langle \hat{O}(t)\rangle. \qquad \text{(A.9-4)}$$

The Heisenberg equations and associated expectation values offer a picture of system behavior that, while incorporating exactly the same physics as the TDSE, is closer to that we deal with in considering classical systems [Dir65; Ack73; Sen73; Ack74; Mil76; Dal82; Sho90; Bar88; Sho93; Luk03; Har06; Lam06; Mil19]. In particular, the emission and absorption of photons is manifest only through alteration of atomic properties such as stationary-state populations and dipole moments, rather than with quantum jumps.

A.9.3 The semiclassical Hamiltonian

When we regard a discrete-state quantum system as being affected by experimentally controlled radiation (say, from a laser beam) then the Hamiltonian energy matrix is the sum of two matrices,

$$\mathsf{H}(t) = \mathsf{H}^0 + \mathsf{H}^{\text{int}}(t). \qquad \text{(A.9-5)}$$

That for an undisturbed Hamiltonian H^0 has the basis vectors ψ_n as its eigenvectors,

$$\mathsf{H}^0 \psi_n = E_n \psi_n, \qquad \text{(A.9-6)}$$

and its eigenvalues E_n are the energies observable for the undisturbed system. The time dependence occurs through an interaction energy $\mathsf{H}^{\text{int}}(t)$ that is controlled by an experimenter and which, in turn, is responsible for controlling changes to the quantum-state structure of the system.

Throughout the following discussion I shall assume that the fixed Hilbert-space unit vectors ψ_n with which the statevector is described in eqn (A.8-8) are chosen from the eigenvectors of an appropriate undisturbed Hamiltonian matrix H^0. The interaction energy $\mathsf{H}^{\text{int}}(t)$ originates ultimately with the Lorentz force between charges and electromagnetic fields, but for systems of bound charges, such as atoms and molecules, it is commonly expressed as the effect of electric and magnetic multipole moments, discussed in Appendix A.9.5.

A.9.4 The coupled equations for the TDSE

To turn the TDSE of eqn (A.9-2) into useful form we use the statevector expansion of eqn (A.8-8) and obtain,[18] for the N probability amplitudes, the set of N coupled equations

$$i\frac{d}{dt}C_n(t) = \sum_m W_{nm}(t)C_m(t) \quad \text{or} \quad i\frac{d}{dt}\mathbf{C}(t) = \mathsf{W}(t)\mathbf{C}(t), \qquad (A.9\text{-}7)$$

where the diagonal and off-diagonal elements of the $N \times N$ coefficient matrix $\hbar\mathsf{W}(t)$ are, respectively, the energies

$$\hbar W_{nn}(t) = E_n - \hbar\dot{\zeta}_n(t), \qquad (A.9\text{-}8a)$$
$$\hbar W_{nm}(t) = -H_{nm}^{\text{int}}(t)\exp[i\zeta_n(t) - i\zeta_m(t)]. \qquad (A.9\text{-}8b)$$

These formulas are an exact transcription of the multistate Hamiltonian subject to monochromatic radiation. They generalize to treatments of other frequency distributions.

With various specifics of multipole interactions and frequencies, the set of linear ODEs presented by eqn (A.9-7) underlie all the physics of coherent quantum-state manipulation—they are a presentation in vector-component form of the time-dependent Schrödinger equation, (A.9-2), that governs all such nonrelativistic quantum physics. The vector $\mathbf{C}(t)$ appearing here is a statevector expressed in a rotating coordinate system defined by the phases $\zeta_n(t)$. The effect of ongoing time expressed in eqn (A.9-7) involves a matrix $\mathsf{W}(t)$ of coefficients whose elements $W_{nm}(t)$ have the dimension of frequencies. In view of their relationship to the underlying physics the diagonal elements are termed *detunings* and the off-diagonal elements are *Rabi frequencies*. The elements of $\mathsf{W}(t)$ differ from the corresponding elements of the Hamiltonian $\mathsf{H}(t)$ not only by a scaling $\hbar$ but by an offset (in detunings), from the rotating coordinates, and so it is appropriate to introduce a new symbol $\mathsf{W}(t)$ for their array. For most applications the expressions for $\mathsf{W}(t)$ are simplified with approximations that assume the radiative interaction energy to be significantly smaller than a single-photon energy—the rotating-wave approximation discussed in eqn (A.9.6).

A.9.5 The multipole Hamiltonian

The interaction of visible, infrared, and radio-frequency radiation with atoms and molecules involves radiation fields acting on objects that are much smaller than the radiation wavelength. Such situations are best treated by expressing the distribution of charges and currents within the particle as a succession of multipole moments centered at the particle center of mass, $\mathbf{R}$. The most important of these are the electric-dipole moment $\mathbf{d}$, built from constituent electric charges, see eqn (A.1-18), and the magnetic-dipole moment $\mathbf{m}$, built from angular momenta, see eqn (A.1-19). The interaction energies of these two moments with electromagnetic fields, evaluated from the Lorentz force, are expressed by the Hamiltonians[19]

---

[18]The needed step is to project $\mathbf{\Psi}(t)$ onto each of the basis vectors ψ_n, here taken to form an orthonormal set.

[19]The $\mathbf{E}$ appearing here and in higher electric multipoles is the solenoidal (transverse) part of the electric field [Pow78; Pow80; Cra84; Pow85; Mil19].

$$\mathsf{H}^{E1}(t) = -\mathbf{d} \cdot \mathbf{E}(\mathbf{R}, t), \qquad \mathsf{H}^{M1}(t) = -\mathbf{m} \cdot \mathbf{B}(\mathbf{R}, t). \qquad \text{(A.9-9)}$$

Each of these interactions is proportional to the projection angle of a vector associated with the underlying particle structure, embodied in $\mathbf{d}$ or $\mathbf{m}$, upon the direction of a field. The center of mass is usually taken as the coordinate origin, $\mathbf{R} = 0$, and it is at this position (moving with the system) that the fields are to be specified. The multipole moments are sources that lend themselves to multipole radiation and to multipole photons when used as Dirac modes for fields.

Classically, the interaction energy is a scalar. Here, with the assumption of matter as obeying quantum mechanics, the dipole moments are operators (matrices) acting upon the statevector. For atomic transitions the M1 interaction strength is typically smaller than the E1 interaction by roughly the ratio of active-electron speed to the speed of light, typically of the order of the Sommerfeld fine-structure constant $\alpha \approx 1/137$. However, the possibility of a transition between two atomic states depends on selection rules, and these may forbid E1 transitions between a particular pair.

Quadrupole interaction. In using this dipole form of the interaction between radiation and matter [Ros55; Pow78; Pow80; Cra84; Ack84; Pow85; Arn05; Bay06; Mil19] we make no reference to any spatial variation of the field around the center of mass—to any direction of photon propagation. Only the directions of the electric and magnetic field vectors matter. Higher-order multipole moments may also occur in radiative interactions between atoms and laser fields. These do depend on propagation directions.

The next term in the multipole expansion is that of a *quadrupole moment*, a tensor Q of order 2, see eqn (A.1-21). The energy of this moment with an electric field involves not only the direction of the field vector (the component j of field $E_j(\mathbf{R}, t)$) but also the direction of the field propagation (the gradient of the field amplitude), expressible as a sum over Cartesian-coordinate products,

$$\mathsf{H}^{E2}(t) = -\frac{1}{6} \sum_{ij} Q_{ij} \frac{\partial}{\partial R_j} E_j(\mathbf{R}, t), \qquad \{i, j\} = \{x, y, z\}. \qquad \text{(A.9-10)}$$

For a traveling plane wave having amplitude $\exp(\mathrm{i}\mathbf{k} \cdot \mathbf{r} - \mathrm{i}\omega t)$ the gradient $\partial/\partial R_i$ brings Cartesian components of the wavevector $\mathbf{k}$ to the energy expression,

$$\mathsf{H}^{E2}(t) = -\frac{\mathrm{i}}{6} \sum_{ij} k_i Q_{ij} E_j(\mathbf{R}, t) = -\frac{\mathrm{i}}{6} \mathbf{k} \cdot \mathsf{Q} \cdot \mathbf{E}(\mathbf{R}, t). \qquad \text{(A.9-11)}$$

It is noteworthy that, unlike the dipole interactions, the energy depends on both the direction of the electric vector (e.g. the helicity) and the direction of propagation. The wavevector is inversely proportional to the wavelength of the light, and so the size of the quadrupole interaction is typically smaller than the electric dipole interaction by the ratio of atomic size to wavelength of light. Quadrupole transitions [Deh54] become important when selection rules (see Appendix A.11.1) forbid E1 and M1 transitions [Bri68; Zar88; Sho90; Lou06].

Moving atoms. The multipole interaction depends upon values of fields at the center-of-mass position $\mathbf{R}$. The time dependence of a traveling electromagnetic field of carrier frequency ω and propagation vector $\mathbf{k}$ appears through factors such as (see Appendix B.3.1)

$$f^+(\mathbf{R}, t) = \exp[-\mathrm{i}\omega t + \mathrm{i}\mathbf{k} \cdot \mathbf{R}]. \tag{A.9-12}$$

When the center of mass is moving steadily with velocity $\mathbf{v}$ the exponential becomes

$$f^+(\mathbf{R}, t) = \exp[-\mathrm{i}(\omega - \mathbf{k} \cdot \mathbf{v})t + \mathrm{i}\mathbf{k} \cdot \mathbf{R}_0], \qquad \text{for} \quad \mathbf{R} = \mathbf{R}_0 + \mathbf{v}t. \tag{A.9-13}$$

Thus the frequency appearing in the exponential differs from from the carrier frequency ω in the laboratory reference frame by the *Doppler shift* (see Appendix A.9.8),

$$\Delta\omega = -\mathbf{k} \cdot \mathbf{v}. \tag{A.9-14}$$

This is the projection of velocity along the propagation axis of the radiation, see eqn (A.9-31); it vanishes for atoms moving along the axis of a radiation beam.

The center of mass appears here as a parameter. When treated as a dynamical variable, paired with a canonical momentum, the resulting Hamiltonian [Bab73; Hea77; Son17; And18] can describe cooling and trapping of molecules by radiation forces [Bla88; Phi98; Coh98; Jav06; Met12; Nev15].

Stark and Zeeman effects. The explicit time dependence of the electric and magnetic fields in the formulas of eqn (A.9-9) allows their connection with radiation and hence with photons. But important energy shifts occur with static, or slowly varying (quasistatic) fields: The *Stark effect* from electric fields [Bon68; Leo04; Van12; Lem13; Cha15] and the *Zeeman effect* from magnetic fields [Sch77; Kox97; Ser02; Van12; Sha15]. Each of these depends upon the relative orientation of an atomic or molecular vector (a dipole moment) with an externally controlled vector, $\mathbf{E}$ or $\mathbf{B}$. For quantum systems these orientations are discrete, and so there occurs a splitting of the original system energy into a set of discretely shifted energies. The spatial variation of this interaction energy produces a force, usable for separating constituents of molecular beams. When the electric field is associated with radiation, and hence is time varying, the effect is known as the *dynamic Stark effect.*

The TDSE with electric-dipole interaction. In much of the discipline of quantum optics, and in the use of lasers for inducing coherent quantum change at optical frequencies, the interaction parametrized by a Rabi frequency is the electric-dipole interaction. For definiteness, that will be the default choice assumed in the present Memoir: $\mathsf{H}^{\mathrm{int}}(t) = \mathsf{H}^{E1}(t)$.

Appendix B.1 discusses a number of simple mathematical descriptions of electric and magnetic fields. For the common idealizations of an isolated atom, having spatial extent much smaller than the wavelength of light, and radiation from a laser source, it is common to idealize the field near the atom as formed of directed plane waves. Typically the coordinate origin is taken to be the center of mass. The electric field at the (stationary) center of mass, $\mathbf{E}(0, t)$, is expressed by means of a complex-valued unit vector $\mathbf{e}$ (defining the field polarization direction, see Appendix B.1.7) paired with a complex-valued field amplitude[20] $\mathcal{E}(t)$ that may vary slowly with time (i.e. changing only over many cycles of the *carrier frequency* ω), for example

---

[20]Some authors omit the factor $1/2$ when defining $\mathcal{E}(t)$.

$$\mathbf{E}(0,t) = \tfrac{1}{2}\mathbf{e}\,\mathcal{E}(t)\,\mathrm{e}^{-\mathrm{i}\omega t} + \tfrac{1}{2}\mathbf{e}^*\mathcal{E}(t)^*\,\mathrm{e}^{\mathrm{i}\omega t} \equiv \mathrm{Re}\;\mathbf{e}\,\mathcal{E}(t)\,\mathrm{e}^{-\mathrm{i}\omega t}. \tag{A.9-15}$$

Because we consider a fixed position $z = 0$ there is no distinction between a traveling wave $\cos(kz - \omega t)$ and a standing wave $\cos(kz)\cos(\omega t)$. A zero-frequency pulse, $\omega = 0$, corresponds to a quasistatic interaction, and is an option compatible with the mathematics that follows, though not with any involvement of what are usually regarded as photons.

With electric-dipole interactions the matrix elements of the semiclassical interaction Hamiltonian between discrete atomic states has the elements electric dipole

$$\begin{aligned} H_{nm}^{E1}(t) &\equiv \langle n|\mathsf{H}^{E1}(t)|m\rangle \\ &= -\tfrac{1}{2}\langle n|\mathbf{d}\cdot\mathbf{e}\,|m\rangle\mathcal{E}(t)\,\mathrm{e}^{-\mathrm{i}\omega t} - \tfrac{1}{2}\langle n|\mathbf{d}\cdot\mathbf{e}^*|m\rangle\mathcal{E}(t)^*\,\mathrm{e}^{\mathrm{i}\omega t}, \end{aligned} \tag{A.9-16}$$

and leads to the formula (for $n \neq m$)

$$\begin{aligned} \hbar W_{nm}(t) &= -\tfrac{1}{2}\langle n|\mathbf{d}\cdot\mathbf{e}\,|m\rangle\mathcal{E}(t)\exp[\mathrm{i}\zeta_n(t) - \mathrm{i}\zeta_m(t) - \mathrm{i}\omega t] \\ &\quad -\tfrac{1}{2}\langle n|\mathbf{d}\cdot\mathbf{e}^*|m\rangle\mathcal{E}(t)^*\exp[\mathrm{i}\zeta_n(t) - \mathrm{i}\zeta_m(t) + \mathrm{i}\omega t]. \end{aligned} \tag{A.9-17}$$

This is an exact representation of the multistate electric-dipole Hamiltonian subject to monochromatic radiation. For most applications it can be simplified with approximations that assume the radiative interaction energy to be significantly smaller than a single-photon energy—the rotating-wave approximation (RWA) discussed in the following paragraphs.

A.9.6   The rotating-wave approximation (RWA)

What is noteworthy about the off-diagonal elements $\hbar W_{nm}(t)$ of the interaction Hamiltonian is the two exponentials, which contain field carrier frequency ω. The phase factors $\zeta_n(t)$ are arbitrary, for us to define, and the equations are simplest to solve when we choose them to nullify one of the two exponentials. The needed choice depends on which of the two quantum states has the larger energy E_n.

The two-state RWA. Suppose we deal with only the couplings of a two-state atom, with state 2 having larger energy, $E_2 > E_1$. A common, useful choice for the phases is

$$\hbar\zeta_1(t) = E_1 t, \qquad \hbar\zeta_2(t) = E_1 t + \hbar\omega t. \tag{A.9-18}$$

These are the phases with which the rotating statevector coordinates $\psi_n'(t)$ change. The energy E_1 is often taken as the zero point of excitation energy, but regardless of its value its presence in $\zeta_1(t)$ nullifies the element $W_{11}(t)$ of the matrix $\mathsf{W}(t)$. With this choice the elements of $\mathsf{W}(t)$ become, without further approximation,

$$W_{11} = 0, \qquad W_{22} = \Delta, \tag{A.9-19a}$$

$$W_{12}(t) = \tfrac{1}{2}\Omega(t)\,\mathrm{e}^{2\mathrm{i}\omega t} + \tfrac{1}{2}\Omega(t)^*. \tag{A.9-19b}$$

Here Δ is the difference between the Bohr frequency and the carrier frequency (the *detuning*)

$$\hbar\Delta = |E_2 - E_1| - \hbar\omega, \tag{A.9-20}$$

and $\Omega(t)$, quantifying the strength of the atom-field interaction energy, is the *Rabi frequency* for dipole-transition moment $d_{12} \equiv \langle 1|\mathbf{d} \cdot \mathbf{e}|2\rangle$,

$$\hbar\Omega(t) = -\langle 1|\mathbf{d} \cdot \mathbf{e}|2\rangle\mathcal{E}(t) \quad \text{or} \quad \Omega \approx 8.04 \times 10^4 \text{ [rad/s]} \left(\frac{d_{12}}{ea_0}\right)\mathcal{E}\text{ [V/m]}. \quad \text{(A.9-21)}$$

When the electric field is that of a traveling plane wave of intensity $I(t)$, as in eqn (B.1-17), the Rabi frequency magnitude is expressible as

$$|\Omega(t)| \approx 2.21 \times 10^6 \text{[rad/s]} \left(\frac{d_{12}}{ea_0}\right) \sqrt{I(t)\text{[W/m}^2\text{]}}. \quad \text{(A.9-22)}$$

(By including the phase φ of the field in the Hilbert-space coordinate for state 2 we can make the Rabi frequency real-valued.) For optical-frequency fields the carrier frequency is typically many orders of magnitude larger than the Rabi frequency. Its effect is to introduce rapid, small-amplitude modulation of more slowly varying population oscillations. It is those slower changes that usually hold interest, and so we wish to have amplitudes that average over (and thereby omit) the oscillations that result from the exponential $\exp[2i\omega t]$—the so-called "counter-rotating" contribution to $\mathsf{W}(t)$. The result is the two-state rotating-wave approximation (RWA) Hamiltonian matrix,

$$\mathsf{W}(t) = \begin{bmatrix} 0 & \frac{1}{2}\Omega(t)^* \\ \frac{1}{2}\Omega(t) & \Delta \end{bmatrix}, \quad \text{(A.9-23)}$$

for use in the basic quantum-mechanical equation eqn (A.9-7) when specialized to treating two-state coherent excitation. Appendix A.10.2 presents commonly used examples of solutions to the TDSE with this Hamiltonian.

The three-state RWA. A three-state system has three possible pairings, with Bohr frequencies $\omega_{12}, \omega_{23}, \omega_{31}$ that form a loop (or triangle). Associated with these are three possible monochromatic fields, expressible, at the center of mass, as

$$\mathbf{E}(0,t) = \mathcal{E}_P(t) \text{ Re}\left\{\mathbf{e}_P \, e^{-i\omega_P t + i\varphi_P}\right\} + \mathcal{E}_S(t) \text{ Re}\left\{\mathbf{e}_S \, e^{-i\omega_S t + i\varphi_S}\right\}$$

$$+ \mathcal{E}_R(t) \text{ Re}\left\{\mathbf{e}_R \, e^{-i\omega_R t + i\varphi_R}\right\}, \quad \text{(A.9-24)}$$

characterized by carrier frequencies ω_λ, phases φ_λ, real-valued pulse envelopes $\mathcal{E}_\lambda$, and complex unit vectors $\mathbf{e}_\lambda$. For a field to have an appreciable effect its carrier frequency must be near to one of the Bohr frequencies. Selection rules based on atomic symmetries and field polarizations (see Appendix A.11.1) may further limit the association of field components with transitions. Such rules prevent three-state loop linkages for electric-dipole transitions in free space—states linked by electric dipole moments should have opposite parity. Thus it is not possible to have a closed loop of three excitation linkages with only electric dipole radiation in a free atom. However, there exist alternative interactions, and alternative systems (e.g. atoms in a crystal), for which three-state closed loops are possible [Buc86; Sho90; Una97; Kor02b; Mal04; Ran08].

The most common three-state systems involve just two fields in a linkage pattern (configuration) that forms a chain of state energies E_1, E_2, E_3. Typically the energies of states 1 and 2 are ranked $E_1 < E_2$ and are associated, by frequency and selection rules, with a controlled P (for pump) field. The second field, S (for Stokes) is associated with the linkage of states 2 and 3. The two fields have single-photon detunings

$$\Delta_P = \omega_{12} - \omega_P, \qquad \Delta_S = \omega_{23} - \omega_S, \qquad \text{with } \hbar\omega_{nm} \equiv |E_n - E_m|. \qquad \text{(A.9-25)}$$

The energies are ranked[21] either as a ladder ($E_3 > E_2$) or, as occurs with stimulated Raman scattering, as a Lambda ($E_3 < E_2$). In each of these linkages it is possible to choose phases so that the two Rabi frequencies are slowly varying [Sho08; Sho11]. When the interaction energy matrix $H^{\text{int}}(t)$ has no diagonal elements a common choice for phases, and the consequent detunings, is [Sho08; Sho11]

$$\begin{aligned} &\hbar\dot{\zeta}_1 = E_1 && \hbar\dot{\zeta}_2 = \hbar\dot{\zeta}_1 + \omega_P, && \hbar\dot{\zeta}_3 = \hbar\dot{\zeta}_2 \pm \omega_S, \\ &\Delta_1 = 0, && \Delta_2 = \Delta_P, && \Delta_3 = \Delta_P \pm \Delta_S, \end{aligned} \qquad \text{(A.9-26)}$$

where the plus sign goes with the Ladder, the minus sign with the Lambda linkage. The third detuning for the ladder system is the *sum* of two single-step detunings whereas for the Lambda linkage it is the *difference* of two detunings.

The preceding considerations fix the phases of all three states. When selection rules allow the construction of a loop linkage the third field of eqn (A.9-24) may be important; there will be a field-generating dipole moment present for the ω_{31} transition, and the extension of eqn (A.9-23) to the three-state matrix $W(t)$ can be given the form

$$W(t) = \frac{1}{2} \begin{bmatrix} 0 & \Omega_P(t) & \Omega_R(t) \\ \Omega_P(t) & 2\Delta_2 & \Omega_S(t) \\ \Omega_R(t) & \Omega_S(t) & 2\Delta_3 \end{bmatrix}. \qquad \text{(A.9-27)}$$

However, unless the carrier frequency ω_R of the third field matches a resonance condition, with either $\omega_P + \omega_S$ (ladder linkage) or $|\omega_P - \omega_S|$ (Lambda linkage), the Rabi frequency $\Omega_R(t)$ will acquire a phase from the difference of $\zeta_1(t)$ and $\zeta_3(t)$ and will not be slowly varying.

The multistate RWA. When the off-diagonal elements of the Hamiltonian matrix form a simple chain linkage, and there is negligible ambiguity about which of many fields is to be responsible for each transition, then it is possible to choose the phases so that, with a generalization of the RWA discussed above, all of the Rabi frequencies are slowly varying and the detunings Δ_n are cumulative values between states 1 and n. This simplification is not necessarily possible with other linkages, such as a loop. Then some of the Rabi frequencies (i.e. the off-diagonal elements of $W(t)$), will oscillate at beat frequencies—sums and differences of the several carrier frequencies.

The common choice of first-state phase $\hbar\dot{\zeta}_1 = E_1$ used above leads to null first detuning, $\Delta_1 = 0$ and $W_{11}(t) = 0$, as suits the usual initial condition $C_1(0) = 1$; it means that, whatever the linkage pattern, states 1 and n are resonant if $\Delta_n = 0$.

A.9.7 Probability loss

The common approaches to coherent excitation discussed in this appendix, and the generation of photons, idealize the atom (or other photon generator) as being describable by a finite number of discrete quantum states—an N-dimensional Hilbert space

---

[21]When $E_1 > E_2$ and $E_3 > E_2$ the linkage is a letter Vee.

in which a defining statevector moves or an $N \times N$ density matrix. Often it is sufficient to consider only two or three quantum states.

Although this idealization suffices to treat much of the phenomena of quantum optics, it is understood that real quantum systems have many more quantum states. The simple classroom model of the hydrogen atom, for example, has a denumerable infinity of discrete energy states $E_n = -E_1/n^2$ that converge to a limit $E_n \to 0$ as the electron orbitals grow increasingly large with increasing principal quantum number $n = 1, 2, 3, \cdots$. But beyond that limit there lies a nondenumerable continuum of positive-energy scattering states.

To affect populations appreciably quantum states must be linked to the initially populated states by near-resonant transitions. Non-resonant states are not totally without influence: They contribute to the polarizability and are thereby responsible for dynamic Stark shifts [Arm62; Sho90; Sho17].

However, two common situations occur when the idealized N-state model may require modification: Photoionization and spontaneous emission.

Photoionization. When near-resonant excitation of an N-state quantum system brings population into a state that lies within reach of a photoionization (or photodissociation) continuum, there will occur transitions between the N states of the model and quantum states that have not been included in the limited Hilbert space. Such transitions become particularly important when the goal of excitation is to produce selective ionization—choosing laser frequencies that will photoionize only predetermined isotopes of chosen atomic and molecular species—as is done for a variety of commercial purposes [Let79; Gre88; Sho01].

Spontaneous emission. Laser-induced coherent excitation relies on the same dipole-transition moments as does spontaneous emission. It is therefore common to have competition between a laser-activated linkage between states n and m and other states that are linked to these by the possibility of spontaneous emission. Those transitions will occur (unless restricted by cavity geometry) to a set of lower energy states regardless of their transition frequency. The use of density matrices allows the inclusion of such incoherent transitions, but the TDSE cannot treat them within its discrete Hilbert space.

Non-Hermitian Hamiltonian. These two mechanisms, of deliberately applied photoionization and unavoidable spontaneous emission, can each produce quantum states that are not part of the small set of N states whose TDSE is being modeled. These transitions remove population from the considered Hilbert space. They represent probability loss: The N summed probabilities no longer add to unity.

Probability-loss processes that take place incoherently from state n are most often treated by adding an imaginary component to the undisturbed (bare) energy E_n,

$$E_n \to E_n - i\hbar\Gamma_n/2, \qquad (A.9\text{-}28)$$

giving a non-Hermitian RWA Hamiltonian. In the absence of excitation-producing interactions (i.e. absent Rabi frequencies) the TDSE for each state reads

$$i\frac{d}{dt}C_n(t) = (\Delta_n - i\Gamma_n/2)C_n(t), \qquad (A.9\text{-}29)$$

and the resulting probabilities fall away exponentially from their values at $t = 0$,

$$P_n(t) = \exp(-\Gamma_n t) P_n(0). \tag{A.9-30}$$

When loss rates are used in this way in the RWA Hamiltonian $W(t)$ they alter the Rabi oscillations that occur with constant illumination. In a two-state system a small loss from state 2 will act to dampen the periodic oscillation. As Γ_2 becomes comparable to the Rabi frequency the population flows out of state 1 without oscillations. When Γ_2 is very large (and the oscillation is thereby overdamped) the population remains, without appreciable loss, in the initial state [Sho06].

A.9.8 Detuning shifts

Several effects, partially under control of an experimenter, introduce alteration of the detunings that appear in the TDSE. These often require ensemble averaging; see Appendix A.11.2.

Doppler shift. The electric and magnetic fields that appear in the Hamiltonian of eqn (A.9-9) are those at the center of mass for the system. When the system is moving with velocity $\mathbf{v}$ then the field frequencies experienced in the center-of-mass rest frame of the atom are Doppler shifted by an amount proportional to $v_{\parallel}$, the component of system velocity along the local propagation vector $\mathbf{k}$ that defines the traveling-wave components of the field; see eqn (A.9-14). For nonrelativistic motion ($v_{\parallel} \ll c$) the Doppler shift requires the replacement

$$\omega \to \omega[1 + v_{\parallel}/c], \qquad v_{\parallel} = \mathbf{v} \cdot \mathbf{k}/k, \tag{A.9-31}$$

in the formulas for phases $\zeta_n(t)$ and the consequent detunings Δ_n: Each detuning acquires an additional velocity-dependent Doppler increment $\Delta_v = \omega v_{\parallel}/c$. When an experiment involves a beam of atoms then the results must be averaged over Doppler shifts; see eqn (A.11-15) in Appendix A.11.2. When the excitation is produced by several fields, each associated with a pair of states and each traveling in a different direction, then each detuning acquires an appropriate Doppler shift from each field.

Zeeman and Stark shifts. Expressions for phases and detunings involve the energies E_n defined as quantum-state energies in the absence of the particular fields eqn (A.9-9) that produce the changes of interest.

Often an experimenter deliberately introduces controlled static magnetic fields that shift these energies—static *Zeeman shifts* that break the symmetry of the atomic environment and eliminate the degeneracy associated with atom orientation. The resulting discrete energies are associated with *Zeeman sublevels*; see eqn (A.11-10).

Static or slowly varying (quasistatic) electric fields induce *Stark shifts* that only partially reduce the degeneracy. When the electric field originates with a pulsed oscillatory field (a laser or maser) the shifts are known as *dynamic Stark shifts*.

For an atom held within a crystalline environment the neighboring atoms produce static *crystal field* energy shifts characterized by the local symmetry. When experimental conditions cannot pick out a single atom, then results must be averaged over the possible environments; see Appendix A.11.2.

All of these effects are regarded in this section as part of the base Hamiltonian H^0 whose eigenvalues are taken as the energies E_n. When an experimenter controls any of these as slowly varying fields the result is slowly varying E_n and a consequent variation of phases $\zeta_n(t)$. The detunings then become time dependent, $\Delta_n(t)$, in accord with experimental design.

A.9.9   The propagator

Solutions to the N-state TDSE of eqn (A.9-7) that incorporate initial conditions at $t = 0$ can be written as resulting from a time-dependent $N \times N$ time-evolution (or *propagator*) matrix $U(t,0)$ acting upon the initial components of the statevector,

$$\mathbf{C}(t) = U(t,0)\mathbf{C}(0). \tag{A.9-32}$$

The propagator matrix becomes the unit matrix at the starting time and satisfies the TDSE at all later times,

$$i\frac{d}{dt}U(t,0) = W(t)U(t,0), \qquad U_{nm}(0,0) = \delta_{mn}. \tag{A.9-33}$$

The absolute square of element $U_{nm}(t)$ expresses the conditional probability for finding the system in state n at time t, given that it was initially in state m:

$$P_n(t|m) = |U_{nm}(t)|^2. \tag{A.9-34}$$

The propagator is a matrix version of a Green's function for a differential equation, in this case for the TDSE of eqn (A.9-33). It presents the statevector equation of motion in integral rather than differential form.

Steady illumination. When the coefficient matrix $W(t)$ is constant, as occurs for monochromatic illumination, the propagator can be written as an exponential of a matrix, as a power series (using a definition of the exponential function) or, because $t > 0$, as a Laplace-transform complex-plane integral of the inverse of a matrix. The three forms, mathematically equivalent, are

$$U(t,0) = \exp[-iWt], \tag{A.9-35a}$$

$$U(t,0) = \sum_{n=0}^{\infty} \frac{(-iWt)^n}{n!}, \tag{A.9-35b}$$

$$U(t,0) = \frac{1}{2\pi} \int_{\mathcal{C}} ds\, e^{st}\, [W + is1]^{-1}. \tag{A.9-35c}$$

Such constructs are valid for any set of coupled ODEs that have constant coefficients, not just the TDSE for quantum changes.

Application to models of quantum excitation leads to particular interpretations of the several forms for the propagator. For example, the series expression eqn (A.9-35b) allows interpretation of the sines and cosines of Rabi oscillations as infinite sums of multiphoton interactions, each term involving an increasing numbers of photons. In the integral expression eqn (A.9-35c), with the variable s taken as iz, the matrix $[W+z1]^{-1}$

is known as the *resolvent* of the matrix W; its structure offers useful opportunity to exhibit resonance behavior such as occurs with autoionization [Sho67].

Pulse successions. Often we wish to subject a quantum system to a succession of pulses, or to treat a single pulse as a succession of separately defined interactions. The required propagator can be expressed as a product,

$$U(T,0) = U^{(M)}(T, t_{M-1}) \cdots U^{(2)}(t_2, t_1) U^{(1)}(t_1, 0), \qquad (A.9\text{-}36)$$

where successive propagator matrices are obtained from a succession of possibly time-dependent interaction matrices, an extension of eqn (A.9-33):

$$i\frac{d}{dt}U^{(n)}(t, t_{n-1}) = W^{(n)}(t) U^{(n)}(t, t_{n-1}), \qquad U^{(n)}(t_{n-1}, t_{n-1}) = 1. \qquad (A.9\text{-}37)$$

These formulas provide a very useful algorithm for producing either analytic expressions or numerical procedures.

Composite pulses. At times it is useful to craft a single pulse from a succession of rectangular pulse-segments eqn (B.5-3a), or other simple form, either with amplitudes or phases (or both) adjusted in accord with some plan. Such *composite pulses* [Lev79; Lev86; Gen11; Tor11; Tor11b; Vit11; Gen14; Tor15; Tor18] can offer advantages over simple pulses, such as Gaussians, for a variety of objectives.

A.10 Two-state coherent excitation

For two nondegenerate states the TDSE is typically presented as the pair of coupled ODEs

$$i\frac{d}{dt}C_1(t) = \tfrac{1}{2}\Omega(t)C_2(t), \qquad (A.10\text{-}1a)$$

$$i\frac{d}{dt}C_2(t) = \Delta(t)C_2(t) + \tfrac{1}{2}\Omega(t)C_1(t). \qquad (A.10\text{-}1b)$$

The strength of the radiation is parametrized by the Rabi frequency $\Omega(t)$, which here is assumed to be real valued, though not necessarily positive. Appendix A.9.5 discusses this parameter. The detuning $\Delta(t)$ is the difference between the Bohr transition frequency and the carrier frequency of the radiation,

$$\hbar\Delta(t) = E_2 - E_1 - \hbar\omega \equiv \hbar\omega_{21} - \hbar\omega, \qquad (A.10\text{-}2)$$

either or both of which may vary with time. It is real valued and may be of either sign, or zero.

From the absolute squares of solutions to these equations one evaluates the probability, eqn (A.3-15), of finding the system in state n at time t. The other important two-state observable is the *coherence* $\rho_{12}(t) = C_1(t)C_2(t)^*$, needed to evaluate the time-varying dipole moment of the atom.

The presence of the imaginary unit $i \equiv \sqrt{-1}$ in the equations eqn (A.10-1) leads to oscillatory solutions when the radiation is steady, see Figure A.1, in contrast to the behavior of solutions to the radiative rate equations, eqn (A.16-16), which monotonically approach steady equilibrium values, see Figure A.3.

These equations, like those for incoherent radiation in Appendex A.16.3, describe the effect of a given radiation field. They say nothing about how that field, passing by an ensemble of atoms, is altered by atom transitions. They make no reference to photons nor to the quantum nature of the electromagnetic field. Appendix C.6 discusses a simple example of the interplay between atom excitation and photons.

A.10.1 Approximations

The equations presented here as eqn (A.10-1) embody a number of approximations:

- Foremost is the idealization of treating a single isolated atom (or other discrete-state quantum system), free from any uncontrollable environmental influences such as collisions or field fluctuations. Unlike the radiative rate equations eqn (A.16-16), the equations do not apply to ensembles of atoms that have different orientations.

- Notable too is the restriction to just two quantum states. All atoms and molecules have numerous discrete states, evident from their spectra and from theory employed by quantum chemists. The assumption leading to eqn (A.10-1)—the *essential states approximation*—is that those states are unimportant because the Bohr transition frequency linking them to the initially populated state is far from matching the field carrier frequency.

- Finally, it is also assumed that the carrier frequency is much larger than the magnitude of the Rabi frequency—the so-called *rotating-wave approximation* (RWA) that has introduced the time average

$$[1 + e^{\pm i\omega t}]_{\text{av}} \to 1, \qquad (\text{A.10-3})$$

see Appendix A.9.6. Consequently, the equations make no reference to the carrier frequency of the field: Its effect is entirely expressed by the detuning.

Within these approximations the TDSE offers opportunity for exact, closed-form expressions for solutions and for excitation probabilities—and for designing pulsed radiation fields that produce a wide variety of quantum-state changes.

A.10.2 Rabi oscillations

The solution to any TDSE can be conveniently expressed as the effect of a propagator matrix U of Appendix A.9.9, as expressed by eqn (A.9-32). In particular, the (conditional) probabilities of eqn (A.9-34) that grow from an initial population in state 1 at $t = 0$ are expressible as

$$P_1(t|1) = |U_{11}(t,0)|^2, \qquad P_n(t|1) = |U_{n1}(t,0)|^2. \qquad (\text{A.10-4})$$

A variety of pulse shapes lead to known closed-form analytic expressions for the elements of the propagator [Sho90; Kyo06; Sho11]. For a two-state system without loss the elements of the propagator can be expressed by two complex-valued *Cayley-Klein parameters* a, b as

$$\mathsf{U}(t,0) = \begin{bmatrix} a & b \\ -b^* & a^* \end{bmatrix}, \qquad |a|^2 + |b|^2 = 1. \qquad (\text{A.10-5})$$

The Cayley-Klein parameters can be found by numerical integration of the TDSE for any reasonably well-behaved Rabi frequency and detuning. They are known in closed form for a number of expressions for time-varying Rabi frequency and detuning [Kyo06].

Resonant pulses. For the simple case of resonant excitation ($\Delta = 0$) the Cayley-Klein parameters are

$$a = \cos[\mathcal{A}(t)/2], \qquad b = -\mathrm{i}\,\mathrm{e}^{-\mathrm{i}\varphi}\sin[\mathcal{A}(t)/2], \qquad (\text{A.10-6})$$

where φ is the phase of the electric field and $\mathcal{A}(t)$ is the *Rabi angle*,

$$\mathcal{A}(t) = \int_0^t dt'\,\Omega(t'). \qquad (\text{A.10-7})$$

The final value of the Rabi angle, at the conclusion of a pulse, is known as the (temporal) *pulse area*. If (and only if) excitation is resonant the probability amplitudes will be restored to their initial values at any time when $\mathcal{A}(t) = 4\pi n$, where n is an integer. This regularity is twice that of the probabilities, which repeat when $\mathcal{A}(t) = 2\pi n$. The two states interchange populations when $\mathcal{A}(t) = \pi n$ (a *pi pulse*).[22]

Steady illumination. When the excitation of a two-state system is by light that maintains a constant intensity (meaning a constant Rabi frequency) and constant carrier frequency (meaning constant detuning) the Cayley-Klein parameters are

$$a = \cos(\Upsilon t/2) + \mathrm{i}\frac{\Delta}{\Upsilon}\sin(\Upsilon t/2), \quad b = -\mathrm{i}\frac{|\Omega|}{\Upsilon}\,\mathrm{e}^{\mathrm{i}\varphi}\sin(\Upsilon t/2), \quad \Upsilon = \sqrt{\Omega^2 + \Delta^2}, \quad (\text{A.10-8})$$

Here Υ is known as the *flopping frequency*. The populations, for population initially in state 1, are

$$P_2(t|1) = \frac{\Omega^2}{2\Upsilon^2}[1 - \cos(\Upsilon t)], \quad P_1(t|1) = 1 - P_2(t|1). \qquad (\text{A.10-9})$$

Complete population transfer $1 \to 2$ or $2 \to 1$ only occurs when the excitation is resonant, meaning $\Delta = 0$. The frequency Υ then becomes the Rabi frequency Ω of resonant population oscillations. Whatever the detuning Δ may be, there will occur complete population return to the initial state at every time $T_n = 2\pi n/\Omega$ for any integer n.

Examples. The simulation of two-state excitation can make use of a variety of numerical techniques for solving ODEs. Many of these are available as subroutines in commercial apps such as Mathematica and Matlab. The solutions for two-state excitation by constant fields are simply expressible with the formulas above, with detuning as a parameter. Figure A.1 illustrates the periodicity of the populations.

Scattering. With any excitation the population oscillations are accompanied by an oscillatory *coherence*,

$$\rho_{12}(t) = C_1(t)C_2(t)^* = \tfrac{1}{2}\sin(\Omega t). \qquad (\text{A.10-10})$$

This coherence leads to a dipole-moment expectation value (in a rotating frame) that oscillates at the flopping frequency, thereby generating red and blue sidebands on

---

[22]I use the symbol π to denote the number $3.1415\ldots$ and the word "pi" to describe a pulse having Rabi angle π.

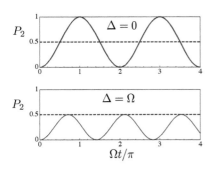

Fig. A.1 *Coherent excitation, without and with detuning.*

The frames of Figure A.1 illustrate the oscillatory two-state excitation population $P_2(t)$ that attends coherent excitation by steady illumination. In the top frame the carrier frequency equals the Bohr frequency, the detuning is zero, and there occur periodic transfers of population from state 1 into state 2 at the Rabi frequency.

In the lower frame there is nonzero detuning, $\Delta = \Omega$. Population oscillations occur more rapidly and less completely than with the resonant excitation of the top frame.

In these plots the time scale is in scaled units of the inverse Rabi frequency, π/Ω: A value of 1 means a *pi pulse* (complete population inversion).

the driving frequency. These are seen in scattered light as a *Mollow spectrum* [Mol69; Mol72]. The sidebands, separated by the flopping frequency, are newly created photons (at the sideband frequencies) of the sort described in [Scu72].

Time reversal. If at some instant t the phase of the electric field φ were to change by π this will produce a change in sign of the Rabi frequency. When the excitation is resonant this change is equivalent to what happens to the equations with the change $t \to -t$, i.e. to time reversal. Thus such a phase change of the electric field will reverse the Rabi oscillation and start the return of population to its initial value. Frequent phase changes of this sort will therefore suppress excitation.

RWA failure. When the carrier frequency ω is comparable to the Rabi frequency Ω the RWA fails, and then the probabilities exhibit modulating oscillations at twice the carrier frequency. These are not seen in eqn (A.10-9) because the populations $P_n(t)$ are here taken to be time averages over many carrier-frequency periods and the carrier frequency is assumed to be much larger than the Rabi frequency. Numerous articles discuss treatments of situations in which the RWA fails, most notably two-state single-mode models without the RWA, the so-called *Rabi model* [Jel90; Bra11; Bra16; Xie17; Rao17; For19].

Perturbation theory and oscillations. Time-dependent perturbation theory and its underlying assumption of minuscule change is quite common throughout quantum theory, most notably in quantum electrodynamics and its use of Feynman diagrams. Although this approach provides a satisfactory description of thermal radiation effects, it gives a poor picture of coherent excitation. Rabi oscillations, for example, involve trigonometric functions to describe the time dependence. An approach involving perturbation theory requires that these functions be replaced by their series expansion. For the sine this means the sum:

$$\sin(x) = \sum_{n=1}^{\infty} (-1)^{n-1} \frac{x^{2n-1}}{(2n-1)!}. \tag{A.10-11}$$

The periodicity of the sine is not evident in this formula. Indeed, to obtain a reasonably accurate picture of the sine for a single period $x = [0, 2\pi]$, requires at least nine terms in the series. Each term can be associated with an emission or absorption of a photon, and so the Rabi oscillation is essentially a multiphoton process.

A.11 Degeneracies and ensembles

The idealized (toy-model) two-state quantum system presented above has provided much insight into the behavior of discrete quantum systems but the atoms and molecules that experimenters must work with have additional states that must be considered. A simple, important, example occurs when the system has angular momentum, so that in the absence of any static electric or magnetic fields each energy value is associated with sublevels that differ by angular-momentum orientation.

A.11.1 Rabi frequency between sublevels

For transitions between magnetic sublevels of angular momentum states the Rabi frequency for an electric-dipole interaction has a dependence upon magnetic quantum numbers expressible, in accord with the *Wigner-Ekart theorem* [Bri68; Zar88; Sho90; Lou06], as proportional to a Wigner *three-j symbol*, (:::), and a reduced matrix element, $(j_1||d||j_2)$:

$$\langle j_1 m_1 | d_q | j_2 m_2 \rangle = (-1)^{j_1 - m_1} \begin{pmatrix} j_1 & 1 & j_2 \\ -m_1 & q & m_2 \end{pmatrix} (j_1||d||j_2). \tag{A.11-1}$$

The connection between the Wigner three-j symbol appearing here and the Clebsch-Gordan coefficients used in constructing coupled states of angular momentum, as in eqn (A.7-29), is

$$(j_1 m_1, j_2 m_2 | JM) = (-1)^{j_1 - J_2 + M} \sqrt{2J + 1} \begin{pmatrix} j_1 & j_2 & J \\ m_1 & m_2 & -M \end{pmatrix}. \tag{A.11-2}$$

Selection rules. The magnetic quantum numbers appearing in eqn (A.11-1) are constrained by the requirement

$$-m_1 + q + m_2 = 0, \tag{A.11-3}$$

and the angular-momentum quantum numbers are constrained by the triangle rule

$$|j_1 - j_2| \leq 1 \leq j_1 + j_2. \tag{A.11-4}$$

This relationship rules out the possibility of a transition between states of angular momentum $j = 0$. These *selection rules*, and their three-j symbol, occur because the matrix element is that of a dipole moment, an example of an irreducible tensor of order 1 [Bri68; Zar88; Sho90; Lou06].

The values of the dipole transition moment, with its factor dependent on angular momentum, determine the details of the angular-momentum photons that are generated by spontaneous emission from a free-space atom; see Appendix B.2.7.

The transition strength. From the summation properties of three-j symbols it follows that the dimensionless *transition strength* $S(1,2)$ or S_{12}, defined as the sum

$$S_{21} = S_{12} = \sum_{m_1 m_2 q} \frac{|\langle j_1 m_1 | d_q | j_2 m_2 \rangle|^2}{(ea_0)^2}, \tag{A.11-5}$$

is proportional to the square of the reduced matrix element of the dipole moment

$$S_{12} = S_{21} = \frac{|(j_1||d||j_2)|^2}{(ea_0)^2}. \tag{A.11-6}$$

The Rabi-frequency magnitude. We can use these results to relate the magnitude of a Rabi frequency between angular momentum states,

$$\hbar\Omega(t) = -\langle j_1 m_1 | d_q | j_2 m_2 \rangle \, \mathcal{E}_q(t), \tag{A.11-7}$$

to a reduced matrix element and thence to a dimensionless transition strength S_{12}, through the formula

$$|\Omega(t)|^2 = |\langle j_1 m_1 | d_q | j_2 m_2 \rangle|^2 \frac{2I(t)}{c\hbar^2 \epsilon_0} = \frac{8\pi a a_0^2}{\hbar} \begin{pmatrix} j_1 & 1 & j_2 \\ -m_1 & q & m_2 \end{pmatrix}^2 S_{12}\, I(t). \tag{A.11-8}$$

Experimental values for the transition strength are often available from measured Einstein A coefficients, through the relationship

$$S_{21} = \frac{3}{4} \frac{1}{\alpha c a_0^2} \left(\frac{\lambda_{12}}{2\pi} \right)^3 \varpi_2 A_{21}. \tag{A.11-9}$$

Here ϖ_2 is the statistical weight (degeneracy) of level 2. It is important to recognize that a Rabi frequency always refers to transitions between nondegenerate quantum states, whereas S_{21} and A_{21} refer to degenerate levels.

Zeeman shifts. Particles that have angular momentum, whether from intrinsic spin or from circling motion, will carry a magnetic moment **m**; see eqn (A.1-19). When there is a static magnetic field **B** present at the center of mass for the particle there will result a magnetic-dipole interaction Hamiltonian H^{M1} akin to the electric dipole interaction; see eqn (A.9-9). Because the magnetic moment is proportional to the angular momentum vector [see eqn (A.1-20)], the interaction energy is most commonly treated by using angular-momentum eigenstates as a basis and taking the quantization axis to lie in the direction of the B field. This static interaction then shifts the energy of every quantum state in proportion to its magnetic quantum number and the resulting expression for the energy shift—the *Zeeman shift*—of electronic origin is a diagonal matrix having elements

$$\langle aJM|\mathbf{m}|bJ'M'\rangle \cdot \mathbf{B} = \gamma\mu_B\, M\, |\mathbf{B}|\, \delta_{ab}\delta_{JJ'}\delta_{MM'}, \quad \text{for } \mathbf{m} = \gamma\mu_B\mathbf{J}, \tag{A.11-10}$$

where μ_B is the Bohr magneton and γ is the gyromagnetic ratio. This interaction breaks the spatial symmetry that underlies the free-space degeneracy of Zeeman sublevels and thereby alters the energy differences that appear in the values of detunings.

Coupled moments. Quite commonly the system of interest can be recognized as involving two magnetic moments. A bound electron has a spin moment and its orbital motion produces another moment (with an associated B field); the resulting interaction produces the fine-structure splitting of atomic energy levels. An atomic nucleus has a spin moment that, in the absence of any external field, will have a discrete alignment with the local field generated by electrons; the energy structure is observed as hyperfine splitting [Sho81b]. An external time-varying magnetic field can induce a transition (a spin flip) between nuclear orientations. This is the origin of the mbox21 cm radio-frequency transition.

A.11.2 Ensemble averages

Unlike changes caused by incoherent mechanisms, which group sets of quantum states into degenerate energy levels for treatment with rate equations such as eqn (A.16-1), coherent excitation with its reliance on statevectors for description, requires a separate TDSE for each quantum system. Each atomic orientation, each center-of-mass velocity (with consequent Doppler shift of radiation frequency) requires a distinct Hamiltonian.

Orientation averages. When angular momentum states are used as the basis for describing a system the Rabi frequencies have an explicit dependence upon the magnetic quantum numbers that appear in the labels of the quantum states. With inclusion of angular momentum a two-state system becomes a two-level system, schematically expressed as

$$|1\rangle, |2\rangle \rightarrow |J_1, M_1\rangle, |J_2 M_2\rangle, \qquad W_{12} = \langle 1|\mathsf{W}|2\rangle \rightarrow \langle J_1 M_1|\mathsf{W}|J_2 M_2\rangle. \qquad \text{(A.11-11)}$$

With the usual assumption of initial population residing equally in the magnetic sub-levels of degenerate level 1 the needed average sums over the set of initial magnetic sublevels, for a prescription that can be written

$$\overline{P}_{nJ}(t) = \frac{1}{2J_1 + 1} \sum_{M_1} |C_{nJM}(t)|^2. \qquad \text{(A.11-12)}$$

Environmental averages. Quite often we must treat sets of Hamiltonians for a variety of environments e (different detunings and Rabi frequencies from different orientations and velocities). For each Hamiltonian we have a separate set of equations,

$$i\frac{d}{dt}\mathbf{C}(e;t) = \mathsf{W}(e;t)\mathbf{C}(e;t). \qquad \text{(A.11-13)}$$

We then evaluate ensemble-averaged probabilities as

$$\overline{P}_n(t) = \sum_e p(e) |C_n(e;t)|^2, \qquad \text{(A.11-14)}$$

where $p(e)$ is the probability for environment e. The evaluation of $\overline{P}_n(t)$ then requires a separate TDSE solution for every different environment.

Doppler averages. If an environment involves a continuum then the sum $\sum_e$ calls for an integral, though for numerical simulation a discretized approximation is needed.

An important example occurs when the atoms that undergo radiation-induced change arrive as a beam, having mean velocity $\bar{\mathbf{v}}$, and intersecting a laser beam. Within the atomic beam there will be a distribution of velocities, both parallel and perpendicular to $\bar{\mathbf{v}}$, typically originating in a source that fits aspects of a thermal distribution at temperature T. The observable population changes will depend upon the duration of the interaction, and hence upon beam velocity $v_\perp$ across the beam, but in addition the detunings must average over Doppler shifts that originate with incremental components of atom velocity $v_\parallel$ parallel to the field propagation direction, as described in eqn (A.9-31). The required average population is defined by the formula

$$\overline{P}_n(t) = \int_{-\infty}^{+\infty} dv_\parallel \, p(v_\parallel) \int_{-\infty}^{+\infty} dv_\perp \, p(v_\perp) \, |C_n(v_\parallel, v_\perp; t)|^2. \qquad \text{(A.11-15)}$$

In thermodynamic equilibrium at temperature T all directions of velocity $\mathbf{v}$ are equally likely. For a particle of mass m the distribution of thermal speed in one direction, say along the x axis, follows the Maxwell-Boltzmann distribution characterized by the probability distribution

$$p(v_x) = \frac{1}{v_p \sqrt{\pi}} \exp\left[-\left(\frac{v_x}{v_p}\right)^2 \right], \qquad v_p \equiv \sqrt{\frac{2 k_B T}{m}}, \qquad \text{(A.11-16)}$$

where k_B is Boltzmann's constant. This is a Gaussian or *normal distribution* about the most probable speed v_p. Each detuning thereby acquires an additional velocity-dependent Doppler increment centered around the value $\Delta_v = \omega v_p/c$.

A further averaging is required to account for different atom paths across the profile of field values of the laser beam.

Average atom. Generally it is not possible to find an averaged RWA Hamiltonian $\overline{\mathsf{W}}(t)$ such that solutions to the averaged TDSE,

$$i \frac{d}{dt} \overline{\mathbf{C}}(t) = \overline{\mathsf{W}}(t) \overline{\mathbf{C}}(t), \qquad \text{(A.11-17)}$$

can be used to obtain the averaged probabilities of degenerate levels as

$$\overline{P}_n(t) = |\overline{C}_n(t)|^2. \qquad \text{(A.11-18)}$$

Typically the contributors to the sum eqn (A.11-14) differ by Rabi frequency or detuning and so the sum combines amplitudes that oscillate with different frequencies. There will be no single "typical" RWA Hamiltonian that will reproduce the superposition.

An exception that allows the use of such averages occurs when we consider various examples of the final outcome $\overline{P}_n(\infty)$ of adiabatic following, because those results are insensitive to details of the RWA Hamiltonian. Then any time history of a representative Hamiltonian will provide a suitable result.

A.12 Adiabatic elimination; Multiphoton interaction

The condition of resonance for a two-state system, eqn (5.2-1), is important for creating population transfer between states. In an N-state chain, eqn (5.9-1), appreciable population transfer can occur between two states whenever selection rules are favorable and the multiphoton resonance condition eqn (8.4-1) holds. A common example is the transition between the end states 1 and N of an N-state chain when only these two states satisfy the resonance condition. In the RWA the TDSE of eqn (A.9-7) comprises a set of equations having the form

$$i\frac{d}{dt}C_n(t) = \tfrac{1}{2}\Omega_{n-1}(t)C_{n-1}(t) + \Delta_n C_n(t) + \tfrac{1}{2}\Omega_n(t)C_{n+1}(t), \tag{A.12-1}$$

where Δ_n is the cumulative detuning between state 1 and state n. When the cumulative detunings of the intermediate states are much larger than their linking Rabi frequencies these intermediate states have small amplitudes that can be *adiabatically eliminated* [Sho79; Sho90; Sho11] to produce a two-state system involving $C_1(t)$ and $C_N(t)$. There results an overall $N-1$-photon Rabi frequency coupling these two states having, apart from factors of $2\hbar$, the general structure [Sho90]

$$\tfrac{1}{2}|\Omega(t)| = \frac{1}{(2\hbar)^{N-1}} \left| \frac{\mathbf{d}_{12}\cdot\mathbf{e}_1\,\mathbf{d}_{23}\cdot\mathbf{e}_2\,\cdots\,\mathcal{E}_1(t)\mathcal{E}_2(t)\cdots}{\Delta_2\Delta_3\cdots\Delta_{N-1}} \right|, \tag{A.12-2}$$

in which there are $N-1$ field amplitudes, each accompanied by a projection of a dipole transition moment upon a field direction. For this Rabi frequency to describe a multiphoton process, rather than a succession of (pulsed) single-photon transitions, the field amplitudes should all have similar time dependence. Taken with monochromatic fields this approximation underlies nonlinear optics [Blo82; Boy08; Gae06; New11; Dru14] and its treatment of field-induced transition moments. Often all amplitudes come from a single field, a single photon mode. When the energies increase monotonically, $E_n < E_{n+1}$ and the Hamiltonian is presented with photon number states, changes to the atom require absorption or emission of $N-1$ photons as a unit.

The presence of detunings in the denominator of the effective transition moment in eqn (A.12-2) makes these notably smaller than single-photon transition moments, and so they seldom provide spontaneous emission events for atoms in free space. An exception occurs with situations in which selection rules negate the possibility of single-photon emission and two-photon spontaneous emission becomes the dominant possibility for radiative decay [Gop31; Lip65b; Mar72; Ota11]. With lasers the intensity can overcome this shortcoming and facilitate two-photon Rabi oscillations [Bre75; Tak71].

A.13   Adiabatic change

Population inversion in a two-state system can be produced not only by resonant excitation that has a Rabi angle of π, it can also be produced by a pulse in which there is a sweeping of detuning (the difference between Bohr frequency and carrier frequency) through zero, either positively or negatively. Such controlled quantum-state manipulation is an example of a broad class of changes known as adiabatic.

Adiabatic inversion.   Analytic expressions for slow (adiabatic) two-state change date from independent works of four authors, known by the appellation Landau, Zener, Majorana, Stückelberg (LZMS) [Lan32; Lan32b; Zen32; Maj32; Stu32]. The conditional probabilities for a two-state system to undergo complete population inversion from either state (*population swapping*), when subjected during the interval

$-\infty < t < +\infty$ to a linearly changing detuning $\Delta(t)$ that is smallest in magnitude at $t = 0$, are expressible as the conditional probabilities of eqn (A.9-34),

$$P_2(\infty|1) = P_1(\infty|2) \approx 1 - \exp(-\pi\eta), \qquad \eta = \frac{\Omega(0)^2}{|\dot{\Delta}(0)|}, \qquad (A.13\text{-}1)$$

where the adiabaticity-determining parameter η is the ratio, at the moment of least detuning, $t = 0$, of the squared Rabi frequency to the rate of detuning change. Analytic expressions are also available for finite time intervals [Vit98d]. System changes will be adiabatic, thereby producing complete population interchange (*population swapping*), when the detuning changes very slowly:

$$\eta \gg 1, \qquad \Omega(0)^2 \gg |\dot{\Delta}(0)|. \qquad (A.13\text{-}2)$$

A.13.1   Adiabatic states

This adiabatic inversion is a special case of the general procedure of crafting an interaction Hamiltonian that, in the RWA, varies sufficiently slowly that the statevector remains aligned with one of its eigenvectors. These N *adiabatic states* $\mathbf{\Phi}^{(\nu)}(t)$ are instantaneous eigenvectors of the N-dimensional RWA Hamiltonian matrix $\mathsf{W}(t)$,

$$\mathsf{W}(t)\mathbf{\Phi}^{(\nu)}(t) = \varepsilon_\nu(t)\mathbf{\Phi}^{(\nu)}(t), \qquad \sum_m W_{nm}(t)\Phi_m^{(\nu)}(t) = \varepsilon_\nu \Phi_m^{(\nu)}(t). \qquad (A.13\text{-}3)$$

The N numbers $\hbar\varepsilon_\nu(t)$ are known as *adiabatic* energies, to be distinguished from the N *diabatic* energies $\hbar\Delta_n(t)$ that form the diagonal elements of $\hbar\mathsf{W}(t)$ and that thereby define the system energies (in the RWA reference frame) in the absence of interactions.

As is the statevector $\mathbf{\Psi}(t)$, the adiabatic eigenvectors are constructed from superpositions of orthogonal Hilbert-space unit vectors $\psi_n(t)$ whose time-varying phases are prescribed to give a TDSE for slowly varying probability amplitudes $C_n(t)$. These bare states are observable quantum states. They are the customary basis for discussion of intended quantum-state manipulation. The number $|\Phi_n^{(m)}|^2$ is the probability that when the system is known to be in the mth adiabatic state it will be found to be in the nth observable basis state.

The eigenvalues $\varepsilon_\nu(t)$ of $\mathsf{W}(t)$ are real-valued and are commonly assigned labels ν to identify their position in an ordering. With that convention the eigenvalues may, at some time t, be degenerate, but their paths with changing time never cross: At all times

$$\varepsilon_n(t) \le \varepsilon_m(t) \qquad \text{if} \quad n < m. \qquad (A.13\text{-}4)$$

By contrast, the *diabatic!energies*, being controllable detunings, may follow paths that cross.

A.13.2   Adiabatic following

The elements of the RWA Hamiltonian—the field amplitudes and frequencies or Rabi frequencies and detunings– are, in principle, under the control of an experimenter. The time-varying construction of the adiabatic eigenvectors—its components $\Phi_n^{(m)}(t)$ at every moment—is therefore also under experimental control. The concept of *adiabatic following* (and its termination, *adiabatic passage*) aims to control these such that an initial alignment of the statevector with one of the adiabatic states, say $\Phi^{(\text{in})}(0)$ with energy $\varepsilon_{\text{in}}(0)$ at $t = 0$, will continue with ongoing time, and that at the end of an excitation procedure the adiabatic state will be the desired final statevector:

$$\Psi(t) \approx e^{i\varphi(t)}\,\Phi^{(\text{in})}(t). \tag{A.13-5}$$

As indicated here, a phase $\varphi(t)$ often accompanies the changing vectors. When such adiabatic following occurs the energy of the system at any time will follow the path of the adiabatic eigenvalue $\hbar\varepsilon_{\text{in}}(t)$. The resulting superposition of bare states may change. Such changes appear as population changes—adiabatic inversions—produced by the controlled changes of the RWA Hamiltonian. Such controlled quantum-state manipulation offers opportunity for controlling the creation of designer photons, as noted in Chapter 7 and Appendix A.15.5.

The condition for the continuation of this alignment of eqn (A.13-5), the *adiabatic condition*, is that the adiabatic eigenvalues associated with possible contaminant adiabatic states should remain well separated from $\varepsilon_{\text{in}}(t)$. This condition may fail at times when the adiabatic states approach degeneracy, as will happen when curves of detunings cross. The system may then follow a course associated with a diabatic energy and retain its bare-state connection.

Two states. For two-state systems the adiabatic states and their eigenvalues can be defined through the constructions [Vit98d; Vit01]

$$\Phi^{-}(t) = \begin{bmatrix} \cos\theta(t) \\ -\sin\theta(t) \end{bmatrix}, \qquad \varepsilon_{-}(t) = \tfrac{1}{2}[\Delta(t) - \sqrt{\Omega(t)^2 + \Delta(t)^2}], \quad \text{(A.13-6a)}$$

$$\Phi^{+}(t) = \begin{bmatrix} \cos\theta(t) \\ \sin\theta(t) \end{bmatrix}, \qquad \varepsilon_{+}(t) = \tfrac{1}{2}[\Delta(t) + \sqrt{\Omega(t)^2 + \Delta(t)^2}], \quad \text{(A.13-6b)}$$

where the structure is entirely determined by the angle $\theta(t)$ defined through

$$\sin 2\theta(t) = \frac{\Omega(t)}{\sqrt{\Omega(t)^2 + \Delta(t)^2}}, \qquad \cos 2\theta(t) = \frac{\Delta(t)}{\sqrt{\Omega(t)^2 + \Delta(t)^2}}. \tag{A.13-7}$$

When the detuning $\Delta(t)$ sweeps from large negative to large positive values, or vice versa, the two populations interchange. This is an example of a diabatic-curve crossing (of diabatic energies 0 and $\hbar\Delta(t)$) and an *avoided crossing* of adiabatic eigenvalues $\varepsilon_{\pm}(t)$. For this change to occur to the system it is only necessary that the adiabatic condition be maintained [Vit01]:

$$\tfrac{1}{2}|\dot{\Omega}(t)\Delta(t) - \Omega(t)\dot{\Delta}(t)| \ll (\Omega(t)^2 + \Delta(t)^2)^{3/2}. \tag{A.13-8}$$

Here the dots denote time derivatives. Population interchange is also possible without a change of sign in the detuning, when variation of $\Omega(t)$ provides the needed change of $\theta(t)$ [Ran10]. Appendix A.14.6 presents a geometric picture of two-state adiabatic following.

Bright and dark states. With systems for which $N > 2$ it is useful to distinguish two classes of adiabatic states. As with the superpositions introduced by the Morris-Shore transformation of Section 5.10, adiabatic states whose components contain no excited states, are known as *dark* states. Adiabatic states whose components can undergo spontaneous emission, visible as fluorescence, are known as *bright* states. Whereas the MS transformation only combines states within the stable set and the excited set, the adiabatic-state bright superpositions combine both stable and excited states.

In the three-state Lambda linkage the single dark state is a superposition of states 1 and 3. In the four-state letter-M linkage and in other examples of odd-N zigzag linkages the dark state includes superpositions of 1 and N [Sho77; Bia77; Ebe77; Sho78; Coo79; Sho79; Sho90; Vit98b; Yat06; Sho11].

A.14   Density matrices and mixed states

The measurable properties of a quantum system are quantified by various products of probability amplitudes, the simplest of which are the absolute squares that appear in the definition of probabilities. A useful formalism for treating the needed numbers obtains as matrix extensions of the statevector. From the statevector of eqn (A.8-8) we form a *density matrix* [Fan57; Ter61; Blu81; Bar06],

$$\tilde{\rho}(t) = |\Psi(t)\rangle\langle\Psi(t)| = \sum_{n,m} C_n(t)\, C_m(t)^*\, e^{-i\zeta_n(t)+i\zeta_m(t)}\, |\psi_n\rangle\,\langle\psi_m|, \quad \text{(A.14-1a)}$$

whose elements, in the coordinate system of fixed basis states ψ_n, comprises products of probability amplitudes and phase factors associated with those,

$$\tilde{\rho}_{nm}(t) = \langle\psi_n|\Psi(t)\rangle\langle\Psi(t)|\psi_m\rangle = C_n(t)\, C_m(t)^*\, e^{-i\zeta_n(t)+i\zeta_m(t)}. \quad \text{(A.14-1b)}$$

The density matrix defined by eqn (A.14-1a) embodies nothing more, and nothing less, than the information contained in the underlying statevector $|\Psi(t)\rangle$ from which its elements come. It describes a single (pure) quantum state.

It proves useful to extract the phases $\zeta_n(t)$ from the definition of eqn (A.14-1a) and define a density matrix $\rho(t)$ built from the probability amplitudes $C_n(t)$ that refer to a rotating Hilbert-space reference frame. It has the elements

$$\rho_{nm}(t) = \langle\psi'_n(t)|\Psi(t)\rangle\langle\Psi(t)|\psi'_m(t)\rangle = C_n(t)\, C_m(t)^*. \quad \text{(A.14-2)}$$

The time dependence of $\rho(t)$ avoids the rapid variation of $\tilde{\rho}(t)$ that accompanies the field carrier frequencies.

The diagonal elements of the density matrix are *populations*,

$$P_n(t) = \rho_{nn}(t) = \tilde{\rho}_{nn}(t). \quad \text{(A.14-3)}$$

When we ignore population losses, so that the basis states form a complete set, meaning that an appropriate measurement will certainly reveal the system to be in one of the basis states, the probabilities must sum to unity and therefore the trace of a density matrix (the sum of its diagonal elements) must be unity:

$$\text{Tr } \tilde{\rho}(t) = \text{ Tr } \rho(t) = \sum_n \rho_{nn}(t) = 1. \quad \text{(A.14-4)}$$

Because probabilities must be non-negative real numbers no greater than unity the individual diagonal elements are constrained by the requirement

$$0 \leq \rho_{nn}(t) \leq 1. \tag{A.14-5}$$

The off-diagonal elements of a density matrix are known as *coherences*. They are numbers (possibly complex) that have the symmetry property

$$\rho_{nm}(t) = \rho_{mn}(t)^{*}, \qquad \widetilde{\rho}_{nm}(t) = \widetilde{\rho}_{mn}(t)^{*}. \tag{A.14-6}$$

By construction the numbers $\rho_{nm}(t)$ avoid the rapidly varying phases $e^{-i\zeta_{n}(t)+i\zeta_{m}(t)}$ and so their display is more readily interpreted as alterations of excitation; see the examples in Figure A.2.

Pure states. For a system that can be described by a statevector, so that the density matrix is expressible as eqn (A.14-1a) or eqn (A.14-2) the density matrix satisfies the relationship

$$\widetilde{\boldsymbol{\rho}}(t)^{2} = \widetilde{\boldsymbol{\rho}}(t), \qquad \boldsymbol{\rho}(t)^{2} = \boldsymbol{\rho}(t). \tag{A.14-7}$$

Such a situation is known as a *pure state*. Every pure-state density matrix has the property of eqn (A.14-7) and conversely, when a density matrix has this property a single-statevector description suffices.

Mixed states.   Not all situations can be described by a single quantum state, a single statevector. Common situations occur when we deal with an ensemble of similar quantum systems for which different initial conditions apply. We must then deal with expectation values of the form

$$\langle \hat{M}(t) \rangle = \sum_{i} p_{i} \langle \Psi_{i}(t) | \hat{M} | \Psi_{i}(t) \rangle, \tag{A.14-8}$$

where p_{i} denotes the probability of observing the system in the ith state of the ensemble. For such situations we require, instead of eqn (A.14-1a), a *mixed-state* density matrix,

$$\widetilde{\boldsymbol{\rho}}(t) = \sum_{i} |\Psi_{i}(t)\rangle p_{i} \langle \Psi_{i}(t)|, \tag{A.14-9}$$

with which to write expectation values still as eqn (A.14-21b).
   Although the trace of a mixed-state density matrix sums to unity,

$$\mathrm{Tr}\,\widetilde{\boldsymbol{\rho}}(t) = \mathrm{Tr}\,\boldsymbol{\rho}(t) = \sum_{i} p_{i} = 1, \tag{A.14-10}$$

the trace of density matrix products for a mixed state is less than unity,

$$\mathrm{Tr}\,\widetilde{\boldsymbol{\rho}}(t)^{2} = \mathrm{Tr}\,\boldsymbol{\rho}(t)^{2} = \sum_{i} p_{i}^{2} < 1. \tag{A.14-11}$$

This property enables us to identify mixed states and to quantify the amount of mixing.

Examples. Consider (without explicitly noting time dependence) a two-state system that can be described by a statevector $|\Psi\rangle = |\psi_1\rangle$. This is a pure state, with the density matrix

$$\boldsymbol{\rho} = |\psi_1\rangle\langle\psi_1| = \begin{bmatrix} 1 & 0 \\ 0 & 0 \end{bmatrix}, \qquad \boldsymbol{\rho}^2 = \begin{bmatrix} 1 & 0 \\ 0 & 0 \end{bmatrix} = \boldsymbol{\rho}. \tag{A.14-12}$$

Next consider a system that is in a 50:50 coherent superposition of states 1 and 2, with statevector

$$|\Psi\rangle = \frac{1}{\sqrt{2}}\Big[|\psi_1\rangle + |\psi_2\rangle\Big], \qquad P_1 = \tfrac{1}{2}, \quad P_2 = \tfrac{1}{2}. \tag{A.14-13}$$

This too is a pure state, with density matrix

$$\boldsymbol{\rho} = |\Psi\rangle\langle\Psi| = \frac{1}{2}\begin{bmatrix} 1 & 1 \\ 1 & 1 \end{bmatrix}, \qquad \boldsymbol{\rho}^2 = \frac{1}{4}\begin{bmatrix} 2 & 2 \\ 2 & 2 \end{bmatrix} = \boldsymbol{\rho}. \tag{A.14-14}$$

These are two examples of a two-state system in a pure state, a situation that can be described by a statevector.

Next consider a two-state system described by the density matrix

$$\boldsymbol{\rho} = \frac{1}{2}|\psi_1\rangle\langle\psi_1| + \frac{1}{2}|\psi_2\rangle\langle\psi_2| = \frac{1}{2}\begin{bmatrix} 1 & 0 \\ 0 & 1 \end{bmatrix}, \qquad \boldsymbol{\rho}^2 = \frac{1}{4}\begin{bmatrix} 1 & 0 \\ 0 & 1 \end{bmatrix} = \tfrac{1}{2}\boldsymbol{\rho}. \tag{A.14-15}$$

This is a mixed-state combination of basis states 1 and 2 in equal amounts—a *maximally-mixed* state that cannot be described by a single statevector.

More generally we can consider situations in which there are equal probabilities of states 1 and 2 but with a variety of pure-state characteristics, as embodied in the density matrix [Sho90; Sho13; Str17]

$$\boldsymbol{\rho} = \frac{1}{2}\begin{bmatrix} 1 & c \\ c^* & 1 \end{bmatrix}, \qquad \boldsymbol{\rho}^2 = \frac{1}{2}\begin{bmatrix} 1 + |c|^2 & 2c \\ 2c^* & 1 + |c|^2 \end{bmatrix}, \qquad P_1 = P_2 = \tfrac{1}{2}. \tag{A.14-16}$$

The determinant of $\boldsymbol{\rho}$ and the trace of $\boldsymbol{\rho}^2$ provide direct measures of the mixed nature, as parametrized by the parameter c,

$$\mathrm{Det}\,\boldsymbol{\rho} = \tfrac{1}{2}\big(1 - |c|^2\big) \qquad \mathrm{Tr}\,\boldsymbol{\rho}^2 = \tfrac{1}{2}\big(1 + |c|^2\big). \tag{A.14-17}$$

When $|c| = 1$ this is a pure state, the maximally coherent superposition $|\Psi\rangle \propto |\psi_1\rangle + c|\psi_2\rangle$. When $c = 0$ this is a (maximally) mixed state.

Expectation values. Given a quantum system described by a statevector $\Psi(t) \equiv |\Psi(t)\rangle$, we evaluate the theoretical expectation value for some observable that is represented by an operator $\hat{M}$ as

$$\langle\hat{M}(t)\rangle = \langle\Psi(t)|\hat{M}|\Psi(t)\rangle = C_n(t)\,C_m(t)^*\,\mathrm{e}^{-i\zeta_n(t)+i\zeta_m(t)}\,M_{mn}, \tag{A.14-18}$$

where M_{nm} is a transition moment, an element of a constant matrix M, evaluated between the two (fixed) basis states,

$$M_{mn} = \langle \psi_m | \hat{M} | \psi_n \rangle. \tag{A.14-19}$$

These numbers are, in principle, obtainable from wavefunctions. For a single, spinless particle the evaluation requires the wavefunction integral over all space

$$M_{mn} = \int dV \ \psi_m(\mathbf{r})^* M(\mathbf{r}) \psi_n(\mathbf{r}). \tag{A.14-20}$$

Using either of these varieties of a density matrix we can express the expectation value as

$$\langle \hat{M}(t) \rangle = \sum_{nm} \rho_{nm}(t) M_{mn} \, e^{-i\zeta_n(t) + i\zeta_m(t)} \tag{A.14-21a}$$

or, using rotating coordinates, as

$$\langle \hat{M}(t) \rangle = \sum_{nm} \tilde{\rho}_{nm}(t) M_{mn} = \ \text{Tr}\ [\tilde{\rho}(t)\mathsf{M}]. \tag{A.14-21b}$$

A.14.1   Time dependence

The density matrix of eqn (A.14-2) defined from statevector components eqn (A.9-7) has time dependence governed by the *Liouville equation*

$$i\frac{d}{dt}\boldsymbol{\rho}(t) = \mathsf{W}(t)\boldsymbol{\rho}(t) - \boldsymbol{\rho}(t)\mathsf{W}^\dagger(t). \tag{A.14-22}$$

Here changes to populations are driven by coherences, as expressed by the equations

$$\frac{d}{dt}\rho_{nn}(t) = 2\ \text{Im}\sum_{j \neq n} W_{nj}(t)\rho_{jn}(t), \tag{A.14-23a}$$

while the equations for coherences are

$$i\frac{d}{dt}\rho_{nm}(t) = W_{nm}(t)\,[\rho_{mm}(t) - \rho_{nn}(t)]$$
$$+ \sum_{j \neq m} W_{nj}(t)\rho_{jm}(t) - \sum_{j \neq n} W_{jm}(t)\rho_{nj}(t). \tag{A.14-23b}$$

The Liouville equation is just a re-expression of the TDSE in terms of a density matrix rather than a statevector. It requires solving a single equation for N^2 elements of a matrix[23] rather than N coupled equations for N vector components, but the incorporated physics is identical with either equation. As examples below will show, the density-matrix equation of motion has a number of advantages when examining consequences of coherent excitation.

The Lindblad equation.   What is missing from the TDSE, and the Liouville equation equivalent, is any recognition of uncontrollable incoherent processes—spontaneous emission and environmental collisions, most notably. When these are treated as random events—Markov processes—the density matrix offers a means to incorporate their

---

[23] The N^2 elements are not all independent; see Appendix A.14.2 below.

effects by including additional terms into the equations of motion. The result, *Lindblad equation* (or quantum master equation), have the form [Haa73; Lin76; Bre02; Die03; Aol15]

$$i\frac{d}{dt}\boldsymbol{\rho}(t) = \mathsf{W}(t)\boldsymbol{\rho}(t) - \boldsymbol{\rho}(t)\mathsf{W}^\dagger(t)$$
$$+ \sum_\mu \left(\mathsf{L}_\mu \boldsymbol{\rho}(t)\mathsf{L}_\mu^\dagger - \tfrac{1}{2}\mathsf{L}_\mu \mathsf{L}_\mu^\dagger \boldsymbol{\rho}(t) - \tfrac{1}{2}\boldsymbol{\rho}(t)\mathsf{L}_\mu \mathsf{L}_\mu^\dagger \right), \qquad \text{(A.14-24)}$$

where the L_μ are Lindblad operators that express the effect on the given system from averaged interactions with constituents of an external bath (including the photon vacuum); see Section 5.1.3. They act to diminish phase coherences of the density matrix (elastic collisions) and to shift populations (inelastic collisions). In a sense, the Lindblad equation, taken with Maxwell's equations and the Lorentz force, is the "equation of everything" that can happen when bound, nonrelativistic electrons (or other discrete quantum systems) interact with photons.[24] In the limit where decoherence effects dominate they lead to radiative rate equations. When the excitation is fully coherent, they are equivalent to the TDSE. Thus the Lindblad equation is an interpolation between these extremes, based on a model of the bath interactions.

A.14.2 Bloch equations

A useful choice of variables with which to describe the statevector of a two-state system, proposed by Feynman, Frank Vernon, and Robert Hellwarth (FVH) [Fey57; Sho11] is constructed from bilinear products of probability amplitudes in a rotating coordinate system (i.e. elements of the density matrix ρ)

$$r_1(t) = 2\,\mathrm{Re}\,C_1(t)C_2(t)^* = 2\,\mathrm{Re}\,\rho_{12}(t), \qquad \text{(A.14-25a)}$$
$$r_2(t) = 2\,\mathrm{Im}\,C_1(t)C_2(t)^* = 2\,\mathrm{Im}\,\rho_{12}(t), \qquad \text{(A.14-25b)}$$
$$r_3(t) = |C_2(t)|^2 - |C_1(t)|^2 = \rho_{22}(t) - \rho_{11}(t). \qquad \text{(A.14-25c)}$$

Here $r_3(t)$ is the *population inversion*, i.e. the difference between excited and ground-state population. The two variables $r_1(t)$ and $r_2(t)$ are known as *coherences*. Together, these three real-valued variables form components of a three-component abstract vector, the *Bloch vector* $\mathbf{r}(t)$ of unit length,

$$|\mathbf{r}(t)|^2 = r_1(t)^2 + r_2(t)^2 + r_3(t)^2. \qquad \text{(A.14-26)}$$

From the TDSE it follows that the Bloch vector, when undergoing coherent change, obeys the *Bloch equation* [Fey57; All87; Sho90]

$$\frac{d}{dt}r_1(t) = -\Delta(t)\,r_2(t) - \Omega(t)\sin\varphi\,r_3(t), \qquad \text{(A.14-27a)}$$

$$\frac{d}{dt}r_2(t) = \Delta(t)\,r_1(t) - \Omega(t)\cos\varphi\,r_3(t), \qquad \text{(A.14-27b)}$$

$$\frac{d}{dt}r_3(t) = \Omega(t)\sin\varphi\,r_1(t) + \Omega(t)\cos\varphi\,r_2(t), \qquad \text{(A.14-27c)}$$

---

[24]The fundamental nature of the Lindblad equation makes it a plausible candidate for the powerful "secret formula" that must be kept from the sci-fi movie villain.

where φ is the phase of the laser, needed when one considers a succession of pulses. These three equations amount to a re-expression of the two-state TDSE in terms of alternative variables: They express no more and no less than those equations. Because there is no probability loss the Bloch vector retains unit length; its tip moves on the surface of a sphere of unit radius, the *Bloch sphere*.

Picturing Bloch-vector motion; Rotating coordinates. The simplest example of Bloch-vector motion occurs when the excitation is resonant, so that $\Delta(t) = 0$, and the illumination is steady, so that the Rabi frequency is constant. Under these circumstances the population undergoes Rabi oscillations, seen as periodic motion of the Bloch vector from the south to the north pole of the Bloch sphere. The path taken by the Bloch vector depends on the choice of Hilbert-space coordinates, as illustrated in Figure A.2. With the rotating-wave (RW) picture traditional in quantum optics the angular-velocity vector for this excitation remains fixed in the equatorial plane, moving the Bloch vector along a great-circle path. With the Schrödinger picture, based on null phases $\zeta_n(t) = 0$, the carrier frequency with which the RW frame moves causes the route to be a spiral.

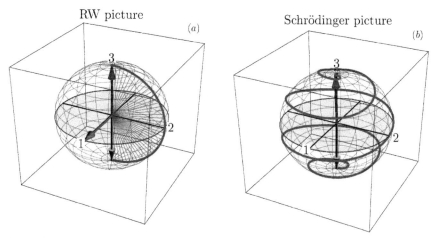

Fig. A.2 *Bloch vector motion for resonant excitation and pi pulse. Left: In rotating-wave picture, with rotating coordinates and fixed angular-velocity vector. Right: In Schrödinger picture, with null phases $\zeta_n(t) = 0$. The two pictures can be regarded as using different gauges. The underlying behavior of the quantum states is identical.*

Relaxation and decoherence. *Homogeneous* decoherence effects (i.e. environmental influences that are the same for all atoms) are traditionally introduced into the Bloch equation by means of two phenomenological *relaxation times*, T_1, and T_2 and an equilibrium value $^{eq}r_3$ for the population inversion. The resulting equations then read, with φ set to zero as is customary,

$$\frac{d}{dt}r_1(t) = -\frac{1}{T_2}r_1(t) - \Delta(t)\,r_2(t), \tag{A.14-28a}$$

$$\frac{d}{dt}r_2(t) = -\frac{1}{T_2}r_2(t) + \Delta(t)\,r_1(t) - \Omega(t)\,r_3(t), \tag{A.14-28b}$$

$$\frac{d}{dt}r_3(t) = -\frac{1}{T_1}\left[r_3(t) - {}^{eq}r_3\right] + \Omega(t)\,r_2(t). \tag{A.14-28c}$$

The parameter ${}^{eq}r_3$ is the equilibrium population inversion that would occur in the absence of the Rabi frequency. Its presence allows the introduction of a thermal equilibrium. The *longitudinal relaxation time* T_1 expresses the time scale for inelastic collisions and spontaneous emission to establish population equilibrium: In the absence of the interaction Ω the population inversion r_3 decays as $\exp(-t/T_1)$ toward the equilibrium value ${}^{eq}r_3$. The term "longitudinal" refers to the choice of the 3 axis of the Bloch vector as the longitudinal direction. The *transverse relaxation time* T_2 expresses the time scale for processes that destroy phase memory without changing population. These include elastic scattering collisions as well as spontaneous emission. T_2 is the decay time for the components $r_1(t)$ and r_2 of the Bloch vector that lie in the plane transverse to the 3 axis: In the absence of Ω these decay as $\exp(-t/T_2)$. In the absence of any other mechanism, spontaneous emission sets a minimum value for the relaxation times, fixed by their inverses as

$$\frac{1}{T_2} > \tfrac{1}{2}A_{21}, \qquad \frac{1}{T_1} > A_{21}. \tag{A.14-29}$$

Additional, *inhomogeneous*, relaxation (i.e. effects that differ from one atom to another) occurs when we average the single-atom Bloch equations over a distribution of detunings.

When the two relaxation times are shorter than the times of interest the Bloch equation describes incoherent excitation. It describes a loss of memory and irreversible photon absorption. To observe Rabi oscillations it is necessary to consider times much shorter than T_1 or T_2. The equations then describe phase-preserving, reversible interchange of atomic and field energy of the sort needed for the storage and retrieval of a single photon.

A.14.3 Rate equation limit

In the limit of short relaxation times the coherences tend to steady values that are tied to the inversion. By setting their time derivatives to zero (adiabatic elimination of the coherences) we obtain the approximations

$$r_2(t) = -\frac{\Omega(t)\gamma}{\Delta^2 + \gamma^2}\,r_3(t), \qquad r_1(t) = -\frac{\Delta}{\gamma}\,r_2(t). \tag{A.14-30}$$

Upon substituting these into the equation for population inversion, and denoting T_2 by $1/\gamma$, we obtain the rate equation

$$\frac{d}{dt}r_3(t) = -\frac{1}{T_1}\left[r_3(t) - {}^{eq}r_3\right] - \frac{\Omega(t)^2\gamma}{\Delta^2 + \gamma^2}\,r_3(t), \tag{A.14-31}$$

The strength of the interaction has a Lorentz profile dependence upon the detuning. For resonant excitation the equation reads

$$\frac{d}{dt}r_3(t) = -\frac{1}{T_1}[r_3(t) - {}^{eq}r_3] - \frac{\Omega(t)^2}{\gamma}r_3(t). \qquad \text{(A.14-32)}$$

A.14.4 The Lorentz atom

Prior to the development of quantum theory Hendrik Lorentz developed a model of atomic structure in which the atom response to weak monochromatic light originated with a "Lorentz electron" of charge e and mass m_e, held by a restoring force to an equilibrium position but responding to the Lorentz force exerted by the radiation [Lor52; Fri65; Mil76; Cra82; Buc01; Sho08]. By including a denumerable set of such electrons the model provides a remarkably good quantitative description of the spectroscopic properties of atoms and their effect upon radiation propagation, as parametrized by a complex-valued index of refraction (a Lorentz medium) [Oug88]. Although Lorentz offered no underlying justification for the model at the time, we can obtain the Lorentz model as a limiting case of a quantum description of two-state excitation when the field is very weak and the excitation is consequently very slight.

The relevant equations draw upon the equation of motion for the density matrix of a two-state atom driven by a monochromatic field and subject to spontaneous emission. Specifically we require the equations for the real and imaginary components of the coherence, because these provide the expectation value of the dipole moment. Expressed in terms of Bloch variables based on stationary Hilbert-space coordinates, rather than the rotating coordinates used above, these equations read [Sho11]

$$\frac{d}{dt}r_1(t) = -\omega_{12}r_2(t) - \gamma r_1(t), \qquad \text{(A.14-33a)}$$

$$\frac{d}{dt}r_2(t) = +\omega_{12}r_1(t) - 2V(t)r_3(t) - \gamma r_2(t), \qquad \text{(A.14-33b)}$$

where $\gamma = A/2$ is half the spontaneous decay rate of state 2 and the interaction $V(t)$ varies at the carrier frequency,

$$V(t) = -d_{12}\mathcal{E}\cos(\omega t) = \tfrac{1}{2}\hbar\Omega\cos(\omega t). \qquad \text{(A.14-34)}$$

To derive an equation descriptive of harmonic motion, as considered by Lorentz, we take the second derivative of r_1, obtaining the acceleration. Taken with the first-order equations this produces the result

$$\frac{d^2}{dt^2}r_1(t) = -[\omega_{12}^2 + \gamma^2]r_1(t) - 2\gamma\frac{d}{dt}r_1(t) + 2\omega_{12}V(t)r_3(t). \qquad \text{(A.14-35)}$$

This is the equation of motion for a driven and damped harmonic oscillator: It has a harmonic restoring force expressed by the Bohr frequency ω_{12} (altered slightly by spontaneous emission), a frictional force proportional to velocity and to the spontaneous emission rate, and a driving force expressed by the final term. To make the connection with a Lorentz oscillator we define the oscillator coordinate to be $x = x_{12}r_1$ where the dipole transition moment is $d_{12} = ex_{12}$. Then the equation reads

$$\frac{d^2}{dt^2}x(t) = -\omega_c^2 x(t) - 2\gamma\frac{d}{dt}x(t) + f_{12}\frac{F(t)}{m_e}r_3(t), \qquad \text{(A.14-36)}$$

where the classical oscillator frequency is

$$\omega_c = \sqrt{\omega_{12}^2 + \gamma^2}, \qquad \text{(A.14-37)}$$

the electron mass is m_e and the force $F(t)$ acting on the electron, of charge e, is

$$F(t) = e\mathcal{E}(t)\cos(\omega t). \qquad \text{(A.14-38)}$$

The dimensionless factor f_{12} is the *oscillator strength*,

$$f_{12} = \frac{2m_e}{\hbar}|x_{12}|^2. \qquad \text{(A.14-39)}$$

This can be written, using atomic units of length and energy [see eqn (2.2-3)] as

$$f_{12} = \frac{2}{3}\frac{\hbar\omega_{12}}{E^{AU}}S(1,2), \qquad \text{(A.14-40)}$$

where $S(1,2) = 3|x_{12}/a_0|^2$ is the spectroscopic transition strength of eqn (A.16-8).

In the absence of the interaction $V(t)$ this oscillator has a natural frequency ω_c differing (slightly) from the Bohr transition frequency ω_{12} by the square of the damping rate $\gamma = A/2$. This damping, due to the loss of coherence due to spontaneous emission, acts like a frictional force upon the oscillator, eventually bringing it to rest at the equilibrium position in the absence of the external force $F(t)$.

The forcing term $F(t)$ here depends upon the population inversion $r_3(t)$. When the excitation is weak, the population remains in the ground state, and so $r_3(t) = -1$. This is the weak-excitation regime of the Lorentz oscillator. When the populations of the two states are equal, so that $r_3(t) = 0$, the force term vanishes.

A.14.5 Torque equations; Adiabatic following

Newton's equation of motion predicts that a constant force $\mathbf{F}$ will cause a change in (linear) velocity that follows a straight path aligned with $\mathbf{F}$. Engineers analyzing mechanical structures must consider not only forces but also the moment of each force, as embodied in the *torque*

$$\boldsymbol{\tau} = \mathbf{r} \times \mathbf{F}. \qquad \text{(A.14-41)}$$

A torque produces a twisting motion in a flexible structure. For a structure to remain at rest and undistorted all net forces and all torques must be zero. Acting on a rigid body a torque imparts a rotation (a spinning motion), a change of angular momentum expressed in terms of dimensionless angular-momentum vector $\mathbf{L}$ as

$$\frac{d}{dt}\hbar\mathbf{L} = \frac{d}{dt}(\mathbf{r}\times m\mathbf{v}) = \mathbf{v}\times m\mathbf{v} + \mathbf{r}\times\frac{d}{dt}m\mathbf{v} = \mathbf{r}\times\mathbf{F} = \boldsymbol{\tau}. \qquad \text{(A.14-42)}$$

Magnetic torques. As an important example, the Lorentz force $\mathbf{F}$ on a charge e that moves with velocity $\mathbf{v}$ in a magnetic field $\mathbf{B}$, is perpendicular to both of those vectors. This force produces an instantaneous torque $\boldsymbol{\tau}$ that is proportional to the instantaneous angular momentum $\hbar\mathbf{L}$,

$$\boldsymbol{\tau} = \mathbf{r} \times \mathbf{F} = e\mathbf{r} \times \mathbf{v}\times\mathbf{B} = \frac{e\hbar}{m}\mathbf{L} \times \mathbf{B}. \tag{A.14-43}$$

The resulting torque equation describes a change to $\mathbf{L}$ that is perpendicular to both $\mathbf{B}$ and $\mathbf{L}$,

$$\frac{d}{dt}\mathbf{L} = -\frac{e}{m}\mathbf{B} \times \mathbf{L}. \tag{A.14-44}$$

This torque occurs in the description of a magnetic moment $\mathbf{m}$ in a constant, uniform magnetic field $\mathbf{B}$, where it appears as

$$\boldsymbol{\tau} = \mathbf{m} \times \mathbf{B}, \quad \text{with} \quad \mathbf{m} = \frac{e\hbar}{m}\mathbf{L}. \tag{A.14-45}$$

In accord with this equation a steady magnetic field acts to rotate the magnetic moment about the fixed $\mathbf{B}$ direction.

Angular velocity vector. As shown in discussion of rigid-body motion, such as spinning tops [Gol50; Lan60b; Kle08b], the alteration of a point $\mathbf{r}$ on the surface of a solid object that is turning about an axis with an *angular velocity*

$$\boldsymbol{\omega} = \frac{\mathbf{r} \times \mathbf{v}}{r^2} \tag{A.14-46}$$

is governed by the equation [Gol50]

$$\frac{d}{dt}\mathbf{r} = \boldsymbol{\omega}\times\mathbf{r}. \tag{A.14-47}$$

The change of the angular momentum vector in eqn (A.14-44) can be written as the action of an angular-velocity vector $\boldsymbol{\Upsilon}$

$$\frac{d}{dt}\mathbf{L} = \boldsymbol{\Upsilon}\times\mathbf{L}. \tag{A.14-48}$$

For the Lorentz force the angular-velocity vector is

$$\boldsymbol{\Upsilon} = -\frac{e\hbar}{m}\mathbf{B}. \tag{A.14-49}$$

Such torque equations occur in several quantum-optics models, and their interpretation there often relies on this magnetic-moment model.

A.14.6 Two-state adiabatic following

For picturing excitation it is often useful to write the Bloch equations eqn (A.14-27) as a torque equation,

$$\frac{d}{dt}\mathbf{r}(t) = \boldsymbol{\Upsilon}(t) \times \mathbf{r}(t), \tag{A.14-50}$$

where the components of the angular-velocity vector (or torque vector) $\boldsymbol{\Upsilon}(t)$ are

$$\Upsilon_1(t) = \Omega(t)\cos\varphi(t), \quad \Upsilon_2(t) = -\Omega(t)\sin\varphi(t), \quad \Upsilon_3(t) = \Delta(t). \tag{A.14-51}$$

The presentation of atomic change as a torque equation makes evident an important class of controllable quantum-state manipulations that occur when the vectors $\boldsymbol{\Upsilon}(t)$

and $\mathbf{r}(t)$ are parallel or antiparallel, so that their cross product vanishes and there is consequently null time derivative of the Bloch vector. This situation underlies the technique of *adiabatic following*, in which the experimenter crafts slow changes to the angular-velocity vector $\Upsilon(t)$ by altering a combination of detuning and Rabi frequency [Vit01; Vit01b; Kra07; Sho08; Sho11; Sho13; Sho17]. When these changes are sufficiently slow (adiabatic) the Bloch vector will remain aligned with the angular-velocity vector. The controlled motion of $\Upsilon(t)$ will thereby carry the Bloch vector $\mathbf{r}(t)$ along any designed path on the Bloch sphere, from an initial state to a chosen final state.

Adiabatic passage. Typically the goal of two-state adiabatic following has been to produce complete population transfer, moving the Bloch vector from one pole to the other on the Bloch sphere. This is accomplished by sweeping the detuning between a very large value (and a consequent initial direction of the angular velocity vector $\Upsilon(-\infty)$ toward one pole of the Bloch sphere), to an oppositely signed large value and $\Upsilon(\infty)$ directed toward the opposite pole. In principle this change of detuning can be produced either by a controlled sweep of Stark or Zeeman shift of the Bohr frequency or by a sweep of the carrier frequency (a chirp). Although the sweep of the detuning must be slow enough to maintain adiabatic following, it must be completed in a time shorter than any relaxation time. The procedure has therefore been termed *rapid adiabatic passage* (RAP) [McC57; Vit01; Sho08; Sho11; Sho13] or adiabatic rapid passage (ARP) [Loy74; Gri75].

A.14.7   Two views of excitation: NMR

Although photon absorption and emission provide a satisfactory model of quantum-state change, other pictures of those changes do not rely on photons. The discipline of nuclear magnetic resonance (NMR) provides an instructive example of complementary views of excitation. The awarding of a Nobel Prize in 1952 for the demonstration of NMR came as my undergraduate education was just beginning to introduce basic quantum theory, and I was puzzled by the seemingly disparate views that were presented by the Harvard and Stanford awardees [Pur46; Blo46; Blo46b; Blo46c; Rig86]. For me it seemed an intellectual challenge rather like that of an earlier generation of physicists who encountered wave and particle pictures of photons, and its resolution is therefore applicable to the present Memoir. The alternate views of NMR have been very clearly explained by John Rigden [Rig86] along the following lines.

Associated with any angular momentum $\mathbf{J}$ is a magnetic moment[25]

$$\mathbf{m} = \gamma \hbar \mathbf{J}, \qquad (A.14\text{-}52)$$

where the proportionality constant γ is termed the *gyromagnetic ratio*. For charged particles (electrons and nuclei) it is proportional to charge and inversely proportional to mass. The basic NMR technique acts on such a proton magnetic moment with two magnetic fields: A relatively strong, static or quasistatic field $\mathbf{B}^0$, and a weaker field

---

[25]This connection is readily understood for charged particles as a circulating electric current, but the connection also applies to the uncharged neutron. Its intrinsic magnetic moment is attributable to currents of constituent charged quarks.

$\mathbf{B}(t)$ at right angles to this, typically varying periodically with time at frequency ω. Their effects are expressed in the two Hamiltonians

$$\mathsf{H}^0 = -\mathbf{m} \cdot \mathbf{B}^0, \qquad \mathsf{H}^{int}(t) = -\mathbf{m} \cdot \mathbf{B}(t), \tag{A.14-53}$$

that combine to give a total Hamiltonian

$$\mathcal{H}(t) = \mathsf{H}^0 + \mathsf{H}^{int}(t). \tag{A.14-54}$$

There are two complementary ways of viewing the effect of such a Hamiltonian on a magnetic moment, reflecting the differing views of the two researchers, Edward Purcell of Harvard and Felix Bloch of Stanford, who shared the 1952 Nobel Prize in physics for the first demonstrations of what became NMR [Rig86].

Spectroscopic description. The viewpoint most in keeping with the present discussion, of a transition between two discrete quantum states, is that used by the Harvard physicists Purcell, Henry C. Torrey, and Robert Pound [Pur46]. In this view the undisturbed Hamiltonian, specialized to nuclear-spin angular momentum $\mathbf{I}$,

$$\mathcal{H}^0 = -\gamma\hbar\,\mathbf{I} \cdot \mathbf{B}^0, \tag{A.14-55}$$

has angular momentum states $|I, M\rangle$ as eigenstates. With this description the magnetic quantum number M takes a discrete set of values $-I \geq M \leq I$. We choose the direction of $\mathbf{B}^0$ as the quantization axis (the z axis); the resulting expression for the eigenvalues of $\mathcal{H}^0$ reads

$$E_M = -\gamma\hbar M B^0, \tag{A.14-56}$$

where B^0 is the magnitude of the field $\mathbf{B}^0$. For a given nuclear spin there are $2I + 1$ distinct equidistant energies. Their frequency separation is the Larmor frequency,

$$\omega_L = \gamma B^0. \tag{A.14-57}$$

The weak field $\mathbf{B}(t)$, lying in the x, y plane, will induce transitions between these states. The excitation will be strongest when this field varies periodically at the Larmor frequency. For a two-state system $(I = \frac{1}{2})$ such as the hydrogen nucleus, and for times shorter than the relaxation time, the populations will undergo periodic Rabi oscillations [at a Rabi frequency proportional to the magnitude of the weak field $\mathbf{B}(t)$]. The two populations correspond to the magnetic moment pointing alternately up and down.

Geometric description. The alternative viewpoint, of magnetic induction taken by Stanford researchers Bloch, William Hansen, and Martin Packard [Blo46; Blo46b; Blo46c], is one based on the behavior of classical spinning tops and magnetic moments. In classical mechanics a static field $\mathbf{B}^0$ exerts on a magnetic moment $\mathbf{m}$ a torque $\mathbf{m} \times \mathbf{B}^0$ that causes the direction of the magnetic moment to trace a cone around the field direction (a precession), always maintaining a fixed angle with respect to the field direction. The precession of the magnetic moment is described by the torque equation

$$\frac{d}{dt}\mathbf{m} = \gamma\,\mathbf{m} \times \mathbf{B}^0. \tag{A.14-58}$$

The frequency of this Larmor precession is the Larmor frequency,

$$\omega_L = \gamma |\mathbf{B}^0|. \tag{A.14-59}$$

Let us take, as above, the static magnetic field to be in the z direction and consider a weaker field in the x, y plane. This will tend to alter the angle between the magnetic moment and the static field. If the weaker field is periodic, with frequency ω equal to the Larmor frequency, then there will occur an orientation change of the magnetic moment.

With sustained weak-field action and negligible relaxation the magnetic moment will undergo periodic reversals of orientation, spin up to spin down, at a rate set by the interaction energy of the weak field—the Rabi frequency for this interaction. Thus the tip of the magnetic moment traces a spiral around the z axis, a combination of precession at the Larmor frequency and slower vertical motion at the Rabi frequency. The picture of this change is greatly simplified if we use a coordinate system that rotates at the Larmor frequency, i.e. at the classical precession frequency. (This is an example of the rotating-wave picture used in quantum optics). With these coordinates we observe just the Rabi oscillation of the magnetic moment, spin up to spin down.

The quantum mechanical counterpart of the classical torque equation is an equation for expectation values,

$$\frac{d}{dt}\langle\mathbf{m}(t)\rangle = \gamma\langle\mathbf{m}(t)\rangle\times\mathbf{B}. \tag{A.14-60}$$

The changes of orientation are observable by detecting the changing magnetic field produced by the magnetic moment.

To summarize the two viewpoints: Whereas the spectroscopic description deals with transitions between populations and with absorption and emission of photons, the geometric description deals with expectation values of moments and hence of coherences.

A.14.8   Multistate adiabatic following

A series of papers during the 1980s [Hio81; Hio82; Hio83; Ore84] established a formalism for recasting the multistate Liouville equation as a torque-like equation for a coherence vector $\mathbf{S}(t)$, constructed from elements of the density matrix, and an angular velocity vector (or torque vector) $\mathbf{\Upsilon}(t)$, built from Rabi frequencies and detunings that define the RWA Hamiltonian matrix $\hbar\mathsf{W}(t)$. Their approach can be presented as follows [Sho17].

Take a complete set of traceless antisymmetric $N \times N$ matrices s_j, generators of the SU(N) Lie group and characterized by commutators

$$[\mathsf{s}_j, \mathsf{s}_k] = 2\mathrm{i}\sum_\ell f_{jk\ell}\,\mathsf{s}_\ell. \tag{A.14-61}$$

Here the $f_{jk\ell}$ are the structure constants of the chosen SU(N) representation. We use these matrices to define the coherence vector $\mathbf{S}(t)$ and torque vector $\mathbf{\Upsilon}(t)$, each of dimension $N^2 - 1$, through their components

$$S_j(t) = \mathrm{Tr}\,\rho(t)\mathsf{s}_j \qquad \Upsilon_j(t) = \mathrm{Tr}\,\mathsf{W}(t)\mathsf{s}_j. \tag{A.14-62}$$

Then the unit-length *coherence vector* $\mathbf{S}(t)$ undergoes a generalized rotation described by the generalized torque equation

$$\frac{d}{dt}S_j(t) = \sum_{k\ell} f_{jk\ell}\, \Upsilon_k(t)\, S_\ell(t). \tag{A.14-63}$$

Adiabatic following occurs when the vectors $\mathbf{S}(t)$ and $\Upsilon(t)$ are parallel or antiparallel and $\Upsilon(t)$ changes sufficiently slowly.

Choices of group representations for the matrices $\mathbf{s}_j$, fitting the symmetry of the Hamiltonian linkage, allow simplification of the description of $\mathbf{S}(t)$; with proper choice of basis the evolution arising from a particular initial condition lies at all times within a subspace of the full Hilbert space [Hio83b; Hio85; Hio87; Hio88]. In particular, for $N = 3$ there can occur motion in a two-dimensional space spanned by the two unit vectors of the dark population-trapping state.

A.15 Three-state pulsed coherent excitation

The extension of two-state excitation to N states is straightforward when the linkage pattern of the Hamiltonian connects the states in a simple chain, starting from state 1 and ending with state N. For three states, excited by a pulse P (for primary or pump, with carrier frequency ω_P) that links states 1 and 2, and a pulse S (for secondary or Stokes, with carrier frequency ω_S) that links states 2 and 3, the three equations introduce two Rabi frequencies and two detunings. The three-state excitation chain is, schematically,

$$|1\rangle_{\text{atom}} \xleftarrow{\ P(t)\ } |2\rangle_{\text{atom}} \xleftarrow{\ S(t)\ } |3\rangle_{\text{atom}}. \tag{A.15-1}$$

These equations are of particular interest when the the two carrier frequencies satisfy a two-photon resonance condition. For the commonly treated linkage pattern appropriate to stimulated-Raman scattering (SRS) [Hel63; She65; Blo67; Ray81; Sir17], in which the energy of state 2 lies above those of states 1 and 3 (a Lambda linkage) the condition is, with absolute values to allow any energies of states 1 and 3:

$$|\hbar(\omega_P - \omega_S)| = |E_1 - E_3|. \tag{A.15-2}$$

The consequent three-state TDSE reads

$$i\frac{d}{dt}C_1(t) = \tfrac{1}{2}\Omega_P(t)C_2(t), \tag{A.15-3a}$$

$$i\frac{d}{dt}C_2(t) = \Delta(t)C_2(t) + \tfrac{1}{2}\Omega_P(t)C_1(t) + \tfrac{1}{2}\Omega_S(t)C_3(t), \tag{A.15-3b}$$

$$i\frac{d}{dt}C_3(t) = \tfrac{1}{2}\Omega_S(t)C_2(t). \tag{A.15-3c}$$

Here the two Rabi frequencies (taken to be real-valued) parametrize the strengths of the two fields and $\Delta(t)$ is the single-photon detuning, of either field from its associated Bohr frequency.

Missing from this set of equations is any direct linkage between states 1 and 3. Such loop-forming linkages [Buc86; Sho90; Una97; Kor02b; Mal04; Ran08] can occur in atoms that are embedded in a solid matrix and as a consequence of a two-photon interaction, e.g. an induced dipole moment.

A.15.1 Multiphoton population cycling

The three-state system is the first of the multiple-state chains in which successive link-ages can move population away from an initially populated state. The first examples of such systems to be considered theoretically treated simultaneous actions of multi-photon fields—linkages in which all Rabi frequencies have the same time dependence and the detunings are constant [Bia77; Ebe77; Coo79; Sho79; Sho81; Sho90; Vit98; Vit98b; Pet10; Sho11]. For the three-state system, and a few others, the system un-dergoes periodic alteration, a generalization of the two-state Rabi oscillations. The simplest N-state behavior typically occurs when all transitions are resonant with their paired states, so that the RWA Hamiltonian has null diagonal elements. With three states, and with particular N-state chains, there occurs periodic complete population transfer between the ends of the chain, customarily taken to be states 1 and N.

A.15.2 Weak probe field: Autler-Townes splitting

When the S field of eqn (A.15-1) is sufficiently strong (i.e. with Rabi frequency much larger than that of the first-step P transition) then the excitation behavior changes dramatically: Once the system is in an excited state, it undergoes many S-field induced Rabi oscillations between the two strongly coupled states 2 and 3 before returning to the ground state via the P-field coupling. These Rabi oscillations affect the fluores-cence signal: Instead of observing the single Lorentz profile of the two-state system, one observes a splitting of the peak into two components, the Autler-Townes doublet separated by the Autler-Townes splitting [Aut55; Gra78; Sho11].

The Autler-Townes doublet is readily explained without resorting to photon num-bers. We begin with the three-state RWA for constant real-valued Rabi frequencies [Sho11]:

$$\mathsf{W} = \begin{bmatrix} 0 & \frac{1}{2}\Omega_P & 0 \\ \frac{1}{2}\Omega_P & \Delta_P & \frac{1}{2}\Omega_S \\ 0 & \frac{1}{2}\Omega_S & \Delta_P - \Delta_S \end{bmatrix} \begin{matrix} \psi_1' \\ \psi_2' \\ \psi_3' \end{matrix}. \tag{A.15-4}$$

The box encloses the portion of the RWA Hamiltonian, the 2×2 matrix $\mathsf{W}^{(2)}$, that involves the strongly coupled states 2 and 3. We now treat these two states using a basis of eigenstates $\Phi_\pm$ of $\mathsf{W}^{(2)}$. These two states are examples of dressed states or, when the RWA Hamiltonian has some slow time dependence, adiabatic states. The two eigenvalues of $\mathsf{W}^{(2)}$ are

$$\varepsilon_\pm = \Delta_P + \tfrac{1}{2}\left[-\Delta_S \pm \delta\right], \qquad \delta \equiv \sqrt{|\Omega_S|^2 + (\Delta_S)^2}, \tag{A.15-5}$$

where δ is the generalized (nonresonant) Rabi frequency of the strongly-coupled tran-sition. The full three-state Hamiltonian in this partially adiabatic basis has couplings between state 1 and each of the adiabatic states,

$$\mathsf{W}^A = \begin{bmatrix} 0 & \frac{1}{2}\Omega_- & \frac{1}{2}\Omega_+ \\ \frac{1}{2}\Omega_- & \varepsilon_- & 0 \\ \frac{1}{2}\Omega_+ & 0 & \varepsilon_+ \end{bmatrix} \begin{matrix} \psi_1' \\ \Phi_- \\ \Phi_+ \end{matrix}. \tag{A.15-6}$$

In general, there will occur significant population transfer only when a diagonal element of the Hamiltonian is equal to the element associated with the initially populated state (a resonance condition). In the present situation, with population starting in state 1, that is the first element on the diagonal. For the matrix of eqn. (A.15-6) the two possible resonance conditions are $\varepsilon_\pm = 0$, leading to the P-field detuning condition

$$\Delta_P = \mp\tfrac{1}{2}\left[\delta - \Delta_S\right]. \tag{A.15-7}$$

In the absence of the S field the two values coincide, requiring $\Delta_P = 0$: Resonance occurs when the P field carrier matches the Bohr frequency. When the P field is tuned to the Bohr frequency ω_{12} and the S field is strong then the P field induces little population change to state 1.

A.15.3 Three-state adiabatic following

Extensive theoretical studies of possible examples of adiabatic following in three-state systems began in 1981 with use of a nine-dimensional coherence vector [Hio81; Hio82; Hio83; Ore84]. Further theoretical studies of three-state adiabatic following that maintained two-photon resonance examined in detail the three components of the adiabatic states, with allowance for temporal separation of the two pulses [Gau90; Few97; Vit97; Vit97b; Vit98c; Vit01]. The first experimental demonstration of three-state adiabatic passage, using a molecular beam passing across monochromatic laser beams, was given the acronym STIRAP (for stimulated-Raman adiabatic passage) [Gau90]. Numerous reviews describe subsequent extensions of the underlying theory and its experimental demonstrations [Ber98; Vit01; Vit01b; Sho13; Ber15; Sho17; Vit17]. Numerous other forms of three-state adiabatic following involve swept detunings [Sho13].

A.15.4 The dark adiabatic state

Counter-intuitive pulse sequences in three-state systems are situations in which the initial population is in state 1 and the S pulse, though unlinked to this state, precedes, but overlaps with, the P pulse—a counter-intuitive pulse sequence, see Section 5.11.1. For pulses that maintain two-photon resonance this sequence aligns the initial statevector $\boldsymbol{\Psi}(t)$ with the two-dimensional *dark state* adiabatic eigenvector, expressible as

$$\boldsymbol{\Phi}^D(t) = \frac{1}{\Omega_T(t)}\begin{bmatrix}\Omega_S(t)\\0\\-\Omega_P(t)\end{bmatrix}, \qquad \Omega_T(t) \equiv \sqrt{\Omega_S(t)^2 + \Omega_P(t)^2}. \tag{A.15-8}$$

Here $\Omega_T(t)$ is $\sqrt{2}$ times the root-mean-square Rabi frequency. This definition is valid as long as there is some field present, so that $\Omega_T(t)$ is not zero. This dark state has no component of state 2 and has the asymptotic connections

$$\boldsymbol{\Phi}^D(t) \to \begin{cases}\boldsymbol{\psi}_1'(t) & \text{when } \Omega_P(t) \to 0\\ -\boldsymbol{\psi}_3'(t) & \text{when } \Omega_S(t) \to 0\end{cases}. \tag{A.15-9}$$

This dark-state construction is often expressed in terms of a mixing angle,

$$\tan\vartheta(t) = \frac{\Omega_P(t)}{\Omega_S(t)}, \qquad \sin\vartheta(t) = \frac{\Omega_P(t)}{\Omega_T(t)}, \qquad \cos\vartheta(t) = \frac{\Omega_S(t)}{\Omega_T(t)}. \tag{A.15-10}$$

Controlled variation of this angle by an experimenter (through suitably crafted pulses) will move the dark state within its two-dimensional Hilbert subspace, as discussed in the next paragraph.

The construction defined by eqn (A.15-9) remains an eigenvector of the instantaneous RWA Hamiltonian for any single-photon detuning, whether constant or time varying, though it does require two-photon detuning. The full set of three-state eigenvectors includes two bright states that incorporate the excited state [Gau90; Few97; Vit97; Vit97b; Vit98c; Vit01].

A.15.5   Stimulated-Raman adiabatic passage (STIRAP)

Traditionally the pulse-induced change went from an initial value $\vartheta \to 0$ to a final value $\vartheta \to \pi/2$, thereby altering the dark state from alignment with state 1 to a complete population transfer into state 3. When the two Rabi frequencies are suitably slowly varying (adiabatic), the statevector remains aligned with the dark state, and the prescription of eqn (A.13-5) becomes

$$\Psi(t) \approx \Phi^D(t). \tag{A.15-11}$$

This means that at the end of the pulse sequence, when the S field is negligible, the dark state will bring the statevector into alignment with state 1, and there will be a corresponding transfer (an adiabatic passage) of population between state 1 and state 3. There will have occurred no transient population in state 2. This process, for fixed-frequency pulses, has been termed STIRAP; see Section 5.11.1.

The adiabatic following of eqn (A.15-11) succeeds in producing complete population transfer for any single-photon detuning. It relies on a counterintuitive pulse sequence, S before P. When the detuning is large, such complete transfer will also occur with the intuitive pulse sequence, P before S, when the statevector follows a bright adiabatic state (termed B-STIRAP) [Vit97b; Kle08; Gri09; Bor10; Sca11; Sca11b; Sho11; Sho17]. With large detuning either pulse sequence will produce population swapping, interchanging populations in states 1 and 3.

The STIRAP coherence. The transitions whose action is parametrized by the two Rabi frequencies typically take place between by means of electric-dipole transitions—the most common of the optical transitions whose absorption and emission continues to serve spectroscopists. For an atom in free space the selection rules for optical transitions 1-2 and 2-3 then imply that, if states 1 and 3 have different energies, the third transition 1-3 is very weak (an example of a "forbidden" optical transition). If states 1 and 3 are degenerate, then transitions between them can be initiated by static fields.

However, an atom that is embedded into a crystalline environment, or some other quantum system that is unrelated to the discrete states of an atomic electron, may very well respond to a field that links states 1 and 3. What matters for this interaction is the coherence $\rho_{13}(t)$. This takes its maximum value when states 1 and 3 have equal magnitudes. This condition occurs when the STIRAP process has proceeded halfway towards its final complete population transfer.

A.16   Radiative rate equations

Prior to the development of coherent, monochromatic light sources from masers and lasers, discussions of radiation effects upon atomic and molecular excitation drew upon principles postulated by Einstein and embodied in three Einstein coefficients. The resulting equations are not applicable to coherent excitation and the physics that is described by the TDSE. Nonetheless they remain widely used in astrophysics and regimes where the radiation is thermal [Ath72; Tuc75; Ryb85]. They are presented here for their historical significance in establishing the usefulness of photons as conveyors of radiation.

Populations and changes.   Changes induced by thermal radiation in ensemble-average populations $\overline{P}_n(t)$ [see eqn (A.3-16)] have the form of first-order ordinary differential equations (rate equations)

$$\frac{d}{dt}\overline{P}_n(t) = \sum_m \overline{P}_m(t)\,\mathcal{R}_{m \to n}, \qquad (A.16\text{-}1)$$

where $\mathcal{R}_{m \to n}$ is the single-atom rate for transition of an atom of (degenerate) energy E_m to energy E_n, taken here to be independent of time. Solutions to these radiative rate equations provide steady-state (equilibrium) populations $\overline{P}_n(\infty)$. Notably, rate equations deal with probabilities, not with the probability amplitudes of quantum theory, and so they do not offer the oscillatory behavior that occurs with coherent excitation. They can be obtained either as postulated, plausible descriptors of change—the route of Einstein—or as the limit of quantum-theory equations when uncontrollable random influences dominate; see Appendix A.14.

A.16.1   The Einstein rates

The Einstein approach to radiation-induced changes dealt with steady, incoherent, isotropic, unpolarized radiation characterized by a broad distribution of frequencies that can be quantified by a spectral energy density $u(\omega)$ in an increment $d\omega$ around angular frequency ω and having dimensions of energy per unit volume per unit angular frequency. In his papers that first dealt with radiation quanta in a thermal-equilibrium enclosure Einstein postulated that three radiative mechanisms contribute to establishment of what we would now call equilibrium populations of discrete energy levels, possibly degenerate:

- Absorption of radiation
- Stimulated emission of radiation
- Spontaneous emission of radiation.

The Einstein coefficients.   Einstein's mechanism of absorption means that a single atom undergoes excitation from energy level E_1 to a level of higher energy E_2 by radiation absorption at a rate proportional to spectral energy density,[26]

---

[26] Here I follow the carefully detailed discussion of Robert Hilborn [Hil82] and indicate by a superscript on B_{12}^{ω} the use of angular frequency ω rather than the circular frequency ν used by Einstein and others as the differential increment. For thermal radiation the energy density, given by the Planck formula eqn (B.7-3), is a function of temperature, $u(\omega, T)$.

$$\mathcal{R}_{1\to 2} = B_{12}^{\omega}\, u(\omega). \tag{A.16-2a}$$

The radiative de-excitation transitions to lower energy, $2 \to 1$, combining stimulated and spontaneous emission, were postulated to occur for each atom at the rate

$$\mathcal{R}_{2\to 1} = B_{21}^{\omega} u(\omega) + A_{21}. \tag{A.16-2b}$$

The connections between the Einstein A and B coefficients appearing in these expressions for radiative rates,[27] deduced originally from thermodynamic arguments applicable to isotropic broadband radiation, are [Hil82]

$$B_{12}^{\omega} = \frac{\varpi_2}{\varpi_1} B_{21}^{\omega}, \qquad B_{21}^{\omega} = \frac{(\lambda_{12})^3}{4\hbar} A_{21}, \tag{A.16-3}$$

where ϖ_n is the statistical weight of level n and $\lambda_{12} = 2\pi c/w_{12}$ is the resonance wavelength of the transition, for Bohr frequency w_{12}. Because of these connections, theory need only provide the A coefficient.

Rates for beams. Although radiative rate equations eqn (A.16-2), and the A and B coefficients, remain widely used, the original equations as presented by Einstein for enclosed radiation require modification for use with radiation beams. For these the quantifying characteristic is the power per unit area per unit frequency—the *irradiance* $I(\omega)$. The removal of energy from a steady beam to produce excitation is quantified by a cross section; see eqn (B.8-5b) in Appendix B.8.2. The frequencies producing excitation of a transition $1 \to 2$ are concentrated around the Bohr frequency for that excitation. It is useful to express that frequency dependence as a line-shape factor $\mathbf{g}(\omega)$ and write the absorption cross-section $\sigma_{12}(\omega)$ and stimulated emission cross-section $\sigma_{21}(\omega)$ as

$$\sigma_{12}(\omega) = \frac{\varpi_2}{\varpi_1}\sigma_{21}(\omega) = \sigma_{12}^{\text{tot}}\, \mathbf{g}(\omega), \quad \text{with} \quad \int d\omega\, \mathbf{g}(\omega) = 1. \tag{A.16-4}$$

The rate equations for beams, like those for isotropic radiation, are most appropriate when the spectral contributions to the radiation are much broader than the profile $\mathbf{g}(\omega)$. Then we can consider the average irradiance surrounding the Bohr frequency, and replace the steady radiative rates for population changes, eqn (A.16-2), by the expressions

$$\mathcal{R}_{1\to 2} = \sigma_{12}^{\text{tot}}\frac{I^{\text{avg}}}{\hbar w_{12}}, \qquad \mathcal{R}_{2\to 1} = A_{21} + \frac{\varpi_1}{\varpi_2}\sigma_{12}^{\text{tot}}\frac{I^{\text{avg}}}{\hbar w_{12}} \quad \text{with} \quad I^{\text{avg}} = I(w_{12}). \tag{A.16-5}$$

These rates are in keeping with the Einstein postulates, but they parametrize the radiation-induced changes without B coefficients. The connection between stimulated and spontaneous emission occurs through the relationship [Hil82]

$$\sigma_{21}^{\text{tot}} = \int d\omega\, \sigma_{21}(\omega) = \frac{(\lambda_{12})^2}{4} A_{21}. \tag{A.16-6}$$

---

[27] The two B coefficients appearing here have dimension of volume × angular frequency over energy × time; the A coefficient has dimension of inverse time.

Spontaneous emission. The spontaneous emission rate for electric-dipole radiation, evaluated from quantum electrodynamics with time-dependent perturbation theory for unoriented atoms in free space, is expressible as [see eqn (A.11-9)]

$$A_{21} = \frac{4}{3} \left(\frac{\omega_{21}}{c} \right)^3 \frac{(ea_0)^2}{4\pi\epsilon_0 \hbar} \frac{S_{21}}{\varpi_2},$$ (A.16-7)

where, for transitions from sublevel m_2 of level 2 to all possible sublevels of level 1 the dimensionless *transition strength* [Con53; Sho65; Hil82; Lar83; Sho90] S_{21} is a sum over squared dipole-transition moments between sublevels,

$$S_{21} = S_{12} = \sum_{m_1 m_2} \frac{|\langle 1m_1 | \mathbf{d} | 2m_2 \rangle|^2}{(ea_0)^2},$$ (A.16-8)

and the statistical weights are counts of the number of sublevels,

$$\varpi_1 = \sum_{m_1} 1, \qquad \varpi_2 = \sum_{m_2} 1.$$ (A.16-9)

When the source atoms are not oriented, as these averages assume, the spontaneously emitted radiation spreads equally in all directions. In keeping with the classical Larmor formula, eqn (C.1-3), the power radiated spontaneously scales as the fourth power of the frequency.

A.16.2 Fermi's Golden Rule; The Purcell effect

Calculations of the Einstein coefficients, quantifying the rate at which an atom will exchange excitation energy with radiation, are presentable as an example of *Fermi's Famous Golden Rule* for evaluating transition rates. With this formula, based on time-dependent perturbation theory for a steady interaction, the transition rate from a discrete initial state i to an energy-conserving final state f (from an energy continuum) is expressed as the absolute square of an interaction energy V times the density of states ϖ_f associated with the final-state energy continuum [see eqn (C.5-11)]:[28]

$$\mathcal{R}_{i \to f} = \frac{2\pi}{\hbar} |\langle f | V | i \rangle|^2 \varpi_f.$$ (A.16-10)

When treating spontaneous emission the density of states is a count of available radiation modes. For an isolated atom in unconstrained free space (idealized as extending indefinitely), with integration over all emission angles, this is [Kle81]

$$\varpi_f = 2 \frac{\omega^2}{\pi c^3},$$ (A.16-11)

where ω is the frequency of the energy-conserving photon. Taken with the electric-dipole interaction averaged over orientation, this yields the traditional expression,

---

[28]The derivation first produces an energy-conserving Dirac delta, $\delta(E_i - E_f)$. Integration over the continuum of final energies E_f replaces this with the density of energy states ϖ_f.

eqn (A.16-7), for the Einstein-A coefficient [Pow64]. When confining surfaces are present, as they are between mirrors, or in an enclosed cavity, or in a waveguide, then the available modes become restricted and the density of states becomes correspondingly redistributed [Kle81]. If the Bohr frequency coincides with one of the cavity-mode frequencies, and the electric field direction meets the requirements for angular-momentum selection rules, the spontaneous emission rate will be increased; otherwise the rate will be slowed. The change in free-space rate (a multiplier of the Einstein A) is known as the *Purcell factor* [Pur46b; Kle81; Sal19]. In treatments of cavity QED the factor is expressed as [Kuh10]

$$f_P = \frac{g^2}{\kappa\gamma},$$ (A.16-12)

where g is the coupling rate between atom and cavity field (half the single-photon Rabi frequency), κ is the rate at which cavity energy decays (loss of photon number) and 2γ is the spontaneous emission rate of the free atom (the loss of atom number). Rabi oscillations are observable in the strong-coupling regime, when $g \gg \kappa, \gamma$.

Photon rates. In setting up radiative rate equations for beams a useful scaling parameter is the *saturation intensity* I^{sat}. To introduce a photon description of the changes induced by a beam of near-monochromatic radiation we introduce a *photon flux* $\mathcal{F}(\omega)$, dimensionally photons per unit area per unit time. The definitions read

$$I^{\mathrm{sat}} = \frac{\hbar\omega_{12}}{\sigma_{12}^{\mathrm{tot}}}, \qquad \mathcal{F}(\omega) = \frac{I(\omega)}{\hbar\omega}.$$ (A.16-13)

We define $\bar{n}$ to be the mean number of interacting photons, i.e. the photon flux within the integrated atomic absorption cross-section $\sigma_{12}(\omega)$,

$$\bar{n} = \sigma_{12}^{\mathrm{tot}} \frac{I^{\mathrm{avg}}}{\hbar\omega_{12}} = \frac{I^{\mathrm{avg}}}{I^{\mathrm{sat}}} = \sigma_{12}^{\mathrm{tot}} \mathcal{F}(\omega_{12}).$$ (A.16-14)

The steady radiative rates can then be expressed as

$$\mathcal{R}_{1\to2} = \frac{I^{\mathrm{avg}}}{I^{\mathrm{sat}}} = A_{21}\bar{n}, \qquad \mathcal{R}_{2\to1}(\omega) = A_{21}\left[1 + \frac{\varpi_1}{\varpi_2}\bar{n}\right].$$ (A.16-15)

The photon number $\bar{n}$ appearing here does not imply any quantum nature of the radiation. It is simply a useful way of parametrizing the radiation irradiance, a dimensionless parameter that quantifies the quantity of radiation.

A.16.3 The two-level radiative rate equations

These formulas are usually applied with the idealizing approximation $\varpi_1 = \varpi_2$, so that the radiative rate equations can be written without mention of degeneracy as

$$\frac{d}{dt}\overline{P}_1(t) = -\frac{d}{dt}\overline{P}_2(t) = A_{21}\left[1 + \bar{n}\right]\overline{P}_2(t) - A_{21}\bar{n}\overline{P}_1(t).$$ (A.16-16)

When no excitation is present at $t = 0$ that population $\overline{P}_2(t)$ rises monotonically in accord with the formula

$$\overline{P}_2(t) = \frac{\overline{n}}{1 + 2\overline{n}} \Big[1 - \exp[-(1 + 2\overline{n})A_{21}t] \Big], \qquad (\text{A.16-17})$$

to a steady-state value

$$\overline{P}_2(\infty) = \frac{\overline{n}}{1 + 2\overline{n}}. \qquad (\text{A.16-18})$$

For very weak fields ($\overline{n} \ll 1$) the excitation probability approaches the value $\overline{n}$ at the rate A_{21}. With intense radiation ($\overline{n} \gg 1$), at most half the population will be in the excited state, an equilibrium value it approaches at the rate $\overline{n}A_{21}$.

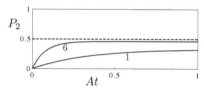

Fig. A.3 *Rate-equation excitation*

Figure A.3 shows examples of two-level rate-equation excitation probability $P_2(t)$, with equal degeneracies $\varpi_1 = \varpi_2$ and mean photon number $\overline{n} = 1$ and $\overline{n} = 6$. The time scale here is in units of the lifetime (the inverse Einstein A coefficient).

In setting up radiative rate equations eqn (A.16-16) I used the word "photon" for a dimensionless measure of radiation intensity $\overline{n}$ defined in eqn (A.16-14). The underlying radiation is purely classical and has been endowed with no explicitly quantum properties; those remain to be discussed. In particular, the "photons" of $\overline{n}$ have not been shown to have granularity—to be discrete indivisible quanta of energy. They are simply a dimensionless parameter that quantifies radiation intensity.

These radiative rate equations deal with given incoherent radiation, such as originates in incandescent lamps and gas discharges, or sunlight. They describe changes of atomic populations but say nothing about the changes the atoms make to the radiation: They describe the effect of photons on atoms but not the creation of photons. Only the discreteness of the atomic energy levels requires quantum theory.

Population inversion. For comparison with the two-state excitation predicted by quantum theory it is useful to express the radiative rate equations in terms of population inversion,

$$w(t) = \overline{P}_2(t) - \overline{P}_1(t). \qquad (\text{A.16-19})$$

For this variable the radiative rate equations eqn (A.16-16) read

$$\frac{d}{dt}w(t) = A_{21} + A_{21}[1 + 2\overline{n}]w(t). \qquad (\text{A.16-20})$$

Two limits have particular significance. In the absence of external radiation ($\overline{n} \to 0$) the equation reads

$$\frac{d}{dt}w(t) = A_{21}[1 + w(t)], \qquad \overline{n} \ll 1. \qquad (\text{A.16-21a})$$

The solution tends (at the spontaneous-emission rate A_{21}) to the steady state limit $w(\infty) \to -1$, with all population in state 1. When there are many photons the inversion equation reads

$$\frac{d}{dt}w(t) = A_{21}2\overline{n}\, w(t), \qquad \overline{n} \gg 1, \qquad (\text{A.16-21b})$$

and the solution tends (at the induced-transition rate $2\bar{n}A_{21}$) to a steady state in which the two populations are equal, $w(\infty) \to 0$.

Transparency and gain. When the radiation produces appreciable excitation then expression eqn (B.8-4) no longer holds; not only will population in level 1 remove radiation but population in level 2 will add radiation, by stimulated emission and by spontaneous emission. The B coefficient for emission, and the corresponding emission cross-section, differs from that for absorption by a ratio of statistical weights, and so the needed alteration of the absorption coefficient, expressed in terms of total number density $\mathcal{N}$ and probabilities $\overline{P}_n$, reads

$$\alpha(\omega) = \mathcal{N} \left[\overline{P}_1 - \frac{\varpi_1}{\varpi_2} \overline{P}_2 \right] \sigma_{12}(\omega). \qquad (\text{A.16-22})$$

As the population in level 2 increases, the absorption decreases. When excitation $\overline{P}_2$ reaches the fraction $\varpi_2/(\varpi_2 + \varpi_1)$ the absorption coefficient vanishes, $\alpha(\omega) = 0$, and the material becomes transparent. However, the radiation will be replenished, in part, by spontaneous emission.

If some mechanism can produce population inversion, $\varpi_2 \overline{P}_2 > \varpi_1 \overline{P}_1$, then the radiation intensity will *increase* with distance; it will undergo *gain* rather than loss [Scu66; Sar74; Sie86; Mil88; Sal19]. All such effects require treatment of coupled partial-differential equations dealing with atom excitation (in time) by photons and with photon change produced (with distance) by atoms, the subject matter for studies of radiative transfer [Ath72; Can85; Cha60; Gal06b].

Failures of rate equations. Coherence of the atom-excitation dynamics and quantum properties of the radiation (the photons) have remarkable effects on the radiation-matter combination. The behavior of a coherently excited two-state system influenced by classical radiation, discussed in Appendix A.10, provides a notable example. Allowance for quantum properties of radiation, discussed in Appendix C.6, introduces still further observable effects. The radiative rate equations presented above, with their underlying limitation to averaged populations and incoherent radiation, cannot predict these effects.

A.17   Algebras

The definition of a vector as an ordered list of numbers (or of words or other recordable entities) provides a useful start for defining an abstract vector space as an ordered set of vectors. This is a useful concept in computer programming and data processing, as well as in engineering and traditional physics.

With the further definition of a scalar product eqn (A.1-6a) between pairs of vectors, and with that the possibility of vector lengths and angles between vectors, we have the useful mathematical structure of a *metric space* (or *Hilbert space*) formed from sets of abstract vectors. Mathematically, we deal with elements that can be added to each other and can be multiplied by scalars (real or complex numbers), to produce other vectors. We have the essential framework for discrete-state quantum theory, as was presented in Appendix A.3.

But the most elementary notion of physical vectors has more to it, as comes to mind when we recall the definition of a vector (in a two-dimensional plane or in three-dimensional Euclidean space) as having *magnitude* and *direction*—two parts, each expressible by numbers. The enlarged mathematics for dealing with such objects is a *division algebra*, meaning the operations of left and right multiplication by any nonzero element have an inverse [Bae01].

Arithmetic. The real and complex numbers are subject to two very familiar operations of arithmetic involving pairs of numbers: Addition + (and its inverse, subtraction) and multiplication, × (and its inverse, division , except by zero). These binary operations are *associative* and *commutative*, meaning that they obey the rules

addition   multiplication

$$\text{associative} \quad a + (b + c) = (a + b) + c, \quad a \times (b \times c) = (a \times b) \times c,$$
$$\text{commutative} \quad a + b = b + a, \quad\quad\quad a \times b = b \times a.$$

The combination of the two operations, multiplication and addition, is *distributive*,

$$\text{left distributive} \quad a \times (b + c) = (a \times b) + (a \times c),$$
$$\text{right distributive} \quad (a + b) \times c = (a \times c) + (b \times c).$$

Generalization of these rules for working with numbers leads to the mathematical structures of an *algebra*.

Algebras. The introduction of *multiplication* operations on pairs of vectors—the scalar (dot) and vector (cross) product—extends the mathematical structure from a *vector space* to an *algebra* [Hes71; Bae01]. With that recognition comes the possibility of expressing physical relationships, and laws of motion, in terms of elements of an algebra and thereby incorporating various symmetries of free space.

Several mathematical algebras occur in discussions of quantum theory, although the mathematical structure is not always mentioned explicitly. These mathematical structures also occur more generally with purely classical description of dynamics and electromagnetism, although they have application to the presentation for the equations of motion for electrons and other quantum particles. The next paragraphs give some idea of the mathematical tools that are becoming recognized.

Hypercomplex numbers: Quaternions and octonions. The extension of the one-dimensional number line for $\mathbb{R}$ to the two-dimensional complex plane for $\mathbb{C}$ can be carried further, into a four-dimensional algebraic base for treating rotations [Alt89; Alt05], the four-dimensional associative division algebra over the real numbers discovered in 1843 by William Rowan Hamilton and known as *quaternions*,[29] denoted $\mathbb{H}$, see [Fin62; Bor66; Ste66; Edm74; Win77; Edm78; Ila81; Alt89; Mar92; Sil02; Alt05]. Whereas complex numbers can be regarded as occupying a two-dimensional space involving two independent coordinate elements, one of which is a square root of -1, a

---

[29]The discovery is famously commemorated by a plaque on the bridge next to the footpath where Hamilton had his inspiration. It reads "Here as he walked by on the 16th October, 1843, Sir William Rowan Hamilton in a flash of genius discovered the fundamental formula for quaternion multiplication "$i^2 = j^2 = k^2 = ijk = -1$, and cut it on a stone of this bridge".

quaternion uses three independent coordinate elements[30] i, j, ℓ that provide such roots. The basic structure of the algebra follows from the equations

$$i^2 = j^2 = \ell^2 = -1, \qquad ij = \ell, \quad j\ell = i, \quad \ell i = j. \tag{A.17-1}$$

A general quaternion $\mathfrak{a}$ is a quartet (or 4-tuple) of real numbers (or a pair of complex numbers) that provide four components a_i

$$\mathfrak{a} = a_0 + a_1 i + a_2 j + a_3 \ell. \tag{A.17-2}$$

These are organized into a single scalar, a_0, and three components a_1, a_2, a_3 of a Euclidean vector $\mathbf{a}$. The product of two quaternions is associative and distributive, but not commutative.

In a 1912 paper [Sil12] Ludwik Silberstein showed, very simply and readably, how the Lorentz transformation of the four-dimensional space of Minkowski, with its hyperbolic metric, used for the special theory of relativity, appeared as part of the intrinsic, built-in mathematical structure of quaternions. He further showed how the electric and magnetic field vectors, regarded as part of a quaternion, allowed the Maxwell equations to be presented as a single quaternion equation. This field combination is what is now regarded as the Riemann-Silberstein vector; see Appendix B.1.11. Here too, the physics of electromagnetism is built into the mathematical structure of quaternions.

The next more complicated generalization, *octonions*, denoted $\mathbb{O}$ [Dix94; Bae01; Con03], have seven basic elements e_i (an eight-dimensional space), and multiplication is neither commutative nor associative. An octonion can be regarded as a pair of quaternions or as an octet (or 8-tuple) of real numbers.

Clifford and geometrical algebras. Quaternions were promoted by Hamilton and Peter G. Tait during the late years of the nineteenth century before losing to the vector calculus of Gibbs as a means of presenting the equations of physics, a rivalry discussed engagingly in [Bor66; Alt89]. After decades of neglect, quaternions (and octonions) are now regarded as a particular example of a *Clifford algebra*, a mathematical construct proposed by William Clifford that includes Pauli algebra and Dirac algebra as special cases. An n-dimensional Clifford algebra [Bay92; Bay93; Las93; Alt96; Tra01] can be based on a set of n anti-commuting basis vectors $\mathbf{e}_i$ that satisfy the multiplication property

$$\mathbf{e}_i \, \mathbf{e}_j + \mathbf{e}_j \, \mathbf{e}_i = 2\delta_{ij}, \quad i, j = 1, \ldots, n, \tag{A.17-3}$$

where δij is the Kronecker delta[31]. Any element of the algebra can be expressed as a scalar and an n-dimensional vector, in the form

$$\mathfrak{a} = a_0 + a_1 \mathbf{e}_1 + \cdots + a_n \mathbf{e}_n. \tag{A.17-4}$$

The product of two elements $\mathfrak{a}$ and $\mathfrak{b}$ of the algebra, or of two quaternions, combines the conventional scalar product and vector product of Euclidean vectors into an inner

---

[30] Here I use an untraditional font $\mathfrak{a}$ to distinguish quaternions, or elements of an algebra, from the numbers a and vectors $\mathbf{a}$ from which they are constructed.

[31] The algebra can be applied to the space-time coordinates of special relativity by including a metric along with the Kronecker delta [Hes03b].

product and an outer (wedge) product that incorporates the geometry of the cross product[32] [Alt89],

$$\mathfrak{ab} = \mathbf{a} \cdot \mathbf{b} + \mathbf{a} \wedge \mathbf{b}. \tag{A.17-5}$$

Instead of being regarded as a form of vector, the cross product is viewed here as an oriented surface. The volume element is the imaginary unit,

$$\mathbf{e}_1 \, \mathbf{e}_2 \, \mathbf{e}_n = \mathrm{i}. \tag{A.17-6}$$

The three Pauli matrices of eqn (A.7-23), together with the unit matrix, form the basis for a Clifford algebra of dimension 3 (the *Pauli algebra*). These mathematical structures, of linear spaces and algebras, occur with purely classical description of vector dynamics, although they have application to the incorporation of quantum dynamics. [Fin62; Mar92].

David Hestenes began popularizing the use of Clifford algebra, now commonly known as *geometric algebra* (GA) and earlier as multi-vector calculus, to the curriculum of physics in mechanics, optics and electrodynamics, as a means of incorporating properties of Euclidean space and special relativity into the fabric of the mathematics being used for physics [Hes66; Hes71; Hes75; Hes86; Hes86b; Hes87; Hes03; Hes03b; Hes12]. Numerous reviews and texts of GA now exist. [Ham84; Gul93; Las93; Vol93; Vol93b; Bay93; Las98; Dor03; Sug04; Per09; Van13; Dre15; Art16]. The use of GA in presenting optics, mechanics, and electrodynamics brings a notable simplification of the equations.

This is a topic that has only recently come to my attention as more than a novelty; it has much to offer as a means of unifying such things as complex numbers, spin and spinors, rotations and reflections, and Maxwell's equations [Maj76; Win77; Edm78]; in that sense photons can be regarded as increments of quaternions. GA has also been used to explain symmetries associated with the Standard Theory of particle physics, [Tra01; Dor03] outlined in Appendix A.19. Octonions have also been used for that purpose [Gun74; Gun96].

Commutators and Lie Algebra. From the postulated commutator properties of quantum mechanical variables it is possible to deduce the values of matrix elements and hence the quantitative behavior of the system. This is the matrix mechanics of Appendix A.7. The needed mathematics, of Lie algebras, is noted in Appendix A.18.4.

A.18   Group theory

Although my own undergraduate education at a small liberal arts college gave me an understanding of what sociologists regard as groups, it was only when I began graduate studies that I was introduced to the astonishingly powerful mathematics known in physical sciences as *group theory*. I was shown by new professor F. Albert Cotton [Cot90] how with only four axioms, presented below, it was possible to create a vast mathematical enterprise that could provide explanations for a remarkable expanse of results in inorganic chemistry—and in turn, throughout physics and mathematics. The

---

[32]The connection is $\mathbf{a} \wedge \mathbf{b} = \mathrm{i}\mathbf{a} \times \mathbf{b}$.

underlying concepts of group theory are touted by advocates as offering the essence of quantum theory as well as useful techniques for calculation.

Group theory was not universally welcomed in the physics departments of the 1950s when I was a student. John Slater, head of the MIT physics department between 1931 and 1951, is famously said to have derided papers that were "incomprehensible to those like me who had not studied group theory" and which relied on what he termed examples of the "Gruppenpest" [Sla75; Wei15] (a name translatable as "group plague" that Eugene Wigner attributed to Pauli [Sza92]). A generation of chemists learned about *Slater determinants* [Sla60] (see page 235) as an alternative to dealing with representations of the symmetric group of degree n, denoted S_n, the group of all permutations of n symbols. Students learned their mathematical physics from the renowned textbook by Philip Morse and Herman Feshbach [Mor53], wherein there was no mention of group theory (or topology, vital now for describing superconductors). Victor Weisskopf lectured from his theoretical nuclear physics textbook [Bla52] without mentioning group theory. But within a decade these attitudes would be reversed, and group theory would become as essential in the education of physicists as it had become to inorganic chemists and molecular spectroscopists.

A.18.1   Coordinate transformations

To motivate the math, I note that it often proves useful to classify solutions to partial differential equations (wavefunctions), and the laws of physics from which they arise, according to their behavior as the three spatial coordinates $\mathbf{x} = \{x, y, z\}$ of Euclidean space and the time variable t are replaced by altered values $\mathbf{x}', t'$ associated with simple static changes: space and time *translations*,

$$\mathbf{x} \to \mathbf{x}' = \mathbf{x} + \mathbf{x}_0, \qquad t \to t' = t + t_0, \tag{A.18-1}$$

where $\mathbf{x}_0$ and t_0 are constants, and with length-preserving spatial *rotations*,

$$\mathbf{x} \to \mathbf{x}' = \mathsf{R}\mathbf{x}, \tag{A.18-2}$$

where R is a 3×3 orthogonal matrix that is parametrizable by three real numbers (often, but not necessarily, taken as the three Euler angles). For equations of motion for point particles based upon nonrelativistic motion, and the nonrelativistic Schrödinger equation, the transformations are termed *Galilean* [Kli15] and they deal separately with space and time. For relativistic physics the transformations are subject to *Lorentzian* constraints in which, as noted below, space and time coordinates are treated as four inter-related coordinates.[33]

It is the existence of these coordinate transformations amongst vector components that distinguishes the vectors of forces and space-time events or electromagnetic fields from the abstract vectors of a general metric space, such as the Hilbert space used to describe quantum states, for which vector components are constrained only by the requirement to maintain unit length. The vectors of a property space have no such constraints.

---

[33]Rotations in four-dimensional spacetime are representations of the Lorentz group; with translations this becomes the inhomogeneous Lorentz group. See Appendix A.18.3.

A.18.2   Mathematical groups

The sets of such transformations are examples of mathematical *groups* [Wey50; Wig60; Ham62; Cot90; Mir95; Cor97; Sze04], a concept that requires only a set of *elements*, say $A, B, C, \cdots$ and an *operation*, here denoted as •, involving pairs of elements.[34] Together the elements and the operation must satisfy four conditions:

- *Closure*: The operation $A \bullet B$ between any two elements must produce a unique element C of the set,

$$A \bullet B \rightarrow C. \tag{A.18-3}$$

- *Associative*: A succession of operations can be pairwise organized (as in multiplication of a set of numbers):

$$A \bullet B \bullet C = A \bullet (B \bullet C) = (A \bullet B) \bullet C. \tag{A.18-4}$$

- *Identity*: One element E (the *identity* element) has no effect on the others:

$$A \bullet E = A. \tag{A.18-5}$$

- *Invertible*: Every interaction can be reversed: Every element A has an *inverse* A^{-1} that reverses the operation action,

$$A \bullet A^{-1} = E. \tag{A.18-6}$$

A *subgroup* is a portion of the set of group elements that by themselves satisfy the four requirements of a group under the same operation • as the full group.

The ordering of group elements in the paired operation may matter. An *Abelian* group is one for which the elements *commute*, $A \bullet B = B \bullet A$. If elements do not commute the group is *non-Abelian*.

As an example, the real numbers, including zero, form an Abelian group under addition, with subtraction as the inverse and zero as the identity. They do not form a group under multiplication, with division as the inverse, because division by zero is undefined.

Representations.   A group *representation* is a mapping between two sets of mathematical objects that incorporates the group operations as successive maps. Coordinate changes, such as the translations and rotations discussed above, provide common examples of representations. Matrix representations are particularly common: Their group operations are associated with multiplications of square matrices.

An *irreducible representation* is one that has no proper subrepresentations, i.e. it forms a closed set under the group actions and cannot be expressed as a sum of smaller representations.

Discrete groups.   As a simple example, the integers form a group (Abelian) under the arithmetic operation of addition (subtraction as the inverse and with 0 as the identity), but not under multiplication. The group elements may form a finite discrete set, such as occurs when treating idealizations of crystals as an infinite distribution of

---

[34]Whereas a *group* requires only a single binary operation, an *algebra* involves two such operations.

pointlike atoms, for which 32 *crystalline point groups* codify the reflections and discrete rotations about symmetry axes that characterize possible structures of crystals. The discrete rotations that occur in these groups are examples of *cyclic groups* formed by multiple operations with a single group element in a multiplication-like operation, $AA^{n-1} = A^n$. The group elements for the cyclic group of order n, are the n elements

$$A, \ AA = A^3, \cdots, A^{n-1}, \ E = A^n. \tag{A.18-7}$$

The nth roots of unity form a cyclic group of order n under multiplication.

The *symmetric group* S_n is the group of $n!$ permutations of n distinguishable elements. Cayley's group theorem states that every group of order n is a subgroup of the symmetric group S_n.

A.18.3   Lie groups

The group elements may form a continuum of elements in which the operations are identifiable by some continuously varying parameter, a *Lie group* [Che46; Wey50; Lip65; Ger01]. Their mathematics has connections not only with the Standard Model photons, but with the quantum-state changes of multiphoton interactions.

The unitary group.   With allowance for vectors that have complex numbers as their elements, as do vectors in Hilbert space, the extension of orthogonality becomes *unitarity.* An $n \times n$ matrix U is unitary if its inverse is equal to the complex conjugate of its transpose (its Hermitian adjoint)

$$\mathsf{U}^{-1} = (\mathsf{U}^T)^* \equiv \mathsf{U}^\dagger, \qquad [U^{-1}]_{ij} = [U]^*_{ji}. \tag{A.18-8}$$

The group of multiplications of unitary matrices of order n is U(n). The group U(1) is the set of phase alterations. The statevector and the electromagnetic field exhibit examples of U(1) symmetry.

The determinant of a unitary matrix has unit magnitude. Restriction to determinant $+1$ gives the special unitary group SU(n). This is the appropriate group for describing the time-dependent Schrödinger equation and its solutions for an n-state system [Hio87].

The orthogonal group.   The *orthogonal group* of dimension n, denoted O(n), is the set of transformations of n real-valued variables $x_1, \cdots, x_n$ that preserve the quadratic form

$$x_1^2 + x_2^2 + \cdots + x_n^2. \tag{A.18-9}$$

It is also the group of multiplications for orthogonal $n \times n$ matrices [Ham62; Itz66]; such a matrix $\mathsf{M} \equiv [M]$ has real-valued numbers as elements M_{ij} and its inverse is its transpose:

$$\mathsf{M}^{-1} = \mathsf{M}^T, \qquad [M^{-1}]_{ij} = [M]_{ji}. \tag{A.18-10}$$

The determinant of an orthogonal matrix is either $+1$ or -1. The *special orthogonal group*, SO(n) is a subgroup of O(n) in which matrices have determinant $+1$.

Rotations in two dimensions (a plane) or three dimensions (Euclidean space) are examples of the groups SO(2) and SO(3) respectively: They maintain lengths of real-component vectors and angles between such vectors. When group operations include inversions the group SO(n) becomes the group O(n).

The Lorentz and Poincaré groups.   A more general orthogonal Lie group, denoted $O(n, m)$, derives from the set of linear transformations (and reflections) involving n real-valued variables x_i and m real-valued variables y_j that preserves the value of the quadratic form

$$(x_1^2 + x_2^2 + \cdots x_n^2) - (y_1^2 + \cdots + y_m^2).$$ (A.18-11)

Of particular interest for invoking special relativity is the connections between the four space-time coordinates in two steadily moving reference frames known as *Lorentz transformations*. These linear relationships between four variables preserve the space-time interval between events—the the four-dimensional Minkowski norms of eqn (8.3-1) involving three space and one time variables – while maintaining a fixed coordinate origin. The symmetries of these relationships fit the generalized orthogonal group $O(1,3)$ or $O(3,1)$, depending on the choice of metric signature, and known as the *Lorentz group*. The Lorentz group is a non-Abelian Lie group having six degrees of freedom: spatial rotations, requiring two real parameters to define the axis of rotation and one to specify the amount of rotation; and *boosts*—linear transformations without rotation of the space coordinates and requiring another three real parameters to define the relative velocity of the two frames. The Lorentz group is a subgroup of the ten-parameter *Poincaré group*, (also known as the inhomogeneous Lorentz group) which encompasses all changes to a space-time point (the full symmetry of special relativity), adding to the six parameters of Lorentz transformations another four that add constants to time and space coordinates. It is this group that describes the free-space photons of Wigner [Wig39; Bia17].

The Wigner little group   In 1939 Wigner considered subgroups of the Lorentz group that preserved the value of momentum, calling them *little groups* [Wig39]. They dictate the internal space-type symmetries of relativistic particles, massive or massless. The little group for massive particles is the three dimensional rotation group, $O(3)$, with its three possible vector directions. The little group for a massless particle is the Euclidean group in two dimensions, $E(2)$, involving cylindrical rotations (about the momentum axis) and translations in a plane (transverse to momentum). This little group offers a useful description of the two vector directions of free-space photons [Han81; Kim88; Vas89; Kim90; Cab03; Lin03; Haw19].

A.18.4   Lie algebras

Mathematicians use the word *algebra* in a technical sense to define a collection of entities (numbers and matrices for example) between which there are two independent binary operations—an extension of the single operation that defines a group; see Appendix A.17. These two operations are taken to be addition (whereby the entities form a vector space) and something akin to multiplication. Addition requires an inverse (subtraction), and therefore forms a group, but multiplication does not, though some divisions may be possible. Imposition of various constraints leads to a great variety of algebras, amongst which is that for quaternions. Algebras are often named for the person who first studied them; examples include Clifford algebras (quaternions are an example) after William Clifford, and Lie algebras, after Sophus Lie.

Lie groups are associated with discrete sets of parameters that can undergo smooth change. The changes induced by an infinitesimal increment ϵ to a single parameter are expressible in operator form as an exponential, definable by a power series

$$\mathcal{U}(\epsilon) = \exp[i\epsilon\mathcal{M}], \qquad \exp(x) = 1 + x + \frac{x^2}{2!} + \cdots . \qquad \text{(A.18-12)}$$

The operator $\mathcal{M}$ is termed the *generator* of infinitesimal transformations. When the group elements are functions of an alterable variable x the operator $\mathcal{M}$ is the differential operator

$$\mathcal{M} = i\frac{\partial}{\partial x}. \qquad \text{(A.18-13)}$$

When the group elements are $n \times n$ unitary matrices the generator $\mathcal{M}$ is an $n \times n$ Hermitian matrix. Groups whose definition requires N parameters will have N generators, say $\mathcal{M}_1, \cdots, \mathcal{M}_N$. The commutators of these operators form a *Lie algebra*, a set of relations having the form

$$[\mathcal{M}_i, \ \mathcal{M}_j] \equiv \mathcal{M}_i\mathcal{M}_j - \mathcal{M}_j\mathcal{M}_i = \sum_k c_{ijk}\mathcal{M}_k, \qquad \text{(A.18-14)}$$

where the *structure constants* c_{ijk} define the particular Lie algebra and, in turn, the underlying Lie group. For example, the commutation properties of angular-momentum operators identify them as representations of the non-Abelian group SO(3) or SU(2).

A.18.5 Constants of motion; Noether's theorem

Apart from constant scaling factors associated with the choice of units the generators found in the Lie groups of most interest to physicists typically have physical significance:

- The generator of time displacements is the Hamiltonian.
- The generator of spatial displacement is the (linear) momentum along the displacement.
- The generator of a rotation around an axis is an angular momentum, represented either by a matrix S, for spin angular momentum acting on components, or by a differential operator **L**, an orbital angular momentum acting on function arguments (coordinates).

From this connection between group-defining parameters and the generators of infinitesimal change in those parameters one deduces that if the basic Hamiltonian (or Lagrangian) is independent of a parameter (time, position, orientation) then the dynamics leaves unchanged the dynamical variable associated with that parameter (energy, linear momentum, angular momentum).

For example, when the Hamiltonian is independent of time, energy is conserved (although it may be reapportioned amongst parts of a system). When the Hamiltonian is independent of location (i.e. there are no bounding walls), momentum is conserved. When the Hamiltonian is cylindrically symmetric the component of total angular momentum along the cylinder axis is conserved.

When the Hamiltonian has no dependence on orientation, the (total) angular momentum **J** is conserved; the corresponding operator, responsible for infinitesimal rotations in three-dimensional Euclidean space, is the angular momentum $\hat{\mathbf{J}}$. It is notable that for any vector field the rotation causes two alterations [Bar10]: To the spatial arguments of the functions that describe field components (induced by orbital angular momentum $\hat{\mathbf{L}}$) and to the three vector components (induced by spin angular momentum $\hat{\mathbf{S}}$) that define the direction of the vectors. The two aspects of rotation combine to give a total angular momentum,

$$\hat{\mathbf{J}} = \hat{\mathbf{S}} + \hat{\mathbf{L}}. \tag{A.18-15}$$

Photons, as carriers of the vector electromagnetic field, always have spin angular momentum unity and may also have orbital angular momentum. For free-space electromagnetic fields the Maxwell equations provide a connection between electric and magnetic vectors that is associated with helicity [Bar10; Bar11; Bar12b; Cam12].

Noether's theorem. These conservation laws are all contained within what is known as *Noether's theorem* [Des77; Cam12; Cam15; Neu17], put simply as:

> *Every differential symmetry in nature has an associated conservation law.*

The symmetry may be visible either in the Hamiltonian (or Lagrangian) or in the equations of motion—the Newton equations for particles and the Maxwell equations for electromagnetic fields. (As several authors have pointed out, the Maxwell equations have an infinite number of conservation-related symmetries [Lip64; Kib65; Cam12; Bar14].) Conversely, a conservation law implies some symmetry of the Hamiltonian, a relationship that has been used in devising Lagrangians for elementary particles.

A.19   The Standard Model of particle physics

Much of this Memoir treats the sort of photons found in Quantum Optics, those that university scientists routinely manipulate in their laboratories. But photons are also of interest in quite different research realms. A very prominent one, garnering Nobel Prizes and influencing university hires, is associated with the photons of the particle-physics community. Researchers there often have very different notions of what a photon is, presented in popular books. Few readers of this present Memoir will not have been aware of this research. Those photons are equally entitled to attention as are those discussed in the other parts of this Memoir. This section aims to to summarize the particle-physics view of the world of particles,[35] for the readers who might willingly view a particle chart on the internet.

Extensions and revisions of the sort of diagrammatic representation of physics interactions that began with Feynman have continued to find application in the description of fundamental interactions at high energies, in the unification of strong, weak, and electromagnetic forces. These are now unified by what has become known

---

[35]The term "particle" is something of a misnomer in this context, because they are here regarded as increments of quantum fields, not as point masses [Hob13].

as *The Standard Model* of particle physics[36] (to be distinguished from standard models in other branches of science and technology, such as the astrophysicists's standard model of the solar interior).

In studies of electromagnetic interactions at very high energies, or in subnuclear spaces or in galactic-scale processes, it is common to work with fields in momentum space, taking into account the constraints of *Lorentz transformations* but without boundary conditions; see Appendix A.18.3. Such photons have been discussed, in Section 8.3.3, as representations of the Poincaré group [Bia17], and, as such, constituents of the Standard Model of particle physics.

A.19.1   The fundamental particles: Leptons and quarks

This unifying model of particles observed in high-energy particle-accelerator collisions involves twelve spin-half (fermion) particles and their antiparticles, organized into three *generations* I, II and III:

- Six different *leptons*: Three with unit negative electron charge (the electron e_-, the mu or muon μ_-, and the tau τ^-, of increasingly larger mass), each of which pairs with its own electrically uncharged neutrino,

$$
\begin{matrix} I \\ \begin{pmatrix} e^- \\ \nu_e \end{pmatrix} \end{matrix} \qquad \begin{matrix} II \\ \begin{pmatrix} \mu^- \\ \nu_\mu \end{pmatrix} \end{matrix} \qquad \begin{matrix} III \\ \begin{pmatrix} \tau^- \\ \nu_\tau \end{pmatrix} \end{matrix}. \tag{A.19-1a}
$$

The generation I pair has the lightest charged lepton, the electron, and so it is the eventual termination of decay processes from generations II and III.

- Six *flavors* of differing-mass *quarks*, each notably more massive than any lepton and carrying a fractional electric charge (either -1/3 or 2/3) and one of three so-called *color charges* R, G, B. (Their antiparticles carry anticolor $\bar{R}, \bar{G}, \bar{B}$ and electric charge +1/3 or -2/3). Like the leptons, the quarks are organized into three generations of pairs, of increasing mass,

$$
\begin{matrix} I \\ \begin{pmatrix} \text{up } (u) \\ \text{down } (d) \end{pmatrix} \end{matrix} \begin{matrix} II \\ \begin{pmatrix} \text{charm } (c) \\ \text{strange } (s) \end{pmatrix} \end{matrix} \begin{matrix} III \\ \begin{pmatrix} \text{top } (t) \\ \text{bottom } (b), \end{pmatrix} \end{matrix}. \tag{A.19-1b}
$$

Antiquarks for each of these are denoted by an overbar, for example $\bar{u}, \bar{d}, \bar{c}, \dots$.

A.19.2   Quark composites: Hadrons, mesons, baryons

Composites of quarks are known collectively as *hadrons*. The lightest of these, pairings of a quark and antiquark, are termed *mesons*.[37] The pion, for example, is the first-generation combination $(u\bar{d})$. Composites of three quarks are known as *baryons*. From triplets of generation-I quarks (the lightest of the quark pairs) are constructed the

---

[36]This is under the supervision of the Particle Data Group (PDG) collaboration, consisting in 2018 of 223 authors from 148 institutions in 24 countries and led by a small team at Lawrence Berkeley National Laboratory (LBNL), who publish a comprehensive review every two years. Their web site is pdg.lbl.gov/index.html.

[37]The mu and tau leptons were originally referred to as "mesons", but no longer.

proton, $p^+ = (uud)$, and the neutron $n^0 = (udd)$ baryons. Numerous unstable heavier hadrons are constructed from other quark triplets.

These constituents, taken with rules for combining quarks into composite particles, have provided a consistent explanation for the assortment of unstable particles and scattering resonances that populated the necessarily descriptive courses on particle physics that were available when I was a graduate student. It was during those years that experiments revealed that neutrinos are intrinsically right handed and anti-neutrinos are left handed, an unexpected property that was revealed when experiments demonstrated that parity (reflection symmetry) was not preserved in beta decay.

A.19.3 The interactions: Photons, mesons, and gluons

The forces between these fermionic particles are attributed to the exchange of various elementary spin-one bosonic fields, each with its particle quantum, a set that includes the photon. These are termed *gauge fields* because they are associated with Lie-group symmetries in which conservation of a current leads to the need for an associated field, see Appendix A.18.3. The Standard Model incorporates three interactions associated with gauge fields, discussed in the following paragraphs.

The electromagnetic interaction; photons. The electromagnetic interactions, affecting all but the neutrinos, are regarded as originating with exchanges of the massless but electrically neutral photon. Its gauge symmetry, described by the one-dimensional unitary group U(1), is associated with invariance under local phase change. The conserved current is that of electric charge and the associated gauge field is the vector potential found in formulations of the physics of electric and magnetic fields. The quantum dynamics of charges and photons is governed by *quantum electrodynamics* (QED).

The weak interaction; mesons. Between leptons and quarks there occurs the weak interaction responsible for beta decay of radioactive nuclei – the decay of a neutron into a proton, an electron, and an electron-antineutrino for example. This behavior is a process involving generation-I particles,

$$\underset{\text{neutron}}{(uud)} \rightarrow \underset{\text{proton}}{(udd)} + \underset{\text{electron}}{e^-} + \underset{\text{anti-neutrino}}{\bar{\nu}_e} . \tag{A.19-2}$$

The weak interaction, often termed *quantum flavordynamics* (QFD), is mediated by exchange of the electrically charged meson pairs W^+, W^- and the neutral meson Z^0. The nonzero mass of these particles leads to the very short range of the weak interaction. They occur as three gauge fields associated with the symmetries of SU(2) ×U(1), the special unitary group in two dimensions (for the electron-neutrino pairing as an example) and the unitary group in one dimension.

The strong interaction; gluons. Between the quarks there occurs the strong interaction that binds neutrons and protons together within atomic nuclei. That force is attributed to exchange of electrically uncharged spin-one *gluons*. There are eight of these gluon gauge-fields. The gluons share the quark symmetry of SU(3). The dynamics of quarks and gluons, with their reliance on three color charges as an extension of electrical charge, is known as *quantum chromodynamics* (QCD).

Summary. The three basic interactions associated with the foregoing three fundamental forces have the following organization:

$$
\begin{matrix}
QED & QFD & QCD \\
\begin{pmatrix} \text{U(1)} \\ \text{1 photon} \end{pmatrix} & \begin{pmatrix} \text{SU(2)} \\ \text{3 mesons} \end{pmatrix} & \begin{pmatrix} \text{SU(3)} \\ \text{8 gluons} \end{pmatrix}
\end{matrix} \, .
\tag{A.19-3}
$$

In addition to these bosonic gauge fields the theory includes a pair of scalar (spin-zero)
Higgs fields, whose interaction with quarks leads to spontaneous symmetry breaking of
an otherwise degenerate ground state and thereby to nonzero masses for boson fields
other than the photon.

Gravity, with its spin-2 gravitons, remains to be incorporated satisfactorily into a
quantum formalism, though not for lack of trying [Roc13; Pla16].

Appendix B
Radiation and photons

The toil of research can seem a swamp.

The previous appendix discussed the physics of atoms (or other discrete-state quantum systems) acted on by specified, controlled (classical) radiation. This next appendix discusses the physics of the radiation, regarded as electromagnetic fields in free space. Although the radiation is here regarded as a quantum system, it is treated as not being appreciably affected by the atoms.

Appendix B.1 presents the Maxwell equations that govern electromagnetic phenomena in the absence of matter. Appendix B.2 discusses examples of traveling-wave modes of radiation that arise from solutions to these free-space Maxwell equations. Appendix B.3 applies quantization conditions to these free-space modes, thereby presenting examples of Dirac photons. Appendix B.5.3 discusses the correlation characteristics that mark non-classical (i.e. quantum) fields. The discussion shifts, in Appendix B.7, to a summary of the thermal physics that underlies what I have called the Planck (or thermal) photon. The final section, Appendix B.8, defines the various parameters—absorption coefficients, cross sections, and optical depth—with which one traditionally quantifies the alterations matter produces on crowds of incoherent photons—the pre-laser (and astronomical) photons.

B.1   Electromagnetic equations in free space

Although the equations of motion for terrestrial particles have undergone two significant alterations since the time of Newton—first to incorporate the limitations of the special theory of relativity and then to incorporate such quantum-mechanical effects as the uncertainty relations—the content of the foundation equations governing electromagnetism (EM), the classical Maxwell equations discussed here, have remained relatively unaffected since they were presented by Maxwell in 1881[1] [Max81]. The equations, presented to unify the eighteenth and nineteenth century observations of

---

[1]Presentation of the equations using vector calculus and notation ∇ came later.

electricity and magnetism, have remained a valid description not only for their predictions of free-space radiation fields but also to describe atomic and subatomic phenomena where quantum theory rules. The following material provides a basic summary of electromagnetic theory as embodied in Maxwell's equations.

B.1.1 The free-space Maxwell equations

The starting point for a description of radiation is the set of Maxwell equations for the time-varying electric field $\mathbf{E}(\mathbf{r}, t)$ and magnetic field[2] $\mathbf{B}(\mathbf{r}, t)$ in free space, i.e. in the absence of any electric charges or currents and with no boundaries. It is through their action upon charges and currents in accord with the Lorentz force eqn (A.1-11) in Appendix A.1.2 that these two classical fields become detectable. To simplify the typography I shall omit explicitly showing the independent variables $(\mathbf{r}, t)$ as arguments.

The Maxwell equations comprise, in part, two dynamical equations, written in vector form and with SI units as

$$\frac{\partial}{\partial t} \epsilon_0 \mathbf{E} = \nabla \times \frac{\mathbf{B}}{\mu_0}, \qquad \frac{\partial}{\partial t} \mathbf{B} = -\nabla \times \mathbf{E}. \tag{B.1-1}$$

These equations involve two universal constants, the *vacuum permittivity* (or electric constant) ϵ_0, and the *vacuum permeability* (or magnetic constant) μ_0, discussed below; see eqn (B.1-4). The Maxwell equations also include two static equations of constraint that impose the condition that the free-space fields be divergenceless (solenoidal),

$$\nabla \cdot \mathbf{E} = 0, \qquad \nabla \cdot \mathbf{B} = 0. \tag{B.1-2}$$

The differential operations of *curl*, $\nabla \times \mathbf{F}$, appearing in eqn (B.1-1), and *divergence*, $\nabla \cdot \mathbf{F}$, appearing in eqn (B.1-2), use the nabla vector differential operator ∇ whose Cartesian-coordinates representations are

$$\nabla_x \equiv \frac{\partial}{\partial x}, \qquad \nabla_y \equiv \frac{\partial}{\partial y}, \qquad \nabla_z \equiv \frac{\partial}{\partial z}, \tag{B.1-3}$$

and whose action is shown in eqn (A.1-4a) and eqn (A.1-4b). The two universal constants appearing in eqn (B.1-1) have the values

$$\epsilon_0 \equiv 8.854 \times 10^{-12}\,\text{F/m}, \qquad \mu_0 = 4\pi \times 10^{-7}\,\text{H/m}. \tag{B.1-4}$$

These combine to give the speed of light in vacuum[3] c, and the *vacuum impedance* or impedance of free space, Z_0,

$$c = \frac{1}{\sqrt{\epsilon_0 \mu_0}} \approx 300 \times 10^6\,\text{m/s}, \qquad Z_0 = \sqrt{\frac{\mu_0}{\epsilon_0}} = 120\pi\ \text{ohm} \approx 376.7\text{ohm}. \tag{B.1-5}$$

From the assumption that the Maxwell equations must appear to all observers as in eqn (B.1-1) it follows that ϵ_0 and μ_0 and the speed of light must be the same to all observers, a key ingredient of Einstein's special theory of relativity.

---

[2] The field $\mathbf{B}$ is commonly termed magnetic flux density or magnetic induction.

[3] By international agreement c is defined to be exactly 299 792 458 m / s.

The first of the dynamical Maxwell equations is often expressed with the use of auxiliary fields $\mathbf{D}(\mathbf{r}, t)$ and $\mathbf{H}(\mathbf{r}, t)$ that, in vacuum, have the definition

$$\mathbf{D} = \epsilon_0 \mathbf{E}, \qquad \mathbf{H} = \frac{\mathbf{B}}{\mu_0}, \qquad \text{so} \qquad \frac{\partial}{\partial t} \mathbf{D} = \nabla \times \mathbf{H}. \tag{B.1-6}$$

When treating fields in bulk matter, as discussed in Appendix C.2.1, these definitions generalize with alternative (but uniform and constant) ϵ and μ.

Electromagnetic units. Generations of students struggled with the separate units of electricity and magnetism that now are commonly accepted as combined into the Systèm International (SI) metric units of electromagnetism, an upgrade of the mks (meter-kilogram-second) and Georgi systems. Relationships amongst SI electromagnetic units A = ampere, C = coulomb, F = farad, H = henry, N = newton, T = tesla, V = volt, W = watt, m = meter, s = second include

$$1\mathrm{F} = \frac{\mathrm{A\,s}}{\mathrm{V}} = \frac{\mathrm{W\,s}}{\mathrm{V^2}} = \frac{\mathrm{s^2}}{\mathrm{H}}, \quad 1\mathrm{T} = \frac{\mathrm{V\,s}}{\mathrm{m^2}} = \frac{\mathrm{W\,s}}{\mathrm{A\,m^2}} = \frac{\mathrm{H\,A}}{\mathrm{m^2}}, \quad 1\mathrm{H} = \frac{\mathrm{V\,s}}{\mathrm{A}} = \frac{\mathrm{W\,s}}{\mathrm{A^2}} = \frac{\mathrm{s^2}}{\mathrm{F}}. \tag{B.1-7}$$

Natural, rationalized units. Theoretical discussions of free-space electromagnetism often employ a *rationalized* or *natural* system of units in which $\epsilon_0 = \mu_0 = c = 1$. The resulting simplified expressions for the Maxwell equations, and for such derived quantities as energy and momentum, exhibit an often-noted symmetry of electric and magnetic fields expressed with eqn (B.1-92).

B.1.2 Density of field energy and momentum

Traveling electromagnetic fields carry energy (as is evident from the ability of sunlight to bring warmth) and momentum (as is evident from comet tails blown away from the sun). The energy density $u(\mathbf{r}, t)$ of any electromagnetic field is the sum of electric and magnetic field contributions,

$$u = \frac{\epsilon_0}{2} \left[\mathbf{E}^2 + c^2 \mathbf{B}^2 \right] = \tfrac{1}{2} \left[\mathbf{D} \cdot \mathbf{E} + \mathbf{B} \cdot \mathbf{H} \right]. \tag{B.1-8}$$

The energy-flux density, or instantaneous power flow, is the *Poynting vector*[4] $\mathbf{G}(\mathbf{r}, t)$,

$$\mathbf{G} = \frac{1}{\mu_0} \mathbf{E} \times \mathbf{B} = \mathbf{E} \times \mathbf{H}. \tag{B.1-9}$$

The Poynting vector is c^2 times the momentum density of the field. Poynting's theorem expresses the relationship between these two fields, imposed by the Maxwell equations, as an equation stating that (in the absence of any currents) the temporal change of electromagnetic energy in any volume is attributable to flow into or out of that volume,

$$\frac{\partial}{\partial t} u = -\nabla \cdot \mathbf{G}. \tag{B.1-10}$$

This is an example of a *continuity equation*, eqn (A.1-39), for electromagnetic energy, treated as a fluid. Any definition of energy density and momentum density must be consistent with this relationship.

---

[4]The symbol **S** is traditional for the Poynting vector. To avoid confusion with the equally traditional use of this symbol for Stokes vectors and spin I am using the unconventional symbol **G**.

B.1.3 Plane-wave fields

The simplest examples of moving solutions to the free-space Maxwell equations take the form of planes of constant amplitude and phase traveling steadily along a Cartesian axis. Such a solution is a *plane wave* – a disturbance that has spatial variation in only one direction (that of propagation) while remaining constant in an infinite plane normal to that direction. Let us take the direction of spatial variation to be the z direction. The solenoidal requirement, eqn (B.1-2), then requires that the vectors $\mathbf{E}$ and $\mathbf{B}$ lie in the x, y plane, i.e. the fields are *transverse* waves, in which the field vectors are perpendicular to the propagation axis and to each other. Two sorts of solutions have particular interest: Traveling waves and standing waves.

Example: Traveling wave. An example is the (idealized) traveling plane-wave that is independent of location in the x, y plane, dependent only on the position along the propagation axis z. Let us take the electric field to lie along the x axis. Then the two fields have the form

$$\mathbf{E}(z,t) = \mathbf{e}_x \, \mathcal{E} \cos(kz - \omega t), \qquad \mathbf{B}(z,t) = \mathbf{e}_y \, \mathcal{B} \cos(kz - \omega t), \qquad \text{(B.1-11)}$$

where $\mathbf{e}_x$ and $\mathbf{e}_y$ are unit vectors in the x and y directions respectively and ω is the *carrier frequency* . Such a field is said to be *linearly polarized*: Its electric vector, along the x axis, fixes the direction of polarization. Other choices of unit vectors, discussed in Appendix B.1.6, produce other polarizations. The parameters k and $\mathcal{B}$ are fixed, by the Maxwell equations, in terms of ω and $\mathcal{E}$, all of them real-valued constants, as

$$k = \omega/c, \qquad \mathcal{B} = \mathcal{E}/c. \qquad \text{(B.1-12)}$$

These fields oscillate everywhere with angular velocity ω and with a spatial repetition length (the wavelength)

$$\lambda = 2\pi/k = 2\pi c/\omega. \qquad \text{(B.1-13)}$$

For traveling plane waves the electric and magnetic field energies are equal because

$$|\mathbf{E}|^2 = |c\mathbf{B}|^2. \qquad \text{(B.1-14)}$$

The energy density and Poynting vector for the example fields eqn (B.1-11) are

$$u(z,t) = \epsilon_0 \mathcal{E}^2 \cos^2(kz - \omega t), \qquad \text{(B.1-15a)}$$
$$\mathbf{G}(z,t) = \mathbf{e}_z \, c\epsilon_0 \mathcal{E}^2 \, \cos^2(kz - \omega t) = \mathbf{e}_z \, c\, u(z,t). \qquad \text{(B.1-15b)}$$

This Poynting vector describes steady energy and momentum flow, with velocity c, along the z axis. The time-averaged values of these two fields are independent of position:

$$\bar{u}(z) = \tfrac{1}{2}\epsilon_0 \mathcal{E}^2, \qquad \bar{\mathbf{G}}(z) = \mathbf{e}_z \, \tfrac{1}{2}c\epsilon_0 \mathcal{E}^2 = \mathbf{e}_z \, c\, \bar{u}(z). \qquad \text{(B.1-16)}$$

The magnitude of this cycle average is the *irradiance*, or laser intensity (power per unit area), given for time-varying amplitude $\mathcal{E}(t)$ by the formula

$$I(t) = \tfrac{1}{2}c\epsilon_0 \, |\mathcal{E}(t)|^2 \quad \text{or} \quad \begin{array}{l} I \approx 1.33 \times 10^{-3} \, [\text{W/m}^2] \, |\mathcal{E}[\text{V/m}]|^2 \\ |\mathcal{E}| \approx 27.4 \, [\text{V/m}] \sqrt{I[\text{W/m}^2]} \end{array}, \qquad \text{(B.1-17)}$$

that provides the connection between theory, with $\mathcal{E}(t)$, and experiment, with $I(t)$.

Example: Standing wave. Another example of idealized plane-wave structure along the z axis is the standing wave with spatially stationary fields

$$\mathbf{E}(z,t) = \mathbf{e}_x \, \mathcal{E} \sin(kz) \sin(\omega t), \qquad \mathbf{B}(z,t) = \mathbf{e}_y \, \mathcal{B} \cos(kz) \cos(\omega t). \qquad \text{(B.1-18)}$$

These have energy density and Poynting vector

$$u(z,t) = \tfrac{1}{2}\epsilon_0 \mathcal{E}^2[\sin^2(kz)\sin^2(\omega t) + \cos^2(kz)\cos^2(\omega t)], \qquad \text{(B.1-19a)}$$

$$\mathbf{G}(z,t) = \mathbf{e}_z \, \tfrac{1}{4} c\epsilon_0 \mathcal{E}^2 \sin(2kz) \sin(2\omega t). \qquad \text{(B.1-19b)}$$

This Poynting vector oscillates forward and backward along the z direction, at frequency 2ω. Its time average is therefore zero:

$$\bar{u}(z) = \tfrac{1}{4}\epsilon_0 \mathcal{E}^2, \qquad \bar{\mathbf{G}}(z) = 0. \qquad \text{(B.1-20)}$$

There is no stationary flow of energy or momentum with a standing wave.

Both of these examples are idealizations akin to the notion of frictionless motion of particles: In practice any electromagnetic field must be spatially bounded and of finite duration. But these plane-wave models are satisfactory for descriptions of fields over moderate distances and times, say, over an atom.

B.1.4   The free-space wave equations

The two dynamical free-space Maxwell equations eqn (B.1-1), first-order differential equations that couple electric and magnetic fields, are often combined into two identical second-order equations,

$$\frac{\partial^2}{\partial t^2}\mathbf{E} = -c^2 \, \nabla \times \nabla \times \mathbf{E}, \qquad \frac{\partial^2}{\partial t^2}\mathbf{B} = -c^2 \, \nabla \times \nabla \times \mathbf{B}. \qquad \text{(B.1-21)}$$

These equations are examples of (vector) *wave equations*: They have wavelike solutions described by moving surfaces (*wavefronts*) where the phase of complex-valued field amplitudes, functions of space and time, maintain a constant value. Quite commonly these surfaces are idealized as being associated with changes that maintain a constant value of one of the spatial coordinates ξ combined with increasing time values, in the form $k\xi - \omega t$ that generalizes the plane-wave function arguments $kz - \omega t$ of eqn (B.1-11).

The electromagnetic fields that describe what we consider radiation—from the sun, from lamps and lasers—are all traveling fields. They are physical examples of what, in simplest form, are vector fields that carry energy and momentum along traveling wave fronts. For such free-space fields the solutions to equations (B.1-21) provide the observed wavelike properties.

Solenoidal and lamellar fields. Electromagnetic fields in free space that are far from any sources are divergenceless. This property distinguishes *radiation fields* from other forms of electromagnetic fields (for example, *evanescent waves* that fall exponentially; see Appendix C.3.4), and therefore it must underlie any practical definition of photons as radiation increments found in free space. Any vector field $\mathbf{F}$, with or without time dependence, can be separated into a *solenoidal* (transverse) part $\mathbf{F}_\perp$ that has zero divergence and a *lamellar* (longitudinal) part $\mathbf{F}_\parallel$ that has zero curl,

$$\mathbf{F} = \mathbf{F}_\perp + \mathbf{F}_\parallel, \qquad \nabla \cdot \mathbf{F}_\perp = 0, \qquad \nabla \times \mathbf{F}_\parallel = 0. \tag{B.1-22}$$

Free-space traveling-wave radiation fields are examples of solenoidal vector fields. With this in mind, the electrostatic Coulomb field surrounding a stationary point charge, a lamellar field, would not be regarded as comprising photons. (However, relativistic presentations of electrodynamics sometimes introduce longitudinal and scalar and timelike photons.)

To simplify the wave equations (B.1-21) for solenoidal fields we use the vector identity, valid for any vector field $\mathbf{F}$,

$$\nabla \times \nabla \times \mathbf{F} = \nabla(\nabla \cdot \mathbf{F}) - \nabla^2 \mathbf{F}, \tag{B.1-23}$$

to rewrite the equation in terms of the *Laplacian* operator ∇^2, defined in terms of Cartesian components as

$$\nabla^2 = \frac{\partial^2}{\partial x^2} + \frac{\partial^2}{\partial y^2} + \frac{\partial^2}{\partial z^2}. \tag{B.1-24}$$

Because the two free-space fields are divergenceless (solenoidal) the resulting equations read

$$\frac{\partial^2}{\partial t^2} \mathbf{E} = c^2 \nabla^2 \mathbf{E}, \qquad \frac{\partial^2}{\partial t^2} \mathbf{B} = c^2 \nabla^2 \mathbf{B}. \tag{B.1-25}$$

These are known as wave equations because, as discussed with examples below, they have solutions that describe traveling wavefronts. Although the two wave equations are uncoupled, their solutions are not independent; they must be linked by the Maxwell equations, (B.1-1).

Electromagnetic fields $\mathbf{E}$ and $\mathbf{B}$ in free space each satisfy the same second-order PDE, either eqn (B.1-21) or eqn (B.1-25). Because the equation is linear in the field, it holds for any superposition $\mathbf{F}$ of these fields. Much of the mathematics can therefore be employed for a generic field $\mathbf{F}(\mathbf{r}, t)$, with specialization to $\mathbf{E}$ or $\mathbf{B}$ as needed; see the RS vector in Appendix B.1.11.

B.1.5   Helmholtz equations; Modes

The possible wavelike behavior of the solutions to the Maxwell equation are most evident when the fields are taken to be monochromatic, with time variation fixed by an angular frequency ω. Taking the arbitrary electromagnetic vector field $\mathbf{F}(\mathbf{r}, t)$ to be expressed in terms of complex-valued positive- and negative-frequency parts,

$$\mathbf{F}(\mathbf{r}, t) = \mathbf{F}^{(+)}(\mathbf{r}, t) + \mathbf{F}^{(-)}(\mathbf{r}, t), \tag{B.1-26}$$

we consider a monochromatic field having the factored form

$$\mathbf{F}^{(+)}(\mathbf{r}, t) = \mathbf{F}^{(-)}(\mathbf{r}, t)^* = \mathbf{U}(\mathbf{r}) \exp(-\mathrm{i}\omega t). \tag{B.1-27}$$

With this *ansatz* the time derivative in the vector wave equation (B.1-21) produces a squared angular frequency and leads from the vector wave equation to the (free-space) *vector Helmholtz equation*

$$\nabla \times \nabla \times \mathbf{U}(\mathbf{r}) - (\omega/c)^2\, \mathbf{U}(\mathbf{r}) = 0, \tag{B.1-28}$$

where the static vector field $\mathbf{U}(\mathbf{r})$ may be complex valued.

Alternatively, we can start with the scalar wave equation (B.1-25) for solenoidal fields. Here the Laplacian makes no directional distinction—it acts only on spatial arguments of functions. The *ansatz* eqn (B.1-27) then gives the *scalar Helmholtz equation*

$$\left[\nabla^2 + \frac{\omega^2}{c^2}\right] U(\mathbf{r}) = 0, \tag{B.1-29}$$

applicable to each component of the vector $\mathbf{U}$. These directional equations are independent, and so each vector component may be a different complex-valued solution to eqn (B.1-29).

Helmholtz modes. Like other linear PDEs, the vector and scalar Helmholtz equations (and their underlying wave equations) each allow an infinite set of distinguishable free-space solutions (modes). To distinguish the solutions we require labels, here denoted ν, that fully identify particular solutions—the vectors $\mathbf{U}_\nu(\mathbf{r})$ or scalars $U_\nu(\mathbf{r})$. The labels ν, as indices, may be discrete, possibly integers $\mathbb{Z}$, or may be from a real-valued continuum $\mathbb{R}$.

To pick out particular solutions, and define the labels ν, it is necessary to impose some constraints. These may come from boundary conditions that confine the fields to finite volumes, often with surfaces that coincide with a spatial coordinate; see Appendix B.2. When the full Helmholtz equation is separable into products of independent factors for different coordinates the label ν expresses the several *separation constants*; see Appendix B.2.2. Alternatively, when the fields are to be regarded as occupying all of free space the constraints may come from requiring that the solutions be eigenfunctions of a differential operator, or a set of commuting operators, notably those associated with linear momentum and angular momentum. These eigenvalues provide the set of labels ν that fully identify particular solutions.

The most common choice of modes is that of transverse plane waves characterized by propagation vectors $\mathbf{k}$, see Appendix B.3.1. For these spatial modes the solenoidal (divergenceless) condition (B.1-2) required of free-space electromagnetic fields imposes the requirement that the <u>field direction</u>, associated with a unit vector $\mathbf{e}$, must be transverse—in the plane normal to the <u>propagation direction</u> defined by $\mathbf{k}$, a condition expressed in eqn (B.3-6) as $\mathbf{k} \cdot \mathbf{e} = 0$. The mode-labeling index ν must specify $\mathbf{k}$ and the direction of $\mathbf{e}$.

Whatever the defining constraints, the set of modes enumerated by the label ν can provide a complete set of functions, vector or scalar, with which to describe any electromagnetic field. Commonly these are assumed to be orthogonal, with normalization factor U_ν,

$$\int dV\ \mathbf{U}_{\nu'}(\mathbf{r})^* \cdot \mathbf{U}_\nu(\mathbf{r}) = \delta_{\nu,\nu'} |U_\nu|^2, \tag{B.1-30}$$

where, for discrete indices, the delta is a Kronecker delta and for continuous variables it is a Dirac delta, to be used in integrals. A general spatial field is expressible as a superposition (with summation over discrete indices and integration over continua),

$$\mathbf{U}_a(\mathbf{r}) = \sum_\nu a(\nu)\mathbf{U}_\nu(\mathbf{r}), \quad \text{with} \quad \sum_\nu |a(\nu)|^2 = 1. \tag{B.1-31}$$

The description of field possibilities by modes amounts to an identification of field degrees of freedom and, with the imposition of quantum conditions, for degrees of freedom for photons. When the electromagnetic field is quantized in the manner of Dirac, the mode indices become quantum numbers of the photons: *Each choice of mode is associated with a choice of photon*; see Appendix B.3.

Factoring Helmholtz solutions. Several coordinate systems allow useful separation of the Laplacian for three spatial variables, allowing $U(\mathbf{r})$ to be factored into an independent function for each coordinate [Lev16]. Each of these factors brings the introduction of a separation constant, counterparts of the frequency w that occurs in separation of the time variable in the wave equation. The set of separation constants provides the label ν. The simplest and most commonly used coordinate systems are: Cartesian, x, y, z; circular-cylindrical, r, ϕ, z; and spherical, r, θ, ϕ. Cartesian and cylindrical coordinates are of most use in describing directed radiation and beams. Spherical coordinates offer opportunities for describing monochromatic fields emitted into all directions by a localized source, and so the resulting complete sets of multipole fields find use in classical optics [Lud91] as well as quantum optics [Fiu62; Dev74; Bay06]. Multipole expansions provide useful techniques for rapidly evaluating the (elastic) scattering of radiation from macroscopic objects as large as airplanes [Coi93; Son97].

Helmholtz eigenfunctions; Momentum operators. Because the Helmholtz equations are linear, any superposition of distinct solutions will be another solution. By means of such constructions it is possible to create solutions that have particular properties, such as satisfying boundary conditions along coordinates. To manage such infinite sets of field modes it is desirable to classify them by various symmetries or by eigenvalues of various differential operators, thereby introducing labels whereby the modes may be referenced and used to form superpositions. When quantization is introduced each such mode field provides a possible spatial pattern for a traveling free-space photon. For this purpose it is useful to consider the partial derivatives appearing in the Helmholtz equations as scaled (linear) momentum operators,

$$\hat{\mathbf{p}} = -i\hbar\nabla, \qquad \hat{\mathbf{p}}^2 = -\hbar^2\nabla^2, \tag{B.1-32}$$

and scaled (orbital) angular momentum operators (see see Appendix B.1.10)

$$\hbar\hat{\mathbf{L}} = \mathbf{r}\times\hat{\mathbf{p}}. \tag{B.1-33}$$

All solutions to the scalar Helmholtz equation then appear as eigenfunctions of $\hat{\mathbf{p}}^2$ with eigenvalue $(\hbar w/c)^2 = (\hbar k)^2$,

$$\hat{\mathbf{p}}^2 U(\mathbf{r}) = (\hbar w/c)^2 U(\mathbf{r}). \tag{B.1-34}$$

B.1.6   Unit vectors for fields

To describe any vector field it is necessary to specify, at every position, a field direction. In general, both the direction of a field and its phase may vary with position, as expressed by the construction

$$\mathbf{U}(\mathbf{r}) = \mathbf{e}(\mathbf{r})U(\mathbf{r}) = \mathbf{e}(\mathbf{r})\,|U(\mathbf{r})|\,\exp[i\chi(\mathbf{r})], \qquad \mathbf{e}(\mathbf{r})^* \cdot \mathbf{e}(\mathbf{r}) = 1, \qquad \text{(B.1-35)}$$

where the locally defined unit vector $\mathbf{e}(\mathbf{r})$, possibly complex-valued, expresses the direction of field components at position $\mathbf{r}$. Traveling plane electromagnetic fields in free space have vectors that are perpendicular to their propagation direction, so only two unit vectors are required for their description. The gradient of the phase $\chi(\mathbf{r})$ determines the local wavefront of the field, independent of the vector components.

Local unit vectors; Spin components.   Local reference frames for vectors are needed when we consider fields at boundaries between materials that have different composition, as embodied in electric and magnetic constants ϵ and μ. When we take the local x, y plane to be the surface, we require continuity across the surface of the tangential (x, y) component of $\mathbf{E}$ and the longitudinal (z) component of $\mathbf{B}$ [Str41], see eqn (C.2-8).

To create a set of three orthogonal locally-defined reference vectors we begin by defining a primary local reference axis, say the position vector $\mathbf{r}$ or, for treating vectors in inverse (Fourier) space, a propagation vector $\mathbf{k}$. This axis will serve as the third of the three Cartesian unit vectors,

$$\mathbf{e}_3(\mathbf{r}) = \frac{\mathbf{r}}{|\mathbf{r}|}, \quad \text{or} \quad \mathbf{e}_3(\tilde{\mathbf{k}}) = \frac{\mathbf{k}}{k}. \qquad \text{(B.1-36)}$$

We then choose another direction, not parallel to this, for an ancillary unit vector $\mathbf{e}_0$. From these we construct by vector multiplication two further vectors that are perpendicular to $\mathbf{e}_3$ and that form an orthogonal set in three dimensions:

$$\mathbf{e}_2 = \mathbf{e}_0 \times \mathbf{e}_3, \qquad \mathbf{e}_1 = \mathbf{e}_3 \times \mathbf{e}_1. \qquad \text{(B.1-37)}$$

From the construction procedure it is evident that the three vectors 1,2,3 are mutually orthogonal,

$$\mathbf{e}_i \cdot \mathbf{e}_j = \delta_{ij}. \qquad \text{(B.1-38)}$$

Vector 3 is along the (local) reference direction, which typically is labeled the z axis. Vectors 1 and 2, each transverse to vector 3, then define the x, y plane. The auxiliary vector $\mathbf{e}_0$ merely serves to distinguish vectors 1 and 2.

Global unit vectors.   In treatments of general fields, particularly those with boundaries, it is useful to choose a fixed Cartesian frame of reference and use the vectors $\mathbf{e}_j(\mathbf{r})$ uniformly throughout the region of interest – they have no dependence on $\mathbf{r}$. The three unit vectors are denoted variously as

$$\mathbf{e}_1 \equiv \mathbf{e}_x \equiv \mathbf{e}_H, \qquad \mathbf{e}_2 \equiv \mathbf{e}_y \equiv \mathbf{e}_V, \qquad \mathbf{e}_3 \equiv \mathbf{e}_z \equiv \mathbf{e}_{||}. \qquad \text{(B.1-39)}$$

Using polar cylindrical coordinates we can define, in the transverse plane to $\mathbf{e}_z(\mathbf{r})$, radial (r) and azimuthal (ϕ) components that depend on position, either in Euclidean space or in wavevector (Fourier) space,

$$\mathbf{e}_r = \cos\phi\,\mathbf{e}_x + \sin\phi\,\mathbf{e}_y, \tag{B.1-40a}$$

$$\mathbf{e}_\phi = -\sin\phi\,\mathbf{e}_x + \cos\phi\,\mathbf{e}_y. \tag{B.1-40b}$$

With spherical coordinates, used for treating multipole fields centered on a radiating atom, we have the unit vectors

$$\mathbf{e}_r = \cos\theta\sin\phi\,\mathbf{e}_x + \sin\theta\sin\phi\,\mathbf{e}_y + \cos\phi\,\mathbf{e}_z, \tag{B.1-41a}$$

$$\mathbf{e}_\theta = -\sin\theta\,\mathbf{e}_x \quad + \cos\theta\,\mathbf{e}_y, \tag{B.1-41b}$$

$$\mathbf{e}_\phi = \cos\theta\cos\phi\,\mathbf{e}_x + \sin\theta\cos\phi\,\mathbf{e}_y - \sin\phi\,\mathbf{e}_z. \tag{B.1-41c}$$

Beam vectors. In treatments of radiation beams, the beam axis $\check{\mathbf{k}}$ is taken as the reference axis, z. The electric field, perpendicular to this axis, defines the polarization direction. For linear polarization this is often referred to as horizontal (H) and vertical (V). In that situation the x, y plane is transverse to the z direction, which is longitudinal.

Helicity vectors. When we deal with plane waves the propagation vector $\mathbf{k}$ establishes the direction of momentum flow. With this as z axis the above procedure defines a set of real orthogonal unit vectors $\mathbf{e}_j(\check{\mathbf{k}})$. In place of these three real-valued locally Cartesian unit vectors we can use three complex-valued basis vectors (often called spherical vectors from their connection with angular momentum),

$$\mathbf{e}_q(\check{\mathbf{k}}) = \delta(q,0)\mathbf{e}_z(\check{\mathbf{k}}) - \frac{q}{\sqrt{2}}\left[\mathbf{e}_x(\check{\mathbf{k}}) + iq\,\mathbf{e}_y(\check{\mathbf{k}})\right], \tag{B.1-42}$$

labelled by $q = -1, 0, +1$. These vectors are eigenvectors of the spin-one matrix s_z, defined in eqn (A.7-26), with eigenvalue q,

$$\mathsf{s}_z\mathbf{e}_q(\check{\mathbf{k}}) = q\,\mathbf{e}_q(\check{\mathbf{k}}), \quad \text{with} \quad \mathsf{s}_z = \begin{bmatrix} 0 & -i & 0 \\ i & 0 & 0 \\ 0 & 0 & 0 \end{bmatrix}, \tag{B.1-43}$$

as follows from the representation of the Cartesian unit vectors by column vectors,

$$\mathbf{e}_x(\check{\mathbf{k}}) = \begin{bmatrix} 1 \\ 0 \\ 0 \end{bmatrix}, \quad \mathbf{e}_y(\check{\mathbf{k}}) = \begin{bmatrix} 0 \\ 1 \\ 0 \end{bmatrix} \quad \mathbf{e}_z(\check{\mathbf{k}}) = \begin{bmatrix} 0 \\ 0 \\ 1 \end{bmatrix}. \tag{B.1-44}$$

Their orthogonality properties are expressed as

$$\mathbf{e}_q(\check{\mathbf{k}})^* \cdot \mathbf{e}_{q'}(\check{\mathbf{k}}) = \delta_{q,q'}. \tag{B.1-45}$$

These vectors are defined with respect to the propagation vector $\mathbf{k}$. When the propagation axis $\mathbf{k}$ coincides with the fixed reference axis we have the usual spherical vectors, $\mathbf{e}_q(\check{\mathbf{r}}) \equiv \mathbf{e}_q$.

From the two transverse vectors above we define two complex-valued *helicity vectors* for describing right- and left-circular polarization,

$$\mathbf{e}_R(\check{\mathbf{k}}) \equiv -\mathbf{e}_+(\check{\mathbf{k}}) = \frac{1}{\sqrt{2}}[\mathbf{e}_x(\check{\mathbf{k}}) + \mathrm{i}\mathbf{e}_y(\check{\mathbf{k}})], \qquad (\text{B.1-46a})$$

$$\mathbf{e}_L(\check{\mathbf{k}}) \equiv \mathbf{e}_-(\check{\mathbf{k}}) = \frac{1}{\sqrt{2}}[\mathbf{e}_x(\check{\mathbf{k}}) - \mathrm{i}\mathbf{e}_y(\check{\mathbf{k}})]. \qquad (\text{B.1-46b})$$

Fields constructed with helicity unit vectors have $\pm\hbar$ units of spin angular momentum along the propagation axis. These are examples of more general independent polarization pairs [Col12] corresponding to diametrically opposite points on the Poincaré sphere of Figure B.1,

$$\mathbf{e}_1(\check{\mathbf{k}}) = \sin\theta\,\mathbf{e}_x(\check{\mathbf{k}}) + \mathrm{e}^{\mathrm{i}\phi}\cos\theta\,\mathbf{e}_y(\check{\mathbf{k}}), \qquad (\text{B.1-47a})$$
$$\mathbf{e}_2(\check{\mathbf{k}}) = \cos\theta\,\mathbf{e}_x(\check{\mathbf{k}}) - \mathrm{e}^{\mathrm{i}\phi}\sin\theta\,\mathbf{e}_y(\check{\mathbf{k}}). \qquad (\text{B.1-47b})$$

Photons are said to be spin-one particles because, as traveling waves, they can be expressed with unit angular-momentum helicity vectors. However, the photons emitted and absorbed in particular atomic transitions need not be states of definite helicity.

B.1.7   Polarization and Stokes parameters

The idealized electric field of the simplest traveling, monochromatic plane-wave field can be expressed as a complex-valued product of amplitude and direction prescribers. Taking the propagation direction to be the z axis we can express the positive frequency part of the field as components of two orthogonal unit vectors $\mathbf{e}$ transverse to the propagation direction, either with horizontal and vertical linear polarization,

$$\mathbf{E}(z,t) = \mathrm{Re}\Big[\mathbf{e}_H\mathcal{E}_H(z,t) + \mathbf{e}_V\mathcal{E}_V(z,t)\Big], \qquad (\text{B.1-48a})$$

or by right- and left-circular polarization,

$$\mathbf{E}(z,t) = \mathrm{Re}\Big[\mathbf{e}_R\mathcal{E}_R(z,t) + \mathbf{e}_L\mathcal{E}_L(z,t)\Big]. \qquad (\text{B.1-48b})$$

Here the two unit vectors $\mathbf{e}_\nu$ fully describe the possible directions of the electric vector (an example of a Cartesian vector) in the plane transverse to the propagation direction, while the amplitudes $\mathcal{E}_\nu(z,t)$ fully describe the magnitude of the field. Because this field is a plane wave there is no variation in the transverse x,y plane of either the magnitude or direction of the field. The field is said to be *polarized* if there is an imbalance between the magnitudes of the two vector components so that, on average, the electric-field vector has a perceptible direction.

Generalized polarization. It is noteworthy that, as pointed out by Eberly [Ebe16], everyday usage of the term "polarized" refers to concentration around a given value (or within a given realm, as with political views of voters). With such generalization, polarization can be defined for degrees of freedom other than the spin that is associated with unit vectors in Euclidean space. As Appendix C.8.4 notes, the mathematics traditionally used for describing partially polarized optical beams has application to a variety of other phenomena, both quantum mechanical and classical, in which there may be a dominance of one of two values [Ebe16].

With such separation of variables there is a correlation between polarization and space-time variation of the field: If the polarization direction is determined to be $\mathbf{e}_H$, then the field amplitude is $\mathcal{E}_H(z,t)$. A measurement of field intensity, proportional to $|\mathbf{E}(z,t)|^2$, gives no information about the relative strengths of the two field components. A detailed description of the field requires further information, such as the determination of the four Stokes parameters defined in the following paragraphs. This correlation of polarization with space-time amplitude does not originate with any quantum-mechanical properties of the field; it is present for completely classical fields [Qia11; Qua17].

Stokes parameters. The observable property of a radiation beam known as polarization is a measure of the time-averaged direction of the electric field vector. Polarization of a simple beam of classical radiation (one that is not tightly focused and is locally idealized as a plane wave whose vector properties are uniform across the wave front and therefore expressible by global unit vectors) can be specified by parametrizing the direction of a unit vector in the plane transverse to the beam axis. For this purpose we can specify the components (possibly complex-valued) of the electric vector along any two independent vectors in that transverse space, say those of horizontal (H) and vertical (V) linear polarization or the helicity vectors of right (R) and left (L) circular polarization. The characterization of polarization is conveniently done with use of four filtered-intensity measurements first proposed by Sir George Stokes in 1852 [Sto52; Fan49; Fan54; Wal54; McM54; Cha60; Hec87; Bic85; Sch07b]. The resulting numbers, the so-called Stokes parameters denoted here as S_n, are defined as classical expectation values of the following measurement operators [Bor99]:

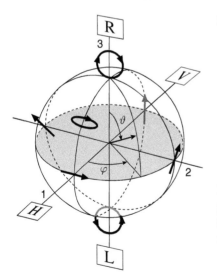

Fig. B.1 *The Poincaré sphere.*

Measurements of Stokes vector components:

S_0 with a filter that transmits 50% of the incident radiation, regardless of its polarization;

S_1 with a polarizer that transmits only horizontally (H) polarized light;

S_2 with a polarizer that transmits only light polarized at 45° to the horizontal; and

S_3 with a polarizer that transmits only (R) right-circularly polarized light.

Figure B.1, known as the *Poincaré sphere*, depicts the connection between polarizations and the Stokes parameters S_1, S_2, S_3 relative to S_0. This figure of the Poincaré sphere is after Fig. 9-3 of [ONe63]; Fig. 1.5-2 of [Sho90]; Fig. 3.2 of [Sho11].

The relationship between these measurements and averages of the electric field ampli-tudes eqn (B.1-48) at a fixed position is

$$S_0 \propto \langle \mathcal{E}_H^2 \rangle + \langle \mathcal{E}_V^2 \rangle = \langle \mathcal{E}_R^2 \rangle + \langle \mathcal{E}_L^2 \rangle = \langle \mathcal{E}_\nearrow^2 \rangle + \langle \mathcal{E}_\searrow^2 \rangle, \qquad \text{(B.1-49a)}$$

$$S_1 \propto \langle \mathcal{E}_H^2 \rangle - \langle \mathcal{E}_V^2 \rangle, \qquad \text{(B.1-49b)}$$

$$S_2 \propto \langle \mathcal{E}_\nearrow^2 \rangle - \langle \mathcal{E}_\searrow^2 \rangle, \qquad \text{(B.1-49c)}$$

$$S_3 \propto \langle \mathcal{E}_R^2 \rangle - \langle \mathcal{E}_L^2 \rangle. \qquad \text{(B.1-49d)}$$

The brackets $\langle \cdots \rangle$ appearing in these definitions refer to averages over time and space. The time average is affected by any random fluctuations in the field direction $\mathbf{e}$, as expressed by fluctuations in any two orthogonal basis vectors. The spatial average becomes important when the light differs significantly from the plane-wave idealization eqn (B.1-48). By contrast with plane waves, beams of radiation have a transverse variation of intensity away from their axis. If they also have direction vectors that vary with position, as do the beams mentioned in Appendix B.2.3, then averaging over the entire beam will effectively average localized Stokes parameters and diminish the polarization.

The classical intensity is proportional to S_0 (and vice versa); The other three Stokes parameters, collected as a vector $\mathbf{S}$ in $\mathbb{R}^3$, provide measures of polarization. When the light is unpolarized only S_0 is nonzero. Generally, a beam may comprise a mixture of polarized and unpolarized light—it is then *partially polarized*. In all cases the four Stokes parameters satisfy the inequality

$$S_0^2 \geq \mathbf{S}^2 \equiv S_1^2 + S_2^2 + S_3^2. \qquad \text{(B.1-50)}$$

The equality $S_0^2 = \mathbf{S}^2$ holds for quasimonochromatic laser light which is completely polarized (having a single direction $\mathbf{e}$ in the entire transverse plane). Then S_0 is a redundant parameter, determined by the other three parameters, and is typically taken to be unity. The ratios of the three real-valued numbers S_1, S_2, S_3 to S_0 can then serve as three coordinates of a unit vector (the *Stokes vector*) in a three-dimensional space, or of coordinates of a point on a sphere of unit radius, the Poincaré sphere shown as Figure B.1. Points on the equator represent linearly polarized light, while the two poles represent circularly polarized light. All other points are associated with elliptically polarized light, in which the electric vector (of a plane wave) traces an ellipse as the beam propagates.

Polarization choices. Characterization of a free-space beam by circular polariza-tion is independent of any arbitrary coordinate system aligned in a laboratory. The contrasting use of linear-polarization states is of value when local conditions intro-duce a natural system of Cartesian coordinates, as occurs when treating propagation of light through birefringent material (the two refractive indices introduce a fast and slow direction for the electric field vector) or for treating reflection from flat surfaces (the $\mathbf{E}$ and $\mathbf{B}$ fields there must satisfy continuity equations). Many lasers emit lin-early polarized light that gives a natural coordinate basis oriented with respect to the laser enclosure. Such fields are not in completely free space; they are constrained by boundary conditions; they can be expressed as coherent superpositions of unfettered free-space fields. In all such situations the labels H and V need not be associated with the direction of local gravity; they are simply useful labels for the two transverse-field components. The properties of beams carrying orbital angular momentum, whose electric fields are directed radially from a beam axis or azimuthally around the beam axis, are not well suited to full description by the single helicity vectors or Cartesian vectors used for traditional parametrization of polarization.

Elaborate beams.   Increasingly there is interest in fields that, though traveling in a definable direction, have elaborate transverse structure. Not only may the field amplitude show node structure, but the direction of the field may be nonuniform. This happens, for example, with tightly focused beams. For such situations there may be differing Stokes vectors (and differing unit vectors) for different points in the transverse plane. Beams whose Stokes vectors are distributed over the entire Poincaré sphere have been termed *full Poincaré* (FP) beams [Bec10; Gal12].

B.1.8   Wave-particle complementarity for photons

Photons, like electrons, are often regarded as exhibiting complementary attributes of waves and particles, manifest with appropriate choice of experimental observables. A common illustrative example is the idealized system of near-monochromatic waves reaching a pointlike detector as beams from two distinguishable point sources (as in the well-studied two-slit demonstration experiment).

Consider steady radiation that arrives at a detector point c from two distinguishable free-space paths, a and b, to produce there the superposition field (expressed here with suppression of space and time arguments)

$$\mathbf{E} = \text{Re}\,[\mathbf{e}_a \mathcal{E}_a + \mathbf{e}_b \mathcal{E}_b]. \tag{B.1-51}$$

Here the unit vectors $\mathbf{e}_a$ and $\mathbf{e}_b$ need not have any particular relationship with each other; they need not be aligned along axes or orthogonal.

At this detector the two arriving field amplitudes $\mathcal{E}_a$ and $\mathcal{E}_b$ are described by a two-dimensional polarization coherence matrix constructed from averages or expectation values of amplitude products, W [Bor99; Wol03; Ebe16; Ebe17; DeZ18] or a dimensionless density matrix ρ [DeZ18],

$$\mathsf{W} = \begin{bmatrix} \langle \mathcal{E}_a^* \mathcal{E}_a \rangle & \langle \mathcal{E}_a^* \mathcal{E}_b \rangle \\ \langle \mathcal{E}_b^* \mathcal{E}_a \rangle & \langle \mathcal{E}_b^* \mathcal{E}_b \rangle \end{bmatrix}, \qquad \rho_{ij} = \frac{W_{ij}}{W_{aa} + W_{bb}}. \tag{B.1-52}$$

The diagonal elements of the matrix W are the intensities I_a and I_b that would occur at the detector from either single path alone,

$$I_i = W_{ii} = |\langle \mathcal{E}_i^* \mathcal{E}_i \rangle|, \quad i = a, b. \tag{B.1-53}$$

Definition of the four Stokes parameters as [McM54; Bor99; Hec70; Kor05b]

$$S_0 = W_{aa} + W_{bb}, \qquad S_1 = W_{aa} - W_{bb}, \tag{B.1-54a}$$
$$S_2 = W_{ab} + W_{ba}, \qquad S_3 = \text{i}[W_{ab} - W_{ba}], \tag{B.1-54b}$$

enable display of the properties of W on a three-dimensional Poincaré sphere. Corresponding quantities defined from the density matrix define a Bloch vector on a Bloch sphere.

Polarization. The field is said to be polarized to the extent that one of the two amplitudes dominates the field. The *degree of polarization* $\mathcal{P}$, a traditional numerical measure of the dominance of one field component, has the definition [Wol59; Man95; Wol03; Wol07],

$$\mathcal{P} = \sqrt{1 - \frac{4\mathrm{DetW}}{(\mathrm{TrW})^2}} = \frac{\sqrt{(W_{aa} - W_{bb})^2 + 4|W_{ab}|^2}}{W_{aa} + W_{bb}}, \tag{B.1-55}$$

leading to the expressions

$$\mathcal{P}^2 = \frac{(I_a - I_b)^2}{(I_a + I_b)^2} + \frac{4|\langle \mathcal{E}_a^* \mathcal{E}_b \rangle|^2}{(I_a + I_b)^2} = \left(\frac{S_1}{S_0}\right)^2 + \left(\frac{S_2}{S_0}\right)^2 + \left(\frac{S_3}{S_0}\right)^2. \tag{B.1-56}$$

The radiation is completely unpolarized if $\mathcal{P}^2 = 0$. Complete polarization, $\mathcal{P}^2 = 1$, requires that

$$\mathcal{P}^2 = 1 \leftrightarrow |W_{ab}|^2 = W_{aa} W_{bb}. \tag{B.1-57}$$

Fringe visibility. The overlap of the two field components produces an intensity modulation at the detector, an interference pattern, [DeZ18]

$$I_c = I_a + I_b + 2|\langle \mathcal{E}_a^* \mathcal{E}_b \rangle| \cos \theta_{ab}, \tag{B.1-58}$$

where θ_{ab} is the phase difference of the two complex-valued field amplitudes at the detector. The maximum intensity $I_c^{\max}$ occurs when the two field components are in phase so the cosine is $+1$; the minimum $I_c^{\min}$ occurs when the cosine is -1. From these extreme values of the detector intensity, and the elements of W, we define the fringe visibility $\mathcal{V}$, a number between 0 and 1, as a measure of the interference (and hence the waviness) [Wol59; Man65; Man91; DeZ18],

$$\mathcal{V} = \frac{I_c^{\max} - I_c^{\min}}{I_c^{\max} + I_c^{\min}} = \frac{2|W_{ab}|}{W_{aa} + W_{bb}}, \qquad \mathcal{V}^2 = \left(\frac{S_2}{S_0}\right)^2 + \left(\frac{S_3}{S_0}\right)^2. \tag{B.1-59}$$

Distinguishability. A useful measure of the possibility of identifying the specific path taken between source and detector (hence of the particle-like localization of beams) is the Englert distinguishability [Eng96],

$$\mathcal{D} = \frac{|I_a - I_b|}{I_a + I_b} = \frac{2|W_{aa} - W_{bb}|}{W_{aa} + W_{bb}}, \qquad \mathcal{D}^2 = \left(\frac{S_1}{S_0}\right)^2. \tag{B.1-60}$$

This varies between 0 (complete indistinguishability) and 1 (completely certain path localization).

Complementarity relations. The quantities $\mathcal{V}$ and $\mathcal{D}$ express numerically the wave-like and particle-like aspects of the two-path system. The so-called polarization coherence theorem (PCT) [Ebe17; DeZ18; DeZ18b; Qia18] expresses the relationship between the three quantities $\mathcal{D}, \mathcal{V}, \mathcal{P}$:

$$\mathcal{D}^2 + \mathcal{V}^2 = \mathcal{P}^2. \tag{B.1-61}$$

Because the degree of polarization is a positive number bounded by 0 and 1 the relationship eqn (B.1-61) is expressible as a complementarity relation between wavelike and particle-like aspects of the field [Eng96],

$$\mathcal{D}^2 + \mathcal{V}^2 \le 1. \qquad (\text{B.1-62})$$

Concurrence. The various relationships expressed above fit readily into formulas involving eigenvalues of the density matrix,

$$\lambda_{\pm} = \tfrac{1}{2}\left[(\rho_{aa} + \rho_{bb}) \pm \sqrt{(\rho_{aa} - \rho_{bb})^2 + 4|\rho_{ab}|^2}\right], \qquad (\text{B.1-63})$$

in which

$$\lambda_+ + \lambda_- = \rho_{aa} + \rho_{bb} = 1. \qquad (\text{B.1-64})$$

The previously defined degree of polarization is also definable as the difference between the two eigenvalues,

$$\mathcal{P} = \lambda_+ - \lambda_-. \qquad (\text{B.1-65})$$

The product of the two eigenvalues provides a definition of the Wootters concurrence $\mathcal{C}$ [Ebe16],

$$\mathcal{C} = 2\sqrt{\lambda_+ \lambda_-} = \frac{2\sqrt{W_{aa}W_{bb} - |W_{ab}|^2}}{W_{aa} + W_{bb}}, \qquad (\text{B.1-66})$$

a measure of the correlation of the two field components. From these definitions there follows the relationship

$$\mathcal{C}^2 + \mathcal{P}^2 = 1 \qquad (\text{B.1-67})$$

connecting the degree of polarization $\mathcal{P}$ (a single-party property) and the concurrence $\mathcal{C}$ (a bipartite property) [DeZ18]. This relationship can be expressed as the maintenance of unit length to a three-dimensional vector,

$$\mathcal{V}^2 + \mathcal{D}^3 + \mathcal{C}^2 = 1, \qquad (\text{B.1-68})$$

an extension of eqn (B.1-61).

Generalizations. Although the preceding discussion builds upon the traditional model of radiation-beam polarization, in which the $\mathbf{e}_i$ are Euclidean vectors transverse to the beam (spin vectors) and $\mathcal{E}_i$ are space-time functions, the formalism only requires that these two mathematical objects be associated with independent degrees of freedom for the object $\mathbf{E}$. Such generalizations have been considered at length by Joe Eberly, Xiao-Feng Qian, and their co-workers [Qia11; Qia13; Qia15; Ebe15; Qua17; Ebe17; Qia18; Qia18b]

B.1.9 Vector potentials; Gauge

Many of the articles and textbooks that discuss relativistic quantum electrodynamics derive the electric and magnetic fields $\mathbf{E}(\mathbf{r}, t)$ and $\mathbf{B}(\mathbf{r}, t)$ from a (magnetic) vector potential $\mathbf{A}(\mathbf{r}, t)$. More generally the electric field can include the gradient of a scalar-potential field $\phi(\mathbf{r}, t)$, used when the Coulomb field is to be included explicitly [Str41; Mil19]:

$$\mathbf{B} = \nabla \times \mathbf{A}, \qquad \mathbf{E} = -\frac{\partial}{\partial t}\mathbf{A} - \nabla\phi. \tag{B.1-69a}$$

An alternative (electric) vector potential, $\mathbf{C}(\mathbf{r}, t)$ [Str41; Bar10c; Cri19] , has been used to maintain the free-space electric-magnetic field symmetries noted in eqn (B.1-92),

$$\mathbf{E} = -\nabla \times \mathbf{C}, \qquad \mathbf{B} = -\frac{1}{c^2}\frac{\partial}{\partial t}\mathbf{C}. \tag{B.1-69b}$$

The use of potentials simplifies the theoretical description of electromagnetic theory by incorporating all of the needed mathematical properties into a single field, from which the two fields $\mathbf{E}$ and $\mathbf{B}$ derive, assuredly fulfilling the needed relationships specified by the dynamical Maxwell equations. When particles are present the potentials are to be constructed so as to incorporate any specified distribution of charges and currents.

In free space the vector potentials, like the fields $\mathbf{E}$ and $\mathbf{B}$, satisfy the homogeneous wave equation, eqn (B.1-21), and so with the mode construction of Appendix B.3.1 they are associated with harmonic oscillators and, upon introducing quantization, with photons.

Guage choices. For application to quantum optics, with its treatments of electrons bound into atoms, the vector potentials are usually taken to be solenoidal, meaning

$$\nabla \cdot \mathbf{A} = 0, \qquad \nabla \cdot \mathbf{C} = 0. \tag{B.1-70}$$

This condition defines the *Coulomb gauge* [And18; Mil19] (also known as radiation or transverse gauge) wherein instantaneous Coulomb interactions between charges are incorporated into the atom portion of the Hamiltonian, leaving the field portion, transverse, to treat radiation and its photons. In this gauge only the transverse fields are quantized; only transverse photons occur.

In treatments presented within the four-dimensional formalism appropriate to special relativity it is common to work with a vector potential $\mathbf{A}$ and a scalar potential ϕ that together form elements of a relativistic four-vector, and to require these to satisfy the *Lorentz condition* (here with SI units),

$$\nabla \cdot \mathbf{A} + \frac{1}{c^2}\frac{\partial}{\partial t}\phi = 0. \tag{B.1-71}$$

Quantization in this *Lorentz gauge*[5] [Jac01; Mil19; Jac01; Jac02] leads to pairs of orthogonally-polarized transverse photons (as with the Coulomb gauge) but also to longitudinal photons and scalar (timelike) photons [Mil94; Fra11]. Well outside any localized source, in the far-field region ($r \gg \lambda$), the field becomes transverse (radiative); the properties of the longitudinal and scalar photons there cancel, so they are traditionally regarded as *virtual*. However, recent articles suggest ways in which they may affect observations [Fra11; Bob16].

---

[5] The review [Jac01], and some of its references, notes that what has long been called the Lorentz gauge, after H. A. Lorentz, should more properly be credited Ludvig Valentin Lorenz.

B.1.10 Field angular momentum

For a point particle the (orbital) angular momentum $\hbar\mathbf{L}$ is the cross product of the position $\mathbf{r}$ with the linear momentum $\mathbf{p}$,

$$\hbar\mathbf{L} = \mathbf{r} \times \mathbf{p}. \tag{B.1-72}$$

When the particle has an intrinsic (spin) angular momentum $\mathbf{S}$ the two combine to give total momentum $\mathbf{J}$,

$$\mathbf{J} = \mathbf{L} + \mathbf{S}. \tag{B.1-73}$$

This relationship holds both for classical, structured particles, such as a spinning planet orbiting a sun, or for quantum structured particles, such as an electron.

Field density of angular momentum. For a classical field the linear momentum is distributed over space and so there is a corresponding distribution of angular momentum. The total angular momentum of a field is the spatial integral of an angular-momentum density $\mathbf{N}(\mathbf{r}, t)$, customarily defined as the cross product of position with the density of linear momentum:

$$\mathbf{N} = \mathbf{r} \times \frac{\mathbf{G}}{c^2} = \epsilon_0\, \mathbf{r} \times (\mathbf{E} \times \mathbf{B}). \tag{B.1-74}$$

Field density of spin, helicity, and chirality. It is possible to show that the spatial integral of this angular momentum density, the angular momentum $\mathcal{J}$, can be separated into an intrinsic spin $\mathcal{S}$ and an orbital angular momentum (OAM) $\mathcal{L}$ [Bar11; Bli11; Bia13; Cam13; Bar16]:

$$\int dV\, \mathbf{N} = \mathcal{J} = \mathcal{L} + \mathcal{S}. \tag{B.1-75}$$

The properties of OAM beams discussed in Section 2.8.5 are a special case.

For fields the separation proceeds by introducing electric and magnetic vector potentials $\mathbf{A}$ and $\mathbf{C}$, as in eqn (B.1-69a) and eqn (B.1-69b). These potentials are taken to be solenoidal, see eqn (B.1-70). The definition of field angular momentum becomes

$$\mathcal{J} = -\int dV\, \mathbf{r} \times \epsilon_0 \Big[(\nabla \times \mathbf{C}) \times (\nabla \times \mathbf{A}) \Big]. \tag{B.1-76a}$$

After integrating by parts we separate the result into two contributions, identified as orbital and spin angular momentum of light [Bar10; Cam12; Bli13; Bar16]:

$$\mathcal{L} = \tfrac{1}{2} \int dV\, \epsilon_0 \Big[\sum_i E_i(\mathbf{r} \times \nabla)A_i + \sum_i B_i(\mathbf{r} \times \nabla)C_i \Big], \tag{B.1-76b}$$

$$\mathcal{S} = \tfrac{1}{2} \int dV\, \epsilon_0 \Big[\mathbf{E} \times \mathbf{A} + \mathbf{B} \times \mathbf{C} \Big]. \tag{B.1-76c}$$

From this expression we identify the *spin density* $\mathfrak{s}(\mathbf{r}, t)$ as the portion of angular-momentum density that has no explicit presence of the position $\mathbf{r}$:

$$\mathfrak{s} = \frac{\epsilon_0}{2}(\mathbf{E} \times \mathbf{A} + \mathbf{B} \times \mathbf{C}). \tag{B.1-77}$$

Helicity. A related quantity is the *helicity density*, $\mathfrak{h}(\mathbf{r}, t)$, defined as

$$\mathfrak{h} = \frac{\epsilon_0}{2}\left[c\mathbf{A} \cdot \mathbf{B} - \frac{1}{c}\mathbf{C} \cdot \mathbf{E}\right]. \tag{B.1-78}$$

The two angular-momentum densities $\mathfrak{h}$ and $\mathfrak{s}$ satisfy a continuity equation,

$$\frac{\partial \mathfrak{h}}{\partial t} = \nabla \cdot c\mathfrak{s}, \tag{B.1-79}$$

and so c times the spin density is identifiable as the flux of helicity.

Helicity arrays. Cameron et al. [Cam12] introduced three-component space- and time-dependent entities $N^{ijk}(\mathbf{r}, t)$ they termed helicity arrays, having dimensions of angular momentum density, whose components incorporated spatial derivatives of electric and magnetic vector potentials:

$$N^{ijk} = \delta_{ij}N^{00i} + \frac{\epsilon_0}{2}\left[-A_i\partial_k C_j - A_j\partial_k C_i + C_i\partial_k A_j + C_j\partial_k A_i\right]. \tag{B.1-80}$$

They identify the helicity density and spin-component density as

$$\mathfrak{h} = N^{000}, \qquad \mathfrak{s}_i = N^{00i}. \tag{B.1-81}$$

These quantities satisfy, for any α, β, the continuity equation

$$\partial_t N^{\alpha\beta 0} + \sum_i \partial_i cN^{\alpha\beta i} = 0. \tag{B.1-82}$$

Chirality. The quantity related to angular momentum is the chirality density $\chi(\mathbf{r}, t)$ for a general field introduced by [And12; Col12] as

$$\chi = \frac{\epsilon_0}{2}\mathbf{E} \cdot \nabla \times \mathbf{E} + \frac{1}{2\mu_0}\mathbf{B} \cdot \nabla \times \mathbf{B}. \tag{B.1-83}$$

The flux of chirality density is the optical chirality flow $\varphi(\mathbf{r}, t)$, defined as,

$$\varphi = \frac{\epsilon_0 c^2}{2}\left[\mathbf{E} \times (\nabla \times \mathbf{B}) - \mathbf{B} \times (\nabla \times \mathbf{E})\right]. \tag{B.1-84}$$

These quantities satisfy a continuity equation first given by Daniel Lipkin without interpretation [Lip64]:

$$\frac{\partial \chi}{\partial t} + \nabla \cdot \varphi = 0. \tag{B.1-85}$$

This is the first of an infinite series of continuity equations for rank-three pseudotensors Lipkin called *zilches*.

B.1.11   The Riemann-Silberstein (RS) field

In examining the properties of electromagnetic radiation it can be useful to combine the electric and magnetic fields into a single complex-valued field, the *Riemann-Silberstein vector* $\mathbf{F}(\mathbf{r}, t)$ [Pow64; Bia96; Bia03; Bia06; Bia06c; Bar11; Bia13; Dre15; Hoy15; Bia17] defined equivalently as

$$\mathbf{F} = \sqrt{\frac{\epsilon_0}{2}} \left[\mathbf{E} + ic\mathbf{B} \right] = \frac{1}{\sqrt{2}} \left[\sqrt{\epsilon_0}\,\mathbf{E} + i\sqrt{\mu_0}\,\mathbf{H} \right] = \frac{1}{\sqrt{2}} \left[\frac{\mathbf{D}}{\sqrt{\epsilon_0}} + i\frac{\mathbf{B}}{\sqrt{\mu_0}} \right], \qquad \text{(B.1-86)}$$

from which we obtain the electric and magnetic fields as real and imaginary parts. The two dynamical Maxwell equations and the divergenceless constraint then read

$$i\frac{\partial}{\partial t}\mathbf{F} = c\nabla \times \mathbf{F}, \qquad \nabla \cdot \mathbf{F} = 0, \qquad \text{(B.1-87)}$$

and the wave equation reads

$$\frac{\partial^2}{\partial t^2}\mathbf{F} = c^2\nabla^2\mathbf{F}. \qquad \text{(B.1-88)}$$

When expressed in terms of this field the energy density $u(\mathbf{r}, t)$ and Poynting vector $\mathbf{G}(\mathbf{r}, t)$ are

$$u = \mathbf{F}^* \cdot \mathbf{F}, \qquad \mathbf{G} = -ic\mathbf{F}^* \times \mathbf{F}. \qquad \text{(B.1-89)}$$

It is also possible to define an angular-momentum density as [Bia11; Bia13]

$$\mathbf{N} = \mathbf{r} \times \frac{\mathbf{G}}{c^2} = \frac{1}{ic}\,\mathbf{r} \times (\mathbf{F}^* \times \mathbf{F}). \qquad \text{(B.1-90)}$$

Null fields. The square of the RS vector is a complex combination of two real-valued scalar invariants Q_R and Q_I of the electromagnetic field [Bia13],

$$\mathbf{F}^2 = Q_R + iQ_I, \qquad Q_R = \tfrac{1}{2}\epsilon_0(\mathbf{E}^2 - c^2\mathbf{B}^2), \qquad Q_I = c\mathbf{E} \cdot \mathbf{B}. \qquad \text{(B.1-91)}$$

Fields for which this square vanishes have been termed *null fields* [Bia03; Irv10; Ked13; Van13; Arr17]. Plane-wave fields are an example; see eqn (B.1-14).

Electric-magnetic symmetry.   It is evident that the Maxwell equations (B.1-87) are unaffected by a change in the RS vector by a constant phase, $\mathbf{F}' = e^{i\Theta}\mathbf{F}$. The resulting electric and magnetic fields, obtained from the real and imaginary parts of this rephased RS vector, are [Bia13]

$$\mathbf{E}' = \mathbf{E}\cos\Theta - c\mathbf{B}\sin\Theta, \qquad c\mathbf{B}' = c\mathbf{B}\cos\Theta + \mathbf{E}\sin\Theta. \qquad \text{(B.1-92)}$$

This alteration of the electric and magnetic fields exhibits an intrinsic symmetry of the Maxwell equations, independent of the general requirements for Lorentz invariance. The conserved quantity associated with this symmetry (sometimes referred to as Heaviside-Larmor symmetry [Bar10; Bar11; Bar12b; Cam12], "*duplex symmetry*" [Bar10; Bar11; Cam12b; Bli13; Bar14; Bar16], and as "*electric-magnetic democracy*" [Ber09; Bar11; Cam13; Bar16]) is the helicity—the projection of total angular momentum onto the propagation axis of a momentum-defined field mode [Cam12; Bia13].

Continuous transformations of the Poincaré group do not superpose waves that have different helicities and so free-field modes that have a definite helicity are natural choices for defining free-space photons.

Photon wavefunction. Because the free-space electromagnetic field moves always with velocity c, its photons cannot be brought to rest. Although the creation and destruction of photons as energy, momentum, and angular-momentum increments can be localized within a small volume (that of an atom, for example), there is no field-operator counterpart to the nonrelativistic position operator of a massive particle. Nevertheless, a variety of possibilities have been suggested for defining position-representation wavefunctions from the well-defined momentum-representation (plane wave) wavefunctions of photons [Coo82; Sip95; Mut05; Haw07; Bar12]. The RS vector of eqn (B.1-86), whose absolute square provides the energy density of an electromagnetic field, offers a consistent means for defining a photon wavefunction in position space [Bia96; Bia00; Smi07]. As remarked in [Bia96], the expressions for energy, momentum and angular momentum have the appearance of expectation values for a particle that has the RS vector field $\mathbf{F}(\mathbf{r}, t)$ eqn (B.1-86) as its wavefunction. The following paragraphs present the resulting particle-like equation, a counterpart to the Dirac equation for an electron shown in eqn (B.1-96).

A Schrödinger equation for the Riemann-Silberstein vector. The replacement of the traditional dynamical Maxwell equations for two vector fields $\mathbf{E}$ and $\mathbf{B}$ by a single first-order PDE for a complex field brings an expression, eqn (B.1-87), for a classical field that has much in common with the first-order Dirac equation used for a relativistic electron wavefunction, a similarity that has been usefully exploited [Bar14]. Toward that goal it is useful to note that the three-dimensional representations of the rotation group, the three spin-one matrices s in eqn (A.7-26) of Appendix A.7.2, acting on Cartesian coordinates (rather than on eigenvectors of angular momentum), allow the vector product of two three-dimensional vectors to be written

$$\mathbf{a} \times \mathbf{b} = -i(\mathbf{s} \cdot \mathbf{a})\mathbf{b}. \tag{B.1-93}$$

This means that the Maxwell equations for the RS vector can be written in the form of a time-dependent Schrödinger equation

$$i\hbar \frac{\partial}{\partial t}\mathbf{F}(\mathbf{r}, t) = \mathcal{H}\mathbf{F}(\mathbf{r}, t), \quad \text{with} \quad \mathcal{H} = -i\hbar c(\mathbf{s} \cdot \nabla). \tag{B.1-94}$$

Now partition the RS vector into a two-component wavefunction, with electric and magnetic fields (three components each) as the two parts. The Maxwell equations then read

$$i\frac{\partial}{\partial t}\mathbf{\Psi}(\mathbf{r}, t) = \begin{bmatrix} 0 & -ic(\mathbf{s} \cdot \nabla) \\ -ic(\mathbf{s} \cdot \nabla) & 0 \end{bmatrix} \mathbf{\Psi}(\mathbf{r}, t), \quad \mathbf{\Psi} = \sqrt{\frac{\epsilon_0}{2}} \begin{bmatrix} \mathbf{E} \\ ic\mathbf{B} \end{bmatrix}. \tag{B.1-95}$$

The similarity of this equation with the Dirac equation (discussed next) has been noted by several authors [Bar14]; it is known as the *Weyl equation* [Bia96]. The absolute square of its solutions express the distribution of electromagnetic energy density –

interpretable as photon probability distribution, definable as a *photon wavefunction* [Bia96; Bia98; Bia06c; Smi07; Sal11].

It has long been thought that, by contrast with massive particles, it was not possible to construct for photons a position operator and its eigenfunctions (as wavefunctions) [New49]. However, in recent years there have been a number of reasonable suggestions for defining photon-position operators [Haw99; Haw01; Haw07; Haw19], and discussions of wavefunctions and localization for free-space photons [Kel05; Haw07; Smi07; Bia09; Bia14].

The Dirac equation. The *Dirac equation* for a particle of mass m_e is the counterpart of a time-dependent Schrödinger equation that is linear (rather than quadratic) in the particle momentum **p**:

$$i\hbar\frac{\partial}{\partial t}\Psi = \mathcal{H}^D\Psi \quad \text{with} \quad \mathcal{H}^D = c\boldsymbol{\alpha}\cdot\mathbf{p} + m_ec^2\beta. \tag{B.1-96}$$

The Dirac Hamiltonian $\mathcal{H}^D$ involves two matrix operators, $\boldsymbol{\alpha}$ with vector elements and β with scalar elements. The Dirac wavefunction Ψ has four components, and so the Dirac operators can be represented by 4×4 matrices. It is customary to group the wavefunction components into a pair of two-dimensional spinors, ψ^+ and ψ^-, and to write the resulting Dirac matrices in the form

$$\alpha_j = \begin{bmatrix} 0 & \sigma_j \\ \sigma_j & 0 \end{bmatrix}, \quad \beta = \begin{bmatrix} 1_2 & 0 \\ 0 & -1_2 \end{bmatrix}, \tag{B.1-97}$$

where 1_2 is the two-dimensional unit matrix, 0 is the two-dimensional null matrix, and the σ_j are the three Pauli matrices eqn (A.7-23) in Appendix A.7.2. In the position representation the momentum operator is $\mathbf{p} = -i\hbar\nabla$ and so the Dirac equation for an electron wavefunction becomes a first-order PDE coupling two two-component electron wavefunctions that can be written

$$i\frac{\partial}{\partial t}\Psi(\mathbf{r},t) = \begin{bmatrix} m_ec^2 1_2 & -ic(\boldsymbol{\sigma}\cdot\nabla) \\ -ci(\boldsymbol{\sigma}\cdot\nabla) & -m_ec^2 1_2 \end{bmatrix}\Psi(\mathbf{r},t), \quad \Psi = \begin{bmatrix} \psi^+ \\ \psi^- \end{bmatrix}. \tag{B.1-98}$$

The wavefunction $\psi^+(\mathbf{r},t)$ represents positive-energy solutions (electrons); the negative-energy solutions $\psi^-(\mathbf{r},t)$ represent anti-electrons (positrons). In turn, each of these has two components, associated with the two possible spin orientations of the electron. Here those directions are taken to be along the momentum.

The Weyl equation for photons differs from the Dirac equation for electrons in two respects: It has zero rest mass rather than the electron mass m_e, and it has matrices of spin-half (for electrons) rather than spin-one (for photons).

B.2 Classical field modes; Examples

The next sections discuss several examples of solutions to the vector Helmholtz equation, from a vast and still growing list, with particular attention to idealizations appropriate to radiation beams. What were, during my graduate studies, presented as exercises for the diligent student [Str41; Mor53], have become important experimental tools in the world of photonics. It is these field modes that become the Dirac photons of Appendix B.3.2 and B.3. I have organized the presentations by the coordinate systems used in their definition. Discussion of experimental techniques for their actual creation I leave to others [Pad04; McG05; Bor09; Lev16b; Rub17].

B.2.1 Simple beams

The textbook discussions of radiation that were part of my own graduate education dealt primarily with plane waves—solutions to wave equations in Cartesian coordinates—and their use in Fourier analysis of traveling-wave fields; see Appendix B.2.2. Over distances much smaller than a wavelength, say over an atom or small molecule, a plane wave provides a suitable approximation to most light sources.

Once lasers became commonplace laboratory tools the mathematics of Gaussian beams was regularly used to describe laser beams [Sie73; Sie86; Lev16b]; see Appendix B.2.5. The intensity profile over the transverse plane of the simplest of such a beam follows a Gaussian pattern on any diameter across the beam center. The beams have curved wave-fronts that bring them into and out of focus along the beam axis. More elaborate common patterns involve rectangular or circular node structures; see Appendix B.2.5.

Light beams that have coherence can be created with a variety of spatial variations in the plane transverse to the beam axis, patterns distinguished in part by their node structure [Koe96; Lev16b]. Wave fronts of beams can now be routinely customized by spatial light modulators (SLMs) [Bor09; Mon10; Wei11; Pad11; Orb16], pixelated liquid-crystal devices that can produce computer-controlled holograms to convert a simple laser beam into a variety of novel beams.

The simple picture of a light beam as being associated, at each transverse slice, with a single direction of the electric field vector requires revision for many of the beams that are now routinely employed in laboratory manipulation of atoms and molecules. The electric-vector direction (the polarization) may not just be along a Cartesian axis but may be radial or azimuthal with respect to the beam axis [Tid90; Oro00].

B.2.2 Cartesian coordinates: Plane waves

Cartesian coordinates express a position **r** with variables x, y, z. The separation of the Laplacian into independent derivatives, as shown in eqn (B.1-24), allows factorization of the Helmholtz solutions into three independent Cartesian-variable functions,

$$U(\mathbf{r}) \equiv U(x, y, z) = U^{(1)}(x)\, U^{(2)}(y)\, U^{(3)}(z), \tag{B.2-1}$$

and leads from the scalar Helmholtz equation eqn (B.1-29) to the three independent equations

$$\left[\frac{d^2}{dx^2} + k_x^2\right] U^{(1)}(x) = 0, \quad \left[\frac{d^2}{dy^2} + k_y^2\right] U^{(2)}(y) = 0, \quad \left[\frac{d^2}{dz^2} + k_z^2\right] U^{(3)}(z) = 0, \tag{B.2-2}$$

where the *separation constants* k_i are constrained by the condition expressed below in eqn (B.2-4). When the k_i are positive or negative real numbers these solutions are trigonometric functions (or exponentials) whose magnitude remains bounded. If any

k_i has an imaginary part the resultant U_i will either grow indefinitely—an unacceptable solution—or will fall to zero—an *evanescent* wave; see Appendix C.3.4. To treat traveling or standing waves we require real k_i. This factorization then leads, with the choice of complex exponentials as solutions (sines or cosines would do equally well), to the structure

$$U_\nu(x, y, z) = \exp(\mathrm{i}k_x x)\, \exp(\mathrm{i}k_y y)\, \exp(\mathrm{i}k_z z), \qquad (\text{B.2-3})$$

where the mode label ν on the function U denotes, in addition to the frequency ω, the set of three separation constants, regarded as components of a three-dimensional propagation vector $\mathbf{k}$ constrained in magnitude by the frequency:

$$\nu = \{\omega, k_x, k_y, k_z\}, \qquad k_x^2 + k_y^2 + k_z^2 \equiv |\mathbf{k}|^2 = (\omega/c)^2. \qquad (\text{B.2-4})$$

This solution to the Helmholtz equation is a single mode of the field, and ν is the set of four *quantum-number* mode identifiers. The constraint in eqn (B.2-4) limits the descriptors ν to two degrees of freedom: Only two of the wave vectors can be chosen independently.

The mode field shown factored in eqn (B.2-3) is an eigenfunction of the three scaled linear-momentum operators eqn (B.1-32),

$$\hat{p}_x U_\nu = \hbar k_x U_\nu, \qquad \hat{p}_y U_\nu = \hbar k_y U_\nu, \qquad \hat{p}_z U_\nu = \hbar k_z U_\nu, \qquad (\text{B.2-5})$$

where, specifically,

$$\hat{p}_x = -\mathrm{i}\hbar \frac{\partial}{\partial x}, \qquad \hat{p}_y = -\mathrm{i}\hbar \frac{\partial}{\partial y}, \qquad \hat{p}_z = -\mathrm{i}\hbar \frac{\partial}{\partial z}. \qquad (\text{B.2-6})$$

Other solutions may be constructed as superpositions of these elementary modes, subject to the wave vector constraint. For example with fixed k_z a general construction eqn (B.1-31) reads

$$U_a(x, y, z) = \int \frac{dk_x}{2\pi} \int \frac{dk_y}{2\pi}\, a(k_x, k_y)\, U_\nu(x, y, z), \qquad (\text{B.2-7})$$

where the two-variable function $a(k_x, k_y)$ is arbitrary.[6] This formula is just a version of a two-dimensional Fourier transform. The three-dimensional transform is constrained by the frequency, as expressed in eqn (B.2-4).

Example: Directed vector plane waves. In Cartesian coordinates the wave fronts form sets of equidistant flat, parallel planes of infinite extent, separated by a wavelength, all propagating in the direction perpendicular to their planes—traveling *plane waves* in which the vectors $\mathbf{E}$ and $\mathbf{B}$ are perpendicular to each other and to the propagation direction.[7]

The general solution to the free-space scalar Helmholtz equation in Cartesian coordinates is a plane wave moving with velocity c. For consideration of directed radiation

---

[6] If U_a is to retain the magnitude of U_ν then the absolute squares of $a(k_x, k_y)$ must integrate to unity.

[7] With general coordinate systems, or when matter is present (as bounding surfaces or near a source) the vectors $\mathbf{E}(\mathbf{r}, t)$ and $\mathbf{B}(\mathbf{r}, t)$ need not be perpendicular to each other [Lev15; Lev16].

we assume the vector field is predominantly a plane wave traveling in the z direction, for example

$$\mathbf{F}^{(+)}(\mathbf{r}, t) = \mathbf{e}_{\perp} U_{\nu}(x, y) \exp(\mathrm{i}k_z z - \mathrm{i}\omega t), \tag{B.2-8}$$

where $\mathbf{e}_{\perp}$ is a unit vector in the x, y plane, perpendicular to the dominant propagation direction along z. The scalar field $U_{\nu}(x, y)$ satisfies the two-dimensional scalar Helmholtz equation,

$$\left[\nabla_{\perp}^2 + k_{\perp}^2\right] U_{\nu}(x, y) = 0, \tag{B.2-9}$$

where $\nabla_{\perp}^2$ is the (transverse) Laplacian in the two dimensions perpendicular to z. With Cartesian coordinates this is

$$\nabla_{\perp}^2 = \frac{\partial^2}{\partial x^2} + \frac{\partial^2}{\partial y^2}. \tag{B.2-10}$$

The transverse wave number $k_{\perp}$ appearing after the removal of the longitudinal wave number k_z is constrained by the frequency ω through the relationship

$$k_{\perp} = \sqrt{k_x^2 + k_y^2} = \sqrt{k^2 - k_z^2}, \qquad k = \omega/c. \tag{B.2-11}$$

With that choice a general expression for the two-dimensional field amplitude is

$$U_{\nu}(x, y) = U_{\nu}(0, 0) \left(X_+ \, \mathrm{e}^{\mathrm{i}k_x x} + X_- \, \mathrm{e}^{-\mathrm{i}k_x x}\right) \left(Y_+ \, \mathrm{e}^{\mathrm{i}k_y x} + Y_- \, \mathrm{e}^{-\mathrm{i}k_y x}\right), \tag{B.2-12}$$

bearing the identifying quantum-number label

$$\nu = \{\omega, \sigma, k_x, k_y\}. \tag{B.2-13}$$

where σ identifies the vector direction in the x, y plane. Full identification of the coherent field $\mathbf{F}^{(+)}(\mathbf{r}, t)$ requires, in addition to the three labels σ, k_x, k_y, identification of the frequency. That is, four numbers are required to identify a field mode (and hence a photon). For these transverse-field Cartesian components the components of the propagation vector $\mathbf{k}$ are restricted by the constraint eqn (B.2-12). To produce a field of constant magnitude $|U_{\nu}(0, 0)|^2$ from this construction we require that

$$|X_+|^2 + |X_-|^2 = |Y_+|^2 + |Y_-|^2 = 1. \tag{B.2-14}$$

These constants determine the nature of the field in directions transverse to the z axis. When $X_- = 0$ or $Y_- = 0$ the field has traveling-wave behavior in the x or y direction and is an eigenfunction of the scaled linear-momentum operators $\hat{p}_x$ or $\hat{p}_y$ associated with that motion:

$$X_- = 0: \quad \hat{p}_x U_{\nu}(x, y) = k_x U_{\nu}(x, y), \tag{B.2-15a}$$
$$Y_- = 0: \quad \hat{p}_y U_{\nu}(x, y) = k_y U_{\nu}(x, y). \tag{B.2-15b}$$

With the choices $X_+ = X_-$ or $Y_+ = Y_-$ the field is a standing wave along the x or y axis, a coherent superposition of two oppositely-traveling waves that have equal amplitudes. For any choice of $X_{\pm}$ or $Y_{\pm}$ the field $\mathbf{F}^{(+)}(\mathbf{r}, t)$ of eqn (B.2-8) is an eigenfunction of the operator $\hat{p}_z$ associated with linear momentum in the z direction; its eigenvalue is $\hbar k_z$:

$$\hat{p}_z \mathbf{F}^{(+)}(\mathbf{r}, t) = \hbar k_z \mathbf{F}^{(+)}(\mathbf{r}, t). \tag{B.2-16}$$

If the unit vector is either $\mathbf{e}_+$ or $\mathbf{e}_-$ the vector field is also an eigenfunction of the z component of spin, i.e. it has helicity ± 1.

More general solutions to the Helmholtz equation obtain as superpositions of this basic solution for different values of the transverse wavenumbers k_x, k_y. Superpositions of such waves that have the same value of $k_\perp$ do not spread or diffract.

The plane-wave fields associated with solutions to the Helmholtz equation in Cartesian coordinates provide a mathematically simple idealization of waves traveling past atoms or other small particles, but a single monochromatic plane wave, by extending indefinitely in the directions transverse to the propagation axis, carries an unbounded amount of energy. To create a description of a model of localized finite energy superpositions—Fourier integrals—are needed.

B.2.3   Elaborate beams: Vortices and orbital angular momentum

In 1987 a brief paper by J. Durnin [Dur87] (see also [Dur87b; Dur88; Arl01]) drew attention to the remarkable nondiffracting nature of idealized beams whose transverse fields were described by Bessel functions. Such a beam may have a concentrated bright center or may exhibit a central node—a dark core surrounded by a pattern of bright rings; see Appendix B.2.4. These papers initiated an extensive, ongoing investigation of novel radiation beams.

A 1992 paper by Les Allen and his colleagues [All92] pointed out how laser-beam fields could be given *orbital angular momentum* (OAM) [Bar94; All99; Pad00; Bar02b; Pad04; Yao11; Cer11; Wil13; Bar16; Fra17]. As has since become well understood, hollowed beams, with a central node, may exhibit spiraling, helical phase fronts in which the beam axis has a singularity in the phase—a line of null optical intensity, an optical *vortex* associated with orbital angular momentum; see Appendix B.2.5 and B.2.5. Such vortex-endowed cylindrical beams, with an azimuthal phase variation $\exp(i\ell\varphi)$, carry an orbital angular momentum equivalent to $\ell\hbar$ per photon, compared with the value $1\hbar$ carried by the projection of photon spin responsible for the usual polarization.

The quantum and photon nature of OAM fields has been discussed [Lea02; Cal06; Aie10; Pli13] and electron-beam analogs of these optical-vortex OAM beams have been extensively studied [Llo17].

All of these beam properties are purely classical, dealing only with solutions to the Maxwell equations of Appendix B.1.1. However, the field properties are inherited by the photons derived from increments of these classical fields. Single Laguerre-Gauss mode photons with OAM as large as 300 $\hbar$ have been created [Fic12]. Such ongoing studies have therefore notably extended the simplistic view of photons held by earlier generations of physicists. Not only have studies of traveling-wave fields (and photons) greatly enlarged, but standing-wave cavity fields have became commonplace sources of photons, leading to studies of Cavity Quantum Electrodynamics (CQED) [Har89; Kim98; Wal06; Mes06; Koc19]. The following sections describe a number of simple beam-like mode structures. As noted in Appendix B.1.5, each of these types of field is associated with a type of Dirac photon.

B.2.4 Cylindrical coordinates: Bessel beams

The commonly used cylindrical coordinates treats the z direction as an axis for beam propagation and treats the directions transverse to this, the x, y plane, with polar coordinates[8] r, ϕ. The function $U(\mathbf{r}) \equiv U(r, \phi, z)$ factors as $U^{(1)}(r)U^{(2)}(\phi)U^{(3)}(z)$, where the separate equations are

$$\left[\frac{d^2}{dz^2} + k_z^2\right] U^{(3)}(z) = 0, \tag{B.2-17a}$$

$$\left[\frac{d^2}{d\phi^2} + \ell^2\right] U^{(2)}(\phi) = 0, \tag{B.2-17b}$$

$$\left[\frac{d^2}{dr^2} + \frac{1}{r}\frac{d}{dr} + \frac{\ell^2}{r^2} - k_\perp^2\right] U^{(1)}(r) = 0. \tag{B.2-17c}$$

To treat traveling-wave solutions we require $k_\perp$ and k_z be real and ℓ be an integer (positive, negative, or zero). The Laplacian for angle variable ϕ has exponential (or trigonometric) solutions and the equation for variable r has Bessel-functions J_ℓ (or more general cylinder-function) as solutions. The possible field solutions $U(r, \phi, z)$ accompanying the time dependence $\exp[-i\omega t]$ can therefore be taken to be the modes

$$U_\nu(r, \phi, z) = J_\ell(k_\perp r)\,\exp[i\ell\phi]\,\exp[ik_z z], \tag{B.2-18}$$

where the mode-identifying quantum-number label ν includes an integer ℓ (for OAM) and two continuous but constrained numbers $k_\perp, k_z$ (for propagation-vector components):

$$\nu = \{\omega, \ell, k_\perp, k_z\}, \qquad k_\perp^2 + k_z^2 \equiv |\mathbf{k}|^2 = (\omega/c)^2. \tag{B.2-19}$$

As with Cartesian coordinates, this magnitude constraint limits the mode label ν to two degrees of freedom. The mode field eqn (B.2-18) is an eigenfunction of $\hat{p}_z$ and of the z component of orbital angular momentum $\hbar \hat{L}_z$,

$$\hat{p}_z U_\nu = \hbar k_z U_\nu, \qquad \hat{L}_z U_\nu = \ell\, U_\nu. \tag{B.2-20}$$

Superpositions of these elementary monochromatic modes take the form of a cylindrically symmetric function,

$$U_a(r, \phi, z) = \sum_\ell \int \frac{dk_\perp}{2\pi}\, c(\ell, k_\perp)\, U_\nu(r, \phi, z). \tag{B.2-21}$$

Bessel waves. The closed-form solutions to the 2D scalar Helmholtz equation in circular cylindrical coordinates involve Bessel functions for the axial r variation and an exponential $\exp(i\ell\phi)$ with integer ℓ for the azimuthal ϕ variation. Bessel waves, as solutions to the Helmholtz equation, are pencil-shaped field amplitudes moving along the z axis without spreading in the transverse direction. They may be associated with any unit vector, i.e. with any choice of polarization.

---

[8]Elliptical and parabolic coordinates also find use for optical beam structure [Lev16].

The intensity pattern of $J_\ell(k_\perp r)$ forms a set of circular nodes that surround the central axis. For $\ell = 0$ the central axis is a bright antinode, but for $\ell > 0$ the central axis is a node, and the center of the beam is a line of null field. Each bright ring carries the same power and so, because the ring circumferences increase with ℓ, the intensity of each ring diminishes as the number of rings increases.

Like directed plane waves, a beam with Bessel-function amplitude extends without limit in the transverse direction. Like plane waves, the energy per unit longitudinal element of a Bessel wave is unbounded. Experimentally created beams can only be idealizations of the Bessel beam defined above; they will not propagate indefinitely without spreading transversely. An approximation to a Bessel beam can be created by using an *axicon*, a type of conical lens [Arl00; Dav16].

B.2.5   The paraxial wave equation; Gaussian beams

A variety of solutions to the Helmholtz equation in cylindrical coordinates z, r, ϕ offer closed-form expressions for fields with beamlike character [Str41; Bia06b; Zha09; Lev16b]. These have gained interest, not only in optics but in acoustics, for offering descriptions of radiation beams: Disturbances that travel predominantly along the z axis of a cylinder with intensity that diminishes with distance r from this axis—the plane transverse to this axis. Examples include Gaussian beams [Sag98; Por01; Wun04; Lev16; Lev16b], in which the magnitude of the field varies in the transverse plane as $\exp[-(r/R)^2]$, and Bessel beams [Dur87; Dur88; Mis91; McG05; Bia06; Now12; Kot12; Sim16; Wan16], in which the dependence on r involves a Bessel function. Such sets of mathematical functions, involving differing numbers of nested cylindrical nodes or planar nodes in ϕ, provide examples of spatial modes that can serve to define beam-like photons.

To describe a traveling-wave field that is localized around an axis (the z axis, say) it is useful to regard the spatial variation along that axis to be much slower than any in the transverse plane, and with this beam-like approximation to neglect the second derivative in the propagation direction. This *slowly-varying envelope approximation* (SVEA), taken for any vector component and with the ansatz

$$F^{(+)}(\mathbf{r}, t) = U(\mathbf{r})\, \exp[\mathrm{i}(kz - \omega t)], \qquad\qquad (\text{B.2-22})$$

leads to the free-space *paraxial wave equation* for a beam-like field,

$$\left[\nabla_\perp^2 + \mathrm{i}2k\frac{\partial}{\partial z}\right] U(\mathbf{r}) = 0. \qquad\qquad (\text{B.2-23})$$

Solutions to this monochromatic equation are eigenfunctions of $\hat{p}_z$ having eigenvalue $k = \omega/c$. There remain second derivatives in the transverse directions, here the x, y plane.

Gaussian beams. The most commonly used solution to the paraxial wave equation is that of a monochromatic focused Gaussian beam (GB). These beams are cylindrically symmetric about a propagation axis, traditionally taken as z, with intensity that follows a Gaussian form in the transverse radial coordinate r, away from a peak value at $r = 0$, independent of the azimuthal angle ϕ. The Gaussian-beam solution to the paraxial wave equation reads [Gre98]

$$U(r, \phi, z) = \frac{w_0}{w(z)} \exp\left(-\frac{r^2}{w^2(z)}\right) \exp\left(-i\frac{kr^2}{2R(z)} - i\psi_G(z)\right). \qquad \text{(B.2-24)}$$

Two parameters fix the Gaussian profile: The wavelength of the light, $\lambda = 2\pi c/w$, and a minimum radius (spot size) w_0. These combine to define a *Rayleigh range*,

$$z_R = \pi w_0^2/\lambda, \qquad \text{(B.2-25)}$$

descriptive of the propagating distance over which focusing occurs. The z-dependent controls of the beam focus are the beam waist $w(z)$, the wavefront curvature $R(z)$, and the *Gouy phase* $\psi_G(z)$,

$$w(z) = w_0\sqrt{1 + (z/z_R)^2}, \quad R(z) = z[1 + (z_R/z)^2], \quad \psi_G(z) = \tan^{-1}(z/z_R), \quad \text{(B.2-26)}$$

all defined in terms of the Rayleigh range. The spherical wavefront brings the beam to a focus at $z = 0$, where the radius is w_0 and the wavefront is flat, and expands the beam into a spherical front as z increases. The Rayleigh length z_R is the distance over which the beam waist $w(z)$ expands to $\sqrt{2}w_0$ of its smallest size w_0. In passing through the focus the beam front acquires the Gouy phase.

The full field of a Gaussian beam includes a globally-defined unit vector and a plane-wave factor

$$\mathbf{F}^{(+)}(\mathbf{r}, t) = \mathbf{e}\, U(\mathbf{r}) \exp[i(kz - \omega t)]. \qquad \text{(B.2-27)}$$

Many traditional lasers emit some form of near-monochromatic Gaussian beam, and the Gaussian intensity profile has been commonly used to describe the field acting on beams of atoms passing through a laser beam.

Hermite-Gaussian beam. With Cartesian coordinates it is possible to factor the solutions of the paraxial wave equation and obtain the general form [Lev16]

$$U_{mn}(x, y, z) = C_{mn}\frac{w_0}{w(z)}H_m\left(\frac{2x}{w(z)}\right)\exp\left(-\frac{x^2}{w(z)^2}\right)$$
$$\times H_n\left(\frac{2y}{w(z)}\right)\exp\left(-\frac{y^2}{w(z)^2}\right)$$
$$\times \exp\left(+i\frac{k(x^2 + y^2)}{2R(z)} - i(m + n + 1)\psi_G(z)\right). \qquad \text{(B.2-28)}$$

Here $H_n(\xi)$ is a Hermite polynomial of order n and argument ξ and C_{mn} is a normalizing constant.

For a Hermite-Gaussian (HG) mode, the surfaces of constant phase are a series of discs separated by the wavelength of the light. Viewed in a transverse (x, y) plane the HG beams have several intensity maxima, separated by a rectilinear pattern of nodes: There are n vertical node lines and m horizontal node lines.

The direction properties of such a wave can be taken as any of the globally-defined unit vectors $\mathbf{e}_\sigma$.

Laguerre-Gaussian beam. The simple Gaussian beam generalizes to include nodal patterns as present with the Laguerre-Gaussian (LG) beam [All92; Pad95; San04; Lev16]

$$U_{m\ell}(r, \phi, z) = C_{m\ell} \frac{w_0}{w(z)} \left(\frac{r\sqrt{2}}{w(z)} \right)^{|\ell|} \exp\left(\frac{-\rho^2}{w(z)^2} \right) L_m^{|\ell|} \left(\frac{2r^2}{w(z)^2} \right)$$

$$\times \exp\left(i\ell\phi + i(2m + |\ell| + 1)\psi_G(z) + i\frac{kr^2}{2R(z)} \right). \qquad \text{(B.2-29)}$$

Here $L_m^{|\ell|}(\xi)$ is a generalized Laguerre polynomial of argument ξ.

As with Hermite-Gauian beams the direction properties of such a wave can be taken as any of the unit vectors $\mathbf{e}_\sigma$ or they can be taken as a radial unit vector or an axial unit vector. When the unit vector has helicity ± 1, the beam carries total angular momentum $(\ell \pm 1)$ per photon.

For an LG mode with $\ell \neq 0$, the surfaces of constant phase have helical form, with a *winding number* ℓ. There is an optical phase change of $2\ell\pi$ upon integration along any closed path in the transverse plane surrounding the beam axis. There is a phase singularity along the z axis, at $r = 0$. This is a topological characteristic termed, in the present context, a *topological charge*, having value ℓ.

The helical nature of the phase surface implies that, even in free space, the wave-fronts are not perpendicular to the direction of propagation of the beam and so the Poynting vector is not parallel to the direction of propagation. These beams carry orbital angular momentum, OAM, around the beam axis.

B.2.6   Spherical coordinates: Multipoles

The third common coordinate system for treating wave equations in free space is the set of spherical coordinates r, θ, ϕ involving two angles and a radial coordinate r that extends from the coordinate origin $r = 0$ to infinity. The scalar Helmholtz equation has solutions whose angular dependence is expressible as a complex-valued spherical harmonic and whose radial variation is expressible as a cylinder function. When we consider fields that are regular at the coordinate origin, $r = 0$, we require the spherical Bessel functions $j_\ell(kr)$ as the radial functions and the solutions can be expressed as

$$U_\nu(r, \theta, \phi) = N_k \, j_\ell(kr) \, Y_{\ell,m}(\theta, \phi), \qquad \text{(B.2-30)}$$

where N_k is a normalization factor. The quantum-number label ν involves two integers and a wavevector magnitude (fixed by the frequency),

$$\nu = \{k, \ell, m\}, \qquad k^2 = (\omega/c)^2, \qquad \ell \geq 0, \qquad -\ell < m < \ell. \qquad \text{(B.2-31)}$$

The spherical harmonics appearing here as angular modes are eigenfunctions of orbital angular momentum, see eqn (A.7-22), and provide a complete orthonormal set of functions for angular variables,

$$\int d\hat{\mathbf{r}} \, Y_{\ell m}(\hat{\mathbf{r}})^* Y_{\ell' m'}(\hat{\mathbf{r}}) = \delta_{\ell\ell'} \, \delta_{mm'}. \qquad \text{(B.2-32)}$$

Arbitrary scalar fields of given frequency $\omega = ck$ are constructible from superpositions of these spherical modes. To create vector fields we must incorporate unit vectors as factors of the fields. This can be done in many ways. To fit the fields that emerge from idealized isolated atoms it is customary to deal with states of well-defined angular momentum for the atom and for the field.

Confined multipoles. Radiations emanating from dipoles and quadrupoles have long been studied as sources of spontaneous emission. Recent years have seen interest in the stationary fields confined within dielectric spheres and waveguides. These bring descriptions with very high-order multipoles; see the mention of whispering-gallery modes in Appendix C.3.4.

Vector harmonics. Consider combinations of the spherical harmonic $Y_{\ell m}(\hat{\mathbf{r}})$ with a complex spherical unit vector $\mathbf{e}_q(\hat{\mathbf{r}})$ whose reference axis coincides with the direction $\hat{\mathbf{r}}$ occurring as argument of the spherical harmonic. Such a vector is an example of an uncoupled state of two angular momenta. The spherical harmonic is an eigenfunction of the orbital angular momentum operators $\mathbf{L}^2$ and L_z, while the spherical unit vectors are eigenvectors of the spin operator $\mathbf{S}^2$ and $\mathbf{S}_z$. We can combine these quantities into coupled angular momentum functions, eigenstates of total angular momentum $\mathbf{J}^2 = (\mathbf{L}+\mathbf{S})^2$, by using the Clebsch-Gordan coefficients, as in eqn (A.7-30). The resultant vectors,

$$\mathbf{Y}_{\ell J M}(\check{\mathbf{r}}) = \sum_{mq}(\ell m, 1q|JM)\,Y_{\ell m}(\check{\mathbf{r}})\mathbf{e}_q(\check{\mathbf{r}}), \qquad J = \ell-1, \ell, \ell+1, \qquad (\text{B.2-33})$$

are known as *vector harmonics* or *vector spherical harmonics* [Edm57; Bie81; Sho90; Arn05]. These vector harmonics are vector functions of an angular argument $\hat{\mathbf{r}}$. They are eigenstates of $\mathbf{S}^2$, $\mathbf{L}^2$, $\mathbf{J}^2$ and J_z, with respective eigenvalues 2, $\ell(\ell+1)$, $J(J+1)$, and M. Because a vector harmonic incorporates triangle coupling rules for unit spin, any given value of the total angular momentum J may accompany at most three choices of the orbital angular momentum ℓ, namely $\ell = J-1$, J, and $J+1$. (But $J = 0$ occurs only with $\ell = 1$.) The vector harmonics $\mathbf{Y}_{\ell\pm1,\ell m}(\hat{\mathbf{r}})$ are each perpendicular to the vector $\mathbf{Y}_{\ell\ell m}(\hat{\mathbf{r}})$ for any choice of direction $\hat{\mathbf{r}}$, but these two vectors are orthogonal to each other only when integrated over all angles. Only the harmonics with $J = \ell$ have no radial component (are fully transverse) [Arn05]:

$$\mathbf{Y}_{\ell\ell M}(\hat{\mathbf{r}}) = \frac{-1}{\sqrt{\ell(\ell+1)}}\left[\frac{M}{\sin\theta}\mathbf{e}_\theta Y_{\ell M}(\theta,\phi) + \mathrm{i}\mathbf{e}_\phi\frac{\partial}{\partial\theta}Y_{\ell M}(\theta,\phi)\right]. \qquad (\text{B.2-34})$$

Coupled vector basis fields. We can supplement these vector angular functions with radial functions and construct vector basis fields. To ensure that the resulting fields satisfy the Helmholtz equation the radial functions must be cylinder functions. Let us consider fields that are regular at the coordinate origin, $r = 0$. This restriction selects the spherical Bessel functions $j_\ell(kr)$ as the radial functions. Then the construction

$$\mathbf{U}_{k\ell J M}(\mathbf{r}) = \sqrt{\frac{2k^2}{\pi}}\,j_\ell(kr)\sum_{mq}(\ell m, 1q|JM)\,Y_{\ell m}(\hat{\mathbf{r}})\,\mathbf{e}_q(\check{\mathbf{r}}) \qquad (\text{B.2-35})$$

provides fields that have the four quantum-number labels $\nu = k\ell J M$. These fields have the normalization

$$\langle k\ell JM|k'\ell' J'M'\rangle \equiv \int dV \; \mathbf{U}_{k\ell JM}(\mathbf{r})^* {\cdot} \mathbf{U}_{k'\ell' J'M'}(\mathbf{r})$$

$$= \delta(k - k')\, \delta_{\ell\ell'}\, \delta_{JJ'}\, \delta_{MM'}. \tag{B.2-36}$$

For the discrete quantum numbers ℓ, J, M this normalization involves the Kronecker delta, δ_{ij}. This generalizes, for the continuum variable k, to the *Dirac delta*, $\delta(k - k')$. The two deltas are definable by the requirements of picking out particular values of components from a sum or argument values from an integral,

$$\sum_j \delta_{ij}\, F_j = F_i, \qquad \int dk \; \delta(k - k')F(k'). \tag{B.2-37}$$

The vector fields $\mathbf{U}_{k\ell JM}(\mathbf{r})$ are eigenfunctions of the operators ∇^2, $\mathbf{S}^2$, $\mathbf{L}^2$, $\mathbf{J}^2$, and J_z, where $\mathbf{J} = \mathbf{L} + \mathbf{S}$ is the total angular momentum operator. It is this angular momentum, not the orbital portion ℓ or the spin portion $\mathbf{S}$, that will be conserved when angular momentum is transferred between atom and field. These fields do not generally have definite helicity, but they do have sharply defined parity. The parity of the vector field is determined by the spherical harmonic portion, and is $(-1)^\ell$.

Hansen multipole fields. To construct a set of vector basis fields we only need to have solutions to the scalar Helmholtz equation in some coordinate system. From such a solution $U(\mathbf{r})$ one can construct a triad of independent, orthogonal vector fields by applying the operators ∇, $\mathbf{L}$, and curl (or, equivalently, the operator $\mathbf{S} \cdot \nabla$). For any given scalar field $U(\mathbf{r})$ we define a triad of orthogonal vector fields, a lamellar field $\mathbf{U}_Z$, and two solenoidal fields $\mathbf{U}_X$ and $\mathbf{U}_Y$,

$$\mathbf{U}_Z(\mathbf{r}) = -\frac{i}{k}\nabla U(\mathbf{r}), \tag{B.2-38a}$$

$$\mathbf{U}_X(\mathbf{r}) = \frac{\mathbf{L}}{|\mathbf{L}|}U(\mathbf{r}), \tag{B.2-38b}$$

$$\mathbf{U}_Y(\mathbf{r}) = -\frac{i}{k}\nabla\times\frac{\mathbf{L}}{|\mathbf{L}|}U(\mathbf{r}) = -\frac{1}{k}\mathbf{S}\cdot\nabla\mathbf{U}_X(\mathbf{r}). \tag{B.2-38c}$$

The resulting fields are, apart from normalization factors, the *Hansen multipole fields* [Han35]. Their full definition requires three additional quantum numbers to specify the scalar $U(\mathbf{r})$.

Orbital angular momentum. To incorporate orbital angular momentum we take $U(\mathbf{r})$ to have the angular dependence of a spherical harmonic $\ell = J$ and, for fields that are regular at the coordinate origin, the radial dependence of a spherical Bessel function:

$$U_{kJM}(\mathbf{r}) = \sqrt{\frac{2k^2}{\pi}}\, j_J(kr)Y_{JM}(\check{\mathbf{r}}). \tag{B.2-39}$$

The triad of vector fields may then be taken as [Bie81; Sho90]

$$\mathbf{U}_{ZkJM}(\mathbf{r}) = -\frac{\mathrm{i}}{k}\nabla U_{kJM}(\mathbf{r}), \tag{B.2-40a}$$

$$\mathbf{U}_{XkJM}(\mathbf{r}) = \frac{\mathbf{L}}{\sqrt{J(J+1)}} U_{kJM}(\mathbf{r}), \tag{B.2-40b}$$

$$\mathbf{U}_{YkJM}(\mathbf{r}) = -\frac{1}{k}\mathbf{S}\cdot\nabla \mathbf{U}_{XkJM}(\mathbf{r}). \tag{B.2-40c}$$

These fields are all eigenfunctions of the operators ∇^2, $\mathbf{S}^2$, $\mathbf{J}^2$, and J_z, but not generally of $\mathbf{L}^2$ nor of helicity. Each field has well defined parity, and can be identified as lamellar or solenoidal.

Transverse Properties. We can rewrite these definitions, for any $U \equiv U(\mathbf{r})$ that solves the scalar Helmholtz equation, in a form that exhibits the radial and transverse directional properties of the fields (here $\hat{\mathbf{r}}$ is a unit vector in the radial direction):

$$\mathbf{U}_Z(\mathbf{r}) = -\mathrm{i}\hat{\mathbf{r}}\left[\frac{\partial U}{\partial kr}\right] - \hat{\mathbf{r}}\times\mathbf{L}\left[\frac{U}{kr}\right], \tag{B.2-41a}$$

$$\mathbf{U}_X(\mathbf{r}) = \frac{\mathbf{L}}{|\mathbf{L}|}U, \tag{B.2-41b}$$

$$\mathbf{U}_Y(\mathbf{r}) = \hat{\mathbf{r}}\frac{\mathbf{L}^2}{|\mathbf{L}|}\left[\frac{U}{kr}\right] - \mathrm{i}\frac{\hat{\mathbf{r}}\times\mathbf{L}}{|\mathbf{L}|}\left[\frac{U}{kr} + \frac{\partial U}{\partial kr}\right]. \tag{B.2-41c}$$

Applied to multipole fields the equations (B.2-40) show that only the (magnetic multipole) field $\mathbf{U}_{XkJM}$ is exactly transverse. The other solenoidal field, the electric multipole $\mathbf{U}_{YkJM}$, has radial components that, by comparison with the transverse part, diminish in inverse proportion to the scaled distance kr. At large distances from the origin it too can be regarded as transverse.

The first of the two solenoidal fields, here bearing subscript X, may be written in terms of a single vector harmonic as

$$\mathbf{U}_{XkJM}(\mathbf{r}) = \mathbf{U}_{kJJM}(\mathbf{r}) = \sqrt{\frac{2k^2}{\pi}}j_J(kr)\mathbf{Y}_{JJM}(\check{\mathbf{r}}). \tag{B.2-42}$$

The other two fields are linear combinations of the fields $\mathbf{U}_{k\ell JM}$ with orbital angular momentum $\ell = J \pm 1$,

$$\mathbf{U}_{ZkJM}(\mathbf{r}) = -\mathrm{i}\sqrt{\frac{J}{2J+1}}\mathbf{U}_{k,J-1,JM}(\mathbf{r}) - \mathrm{i}\sqrt{\frac{J+1}{2J+1}}\mathbf{U}_{k,J+1,JM}(\mathbf{r}), \tag{B.2-43a}$$

$$\mathbf{U}_{YkJM}(\mathbf{r}) = +\sqrt{\frac{J+1}{2J+1}}\mathbf{U}_{k,J-1,JM}(\mathbf{r}) - \sqrt{\frac{J}{2J+1}}\mathbf{U}_{k,J+1,JM}(\mathbf{r}). \tag{B.2-43b}$$

Fields near the origin. For small radial distances, $kr \ll 1$, we can express the spherical Bessel function by the first term of a series expansion,

$$j_\ell(x) \simeq \frac{x^\ell}{(2\ell+1)!!} \equiv \frac{x^\ell}{1\times 3\times 5\times\ldots\times(2\ell+1)}. \tag{B.2-44}$$

At the origin, $r = 0$, only the spherical Bessel function $j_0(0) = 1$ is nonzero. We therefore find that the only multipole basis fields that do not vanish at the coordinate origin are those with $J = 1$. These comprise a lamellar dipole field,

$$\mathbf{U}_{Zk1M}(0) = -\frac{ik}{\pi\sqrt{6}}\,\mathbf{e}_M(\check{\mathbf{r}}), \tag{B.2-45a}$$

and one solenoidal field (the electric dipole field),

$$\mathbf{U}_{Yk1M}(0) = \frac{k}{\pi\sqrt{3}}\,\mathbf{e}_M(\check{\mathbf{r}}). \tag{B.2-45b}$$

Helicity multipoles. The combinations $\mp\mathbf{U}_X - i\mathbf{U}_Y$ correspond to fields with helicity ± 1. Specifically, we have the construction

$$\mathbf{U}_{kJMq}(\mathbf{r}) = \begin{cases} \mathbf{U}_{ZkJM}(\mathbf{r}) & \text{for } q = 0 \\ \frac{1}{\sqrt{2}}\left[-q\mathbf{U}_{XkJM}(\mathbf{r}) - i\mathbf{U}_{YkJM}(\mathbf{r})\right] & \text{for } q = \pm 1, \end{cases} \tag{B.2-46}$$

labeled by the four quantum numbers $\nu = kJMq$.

Each helicity multipole with $q = \pm 1$ is a blend of two parity multipoles. Because atomic states in the absence of external perturbations are usually states with well-defined parity, parity conservation produces a single parity multipole rather than a single helicity multipole.

Connection between multipoles and plane waves. Conventional radiation detectors are small field-sensitive areas that identify directional traveling waves. That is, they record field modes (photons) having well-defined linear momentum and polarization. Hansen multipole fields are coherent superpositions of such fields,

$$\mathbf{U}_{jkJM}(\mathbf{r}) = \int d\check{\mathbf{k}}\,\mathbf{U}_{kj}(\hat{\mathbf{r}})\langle kj|jkJM\rangle, \tag{B.2-47}$$

where j denotes Hansen multipole type, X, Y, or Z. That is, a single multipole source produces plane waves in all directions, with intensities governed by the absolute square of the mode-transformation coefficient

$$\langle jkJM|kj\rangle = \frac{i^J}{k}Y_{JM}(\check{\mathbf{k}})^*. \tag{B.2-48}$$

B.2.7   Spontaneous-emission dipole fields

Spontaneous emission is a random process, differing from the coherent process responsible for the controlled excitation. The loss of energy from one excited atom typically results in an emitted radiation field that is a mixture of three dipole fields traveling spherically outward from the atom center of mass. Each of these fields is associated with a distinct dipole orientation of the source atom (labeled either x, y, z or $-1, 0, +1$), and a resulting dipole radiation pattern around the quantization axis. Figure B.2, from [Sho90; Sho11], shows these patterns for the spherical unit vectors as sources. Their (randomized) mixture produces an unpolarized field carrying, in total, one photon of excitation energy from the single excited atom.

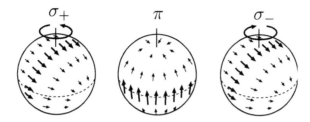

Fig. B.2 *The three dipole fields associated with sources* $\mathbf{e}_q$. *Mixtures of these elementary fields emerge as fluorescence from excited states in free space. After Fig. 2.7 of* [Sho68]; *Fig. 19.5-1 of* [Sho90]; *Fig. 34 of* [Sho08]; *Fig. 12.2 of* [Sho11].

A procedure in which a controlled, coherent field induces atomic excitation and the always-possible spontaneous emission produces loss of excitation energy is known as an example of *optical pumping* [Kas57; Hap72; Ima03; Auz09; Auz10]. It produces a mixed state of the atom and a mixed state of the radiation field.

B.2.8 Multiple-path modes

The free-space field modes discussed in the preceding examples are relatively simple structures, appropriate for describing idealizations of radiation beams by closed-form mathematical expressions. For many uses of radiation it is necessary to supplement the free-space wave equations or the Helmholtz equations with various boundary conditions. The simplest examples are when radiation is confined by surfaces over which components of the electric or magnetic fields can be idealized as vanishing. Such surfaces serve to create enclosures or cavities that restrict possible values of wavevectors and frequencies, converting a continuum into a discrete set.

Some optical designs involve opaque surfaces in which there are openings. The various forms of multi-slit apertures are examples. The fields for such situations can be followed, mathematically, along multiple paths between any given source point and any other point. The requirement that the field be uniquely defined at every point leads to various methods for describing the field as a coherent superposition of beams or simple modes. In addition to the often displayed examples of slit-produced patterns, ongoing development of techniques for imposing remarkably varied wavefronts on traveling fields leads to patterns of node lines and "knotted" fields that lend themselves to topological analysis.

Such situations are to be kept in mind when field quantization, and photon creation, takes place. However elaborate may be the node structure of the field, as long as it is connected the strength of the energy density gives the relative probability of detecting a discrete interaction with a probe (detecting a photon) at every location.

B.2.9 Summary

The great variety of solutions to the scalar and vector Helmholtz equations in various coordinate systems permits a corresponding variety of analytically-expressed mode functions with which to express an electromagnetic field. Each set of modes leads to a class of photons. For classical fields the coefficients of these mode functions, in a mode expansion of the field, must be determined to fit boundary conditions. In simple cases, when the boundary surfaces coincide with surfaces defined by constant values of coordinates, the field will be a single mode, but in general it will appear as a superposition of modes.

Fields generated by quantum-mechanical interaction with an atom have similar properties. For these fields matrix elements of the interaction take the place of boundary conditions in prescribing the field. In general the interaction produces a coherent superposition of modes—the prominent example is the representation of the field by plane waves. By suitably choosing the modes it often becomes possible to identify a single mode, typically a dipole field, as the field produced by the atom. This is the utility of the parity (or Hansen) multipole fields: They provide a compact analytic expression for the field produced by a multipole interaction. When quantized they describe the photons produced in spontaneous emission from an atom in free space.

B.3   Quantized field modes; Dirac photons

The most common approach to establishing a quantum theory of radiation is that first taken by Dirac and Fermi: Introducing field spatial modes whose time variation can be treated as harmonic oscillators and then treaing these as quantum mechanical.

Vector potentials. Much of the literature on quantum electrodynamics employs vector potentials as the fields to be subject to quantum conditions that lead to the definition of photons; see Appendix B.1.9. In quantum optics, where the matter states of interest have been those of bound electrons and molecular motion, it has been customary to deal with a multipole description of bound charges and currents, and to introduce a Hamiltonian based on electric and magnetic fields interacting with these [Ros55; Pow78; Pow80; Cra84; Ack84; Pow85; Arn05; Bay06; Mil19]. In keeping with this quantum optics approach I shall deal with direct quantization of the electric and magnetic fields $\mathbf{E}(\mathbf{r}, t)$ and $\mathbf{B}(\mathbf{r}, t)$ rather than with vector potentials [Van77; Pow78; Pow80; Pow85; Sho90; Sho11].

B.3.1   Spatial modes

We begin by considering an electric field[9] that satisfies the vector wave equation

$$\frac{\partial^2}{\partial t^2} \mathbf{E}(\mathbf{r}, t) = -c^2 \, \nabla \times \nabla \times \mathbf{E}(\mathbf{r}, t). \tag{B.3-1}$$

We proceed by separating the real-valued electric field into positive- and negative-frequency complex-valued parts,

$$\mathbf{E}(\mathbf{r}, t) = \mathbf{E}^{(+)}(\mathbf{r}, t) + \mathbf{E}^{(-)}(\mathbf{r}, t). \tag{B.3-2}$$

We then separate the spatial and temporal variation by introducing a set of spatial-mode vector fields $\mathbf{U}_\nu(\mathbf{r})$ that are complex-valued solutions to the vector Helmholtz equation, as discussed in Appendix B.1.5:

---

[9] All of the following discussion, and the quantization, apply equally well to the **B** field, the RS vector of Appendix B.1.11 and the vector potentials of Appendix B.1.9. Each of these fields will have a variant of the photons discussed here.

$$\nabla \times \nabla \times \mathbf{U}_\nu(\mathbf{r}) = (\omega_\nu/c)^2 \, \mathbf{U}_\nu(\mathbf{r}), \qquad \nabla \cdot \mathbf{U}_\nu(\mathbf{r}) = 0. \tag{B.3-3}$$

The index ν identifies the various possible solutions to this equation.

To complete the definition of the mode fields we need constraints, such as are provided by boundary conditions. A common approach is to require that the field (or its directional derivative) vanish over the surface of a confining volume, regarded as a cavity. With a cubic box the requirement of vanishing field means that an integer number of half wavelengths must fit into each of the three orthogonal box directions. With such modes the possible frequencies ω_ν are discrete. An alternative approach is to require that the mode fields be eigenfunctions of various differential operators, such as those associated with momentum and angular momentum in quantum descriptions. With whatever approach to the definition of modes, boundary conditions, or operator eigenfunctions, the index ν identifies the various independent spatial modes.

Examples of these mode fields appear in the discussion of Appendix B.1. Let us take these basis fields to be orthogonal and to have the normalization $|U_\nu|$, as in eqn (B.1-30). These spatial-mode fields we pair with complex-valued time dependences $a_\nu(t)$ to write the complex-valued electric fields of eqn (B.3-2) as a superposition (with summation over discrete indices and integration over continua) of well-defined spatial modes $\mathbf{U}_\nu(\mathbf{r})$ each having a time-dependent amplitude $a_\nu(t)$ to be determined:

$$\mathbf{E}^{(+)}(\mathbf{r},t) = \tfrac{1}{2} \oint_\nu \mathcal{E}_\nu \, \mathbf{U}_\nu(\mathbf{r}) \, a_\nu(t), \qquad \mathbf{E}^{(-)}(\mathbf{r},t) = \tfrac{1}{2} \oint_\nu \mathcal{E}_\nu^* \, \mathbf{U}_\nu(\mathbf{r})^* \, a_\nu^*(t). \tag{B.3-4}$$

The spatial mode functions $\mathbf{U}_\nu(\mathbf{r})$ incorporate all of the classical characteristics of the field—the polarization, the wavelike interference, the reflections at edges, the dispersion by gratings, etc. It is with the amplitudes $a_\nu(t)$ that the quantum nature of the field, and photons, makes its appearance.

Plane waves and spatial Fourier components. For mathematical manipulations the most common choice for the spatial modes is a set of unit-amplitude solenoidal traveling plane-waves, each characterized by a wavevector $\mathbf{k}$ whose three Cartesian components define the propagation direction $\check{\mathbf{k}}$, and by a unit vector $\mathbf{e}_\nu(\check{\mathbf{k}})$ (possibly complex-valued) required to be transverse to this direction,

$$\mathbf{U}_\nu(\mathbf{r}) = \mathbf{e}_\nu(\check{\mathbf{k}}) \exp(i\mathbf{k}_\nu \cdot \mathbf{r}), \qquad \mathbf{e}_\nu(\check{\mathbf{k}}) \cdot \mathbf{k}_\nu = 0. \tag{B.3-5}$$

The superposition of these modes, allowing for possible propagation directions, takes the form of a spatial Fourier integral summed over two field directions $\mathbf{e}_\mu(\check{\mathbf{k}})$ as specified by label μ,

$$\mathbf{E}^{(+)}(\mathbf{r},t) = \tfrac{1}{2} \int d^3k_\nu \sum_\mu \mathcal{E}_\nu \, \mathbf{e}_\mu(\check{\mathbf{k}}) \, \exp(i\mathbf{k}_\nu \cdot \mathbf{r}) \, a_\nu(t), \tag{B.3-6}$$

where $a_\nu(t)$ is a unit-amplitude solution to eqn (B.3-8). Each choice of propagation vector $\mathbf{k}_\nu$ is associated with a frequency ω_ν that is fixed by the three wavevector components, in accord with the Helmholtz-equation requirements,

$$\omega_\nu = c|\mathbf{k}_\nu|. \tag{B.3-7}$$

To create a field that has a specified frequency the Fourier integral eqn (B.3-6) must be restricted to two dimensions. It is important to recognize that boundary conditions fix the possible discrete values for $\mathbf{k}_\nu$. These, in turn, determine the possible values of the frequencies ω_ν.

B.3.2   Field-mode quantization; Photons

The requirement that the electric field obey the vector wave equation means that the mode functions, in turn, satisfy the vector Helmholtz equation and leads to the conclusion that for each independent field mode the amplitude $a_\nu(t)$ must have the time dependence of an harmonic oscillator,

$$\frac{d^2}{dt^2}\, a_\nu(t) = \omega_\nu^2\, a_\nu(t). \tag{B.3-8}$$

The contribution of each mode to the electric field is established by these oscillator amplitudes: With every time-dependent amplitude $a_\nu(t)$ there is paired a specified spatial vector field $\mathbf{U}_\nu(\mathbf{r})$, a solution to the vector Helmholtz equation. Unlike oscillators that have nonzero rest mass m and a frequency $\omega = \sqrt{\kappa/m}$ fixed by a spring constant κ, here there is no rest mass and no spring constant: The oscillator frequency ω_ν comes from a Fourier analysis of the field (with its boundary conditions), as in eqn (B.3-7).

We turn these expressions of classical electrodynamics into ones associated with quantum theory by assuming that the oscillator variables appearing here are noncommuting operators, making the replacements appropriate to a monochromatic field,

$$a_\nu(t) \to \hat{a}_\nu\, \mathrm{e}^{-\mathrm{i}\omega_\nu t}, \qquad a_\nu^*(t) \to \hat{a}_\nu^\dagger\, \mathrm{e}^{+\mathrm{i}\omega_\nu t}, \tag{B.3-9}$$

and postulating the traditional commutator relationships of quantum oscillators,

$$[\hat{a}_\nu, \hat{a}_{\nu'}^\dagger] = \delta(\nu, \nu'). \tag{B.3-10}$$

The delta appearing here is a product of Kronecker deltas or Dirac deltas, see eqn (B.2-37), depending on the discreteness of the mode-defining quantum numbers. It is with the introduction of these formulas that the electromagnetic field becomes quantized, taking on non-classical characteristics,

$$\hat{\mathbf{E}}^{(+)}(\mathbf{r}, t) = \tfrac{1}{2} \sum_\nu \mathcal{E}_\nu\, \mathbf{U}_\nu(\mathbf{r})\, \hat{a}_\nu(t), \qquad \hat{\mathbf{E}}^{(-)}(\mathbf{r}, t) = \tfrac{1}{2} \sum_\nu \mathcal{E}_\nu^*\, \mathbf{U}_\nu(\mathbf{r})^*\, \hat{a}_\nu^\dagger(t), \tag{B.3-11}$$

with summation or integration as appropriate for the mode-defining quantum numbers of the label ν. It is here that *Dirac photons* make their appearance: A single photon in mode ν is associated with a measurable electric field that has the spatial structure of mode field $\mathbf{U}_\nu(\mathbf{r})$. The choice of traveling plane waves for the mode is by no means a necessity; any of the example modes of Appendix B.2 would do equally well, as would the standing-wave modes of an arbitrarily shaped enclosure: All of these solutions to the free-space vector Helmholtz equation would provide Dirac photons.

Continuum photons.   The traditional approach to quantizing the free-space electromagnetic field [Blo90] imagines the field to be contained in a large cavity, say a cube of edge L, that has node-imposing surfaces so that the possible modes have discrete wavevectors along each Cartesian axis and discrete frequencies. In one dimension their spacing is

$$\Delta k = \frac{2\pi}{L}, \qquad \Delta\omega = \frac{2\pi c}{L}. \tag{B.3-12}$$

Such possible discretization occurs with all the coordinate systems for which the Helmholz equation is separable. The discreteness allows us to introduce an integer index ν to label the modes, and leads to the commutation relations

$$[\hat{a}_\nu, \hat{a}_{\nu'}^\dagger] = \delta_{\nu\nu'} \tag{B.3-13}$$

involving the Kronecker delta. We then imagine the enclosing cavity to grow large, $L \to \infty$, and the mode spacing to become correspondingly small, forming a continuum of values—an example of the mathematician's replacement $\mathbb{Q} \to \mathbb{R}$. The discrete sum over modes becomes an integral,

$$\sum_\nu \to \frac{1}{\Delta\omega} \int d\omega. \tag{B.3-14}$$

The discrete set of photon operators becomes a continuum, the Kronecker delta becomes the Dirac delta, [Blo90]

$$\hat{a}_\nu \to \sqrt{\Delta\omega}\,\hat{a}(\omega), \qquad \delta_{\nu\nu'} \to \delta(\omega - \omega'), \tag{B.3-15}$$

and the commutation relation becomes expressed by the Dirac delta,

$$[\hat{a}(\omega), \hat{a}^\dagger(\omega')] = \delta(\omega - \omega'). \tag{B.3-16}$$

The original underlying mode structure dictated by the coordinate system has here been replaced by a continuum that has no connection with bounding surfaces. When applied to vector fields in three dimensions, for photons specified by momentum $\hbar\mathbf{k}$ and helicity μ this equation generalizes to

$$[\hat{a}(\mathbf{k}\mu), \hat{a}^\dagger(\mathbf{k}'\mu')] = \delta(\mathbf{k} - \mathbf{k}')\delta_{\mu\mu'}. \tag{B.3-17}$$

Time-domain photons. We use a Fourier transform to convert from frequency to time as the independent variable,

$$\hat{a}(t) = \frac{1}{\sqrt{2\pi}} \int d\omega\,\hat{a}(\omega), \tag{B.3-18}$$

and obtain the photon commutation relations

$$[\hat{a}(t)\,\hat{a}^\dagger(t')] = \delta(t - t'). \tag{B.3-19}$$

Varieties of photons. We see from the preceding discussion and the commutators eqn (B.3-13), eqn (B.3-16), and eqn (B.3-19), possible approaches to incorporating quantum theory into wave equations (including, for example, the paraxial equation for a waveguide with Kerr nonlinearity [Lai89; Lai89b]), thereby defining photons in mode space or by frequency variation or by time variation. As with the distinction between emission and absorption photons, the choice is left to the convenience of an experimenter or a theorist.

B.3.3   Photon-number operator; Number states

It is customary when treating quantum oscillators to introduce the number operator $\hat{N}_\nu$ for mode ν,

$$\hat{N}_\nu = \hat{a}_\nu^\dagger \, \hat{a}_\nu \equiv \hat{a}_\nu^\dagger(t) \, \hat{a}_\nu(t). \tag{B.3-20}$$

This operator has eigenstates denoted in Dirac form as $|n_\nu\rangle$,

$$\hat{N}_\nu |n_\nu\rangle = n_\nu |n_\nu\rangle. \tag{B.3-21}$$

It then follows, from the equal-time commutation relation, that

$$\hat{a}_\nu |n_\nu\rangle = \sqrt{n_\nu}\,|n_\nu - 1\rangle, \qquad \hat{a}_\nu^\dagger |n_\nu\rangle = \sqrt{n_\nu + 1}\,|n_\nu + 1\rangle. \tag{B.3-22}$$

The operator $\hat{a}_\nu^\dagger$ adds an increment of energy $\hbar\omega_\nu$ to field mode ν while $\hat{a}_\nu$ removes such an increment. These operators are therefore annihilation and creation operators for field increments, i.e. for photons. The *photon-vacuum state*, for which n_ν is zero, has the property, for all modes ν,

$$\hat{a}_\nu |0_\nu\rangle = 0. \tag{B.3-23}$$

B.3.4   Single-photon superpositions

Although the traditional route to quantization and photons proceeds through discrete modes of the Helmholtz equation subject to boundary conditions, it is possible to express those modes in terms of any arbitrary but complete set of orthogonal functions of frequency (or time). Continuing the restriction to a one-dimensional scalar field, consider such a set of frequency functions $\phi_i(\omega)$ such that the orthogonality and completeness relations read

$$\int d\omega \; \phi_\nu(\omega)\phi_{\nu'}^*(\omega') = \delta_{\nu\nu'}, \qquad \sum_\nu \phi_\nu(\omega)\phi_\nu^*(\omega) = \delta(\omega - \omega'). \tag{B.3-24}$$

For each of these frequency functions we define a photon destruction operator as a superposition,

$$\hat{c}_\nu(\omega) = \int d\omega \; \phi_\nu(\omega)^* \, \hat{a}(\omega), \qquad \hat{a}(\omega) = \sum_\nu \phi_\nu(\omega)\hat{c}_\nu(\omega). \tag{B.3-25}$$

These superposition photons satisfy the traditional commutation relations,

$$[\hat{c}_\nu, \hat{c}_{\nu'}^\dagger] = \delta_{\nu\nu'}, \tag{B.3-26}$$

and so they are annihilation operators, but they have a frequency distribution that is set by the (arbitrary) choice of basis functions $\phi_\nu(\omega)$. Such superpositions of frequencies are a generalization of the superpositions that produce the Morris-Shore photons of Section 5.10 and the superposition of P and S fields of Raman transitions. The photon number operator, when expressed in terms of the chosen functions, reads

$$\hat{N} = \sum_\nu \hat{c}_\nu^\dagger \hat{c}_\nu. \tag{B.3-27}$$

Such a superposition procedure works equally well to introduce a set of basis functions of the time variable, enabling description of any convenient pulse shape as originating with a single photon.

B.3.5   Quantization without photon numbers

The association of electromagnetism with quantum theory occurs with the introduction of commuting operators, here taken to be those associated with electromagnetic field modes. The use of number states, as in eqn (B.3-20), is by no means a required second step. Such states leave field phases undetermined. Other complete basis sets can serve equally well. The overcomplete Glauber *coherent states* of Appendix B.4.1, offering both photon number and phase as descriptors, are a useful alternative.

It is also possible to avoid use of mode fields in the formalism [Hei29; Hei30; Bia96; Bia09; Bia17]. The noncommuting operators then become electric and magnetic field operators. Their commutator relations restrict their simultaneous specification at a common space-time point. The essence of the procedure can be seen from the following simplification. Define a field destruction operator appropriate for scalar plane-wave modes (or any other basis) as

$$\hat{F}(\mathbf{r}) = \sum_{\nu} \exp(i\mathbf{k}_{\nu}\cdot\mathbf{r})\,\hat{a}_{\nu}. \tag{B.3-28}$$

The assumed commutator relationships of the single-mode oscillator operators then gives the field commutators as

$$[\hat{F}(\mathbf{r}),\,\hat{F}(\mathbf{r}')] = 0, \qquad [\hat{F}(\mathbf{r})^{\dagger},\,\hat{F}(\mathbf{r}')^{\dagger}] = 0, \tag{B.3-29a}$$

and

$$[\hat{F}(\mathbf{r}),\,\hat{F}(\mathbf{r}')^{\dagger}] = \delta(\mathbf{r} - \mathbf{r}'). \tag{B.3-29b}$$

These commutation relations replace mode-dependent photon numbers as the basis for predicting the dynamics of electromagnetic fields.

Operator ordering; The vacuum.   The presentation of Maxwell's equations as equations for harmonic oscillators leads to the interpretation of an electromagnetic field as a dynamical system with an infinite number of degrees of freedom—an infinite number of Helmholtz-equation modes (of the Helmholtz equation with boundary conditions)—each of which is associated with an oscillator. To complete the formalism we need to follow a consistent program of operator ordering. The earliest discussions of quantum electrodynamics chose the *symmetric ordering* discussed in Appendix A.4.3. As discussed in Appendix A.7.1 this choice endows each oscillator, and hence each field mode, with a *zero-point energy*. This energy needs to be included for the mechanical oscillators that occur in descriptions of molecular motions involving rest masses and spring constants in the definition of frequencies. But for field oscillators there are no rest masses and there are an indefinite number of frequencies. The resulting total zero-point energy, even for the vacuum in the absence of any photons, will be infinitely large. This result, a consequence of the operator-ordering choice, offers opportunity for interpreting a number of observable effects, notably the Lamb shift and spontaneous emission, as originating in *vacuum fluctuations* of photon numbers [Gla06]. The photons of this vacuum state are not to be expected as having properties of real photons, but their effects are real.

An alternative approach, using *normal ordering* at the outset, introduces no such infinite energy to the quantum vacuum [Mil94]. The observable Lamb shift and spontaneous emission are attributed to radiative reaction, an effect with close ties to classical electrodynamics. This approach is particularly common in quantum optics, and it is the approach I follow.

B.3.6   The radiation-field Hamiltonian with photons

For subsequent introduction of noncommuting operators it is important to establish the ordering of the two complex parts of the field. I shall use normal ordering, writing the (classical) free-space energy as a weighted sum of individual mode energies $\epsilon_0|\mathcal{E}_\nu U_\nu|^2$,

$$H^R(t) = 2\epsilon_0 \int dV \; \mathbf{E}^{(-)}(\mathbf{r},t) \cdot \mathbf{E}^{(+)}(\mathbf{r},t)$$

$$= \tfrac{1}{2}\epsilon_0 \sum_\nu |\mathcal{E}_\nu U_\nu|^2 \, a_\nu^*(t)a_\nu(t). \qquad \text{(B.3-30a)}$$

The quantum version of the normal-ordering field energy is the free-field Hamiltonian, independent of time,

$$\hat{H}^R = 2\epsilon_0 \int dV \; \hat{\mathbf{E}}^{(-)}(\mathbf{r},t) \cdot \hat{\mathbf{E}}^{(+)}(\mathbf{r},t) = \tfrac{1}{2}\epsilon_0 \sum_\nu |\mathcal{E}_\nu U_\nu|^2 \, \hat{a}_\nu^\dagger \hat{a}_\nu. \qquad \text{(B.3-30b)}$$

We express this radiation Hamiltonian as the sum of photon energies,[10]

$$\hat{H}^R = \sum_\nu \hbar\omega_\nu \hat{N}_\nu, \qquad \text{(B.3-31)}$$

by requiring that the single-photon field amplitude $\mathcal{E}_\nu$ associated with the mode function $\mathbf{U}_\nu(\mathbf{r})$ be set by the formula

$$|\mathcal{E}_\nu|^2 = \frac{2\hbar\omega_\nu}{\epsilon_0|U_\nu|^2}. \qquad \text{(B.3-32)}$$

B.3.7   Traveling-waves, standing-waves, and RS photons

The mode functions that accompany photon operators in the expression for an electric field should be chosen to suit boundary conditions that accord with each specific occasion. A simple example occurs when we consider traveling-wave and standing-wave solutions to the Maxwell equations. The positive-frequency field structure for given frequency, polarization, and magnitude for modes that have one-dimensional spatial variation has the form

$$\hat{\mathbf{E}}^{(+)}(z,t) = \mathbf{e}\,\mathcal{E}\Big[U_1(z)\hat{a}_1 + U_2(z)\hat{a}_2\Big]e^{-i\omega t}, \qquad \text{(B.3-33)}$$

where $\mathbf{e}$ is a unit vector in the x,y plane and $\hat{a}_1$ and $\hat{a}_2$ are independent photon annihilation operators that satisfy the standard commutation relations with their conjugates,

---

[10]Normal ordering removes the term $\tfrac{1}{2}\hbar\omega_\nu$ that appears for each mode of mechanical oscillators.

$$[\hat{a}_i, \hat{a}_j^\dagger] = \delta_{ij}. \tag{B.3-34}$$

Field eigenstates that are associated with definite photon numbers (Fock states) have the form of products,

$$|n_1, n_2\rangle \equiv |n_1\rangle_1 |n_2\rangle_2, \tag{B.3-35}$$

in which the ordering of the kets from left to right must be maintained. The action of photon operators on these product states follows the standard pattern. For example, the two independent photon-annihilation operators produce the changes

$$\hat{a}_1 |n_1, n_2\rangle = \sqrt{n_1} |n_1 - 1, n_2\rangle, \tag{B.3-36a}$$

$$\hat{a}_2 |n_1, n_2\rangle = \sqrt{n_2} |n_1, n_2 - 1\rangle. \tag{B.3-36b}$$

Although these, and other, formulas make no distinction between traveling- and standing-wave modes, experimental boundary conditions select only one class, and there are observable differences in the behavior of quantum systems subjected to each class of field.

Traveling waves. Consider the positive-frequency part of a field formed by super-posing two plane waves, of equal frequency, polarization, and magnitude, propagating in opposite directions (counter-propagating fields),

$$\hat{\mathbf{E}}^{(+)}(z, t) = \mathbf{e}\,\mathcal{E}\left[e^{+ikz}\hat{a}_+ + e^{-ikz}\hat{a}_- \right] e^{-i\omega t}. \tag{B.3-37}$$

Here $\hat{a}_+$ and $\hat{a}_-$ are photon annihilation operators for forward- and backward-traveling modes, respectively. Classical traveling waves carry not only energy but momentum (as expressed by the Poynting vector). The photons created by $\hat{a}_\pm^\dagger$ carry not only energy $\hbar\omega$ but momentum of magnitude $\pm\hbar\omega/c$ along the propagation direction. Absorption of such a photon will not only increase the energy of an atom but will also impart a momentum kick. For these traveling-wave modes we use the product states

$$\text{traveling wave:}\quad |n_+, n_-\rangle_T \equiv |n_+\rangle_+ |n_-\rangle_-. \tag{B.3-38}$$

These are to be paired with atom states as products $|\text{atom}\rangle|\text{field}\rangle$ in constructing statevectors that treat atoms interacting with a quantum field.

Standing waves. Rather than use exponentials and traveling waves for the spatial modes we can express this field using sine and cosine standing waves. Doing so we rewrite the construction as

$$\hat{\mathbf{E}}^{(+)}(z, t) = \mathbf{e}\,\sqrt{2}\mathcal{E}\left[\cos(kz)\hat{a}_c + i\sin(kz)\hat{a}_s \right] e^{-i\omega t}, \tag{B.3-39}$$

where the coefficients of the mode functions are now the photon operators

$$\hat{a}_c = \frac{1}{\sqrt{2}}[\hat{a}_+ + \hat{a}_-], \qquad \hat{a}_s = \frac{1}{\sqrt{2}}[\hat{a}_+ - \hat{a}_-]. \tag{B.3-40}$$

Classical standing waves carry no momentum (the Poynting vector vanishes) and hence although the standing-wave photons have energy $\hbar\omega$ they carry no linear momentum.

Absorption of such a photon adds energy to an atom but does not give a momentum kick. To describe changes in these fields we use the standing-wave product states

$$\text{standing wave: } |n_s, n_c\rangle_S \equiv |n_s\rangle_s |n_c\rangle_c. \tag{B.3-41}$$

Mode differences. The difference between traveling- and standing-wave modes (and their photons) has been discussed in some detail in the article [Sho91]. Traveling-wave modes are appropriate when we deal with laser beams or other free-space fields. The two photon-propagation directions are independent degrees of freedom, and the interaction Hamiltonian can act to transfer photons from one mode to the other. In so doing it transfers linear momentum to or from the center-of-mass motion of the atom involved with the interaction.

Standing-wave modes are appropriate when we deal with cavity-confined radiation, such as occurs between parallel confocal mirrors. Such modes do not carry linear momentum and so their effect on atoms is experimentally distinguishable from that of traveling waves. Although a standing-wave photon operator is expressible as a coherent superposition of counter-propagating traveling-wave photons, as in eqn (B.3-40), these two parts must always occur together: They are not independent photons.

Quantization with the RS vector. The quantum version of the RS vector is obtainable by introducing a three-dimensional Fourier decomposition of this field into traveling plane waves and writing, with a single unit vector [Bia13; Bia17],

$$\mathbf{F}(\mathbf{r}, t) = \sqrt{\hbar c} \int \frac{d^3 k}{(2\pi)^{3/2}} \, \mathbf{e}(\check{\mathbf{k}}) \left[U_+(\mathbf{k}) \, e^{i(\mathbf{k} \cdot \mathbf{r} - \omega_k t)} + U_-(\mathbf{k})^* \, e^{-i(\mathbf{k} \cdot \mathbf{r} - \omega_k t)} \right], \tag{B.3-42}$$

where frequency ω_k and wavevector magnitude k are related by

$$\omega = ck = c|\mathbf{k}| = c\sqrt{k_x^2 + k_y^2 + k_z^2}. \tag{B.3-43}$$

To allow a probabilistic interpretation of the amplitudes they should have the normalization [Bia17]

$$\sum_{\lambda = \pm} \int \frac{d^3 k}{k} \, |U_\lambda(\mathbf{k})|^2 = 1. \tag{B.3-44}$$

For the RS vector to be divergenceless the single unit vector $\mathbf{e}(\check{\mathbf{k}})$ that defines the directions of the electric and magnetic field vectors must satisfy the transverse condition (i.e. it must be perpendicular to the propagation vector $\mathbf{k}$)

$$c\mathbf{k} \times \mathbf{e}(\check{\mathbf{k}}) = -i\omega \mathbf{e}(\check{\mathbf{k}}), \quad \text{or} \quad \mathbf{k} \cdot \mathbf{s} \, \mathbf{e}(\check{\mathbf{k}}) = k\mathbf{e}(\check{\mathbf{k}}), \tag{B.3-45}$$

and the normalization condition

$$\mathbf{e}^*(\mathbf{k}) \cdot \mathbf{e}(\check{\mathbf{k}}) = 1. \tag{B.3-46}$$

In order for the two independent expansion coefficients $U_\pm(\mathbf{k})$ to produce fields that are consistent with the inhomogeneous Lorentz group (the Poincaré group) it is useful to take these as one-dimensional irreducible representations of the Poincaré group (without reflections). Identifying labels on the resulting RS vector can then be taken as eigenvalues of the generators of the Poincaré group. A complete specification then requires four labels: The helicity and three eigenvalues of mutually commuting operators.

To quantize the RS vector we replace the Fourier amplitudes by annihilation and creation operators [Bia17],

$$U_\lambda(\mathbf{k}) \to \hat{a}_\lambda(\mathbf{k}), \qquad U_\lambda^*(\mathbf{k}) \to \hat{a}_\lambda^\dagger(\mathbf{k}), \quad \lambda = \pm, \tag{B.3-47}$$

to give the RS field operator [Bia13; Bia17]

$$\hat{\mathbf{F}}(\mathbf{r},t) = \sqrt{\hbar c} \int \frac{d^3k}{(2\pi)^{3/2}} \, \mathbf{e}(\mathbf{k}) \left[\hat{a}_\lambda(\mathbf{k}) \, \mathrm{e}^{\mathrm{i}(\mathbf{k}\cdot\mathbf{r}-\omega_k t)} + \hat{a}_\lambda^\dagger(\mathbf{k}) \, \mathrm{e}^{-\mathrm{i}(\mathbf{k}\cdot\mathbf{r}-\omega_k t)} \right]. \tag{B.3-48}$$

The field operators defined in this way obey the commutation relations [Bia13; Bia17]

$$[\hat{a}_\lambda(\mathbf{k}), \, \hat{a}_{\lambda'}^\dagger(\mathbf{k}')] = k\delta_{\lambda\lambda'}\delta^3(\mathbf{k} - \mathbf{k}'). \tag{B.3-49}$$

The single-photon state is a superposition of Fourier components [Bia13],

$$|f\rangle = \hat{a}_f^\dagger |0\rangle, \qquad \hat{a}_f^\dagger = \sum_{\lambda=\pm} \int \frac{d^3k}{k} \, f_\lambda \, \hat{a}_\lambda^\dagger(\mathbf{k}). \tag{B.3-50}$$

With the RS-vector description of electromagnetic fields and the photon operators eqn (B.3-49) the radiation Hamiltonian appears as [Bia13]

$$\hat{H}^R = \sum_{\lambda=\pm} \int \frac{d^3k}{k} \, \hbar\omega_k \, \hat{a}_\lambda^\dagger(\mathbf{k})\hat{a}_\lambda(\mathbf{k}), \qquad \omega_k = c|\mathbf{k}|, \tag{B.3-51}$$

with an integral over photon directions in momentum space.

B.3.8 Vector properties

The vector nature of the electromagnetic field adds an additional degree of freedom to the three spacial dimensions and requires inclusion in the mode-defining quantum numbers of a photon. A natural extension of classical vector fields to quantum fields (photons) follows from the discussion of polarization and its Poincaré-sphere display in Appendix B.1.7.

Quantum Stokes operators. To treat quantum fields the four Stokes parameters can be regarded as expectation values $S_n = \langle \hat{S}_n \rangle$ of four operators that are constructed from paired creation and annihilation operators for two independent polarization directions, say H and V [Kor02; Bow02; Lui02; Lui06; Bjo10],

$$\hat{S}_0 = \hat{a}_H^\dagger \hat{a}_H + \hat{a}_V^\dagger \hat{a}_V \qquad \equiv \hat{n}_H + \hat{n}_V, \tag{B.3-52a}$$

$$\hat{S}_1 = \hat{a}_H^\dagger \hat{a}_H - \hat{a}_V^\dagger \hat{a}_V \qquad \equiv \hat{n}_H - \hat{n}_V, \tag{B.3-52b}$$

$$\hat{S}_2 = \hat{a}_H^\dagger \hat{a}_V \, \mathrm{e}^{\mathrm{i}\theta} + \hat{a}_V^\dagger \hat{a}_H \, \mathrm{e}^{-\mathrm{i}\theta}, \tag{B.3-52c}$$

$$\hat{S}_3 = \mathrm{i}\hat{a}_V^\dagger \hat{a}_H \, \mathrm{e}^{-\mathrm{i}\theta} - \mathrm{i}\hat{a}_H^\dagger \hat{a}_V \, \mathrm{e}^{\mathrm{i}\theta}, \tag{B.3-52d}$$

where θ is the phase shift between H and V modes [Bow02] (often taken as zero) and $\hat{n}_k$ is the photon-number operator for mode k. The operator $\hat{S}_0$ is the total photon number for both polarization modes, the quantum counterpart for the summed intensity of the two polarizations. The commutators of photon creation and annihilation for individual modes,

$$[\hat{a}_k, \hat{a}_l^\dagger] = \delta_{kl}, \quad k, l \in \{H, V\}, \tag{B.3-53}$$

leads to the two-mode commutators

$$[\hat{S}_1, \hat{S}_2] = 2\mathrm{i}\hat{S}_3, \qquad [\hat{S}_2, \hat{S}_3] = 2\mathrm{i}\hat{S}_1, \qquad [\hat{S}_3, \hat{S}_1] = 2\mathrm{i}\hat{S}_2. \tag{B.3-54}$$

These are commutators found with angular momentum operators [Sch65]; they lead to the recognition that mean values and variances of the Stokes operators are restricted by Heisenberg uncertainty relations.

$$\Delta S_1 \, \Delta S_2 \geq |\langle \hat{S}_3 \rangle|^2, \qquad \Delta S_2 \, \Delta S_3 \geq |\langle \hat{S}_1 \rangle|^2, \qquad \Delta S_3 \, \Delta S_1 \geq |\langle \hat{S}_2 \rangle|^2. \tag{B.3-55}$$

where the variances are defined as

$$(\Delta X)^2 = \langle \hat{X}^2 \rangle - \langle \hat{X} \rangle^2. \tag{B.3-56}$$

Thus, unlike classical radiation beams, it is not possible to characterize quantum radiation with a sharply-defined point on a Poincaré sphere. The counterpart to the classical inequality eqn (B.1-50) is [Has08]

$$\hat{S}_1^2 + \hat{S}_2^2 + \hat{S}_3^2 = \hat{S}_0^2 + 2\hat{S}_0, \tag{B.3-57}$$

and so there is uncertainty in the radius of the Poincaré sphere as well as the angles of the Stokes-vector components. A Stokes parameter is said to be squeezed if its variance falls below the shot noise of an equal power coherent beam [Bow02; Kor02; Sch03; Kor05; Lui06].

Coherency matrix. We can present the information about a localized classical electromagnetic field, either a plane wave or a beam, by means of a 2×2 density matrix obtained by emplacing the expectation values of Stokes operators between orthogonal unit vectors for polarizations, say H and V. The result, with suppression of arguments $(\mathbf{r}, t)$ common to all the variables, is the dimensionless, Hermitian *coherency matrix* [Bor99],

$$\boldsymbol{\rho} \equiv \begin{bmatrix} \rho_{11} & \rho_{12} \\ \rho_{21} & \rho_{22} \end{bmatrix} = \frac{1}{\mathcal{I}} \begin{bmatrix} \langle \mathcal{E}_H \mathcal{E}_H^* \rangle & \langle \mathcal{E}_H \mathcal{E}_V^* \rangle \\ \langle \mathcal{E}_V \mathcal{E}_H^* \rangle & \langle \mathcal{E}_V \mathcal{E}_V^* \rangle \end{bmatrix}, \qquad \mathcal{I} \equiv \langle \mathcal{E}_H \mathcal{E}_H^* \rangle + \langle \mathcal{E}_V \mathcal{E}_V^* \rangle. \tag{B.3-58}$$

This is an example of a density matrix for a classical two-state system: The two states are the two orthogonal directions for the transverse electric field. With this definition, and the assumption that the values of the field amplitudes are normalized such that $\mathcal{I} = 1$, the coherency matrix and the components of the Stokes vector become

$$\boldsymbol{\rho} = \frac{1}{2} \begin{bmatrix} S_0 + S_1 & S_2 + \mathrm{i}S_3 \\ S_2 - \mathrm{i}S_3 & S_0 - S_1 \end{bmatrix}, \qquad \begin{array}{l} S_0 = \rho_{11} + \rho_{22}, \ S_1 = \rho_{11} - \rho_{22}, \\ S_2 = \rho_{12} + \rho_{21}, \ S_3 = \mathrm{i}\rho_{21} - \mathrm{i}\rho_{12}. \end{array} \tag{B.3-59}$$

The determinant and trace of such a matrix, applicable to any 2×2 density matrix, has the following properties:

$$\begin{aligned}
\text{Det } \boldsymbol{\rho} &= \rho_{11}\rho_{22} - \rho_{12}\rho_{21} = \tfrac{1}{2}[S_0^2 - S_1^2 - S_2^2 - S_3^2], \\
\text{Tr } \boldsymbol{\rho} &= \rho_{11} + \rho_{22} = S_0, \\
\text{Tr } \boldsymbol{\rho}^2 &= \rho_{11}^2 + \rho_{22}^2 + 2\rho_{12}\rho_{21} = \tfrac{1}{2}[S_0^2 + S_1^2 + S_2^2 + S_3^2], \\
&= [\text{Tr } \boldsymbol{\rho}]^2 - 2\text{Det } \boldsymbol{\rho}.
\end{aligned} \tag{B.3-60}$$

The eigenvalues of the coherency matrix are readily expressible as

$$\lambda_\pm = \tfrac{1}{2}\left[S_0 \pm \sqrt{S_1^2 + S_2^2 + S_3^2}\right]. \tag{B.3-61}$$

They have the properties

$$\lambda_+ - \lambda_- = \sqrt{S_1^2 + S_2^2 + S_3^2}, \qquad 4\lambda_+\lambda_- = S_0^2 - (S_1^2 + S_2^2 + S_3^2). \tag{B.3-62}$$

Polarization and concurrence. The *degree of polarization* $\mathcal{P}$ at a given position is defined [Bor99] as the ratio of the polarized portion of the intensity to the total intensity. Expressed in terms of Stokes-vector components this means [Qia16]

$$\mathcal{P} = \frac{\sqrt{S_1^2 + S_2^2 + S_3^2}}{S_0} = \frac{(\lambda_+ - \lambda_-)}{(\lambda_+ + \lambda_-)} = \sqrt{1 - \frac{4\text{Det}\boldsymbol{\rho}}{[\text{Tr }\boldsymbol{\rho}]^2}}. \tag{B.3-63}$$

The value $\mathcal{P} = 0$ occurs with completely unpolarized light; the value $\mathcal{P} = 1$ occurs with completely polarized light.

If the quantity $\langle\mathcal{E}_i\mathcal{E}_j^*\rangle$ that defines the elements of $\boldsymbol{\rho}$ is separable, so that it can be written as a simple product of two complex numbers,

$$\langle\mathcal{E}_i\mathcal{E}_j^*\rangle = A_iB_j^*, \tag{B.3-64}$$

then we have the relationship

$$\rho_{ii}\rho_{jj} = A_iA_i^*B_jB_j^* = \rho_{ij}\rho_{ji}, \quad \text{and so} \quad \text{Det } \boldsymbol{\rho} = 0. \tag{B.3-65}$$

The value of the determinant therefore expresses the degree to which the elements of $\boldsymbol{\rho}$ are separable. A second field characteristic, the *concurrence* $\mathcal{C}$, definable variously as [Ebe16; Qia16; Ebe17]

$$\mathcal{C} = \frac{\sqrt{4\text{Det }\boldsymbol{\rho}}}{\text{Tr }\boldsymbol{\rho}} = \sqrt{2}\sqrt{1 - \frac{\text{Tr }[\boldsymbol{\rho}^2]}{[\text{Tr }\boldsymbol{\rho}]^2}} = \frac{2\sqrt{\lambda_+\lambda_-}}{(\lambda_+ + \lambda_-)}, \tag{B.3-66}$$

provides a measure of the separability of the elements of $\boldsymbol{\rho}$. The two numbers $\mathcal{P}$ and $\mathcal{C}$ embody complementary properties of polarization and separability, a relationship quantified at a given position by the expression [Qua17; Ebe17]

$$\mathcal{P}^2 + \mathcal{C}^2 = 1. \tag{B.3-67}$$

Appendix C.8 has more to say about generalizations of these results.

B.3.9   Photon angular momentum

Just as structured point-particles can have angular momentum, so too can fields. For vector fields, as with particles, the angular momentum can be of two sorts, orbital and spin. These combine to give a total angular momentum. For vector fields the spin angular momentum is associated with the three components of the vector, while the orbital angular momentum originates in the field amplitude. Simply stated [Yao11]: If every field vector rotates as it travels, the light has spin; if the phase structure rotates, the light has orbital angular momentum (OAM). This can be many times greater than the spin.

Photons are said to be spin-one particles because as traveling waves they can be expressed with unit angular-momentum helicity vectors. Descriptions of angular momentum properties of radiation fields are facilitated by introducing plane-wave circularly polarized field modes for photons. These have the number operators $\hat{N}_L(\mathbf{k})$ and $\hat{N}_R(\mathbf{k})$. The several quantities introduced earlier have the following quantum counterparts.

Optical helicity. The quantized form for integrated optical helicity is the difference of number operators for right and left circular polarization [Cam12]

$$\int dV \, \hat{\mathfrak{h}} = \sum_{\mathbf{k}} \hbar \left[\hat{N}_L(\mathbf{k}) - \hat{N}_R(\mathbf{k}) \right]. \tag{B.3-68}$$

Spin. The operator for field spin is expressible in terms of photon-number operators as [Cam12; And12; Bar16]

$$\int dV \, \hat{\mathfrak{s}} = \sum_{\mathbf{k}} \frac{\hbar \mathbf{k}}{|\mathbf{k}|} \left[\hat{N}_L(\mathbf{k}) - \hat{N}_R(\mathbf{k}) \right]. \tag{B.3-69}$$

This represents a sum over all modes of the number of photons in each mode multiplied by their helicity and the unit vector $\mathbf{k}/|\mathbf{k}|$ in the direction of propagation.

Chirality.   The integrated chiral density is proportional to the difference between left and right polarized photon numbers [Col12]

$$\int dV \, \hat{\chi} = \sum_{\mathbf{k}} \hbar c k^2 [\hat{N}_L(\mathbf{k}) - \hat{N}_R(\mathbf{k})]. \tag{B.3-70}$$

In all of these expressions the photons need not have circular polarization. Their polarization can be taken as any pair of points on opposite sides of the Poincaré sphere; see eqn (B.1-47). For example [Col12]:

$$\int dV \, \hat{\chi} = \sin(2\theta) \sin \varphi \sum_{\mathbf{k}} \hbar c k^2 \, [\hat{N}_1(\mathbf{k}) - \hat{N}_2(\mathbf{k})]. \tag{B.3-71}$$

Interpretation.   The difference between the helicity and the spin of a beam is well illustrated by a figure in [Bar12; Bar16] showing the effect of mirror reflection upon a monochromatic right-circularly polarized beam directed along the z axis. The beam is characterized by its momentum $k\mathbf{e}_z$, helicity parameter σ, helicity $\mathfrak{H} = \hbar\sigma$, and spin $\mathbf{S} = \hbar\sigma\mathbf{k}/k$. Initially, as the beam travels to the right, it has momentum helicity parameter $\sigma = 1$. Reflection reverses the momentum but retains the circular motion of the electric vector, so it reverses the handedness of the polarization. The result is the set of parameters

$$\text{Initial:} \quad \mathbf{k}/k = +\mathbf{e}_z, \qquad \mathfrak{H} = +\hbar, \qquad \mathbf{S} = +\hbar\mathbf{e}_z, \qquad \text{(B.3-72a)}$$
$$\text{Final:} \quad \mathbf{k}/k = -\mathbf{e}_z, \qquad \mathfrak{H} = -\hbar, \qquad \mathbf{S} = +\hbar\mathbf{e}_z. \qquad \text{(B.3-72b)}$$

B.4   Photon number-state superpositions

Photon-number states were, until laser radiation became plentiful, the standard basis for describing radiation states, and it still is widely used. A photon-number state, with its total neglect of field phase, is a very non-classical state, but superpositions of number states can provide a wide variety of useful field states whose characteristics vary widely [Law96]. The most widely used of these superpositions, a representation of classical fields, are the coherent states of Glauber [Gla63; Gla63b; Per72; Man86; Gaz09; San12; Sal19].

B.4.1   Coherent states

Harmonic-oscillator eigenstates of the number operator $\hat{N}_\nu = \hat{a}_\nu^\dagger\hat{a}_\nu$ begin with the vacuum state. For mode ν this is the state that is annihilated by the number-lowering operator $\hat{a}_\nu$:

$$\hat{a}_\nu|0\rangle_\nu = 0. \qquad \text{(B.4-1)}$$

The excited states of the oscillator can be constructed from this state by repeated action of the number-raising operator $\hat{a}_\nu^\dagger$,

$$|n\rangle_\nu = \frac{(\hat{a}_\nu^\dagger)^n}{\sqrt{n!}}|0\rangle_\nu. \qquad \text{(B.4-2)}$$

Glauber defined *coherent states* $|\alpha_\nu\rangle$ of the field mode ν as eigenstates of the annihilation operator $\hat{a}_\nu$,

$$\hat{a}_\nu|\alpha_\nu\rangle = \alpha_\nu|\alpha_\nu\rangle, \qquad \text{(B.4-3)}$$

where α_ν is a complex-valued eigenvalue associated with mode ν. These states can be expressed, for each mode, in terms of photon-number states as (for typographic simplicity I here omit the mode-specifying subscript)

$$|\alpha\rangle = \exp\left(-\tfrac{1}{2}|\alpha|^2\right) \sum_{n=0}^\infty \frac{\alpha^n}{\sqrt{n!}}|n\rangle. \qquad \text{(B.4-4)}$$

The squared magnitude of α fixes the mean photon number in the mode,

$$\bar{n} = |\alpha|^2, \qquad \text{(B.4-5)}$$

while the phase of α fixes a direction in the complex plane. A coherent state can be written as the action of a unitary displacement operator $D(\alpha)$ acting on the vacuum,

$$|\alpha\rangle = D(\alpha)|0\rangle, \qquad D(\alpha) = \exp[\alpha\hat{a}^\dagger - \alpha^*\hat{a}]. \tag{B.4-6}$$

Weak coherent-state beams. Beams of laser radiation are often idealized as coherent states. Very weak laser beams ($\bar{n} \ll 1$) are dominated by the vacuum—no photons—but the possibility of a single photon always accompanies a (much smaller) probability of two photons:

$$P_0(\alpha) = \exp\left(-\bar{n}\right), \qquad P_1(\alpha) = \bar{n}\, P_0(\alpha), \qquad P_2(\alpha) = \tfrac{1}{2}\bar{n}P_1(\alpha). \tag{B.4-7}$$

Although coherent states have a definition based on quantum theory, their coherences $g^{(n)}$ have the classical-field property of being coherent to all order. A coherent state is thus both quantum and classical. If $|\alpha|$ is very large, it appears classical. If $|\alpha|$ is small the quantum properties are discernible.

These constructs, and many others such as particular superpositions of Fock states, are rather theoretical. How are we to tell, by measurement, where a given field lies on the range of quantum-classical characteristics? The correlation functions offer one approach.

B.4.2   Phase states

A classical monochromatic wave (a possible mode for Dirac quantization as a harmonic oscillator) is characterized by amplitude and phase as well as frequency and direction. With quantization the squared amplitude is associated with mode energy and with the photon number operator, as expressed by paired photon annihilation and creation operators. The wave phase has posed a challenge for theorists [Bar89; Peg97]. If the number of photons in the field can be specified precisely, then there can be no information about the field phase —all phases are equally likely. Dirac and others sought a phase operator that would, in some sense, act as a canonical variable to the number operator, with a commutator relationship that would lead to an uncertainty relationship between phase and photon number. But mathematical implementations of his approach brought considerable criticism [Sus64; Car68; Bar86; Hra90; Lyn95] as well as useful proposals [Lou73].

Prompted by the availability of coherent radiation sources (masers and then lasers) various researchers employed a form of Fourier series, akin to the Fourier integral relationship between position and momentum wavefunctions, to define phase states as particular superpositions of number states. The structure proposed by David Pegg and Stephen Barnett is the limit for $s \to \infty$ of a phased superposition of photon number states [Bar86; Peg88; Bar89; Peg89; Sha91; Peg97]

$$|\theta\rangle^s = \frac{1}{\sqrt{s+1}} \sum_{n-0}^{s} \exp(in\theta)|n\rangle. \tag{B.4-8}$$

Phase states make possible the definition of phase operators and, in turn, phase-number uncertainty relationships [Sha91; Bia93]. They have been an important part of the mathematical tools of Quantum Optics.

B.4.3   Photon distributions

The expectation value of any function $f(\hat{N})$ of photon-number operator $\hat{N}$ obtains from the prescription

$$\langle f(\hat{N}) \rangle = \sum_{n=0}^{\infty} f(n)\, p_n, \qquad \sum_{n=0}^{\infty} p_n = 1, \tag{B.4-9}$$

where p_n is the probability for finding n photons, i.e. the photon distribution function. The mean photon number $\bar{n}$ and the variance of photon number, $(\Delta n)^2$, are defined as the first and second moments of the photon-number distribution,

$$\bar{n} = \langle \hat{N} \rangle, \qquad (\Delta n)^2 \equiv \langle \hat{N}^2 \rangle - \langle \hat{N} \rangle^2. \tag{B.4-10}$$

Example: Coherent state; Laser.   The Glauber definition eqn (B.4-3) of a single-mode coherent state means that for such a field state the probability p_n follows a *Poisson distribution* [Sal19]: The probability of finding n photons in a coherent state $|\alpha\rangle$ is

$$p_n(\alpha) = \frac{x^n}{n!}\, e^{-x} \quad \text{where} \quad x = |\alpha|^2. \qquad\qquad \text{coherent} \tag{B.4-11}$$

This Poisson distribution has the following photon-number moments

$$\langle \hat{N} \rangle \equiv \bar{n} = x, \quad \text{and} \quad \langle \hat{N}(\hat{N} - 1) \rangle = x^2. \tag{B.4-12}$$

Thus the mean photon number $\bar{n}$ is equal to $|\alpha|^2$. For a Poisson distribution the variance is equal to the mean. This implies that for a single-mode coherent state the variance of photon number, or mean-squared deviation, is equal to the mean photon number,

$$(\Delta n)^2 = \langle \hat{N} \rangle \quad \text{or} \quad (\Delta n)^2 = \bar{n}. \qquad\qquad \text{coherent} \tag{B.4-13}$$

Coherent states are the most classical of possible photon distributions. They are commonly used as idealizations of laser fields. Because their mean photon number is usually very large when describing laser radiation, the discreteness of their photons is then hardly perceptible: Industrial lasers are very classical devices.

Example: Fock state.   A state that has smaller variance than $\bar{n}$ is termed *sub-Poisson*. For example, a Fock state (or photon-number state with definite n) has zero variance,

$$(\Delta n)^2 = 0. \qquad\qquad \text{Fock} \tag{B.4-14}$$

Fock states are the most quantum-mechanical of photon distributions.

Example: Thermal radiation.   By comparison, a system of monochromatic photons in thermal equilibrium at temperature T is not in a quantum state. It has a Bose-Einstein (chaotic) distribution of photon numbers,

$$p_n(T) = \frac{1}{1 + \bar{n}} \left(\frac{\bar{n}}{1 + \bar{n}} \right)^n, \qquad\qquad \text{thermal} \tag{B.4-15}$$

where the mean photon number for frequency ω is (Here k_B is *Boltzmann's constant*, see eqn (2.6-2)):

$$\bar{n} = [\exp(\hbar\omega/k_B T) - 1]^{-1}. \qquad (B.4-16)$$

A thermal photon-number distribution has the variance

$$(\Delta n)^2 = \bar{n}^2 + \bar{n}. \qquad \text{thermal} \qquad (B.4-17)$$

The thermal (or chaotic) distribution maximizes the field entropy for fixed mean photon number.

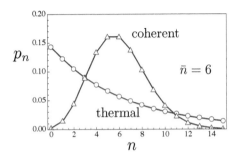

·**Fig. B.3** *Photon distribution functions*

Figure B.3 shows examples of photon-distribution functions p_n for a coherent state and thermal light, as defined in the following paragraphs, each having the mean photon number $\bar{n} = 6$. The distribution for a photon-number state, not shown, is a spike of unit height at $n = 6$.

Photon statistics; The Mandel Q. One of the many ways of characterizing a field is by its distribution of photon numbers. The coherent states are often taken as illustrative of the most classical of coherent stationary fields. Such a single-mode field has a Poisson distribution of photon numbers, eqn (B.4-11), and a variance equal to its mean. The *Mandel Q parameter* [Man79; Lou87; Zou90; Kol99],

$$\mathsf{Q} = \frac{\langle(\Delta\hat{n})^2\rangle - \langle\hat{n}\rangle}{\langle\hat{n}\rangle}, \qquad (B.4-18)$$

has provided a useful measure of the classical nature of a field because it quantifies the deviation of the photon variance from that of a Poisson distribution, the most classical of fields. Values of the Mandel Q parameter distinguish the following categories of photon statistics:

super-Poisson (classical)　if　$\mathsf{Q} > 1$,
Poisson (classical)　if　$\mathsf{Q} = 1$,
sub-Poisson (quantum)　if　$\mathsf{Q} < 1$.

A state with a definite photon number (a Fock state), the most quantum-mechanical of possible states, has a Mandel Q value of $\mathsf{Q} = -1$, the lowest possible value. There is no upper bound on Q.

B.4.4 Quadrature operators and squeezed states

Photon-number states were, until laser radiation became plentiful, the standard basis for describing radiation states, and it still is widely used. But it does not provide any information about field phases. Several alternatives, including coherent states, have been proposed for bases that allow phase description [Bar86; Peg88; Peg89; Hra90; Lyn95; Peg97]. To illustrate treatments of phase it is useful to consider a simple single-mode linearly-polarized monochromatic standing-wave field having spatial variation only along the z axis, and with linear polarization in the x direction. The E and B fields have the form

$$\mathbf{E}(\mathbf{r}, t) = \mathbf{e}_x \, \mathcal{E}_x(z, t), \qquad \mathbf{B}(\mathbf{r}, t) = \mathbf{e}_y \, \mathcal{B}_y(z, t). \qquad \text{(B.4-19)}$$

In the quantized version we deal with single-mode photon creation and annihilation operators

$$\hat{a}(t) = \hat{a} \, e^{-i\omega t}, \qquad \hat{a}^\dagger(t) = \hat{a}^\dagger \, e^{+i\omega t}, \qquad [\hat{a}, \, \hat{a}^\dagger] = 1, \qquad \text{(B.4-20)}$$

with which we write the electric- and magnetic-field-amplitude operators for frequency $\omega = ck$, using sines and cosines for the standing-wave pattern, as

$$\hat{\mathcal{E}}_x(z, t) = \mathcal{N} \sin(kz) \, \tfrac{1}{2}[\hat{a}(t) + \hat{a}^\dagger(t)], \qquad \text{(B.4-21a)}$$

$$c\hat{\mathcal{B}}_y(z, t) = -i\mathcal{N} \cos(kz) \, \tfrac{1}{2}[\hat{a}(t) - \hat{a}^\dagger(t)]. \qquad \text{(B.4-21b)}$$

The normalization constant $\mathcal{N}$, chosen such that the electromagnetic energy integrated over volume $\mathcal{V}$ is $\hbar\omega$, has the value [Lou83]

$$\mathcal{N} = 2\sqrt{\frac{\hbar\omega}{\epsilon_0 \mathcal{V}}}. \qquad \text{(B.4-22)}$$

When the field is in a photon-number state the mean values of these operators are zero,

$$\langle \hat{a} \rangle = \langle \hat{a}^\dagger \rangle = 0, \qquad \text{(B.4-23)}$$

and so the mean values of fields are also zero. However, there occur fluctuations about these values. When the field is in a coherent state the mean photon number is

$$\langle \hat{N} \rangle = |\alpha|^2, \qquad \text{(B.4-24)}$$

and the photon-number variance is

$$\langle (\Delta n)^2 \rangle = \langle \hat{N}^2 \rangle - \langle \hat{N} \rangle^2 = |\alpha|^2. \qquad \text{(B.4-25)}$$

Quadrature operators. A useful way to incorporate some information about phase is by means of time-independent *quadrature operators* [Wal83; Lou87; Kol99; Gar08],

$$\hat{X} = \frac{1}{2}[\hat{a} + \hat{a}^\dagger], \qquad \hat{Y} = \frac{1}{2i}[\hat{a} - \hat{a}^\dagger], \qquad [\hat{X}, \, \hat{Y}] = \frac{i}{2}. \qquad \text{(B.4-26)}$$

With these we rewrite the preceding expressions eqn (B.4-21) for field-amplitude operators as

$$\hat{\mathcal{E}}_x(z, t) = \mathcal{N} \sin(kz)[\hat{X} \cos\omega t + \hat{Y} \sin\omega t], \qquad \text{(B.4-27a)}$$

$$c\hat{\mathcal{B}}_y(z, t) = \mathcal{N} \cos(kz)[\hat{Y} \cos\omega t - \hat{X} \sin\omega t]. \qquad \text{(B.4-27b)}$$

The variances of these operators satisfy the uncertainty relation

$$\langle (\Delta X)^2 \rangle \, \langle (\Delta Y)^2 \rangle \geq \frac{1}{16}, \qquad \text{(B.4-28)}$$

and so these lead to constraints on the relative uncertainty in the two fields. When the field is in a coherent state the relation $\geq$ becomes an equality $=$.

The expressions eqn (B.4-27) are well suited to a field that is in a coherent state, an eigenstate of the annihilation operator with complex-valued eigenvalue α,

$$\hat{a}|\alpha\rangle = \alpha|\alpha\rangle. \tag{B.4-29}$$

In that situation the expectation values of the quadrature operators are the real and imaginary parts of the eigenvalue α,

$$\langle\hat{X}\rangle + i\langle\hat{Y}\rangle = \alpha, \tag{B.4-30}$$

a formula that makes explicit the phase relationship between the electric and magnetic fields. In this situation the two variances are equal,

$$\langle(\Delta X)^2\rangle = \langle(\Delta Y)^2\rangle = \frac{1}{4}. \tag{B.4-31}$$

Squeezing. This minimum uncertainty can be retained while altering the uncertainty in either one of the quadrature variables. Such a field is termed a *squeezed state*—a state in which one variance grows at the expense of the other [Wal83; Lou87; Wod87; Bar91; Dal99; Kol99; Dod02; Wal07; Gar08; Sch17b; Pat18; Sal19]. The resulting relationship is expressible with a real-valued positive or negative squeezing parameter s in the form

$$\langle(\Delta X)^2\rangle = \frac{1}{4}e^{-2s}, \qquad \langle(\Delta Y)^2\rangle = \frac{1}{4}e^{+2s}. \tag{B.4-32}$$

The mean photon number in a squeezed coherent state is [Wal83]

$$\langle\hat{N}\rangle = |\alpha|^2 + \sinh(s)^2. \tag{B.4-33}$$

Squeezed states have become important for devising measurements that avoid the usual uncertainty limit [Pac93; Gio04].

B.4.5   Characterizing fields by density matrix

Just as a general description of the status of an atom or molecule can rely on a density matrix, $\rho(t)$, when evaluating expectation values, so too can the description of a field. A variety of bases have found use for that purpose [Cah69b]. These can often make evident the quantum nature of the system. When used with radiation these exhibit photon properties.

Photon number representation. This is particularly straightforward when we use photon number states as a basis. For a single-mode idealization we then have a set of integer photon numbers (including none, meaning the vacuum) $n = 0, 1, 2, \cdots$, without upper bound. The expression of the density matrix in terms of photon annihilation and creation operators that define number states requires a specification of operator orderings, often done by means of an ordering parameter s [Cah69; Cah69b]:

$$\rho(t) = \sum_{nm} \rho_{nm}(s;t) \times \begin{cases} (\hat{a}^\dagger)^n\,\hat{a}^m, & s = 1, \text{normal ordering} \\ \hat{a}^m\,(\hat{a}^\dagger)^n, & s = -1, \text{antinormal ordering.} \end{cases} \tag{B.4-34}$$

The choice of symmetric ordering, $s = 0$, requires an average over all possible ways of ordering n factors of $\hat{a}^\dagger$ and m factors of $\hat{a}$.

The Husimi $Q(\alpha)$ function. By choosing photon numbers (or field energy) as our basic variable we eliminate the possibility of specifying field phase. A number of other options are available that bring the density matrix more in line with classical field descriptions, which have both amplitude and phase. One of the most intuitive is the Husimi Q function,[11] defined for a single-mode field as the diagonal elements of the density matrix (or a statevector projection for a pure state) in a basis of coherent states:

$$Q(\alpha;t) = \frac{1}{\pi}\langle\alpha|\rho(t)|\alpha\rangle, \quad \text{or} \quad Q(\alpha;t) = \frac{1}{\pi}|\langle\alpha|\Psi(t)\rangle|^2. \tag{B.4-35}$$

This function is non-negative for all α and so its display has offered an intuitive picture of photon distributions [Cah69b; Ste92; Sho93; Wal07; Gar08].

The field Wigner function. The Wigner function of Appendix A.6, measurable by quantum tomography [Ray97], provides a witness to the quantum nature of a given system. The basic single-mode n-photon state has a Wigner function $W_n(q,p)$ that is cylindrically symmetric in its phase space and is expressible in terms of a Laguerre polynomial of order n [Mac03]. For a mode having frequency ω the vacuum ($n = 0$) and the one-photon state have the Wigner distributions [Mac03]

$$W_0(q,p) = \frac{1}{\pi\hbar}\,e^{-\eta}, \qquad W_1(q,p) = \frac{-1}{\pi\hbar}(1 - 2\eta)\,e^{-\eta}, \tag{B.4-36}$$

where the variation is through the parameter

$$\eta = \frac{p^2 + \omega^2 q^2}{\hbar\omega}. \tag{B.4-37}$$

Although the Wigner function for the vacuum state is everywhere positive, those for other photon-number states are marked by circles of negative values, a signature of a quantum system.

B.5   Temporal variations; Quantum character

The Dirac approach to a description of radiation places wavelike characteristics into spatial modes and granular (particle-like) characteristics into temporal parts. It is with the latter, as harmonic oscillators, that quantization is introduced, through photon creation and annihilation operators. This next section discussed that time dependence, both for classical radiation and for quantum radiation.

B.5.1   Fourier components; Coherence and bandwidth

The electric fields that appear in the conventional discussion of Dirac photons, and in modeling idealized laser radiation, are monochromatic, comprising a single frequency and hence a signal of infinite duration. They are to be found with time-independent

---

[11] The use of the coherent-state elements of the density matrix in this way were discussed at length in [Cah69b]. It was only when Stig Stenholm, in 1992 [Ste92], drew attention to poorly accessible work of 1940 by Kodi Husimi that the quantum optics community added his name to the function. The idea of such weighting of expectation values traces back to Schrödinger's Gaussian wavepackets as minimum uncertainty states.

radiation Hamiltonians. Deviations from this idealized approximation occur for several reasons, in addition to the use of finite-duration pulses as descriptors of electric-field amplitudes.

The frequency content of any complex-valued time-varying signal $F(t)$ is defined by means of Fourier analysis, expressible as the complex integral[12]

$$F(t) = \int d\omega \, \exp(-i\omega t) \widetilde{F}(\omega), \tag{B.5-1}$$

where the Fourier components $\widetilde{F}(\omega)$ are obtained as the Fourier transform (FT) of the signal,

$$\widetilde{F}(\omega) = \frac{1}{2\pi} \int dt \, \exp(+i\omega t) F(t). \tag{B.5-2}$$

Such transform pairs, applied to the positive-frequency part $\mathcal{E}(t)$ of the electric-field envelope of a radiation pulse (after factoring carrier frequency), completely characterize its temporal properties. Commonly used models for pulses include:

Rectangular: Unit peak value of duration τ centered at $t = 0$:

$$\mathcal{E}(t) = \begin{cases} 1 & \text{if } |t| \leq \tau/2 \\ 0 & \text{if } |t| > \tau/2 \end{cases}, \qquad \widetilde{\mathcal{E}}(\omega) = \frac{\sin(\omega\tau/2)}{\pi\omega}. \tag{B.5-3a}$$

Gaussian: A Gaussian envelope of unit peak value centered at $t = 0$:

$$\mathcal{E}(t) = \exp[-(t/\tau)^2/2], \qquad \widetilde{\mathcal{E}}(\omega) = \frac{\tau}{\sqrt{2\pi}} \exp[-(\omega\tau)^2/2]. \tag{B.5-3b}$$

Because the Gaussian function $\mathcal{E}(t)$ extends indefinitely in time (it is a function of *infinite support*) it can only represent an idealization of pulses that have finite duration.

Exponential: A model of spontaneous emission, with inclusion of the Bohr frequency ω_B, exponential decay starting from $t = 0$,

$$\mathcal{E}(t) = \begin{cases} 0 & \text{if } t < 0 \\ \mathcal{E}_0 \exp[-i\omega_B t - \Gamma t/2] & \text{if } t \geq 0 \end{cases}, \qquad \widetilde{F}(\omega) = \frac{i\mathcal{E}_0}{2\pi(\omega - \omega_B - i\Gamma/2)}. \tag{B.5-3c}$$

The power spectrum of this spontaneous-emission radiation has a *Lorentz profile* of full-width at half-maximum (FWHM) Γ,

$$I(\omega) = \frac{I_{\max}}{(\omega - \omega_B)^2 + (\Gamma/2)^2}. \tag{B.5-4}$$

When we treat quantum dynamics by means of Heisenberg equations we picture spontaneous emission as the field emerging from a steadily diminishing (decaying) dipole moment [Scu72; Ray12]. This is a spherical wave whose amplitude undergoes exponential decay, modulated sinusoidally. If we regard this outgoing field as that

---

[12] Alternative definitions of Fourier transforms place factors of $1/\sqrt{2\pi}$ with each integral and may use $2\pi\nu$ rather than ω in exponentials.

of a single photon we require this photon to be a coherent blend of monochromatic "photons".

Characterizing pulses. Useful measures of pulse duration and frequency content of an electromagnetic field come from various moments of the radiation power, or absolute square of the pulse envelope, when they exist. From the electric-field moment we define a mean square pulse duration by the ratio

$$(\Delta t)^2 = \int dt\, t^2 |\mathcal{E}(t)|^2 \,/\, \int dt\, |\mathcal{E}(t)|^2. \tag{B.5-5}$$

From the absolute square of the Fourier transform of the electric-field envelope we similarly define the *bandwidth* $\Delta\omega$ from the ratio

$$(\Delta\omega)^2 = \int d\omega\, \omega^2 |\tilde{\mathcal{E}}(\omega)|^2 \,/\, \int d\omega\, |\tilde{\mathcal{E}}(\omega)|^2. \tag{B.5-6}$$

In all cases as any measure of the pulse duration increases the distribution of frequencies becomes proportionally more narrow; the width $\Delta\omega$ of the distribution in frequencies (the *bandwidth*) is inversely proportional to the root-mean-square (RMS) duration of the signal, Δt. For the Gaussian pulses of eqn. (B.5-3b) the two parameters are

$$\Delta t = \frac{\tau}{\sqrt{2}}, \qquad \Delta\omega = \frac{1}{\tau\sqrt{2}} \approx \frac{0.707}{\tau}. \tag{B.5-7}$$

For any pulse shape the time-bandwidth product of these parameters satisfies the inequality

$$\Delta\omega\,\Delta t \geq \tfrac{1}{2}. \tag{B.5-8}$$

This product has its smallest value, $1/2$, for Gaussian pulses.

Bandwidths of pulses are typically used as important parameters for judging effectiveness of incoherent radiation for producing excitation. However, when the system of interest is intended to undergo coherent excitation, the dynamics depends upon details of the field-amplitude variation. Parameters such as Rabi angles, pulse duration, and peak intensity (or Rabi frequency) become the relevant pulse characteristics. It is then misleading to base expectations of excitation solely on pulse bandwidth. Indeed, it is possible to achieve complete excitation with a pulse that has no resonant Fourier component [Vit05].

Photon energy-time uncertainty. The inequality eqn (B.5-8) imposes constraints on how long a time interval Δt must be in order to define a frequency within an uncertainty $\Delta\omega$. When applied to the energy-frequency relationship $E = \hbar\omega$ associated with photons it gives the energy-time uncertainty relationship

$$\Delta E\,\Delta t \geq \tfrac{1}{2}\hbar. \tag{B.5-9}$$

Although this time-frequency uncertainty relationship appears analogous to the position-momentum uncertainty relationship found with quantum particles (see Appendix A.4.2) it differs in a significant way: Although there exist operators whose expectation value is an energy, there is no operator for time—it is just a parameter.

Coherence time. The simplest sort of single-photon field originates with spontaneous emission from individual atoms, possibly supplemented with additional stimulated emission, and so there is an inevitable temporal variation in phase, a departure from a pure sinusoid as each atom makes its contribution. The direction of the polarization axis will vary randomly. For each of these changes there is a characteristic during which phase and direction changes are negligible and the time variation at a fixed location remains expressible as $\exp[-i\omega t + i\varphi]$, but after which there is increasingly less overlap with past values. The inverse of the bandwidth provides one estimate of the coherence time,

$$\tau_{coh} = 1/\Delta\omega. \tag{B.5-10}$$

For times well after the coherence time the polarization state can only be parametrized by statistical measures, such as mean values for two independent components (see eqn (B.1-48) on page 305); the radiation is partially polarized or, when randomization is complete, unpolarized. Even from well-stabilized lasers there is some uncontrollable drift of the phase, although suitable filters can restrict the polarization. It is such fields, of unchanging phase and well-defined polarization, that the theory of coherent excitation deals.

A *bandwidth-limited pulse* (also known as a transform-limited pulse) is one that has the minimum possible bandwidth for a given duration (or maximum duration for a given spectral bandwidth). Such a pulse can have no random fluctuations in phase or amplitude (they would add to the bandwidth); the coherence time cannot be shorter than the pulse duration.

B.5.2 Pulse shaping

As bandwidth-limited pulses become shorter their range of constituent frequencies becomes larger, whatever may be the particular temporal shape of the pulse. The bandwidth of frequencies that occur for a femtosecond pulse (1 fs $= 10^{-12}$ sec) whose carrier has a wavelength of 800 nm spreads over much of the optical spectrum. By passing a pulsed beam through a spectral disperser such as a grating or prism the frequency components rather than the time components become accessible for control. With the use of electronically adjustable phase elements spread over the spectrum it is possible to alter the frequency and polarization structure of the pulse, while still maintaining its coherence [Bri01; Pra03; Wol05; Wol06; Wei11].

In particular, it is possible to extract a pair of frequencies that are fully correlated [Ker17]. They are two components of what began as a single pulse, perhaps a single photon, so they retain that single-photon quantum character. The two electric vectors add to give a single vector that may, with time, follow patterns far more elaborate than the simple elliptical path that occurs with a single frequency.

The shaping of single pulses described here contrasts with the composite-pulse technique discussed on page 249.

B.5.3 Field correlations

Interference is a defining characteristic of waves, made evident by superposing waves from different sources as is done in interferometry. In 1963 Glauber introduced a formalism that has become the basic approach to treating measurements of interference phenomena in any combination of space and time [Gla63]. In applying the theory to radiation he first defined positive- and negative-frequency parts of the field envelope so as to express the real-valued scalar field $\mathcal{E}(\mathbf{r}, t)$, one of the three vector components of the field $\mathbf{E}(\mathbf{r}, t)$, as

$$\mathcal{E}(\mathbf{r}, t) = \mathcal{E}^{(+)}(\mathbf{r}, t) + \mathcal{E}^{(-)}(\mathbf{r}, t), \tag{B.5-11}$$

where the parts are defined from the Fourier decomposition as

$$\mathcal{E}^{(+)}(\mathbf{r}, t) = \int_0^\infty d\omega \, \mathcal{E}(\mathbf{r}, \omega) \, e^{-i\omega t}, \qquad \mathcal{E}^{(-)}(\mathbf{r}, t) = \int_{-\infty}^0 d\omega \, \mathcal{E}(\mathbf{r}, \omega) \, e^{-i\omega t}. \tag{B.5-12}$$

B.5.4 Correlation functions

Using this field decomposition Glauber defined space-time correlation functions of order n with $2n$ space-time points x_i as averaged expectation values of pairs of fields, with all positive-frequency parts $\mathcal{E}^{(+)}$ placed to the right (and hence acting first). For scalar fields, or components of vector fields, the definition reads [Gla63; Gla63b]

$$G^{(n)}(x_1, \ldots, x_n, x_{n+1} \ldots x_{2n})$$
$$= \langle \mathcal{E}^{(-)}(x_1) \ldots \mathcal{E}^{(-)}(x_n) \mathcal{E}^{(+)}(x_{n+1}) \ldots \mathcal{E}^{(+)}(x_{2n}) \rangle_{av}. \tag{B.5-13}$$

More generally, vector fields are treated by including a subscript, say $\mathcal{E}_\mu(\mathbf{r}, t)$, to denote one of three orthogonal vector components, and these are included as subscripts on $G^{(n)}$. For applications we might consider either fields that to be found at a given time spread over a plane, after having traveled various routes from a point source, or time varying fields that fall on a point-source detector. The correlation function exhibits a pattern of nodes and antinodes (wave interference) as the space or time variable changes.

The simplest of this infinite succession of functions, the first-order correlation function evaluated at a single space-time point, $G^{(1)}(x_1, x_1)$, is proportional to the intensity of the radiation.

Density matrix description. An alternative to the expectation value that appears above in the definition of the correlation functions is the trace with a density matrix ρ (see Appendix A.14):

$$G^{(n)}(x_1, \ldots x_{2n}) = \mathrm{Tr} \left[\rho \hat{\mathcal{E}}^{(-)}(x_1) \ldots \hat{\mathcal{E}}^{(+)}(x_{2n}) \right]. \tag{B.5-14}$$

The density matrix incorporates a full statistical description of the field, including the fluctuations that are quantified by correlation functions.

Coherent radiation. A field that is coherent to order n has, by definition, all correlation functions $G^{(j)}$ with j less than or equal to n expressible as factoring into products of first-order functions $G^{(1)}$:

$$G^{(j)}(x_1, \ldots, x_j, x_j \ldots x_1) = G^{(1)}(x_1, x_1) \ldots G^{(1)}(x_j, x_j). \tag{B.5-15}$$

Glauber pointed out that any classical field of nonrandom behavior has correlation functions that have this form. For example, a simple sinusoid (of infinite duration) is coherent to all orders.

Time correlation. The construction and evaluation of these correlation functions has application to the spatial distribution of field nodes, as occurs in observation of radiation patterns from slits, but for treatment of quantum fields we deal more often with a field observed by a single point-like detector at various times. When dealing only with time dependence it is common to write the second-order correlation function for a single amplitude as [Wal07]

$$G^{(2)}(\tau) = \langle \mathcal{E}^{(-)}(t)\mathcal{E}^{(-)}(t+\tau)\mathcal{E}^{(+)}(t+\tau)\mathcal{E}^{(+)}(t)\rangle, \qquad \text{(B.5-16)}$$

where τ is the interval between compared moments, and to introduce the associated first-order single-time correlation function

$$G^{(1)}(t) = \langle \mathcal{E}^{(-)}(t)\mathcal{E}^{(+)}(t)\rangle. \qquad \text{(B.5-17)}$$

This function is proportional to the instantaneous intensity. For steady illumination it is independent of time.

Normalized correlation; Autocorrelation. To compare different field models it is customary to introduce normalized coherence functions [Gla63],

$$g^{(n)}(x_1, \ldots x_{2n}) = \frac{G^{(n)}(x_1, \ldots x_{2n})}{\sqrt{G^{(1)}(x_1, x_1) \cdots G^{(1)}(x_{2n}, x_{2n})}}. \qquad \text{(B.5-18)}$$

The general definition of correlation functions allows multiple positions as well as multiple times. The following discussions apply to *autocorrelation* in which a field is detected at a fixed point for several times. The commonly used first-order and second-order normalized correlation functions for stationary fields are

$$g^{(1)}(t) = \frac{G^{(1)}(t)}{|G^{(1)}(0)|}, \qquad g^{(2)}(\tau) = \frac{G^{(2)}(\tau)}{|G^{(1)}(0)|^2}. \qquad \text{(B.5-19)}$$

If there is second-order coherence for the field then $g^{(2)}(\tau) = 1$.

Quantum correlation functions. In the quantum theory of radiation the field amplitudes become operators involving photon creation and annihilation. The *positive-frequency part*, involving time dependences $e^{-i\omega t}$ for $\omega > 0$ and denoted $\mathcal{E}^{(+)}$, becomes proportional to photon annihilation operators that diminish the field by one photon, while the negative-frequency amplitudes $\mathcal{E}^{(-)}$ become proportional to creation operators. The creation operators commute amongst themselves, as do the annihilation operators, but the creation and annihilation operators for a single mode do not commute. For a single mode this means the replacements

$$\mathcal{E}^{(+)}(t) \to \mathcal{E}_0\, \hat{a}(t), \qquad \mathcal{E}^{(-)}(t) \to \mathcal{E}_0^*\, \hat{a}^\dagger(t), \quad \text{with} \quad [\hat{a}(t), \hat{a}^\dagger(t)] = 1, \qquad \text{(B.5-20)}$$

in the definitions of correlation functions. This leads to a definition for $G^{(n)}$ as the normally ordered expectation value of a succession of n annihilation operators to the right of (and acting before) a balancing set of n annihilation operators for the various photon modes.

For example, the second-order correlation function of a single field is the intensity-intensity correlation function

$$g^{(2)}(\tau) = \frac{\langle \hat{a}^\dagger(t)\hat{a}^\dagger(t+\tau)\hat{a}(t)\hat{a}(t+\tau)\rangle}{\langle \hat{a}^\dagger(t)\hat{a}(t)\rangle\langle \hat{a}^\dagger(t+\tau)\hat{a}(t+\tau)\rangle} = \frac{\langle \hat{N}(t)\hat{N}(t+\tau)\rangle}{\langle \hat{N}\rangle^2}, \tag{B.5-21}$$

involving the photon number operator

$$\hat{N}(t) \equiv \hat{a}^\dagger(t)\hat{a}(t). \tag{B.5-22}$$

In the quantum theory version of correlation functions, where photon operators appear in place of classical amplitudes, a measurement of photon number reduces the photon number n by 1 so that a second measurement finds only the number $n-1$. This measurement-induced field depletion is at the heart of the difference between classical and quantum light, most noticeable for very small photon numbers.

Whether the field is regarded as classical or as originating with photons, the correlation function can exhibit a distribution of nodes and antinodes attributable to interference. Glauber has stressed the proper interpretation of these interference patterns [Gla63; Gla95]:

> It is not photons that interfere, it is probability amplitudes for indistinguishable histories that interfere.

Several simple models of quantum fields have proven useful, notably coherent states (a model of idealized laser light), number states, and thermal radiation. The following paragraphs discuss these.

B.5.5 Characterizing fields by their coherence

The coherent states, being coherent to all orders, are the quantum states whose coherence properties are most classical—most like a pure sine wave. These states contrast with the most quantum-mechanical states, the photon-number states. These two classes of states are distinguishable by their values of $g^{(2)}(0)$:

$$g^{(2)}(0) = \begin{cases} 1 & \text{coherent state} \\ 1 - \frac{1}{n} & \text{number state of } n \text{ photons.} \end{cases} \tag{B.5-23}$$

The connection between the value of $g^{(2)}(0)$ and the Mandel Q parameter is through the formula [Lou87]:

$$Q = \langle \hat{n}\rangle \left[g^{(2)}(0) - 1\right]. \tag{B.5-24}$$

Photons as events; Bunching and antibunching. The detection of a photon, idealized as instantaneous, can be regarded as an event taken from an ongoing stochastic process [Gla06]. The possible sequence of events in a single realization of a stochastic process is characterized, in part, by the mean rate of events and by the correlation function $g^{(2)}(\tau)$. Various photon sources can be distinguished by their event histories: Whether the events tend to cluster (*bunching*), with possible coincidences, or to avoid clustering (*antibunching*). The quantification of this effect is by considering the slope of the curve $g^{(2)}(\tau)$ near $\tau = 0$,

$$
\begin{aligned}
\text{bunching (classical)} \qquad & g^{(2)}(0) > g^{(2)}(\tau), \\
\text{random (classical)} \qquad & g^{(2)}(0) = 1, \\
\text{antibunching (quantum)} \qquad & g^{(2)}(0) < g^{(2)}(\tau),
\end{aligned}
$$

Thermal light sources (stars, for example) have events that tend to cluster, with coincidences quite likely. Photon antibunching is a sign of nonclassical light; it cannot be predicted by a classical description.[13] As has been pointed out [Zou90; Kol99; Mil19], the notion of antibunching is distinct from characterization of light as sub-Poisson; the two effects are independent and need not be found together.

Interpretation. The *bunching* of arrivals is what we expect when radiation is arriving from many independent sources: A peak value of the field is likely to persist. The *antibunching* is typical of what we expect of photon arrivals from spontaneous emission of an excited atomic state that must be re-excited after every emission. The replenishing of excitation requires a time interval that depends on the mechanism of excitation.

Although correlation measurements may be conducted upon a field that is well separated from its source, the information present in the correlation function originates with the source. It is therefore the field source that is responsible for the observable quantum properties of the field embodied in correlation functions.

B.6 Alternative views of photons

The mathematics presented in the preceding sections above summarizes what remains today as the accepted quantum theory of radiation as a unification of Maxwell's equations for electromagnetic fields with the quantum theory of matter that is provided by the Schrödinger equation. It provides a straightforward definition of what a photon is: It is an increment of a mode of the electromagnetic field. But the application of this simple definition to experiments has many practical difficulties. And this approach to a quantum theory of radiation is by no means the only route that has been followed over the last century.

My presentation, in this Memoir, of quantization by way of solutions to the vector Helmholtz equation, allows the possibility of many types of field modes, some of which were discussed in Appendix B.2 above, and thereby many types of photons. In this view, every particular optical system, with its particular Helmholtz equations and boundary conditions, can be equally well associated with a quantum field whose increments are photons.

Appendix B.3.4 carries this diversity still further, by introducing the possibility of photons based upon arbitrary sets of basis functions, each choice leading to a distinguishably different photon.

Not all of my colleagues accept such a broad definition of photons. There are those who, like Einstein, consider the only acceptable definition of a photon to be one that relies on eigenfunctions of linear momentum—traveling plane waves in free space. With such a view any of the other field modes I have discussed, and the discrete field increments created by a single emission or absorption event, or stored within

---

[13]Because it violates the Cauchy-Schwarz inequality eqn (A.3-8) [Kol99; Hen05].

an enclosure, must be regarded as involving something else: A normalized coherent superposition of the free-field basis photons.

B.7 Thermal equilibrium; Planck photons

The origins of quantum theory, and of photons, are often traced to the work of Max Planck that gave a quantitative description of the spectrum of light expected from a thermal source. It is therefore fitting to include a discussion of the radiation increments from these sources that I have referred to as the *Planck photon*.

The notion of thermodynamic equilibrium is an idealization of conditions that hold when there are no flows of energy or material and all portions of connected volumes can be assigned a common temperature as a measure of energy content. Its properties are the subject for courses on chemical thermodynamics as well as statistical physics, for from its general principles follow a great many practical results that deal with the management of heat and chemical reactions.

Equipartition. When a system is in thermodynamic equilibrium at temperature T its properties can only be described statistically: It is an incoherent mixture of possible quantum states. Each degree of freedom that is associated with a momentum that contributes quadratically to energy (velocity components for example) has, on average, the kinetic energy $\frac{1}{2}k_BT$, where k_B is Boltzmann's constant, see eqn (2.6-2). Similarly, a coordinate that contributes quadratically to energy (as does one subject to the Hooke's-law force of an harmonic oscillator) has average energy $\frac{1}{2}k_BT$:

$$\bar{E}^{kin} = \tfrac{1}{2}k_BT, \qquad \bar{E}^{pot} = \tfrac{1}{2}k_BT. \tag{B.7-1}$$

These formulas, the content of the *equipartition theorem* of classical statistical mechanics, endow each oscillator degree of freedom an average energy $\frac{1}{2}k_BT$. However, they fail at low temperature, when the thermal energy k_BT is significantly smaller than the spacing between discrete energy values of a quantum system and is therefore insufficient to produce energy transfer into that degree of freedom. It is then a poor approximation to assume that the energy values form a smooth continuum, and the discreteness of quantum theory is indispensable. (A degree of freedom is said to be "frozen out" when k_BT is much smaller than its energy.)

B.7.1 The Boltzmann formula

In thermodynamic equilibrium at temperature T the ratio of the number-densities $\mathcal{N}_n$ of atoms (or other systems) with discrete energies E_n is given by the *Boltzmann formula*,

$$\frac{\mathcal{N}_2}{\mathcal{N}_1} = \frac{\varpi_2}{\varpi_1}\exp[-(E_2 - E_1)/k_BT], \tag{B.7-2}$$

where ϖ_2 is the statistical weight of energy level n (the number of quantum states sharing the energy E_n). The exponential is known as the *Boltzmann factor*, responsible for placing more atoms in excited states as temperature increases. It does not matter what the system is: Only the temperature T, the energy difference, and the statistical weights enter into the calculation of number densities.

Negative temperature. The Boltzmann formula applies to the idealized situation of steady, local thermodynamic equilibrium. It predicts that, apart from statistical weights, populations decrease monotonically with increasing energy: Higher-energy states have lower population than lower-energy states. When beams of radiation are present the populations often depart significantly from the Boltzmann formula, and population inversions may occur. To ascribe a temperature to such a nonequilibrium population ratio of two states requires an empirically defined *negative temperature*.

B.7.2   The Planck formula

Much has been written about the discovery by Planck of the radiation law that bears his name—a formula for the distribution of wavelengths in radiation emitted by or absorbed by a (black) surface that absorbs all colors equally well: The so-called black-body radiation. In his search for such a formula Planck first introduced a mathematical expression designed to interpolate between formulas that were successful at long wavelengths and short wavelengths. To explain this purely mathematical invention he sought justification from thermodynamics and the statistical mechanics of molecular gases as formulated by Ludwig Boltzmann, applying those results to harmonic oscillators regarded as sources of radiation. Although the use of discrete radiation-energy increments in that work is often credited with taking the first steps toward the concept of a photon, and to the modern quantum theory in which discrete energies occur routinely, there remained critical shortcomings. Concerns over just what Planck had in mind continue to motivate learned discussions by historians of physics [Kra00; Gea02; Gar08; Nau16; Bor17; Sal19]. I will here only attempt a very abbreviated *précis* of that literature.

The formula proposed by Planck for the distribution of thermal-radiation energy density $u(\omega, T)$ at temperature T within an increment $d\omega$ around the continuously distributed angular frequency ω can be written [Mor64; Mil19]

$$u(\omega, T) = \frac{\hbar\omega^3}{\pi^2 c^3} \frac{1}{\exp[\hbar\omega/k_B T] - 1}. \tag{B.7-3}$$

Other commonly found expressions for this Planck formula deal with energy density per wavelength interval or with spectral radiance (power rather than energy).

Density of radiation modes. Planck proposed a foundation for his formula eqn (B.7-3) by using a mathematical description of radiation as a classical gas of massless, non-interacting uncharged particles in thermodynamic equilibrium. To apply the distribution to standing waves in a conducting-wall box of volume $\mathcal{V}$ requires a count of the number of modes of angular frequency ω as $\mathcal{V}$ becomes indefinitely large [Mor64],

$$\varpi(\omega) = \omega^2 \frac{\mathcal{V}}{\pi^2 c^3}. \tag{B.7-4}$$

In that approach it was necessary to enumerate the various distinguishable sorts of radiation of a given frequency that could be found in an enclosed volume—the different directions of propagation and the possible polarizations. This traditional free-space mode density requires alteration—the Purcell effect of Appendix A.16.2—when the radiation is actually enclosed.

The relevant Maxwell-Boltzmann distribution, used by Planck, is the low-density (weak field) limit common to the two possible quantum-mechanical distributions: The Bose-Einstein distribution applicable to bosons (particles with integer spin, such as photons) and the Fermi-Dirac distribution applicable to fermions (particles with half-integer spin, such as electrons). Whereas multiple bosons can occupy the same quantum state, multiple fermions cannot—this is the essence of the Pauli exclusion principle.

The Planck formula eqn (B.7-3) satisfactorily fitted the observations, and became the accepted description of thermal radiation. Although it relies on discrete numbers (quanta) of energy increments, and therefore is often taken as evidence for photons, Planck himself regarded his hypothesis of resonant oscillators as a mathematical device for allowing a derivation of his formula for black-body radiation; he did not attribute any definite physical significance to the oscillators [Kra00; Gea02; Gar08; Nau16; Bor17; Sal19].

Peak of the blackbody distribution. The Planck formula, eqn (B.7-3), for the distribution of frequencies in blackbody radiation has a single maximum, at a frequency ω_{max} for which the photon energy is

$$\hbar\omega_{max} = h\nu_{max} = x\,k_B T, \qquad x \approx 2.82, \tag{B.7-5}$$

in which the constant x is obtained by setting the frequency derivative of $u(\omega)$ to zero and solving numerically the resulting equation

$$(x - 3)\,e^{-x} + 3 = 0. \tag{B.7-6}$$

For practical use we have the formulas

$$\nu_{max} = 58.8\ T\ \text{GHz / K} \qquad h\nu_{max} = 243\ T\ \mu\text{eV / K} \tag{B.7-7}$$

for associating a dominant frequency with a temperature.

The linear relationship between dominant frequency and temperature is a version of the Wien displacement law between wavelength and temperature, applicable to the blackbody distribution expressed in wavelengths. It quantifies the shifting color observed as metal is heated and enables astronomers to evaluate the surface temperatures of stars from measurements of their spectra. The visible solar spectrum is well fitted by a blackbody having temperature $T_{sun} \approx 5.8 \times 10^3$ K, overlain by hundreds of dark absorption lines, the *Fraunhofer lines* associated with elements in the solar atmosphere.

B.8   Incoherent radiation; Photon crowds

Prior to the present popularity of laser light sources and their initiations of coherent excitation, astrophysicists dealt extensively with the flow of incoherent radiation through bulk matter [Ath72; Tuc75; Ryb85]. Spectroscopists, in turn, measured the relevant atomic and molecular characteristics associated with rate-equation excitation. The extensive literature generated by that work, often employing the word "photon" as a handy way of describing the radiation but implying no detailed quantum properties, still has relevance to a variety of situations. The following discussions summarize that regime of matter-radiation interaction, of use in semiclassical treatments of atomic excitation.

B.8.1 Radiation attenuation

Treatment of monochromatic, <u>incoherent</u> radiation beams, rather than broadband enclosed thermal radiation, traditionally begins with the "spectral irradiance", commonly termed the intensity (power per unit area and unit angular frequency), $c\,u(\omega)$, related to the electric field amplitude as

$$I(\omega) = c\,u(\omega) = \tfrac{1}{2}c\epsilon_0\,|\mathcal{E}(\omega)|^2. \tag{B.8-1}$$

Experience has shown that a weak, steady pencil beam of radiation (or a plane wave) passing through a homogeneous sample of matter typically undergoes differential change expressible by a simple rate equation, a first-order ODE that expresses the basic idealization of the theory of radiative transfer for incoherent radiation [Cha60; Sob63; Tuc75; Can85; Ryb85]. If we take z to denote the propagation distance of the beam through homogeneous matter then the change of intensity $I(\omega, z)$ of a monochromatic field with distance can be written either as an ordinary differential equation, expressing the rate of change, or as the integrated solution to that simple equation:

$$\frac{d}{dz}I(\omega, z) = -\alpha(\omega)I(\omega, z), \quad \text{or} \quad I(\omega, z) = \exp[-\alpha(\omega)z]I(\omega, 0). \tag{B.8-2}$$

Most commonly $\alpha(\omega)$ is a positive number, the *attenuation coefficient*, leading to diminution of intensity with distance (what Einstein regarded as absorption). This exponential attenuation is known variously as Beer's law, the Beer-Lambert law or the Beer-Lambert-Bouguer law. The dimensionless product of absorption coefficient and distance, αz, is known as the *optical depth*. The inverse of the attenuation coefficient is termed variously the Beer's length, the absorption length, or the *photon mean free path*. The optical depth is therefore distance measured in mean free paths.

Under suitable conditions $\alpha(\omega)$ may be a negative number, the *gain coefficient*, leading to an <u>increase</u> of intensity with distance—as happens with the formation of an emission line in hot vapor or the growth of a laser beam during its formation (what Einstein regarded as stimulated emission); see Section 5.3.

It can also happen that, at a particular frequency, the attenuation coefficient is zero, or very small; see eqn (B.8-12b). This condition produces a *transparency window*.

Although loss and gain of intensity has obvious interpretation as loss and addition of beam photons, the beam propagation equations makes no requirement on the discreteness of photons; traditional treatises on radiative transfer do not require particulate photons.

It should be noted that, like the rate equations for population change discussed in Section 5.2, a rate equation description of radiation transfer is inappropriate for treatment of coherent excitation. For such situations it is necessary to rely on Maxwell equations for field amplitudes.

The attenuation cross-section. Whether expressing attenuation or gain, the magnitude of $\alpha(\omega)$ increases as the density of matter increases—from vapor to solid, for example. Therefore the assumption of uniform behavior of attenuation with distance z requires revision to deal with spatially varying behavior, either because the density of matter varies or because the traveling field alters the matter behavior. For this purpose we introduce a spatially varying attenuation coefficient $\alpha(\omega, z)$ and express it as a product

$$\frac{d}{dz}I(\omega, z) = -\alpha(\omega, z)I(\omega, z), \quad \text{with} \quad \alpha(\omega, z) = \mathcal{N}(z)\sigma(\omega). \tag{B.8-3}$$

Here $\mathcal{N}(z)$ is the number density of scattering elements (atoms in an absorbing state, for example) and the frequency-dependent $\sigma(\omega)$, dimensionally an area, is the single-atom attenuation cross-section. (The factorization of α in eqn (B.8-3) applies to a single species of scatterers. More generally a weighted sum is required.) The dependence of $\mathcal{N}(z)$ upon distance prevents one from writing a simple exponential solution for the propagated intensity, as in eqn (B.8-2). As illustrated with examples below, the cross section can be evaluated for a variety of models for the radiation-matter interaction. Like the rate equation by which it is defined, a cross section is inappropriate for parametrizing coherent excitation. For those effects Rabi frequencies are more appropriate.

Attenuation and absorption. Several mechanisms act to remove radiation from a pencil beam. The removal of radiation energy may occur by undelayed scattering out of the beam direction, discussed in Appendix B.8.2 below. Alternatively, when the radiation frequency matches, at least approximately, a Bohr frequency of some constituent of the matter, there can occur a (resonant) transfer of energy from the radiation to constituent excitation, with corresponding alteration of some degree of freedom. In a dielectric the excitation typically produces a dipole transition moment that will reradiate as fluorescence, possibly with a change to lower frequency. The absorption may, alternatively, convert the radiant energy into heat. Scattering and absorption both contribute to the attenuation coefficient that appears in the rate equation (B.8-3) for intensity.

The absorption cross-section. Resonant excitation from level 1 to level 2 occurs when a pulse, of carrier frequency ω_{21}, passes through an aggregate of two-level atoms. Let the number density of absorbers (atoms in energy level 1 per unit volume) be $\mathcal{N}_1$. In the absence of any previous excitation (i.e. $\mathcal{N}_2 = 0$), the linear *absorption coefficient* $\alpha(\omega)$, with dimensions of inverse length, is expressible in terms of an absorption cross-section $\sigma_{12}(\omega)$, with dimensions of area (per absorber),

$$\alpha(\omega) = \mathcal{N}_1 \sigma_{12}(\omega). \tag{B.8-4}$$

The frequency dependence can be placed into a single function, either $g(\omega)$ or $g(\nu) = g(\omega)/2\pi$, by writing

$$\sigma_{12}(\omega) = \sigma_{12}^{tot} g(\omega), \quad \text{with} \quad \int_0^\infty d\omega\, g(\omega) = 1. \tag{B.8-5a}$$

The absorption cross-section can be written in terms of Einstein A or B coefficients and the spectral distribution $g(\omega)$ as [Hil82]

$$\sigma_{12}(\omega) = \frac{\hbar\omega_{21}}{c} B_{12}^{\omega} g(\omega) = \frac{\lambda_{21}^2}{4} \frac{\varpi_2 A_{21}}{\varpi_1} g(\omega). \qquad \text{(B.8-5b)}$$

Pair production. A variety of attenuation mechanisms become active as photon energy increases [Eva55]. When photon energies exceed 1.02 MeV (twice the rest-mass energy of an electron) it will happen that, in the neighborhood of a momentum-conserving atomic nucleus, this radiation energy will be converted into an electron and a positron (the anti-particle to the electron), whose kinetic energies and rest masses balance the loss of incident-photon energy. This mechanism of pair production is a dominant contribution to gamma-ray attenuation in matter [Eva55].

B.8.2   Radiation and scattering theory

During the 1970s lasers were replacing other light sources for use in atomic physics, but they continued to be regarded as examples of steady radiation, undergoing absorption and scattering as had traditional light sources. Like those sources, lasers found use in measuring opacities and excited-state lifetimes. Thus the formalism of scattering theory [Gol64; New66; Rod67; Sho67; Tay72; Wat07; Mis10; Fre18], long used in treating particle-induced nuclear reactions and electron impacts upon atoms, a natural companion to the study of stationary states of molecules, atoms, and nuclei, drew attention for describing radiation passage through assemblages of atoms.

Nuclear reactions were presented in my graduate-student days as examples of a two-step scattering process, in which an incident particle (say, a neutron) first enters the nucleus and then, after forming a compound state, either it or some other particle emerges [Fes58b; Fes62]. Between the two steps the nucleus often loses memory of all but the energy and angular momentum of the incident particle. The overall event is well described by a scattering matrix (*S matrix*) which, for specified projectile energy, has elements linking the input to all possible output results.

This simple picture of particle scattering (and an implicit S matrix) occurs, suitably adjusted, in some descriptions of electromagnetic-wave scattering. In that view a photon from a beam is absorbed by an atom, or a composite collection of atoms. There follows re-emission of a photon in some other direction. In particular, description of radiation *attenuation*, implying a sum over all possible final outcomes, is well suited to an S-matrix description [Sho67].

This viewpoint can be misleading; see [Mis09]. To evoke such a particulate view it is necessary that the outgoing photon retains some memory of the incoming photon, in order that the outgoing field be appropriately correlated, in frequency, direction, and polarization, with the incoming photon.

The alternative, non-photon viewpoint, of solutions to Maxwell's equations subject to particular boundary conditions (of a scattering entity), perhaps with a classical decomposition into plane waves and multipole waves, are better suited to depiction and to calculation. Both the classical and quantum theory of scattering treat monochromatic waves—single frequencies of the Maxwell equations for classical fields or single energies of the Schrödinger equation for particle wavefunctions. With each of these formalisms interest lies primarily with averaged squares of amplitudes, parametrized by scattering cross-sections. Such calculations suit well the incoherent behavior of thermal fields, but investigation of the quantum properties of photons are better served by treatments of scattering amplitudes and statevectors. A few particular cases of radiation scattering are often phrased in terms of photons, although discreteness of photons is not needed. These include the following examples.

Scattering theory treats particles—electrons, neutrons, protons—as waves, and so it relies on quantum theory. But its results—formulas for cross sections—do not require that waves should have granular nature: It does not require particulate photons.

Resonance scattering and absorption. The excitation cross-section, or radiation absorption cross-section, associated with resonant excitation is expressible as a constant times a frequency-dependent spectral-profile function $g(\omega - \omega_{12})$ (a probability distribution) centered around the Bohr transition frequency ω_{12},

$$\sigma_{12}(\omega) = \frac{\pi^2 c^2}{\omega^3} \frac{\varpi_2 A_{21}}{\varpi_1} g(\omega - \omega_{12}). \tag{B.8-6}$$

Here ϖ_n is the statistical weight of state n and A_{21} is the Einstein A coefficient for spontaneous emission in the transition $2 \rightarrow 1$. The spectral line profile has the normalization

$$\int d\omega \, g(\omega - \omega_{12}) = 1. \tag{B.8-7}$$

The line profile can take various forms, depending on the surroundings of the absorber and on the structure of its energy levels. The simplest situation is that of an idealized stationary, undisturbed atom in which the only spontaneous emission from state 2 is to state 1. The profile function is then a *Lorentzian* with width parameter γ fixed by the Einstein A coefficient,

$$g(\omega - \omega_{12}) = \frac{1}{\pi} \frac{\gamma}{(\omega - \omega_{12})^2 + \gamma^2}, \qquad \gamma = \tfrac{1}{2} A_{21}. \tag{B.8-8}$$

The absorption is greatest when the monochromatic radiation is tuned to the resonance frequency ω_{12}. For the Lorentz profile of eqn (B.8-8) the maximum absorption cross-section is proportional to the squared resonance wavelength λ_{12},

$$\sigma_{12}^{\max} = \sigma_{12}(\omega_{12}) = \frac{1}{2\pi} (\lambda_{12})^2 \frac{\varpi_2}{\varpi_1}, \tag{B.8-9}$$

a value that can be larger than atomic dimensions by orders of magnitude.

Environmental influences on line profile. In addition to the inevitable broadening of a spectral line from the radiative decay of excitation (the Fourier transform of exponential decay is a Lorentzian), the profile g(ω) typically incorporates a variety of environmental effects, most notably associated with temperature and pressure of the source atoms. The well-developed theory of such effects [Bre61; Muk82; Pea06; Vit09] offers connections between the detected photons and their origin environment. The effects are of two sorts: *homogeneous* broadening (notably decay and interruptions) that has the same effect on every atom, and *inhomogeneous* broadening (notably Doppler shifts) that requires summing over different velocities and environments. The two in-fluences combine most simply in the *Voigt profile* [Lam06; Ste06], a convolution of the basic Lorentz profile with a Gaussian distribution of detunings.

Continuum structure.   Central to a scattering-theory description of radiation pro-cesses was the *Lippmann-Schwinger equation* [Lip79; Sho67] for a scattering wave $\Psi_a^+(E_a)$, an eigenstate of the Hamiltonian $H = H^0 + V$, associated with a particular eigenstate ψ_a of an undisturbed Hamiltonian H^0,

$$\left(H^0 + V - E_a \right) \Psi_a^+(E_a) = 0, \qquad \left(H^0 - E_a \right) \psi_a = 0, \tag{B.8-10}$$

or, in a form that emphasizes the outgoing-wave nature of the eigenstate,

$$\Psi_a^+(E_a) = \psi_a + \lim_{\eta \to 0^+} \left[\left(E_a + i\eta - H^0 \right)^{-1} V \Psi_a^+(E_a) \right]. \tag{B.8-11}$$

The continuum energy E_a appearing here may refer to the kinetic energy of a particle projectile or to the frequency $\omega = E_a/\hbar$ of an electromagnetic wave.

Combined with the Feshbach formalism for nuclear reaction theory [Fes58b; Fes62] it allows a description of scattering in which the photoionization continuum has struc-ture, as occurs with autoionization. Resulting formulas for cross sections typically have a Breit-Wigner dependence on energy [Sho68b],

$$\sigma(E) = C(E) + \frac{(\Gamma/2)B + (E - E_0)\,A}{(E - E_0)^2 + (\Gamma/2)^2}, \tag{B.8-12a}$$

with constant profile parameters A, B and slowly vary background $C(E)$, for a reso-nance having energy E_0 and width Γ. The Fano form for such a cross section, with constant background, is [Fan61; Fan65; Sho68b; Lou92; Sta98]

$$\sigma(\epsilon) = \sigma_b + \sigma_a \frac{(\epsilon + q)^2}{(\epsilon^2 + 1)}, \qquad \epsilon = \frac{E - E_0}{\Gamma/2}, \tag{B.8-12b}$$

with constant Fano-profile parameters σ_c, σ_b, and q.

Notably, this resonance structure, unlike the more common Lorentz profile, has a narrow *transparency window*, a consequence of interference; see eqn (B.8-12b). The structure of the continuum may originate with autoionization of a doubly-excited state or it may be a *laser-induced continuum structure* (LICS) [Kni90; Hal98].

The preceding formulas provide descriptions of electromagnetic fields as they are altered by quantum systems of atoms, but they do not require that the fields be granular: They are examples of semiclassical physics in which atoms obey quantum mechanics but radiation is governed by classical equations.

B.8.3   Radiation scattering examples

A variety of scattering processes, apart from excitation-inducing absorption, occur as radiation passes by scattering centers. Explicit solutions to Maxwell's equations have been obtained for a variety of boundary conditions appropriate to single-center scatterings. These classical equations have no connection with quantum theory but they can be regarded as describing elastic or inelastic scattering of photons.

Mie scattering. The term *Mie scattering* (named for Gustav Mie) refers to solutions that describe the outgoing spherical multipole waves (an infinite series) that emerge from a wavelength-sized homogeneous dielectric sphere as it is illuminated by an incoming plane wave [Mie08; Mis09; Hor09b]. The high-multipole field modes for the sphere interior find use for describing whispering-gallery fields; see Appendix C.3.4.

Compton scattering.   Inelastic scattering of a photon from a charged free particle produces a wavelength change (the Compton shift)

$$\lambda_{\text{after}} - \lambda_{\text{initial}} = \lambda_C (1 - \cos\theta), \qquad (\text{B.8-13})$$

where θ is the deflection angle of the radiation beam and λ_C is the Compton wavelength. For an electron this is

$$\lambda_C = \frac{h}{m_e c} \approx 2.426 \times 10^{-12} \text{ m}, \qquad \lambda\!\!\!\!\lambda_C \equiv \frac{\lambda_C}{2\pi} = \frac{\hbar}{m_e c}. \qquad (\text{B.8-14})$$

Thomson scattering.   The low-energy limit of Compton scattering is *Thomson scattering*, elastic scattering of classical electromagnetic radiation by a classical free particle first presented by J. J. Thomson. In this low-frequency long-wavelength limit the traveling electric field exerts a force on the charged particle. The particle, weakly accelerated by this force (to nonrelativistic velocity), emits radiation at the frequency of field oscillation. The Thomson scattering cross-section for an electron is independent of the radiation frequency and is proportional to the square of the classical radius of the electron, r_e,

$$\sigma_T = \frac{8\pi}{3} (r_e)^2, \qquad r_e = \alpha \lambda\!\!\!\!\lambda_C, \qquad (\text{B.8-15})$$

where α is the Sommerfeld fine-structure constant of eqn (A.2-3) and $\lambda\!\!\!\!\lambda_C$ is the reduced Compton wavelength of the electron. Whereas Thomson scattering scales as the square of the classical electron radius, resonant scattering scales as the square of the wavelength of the light. For optical wavelengths this scaling enhances resonant scattering by eight orders of magnitude.

Rayleigh scattering.   When the medium can be idealized as random distribution of independent molecules whose individual dipole moments are expressible as a polarizability α_{mol} times the electric field, and which have only a small effect on the field in the forward direction, then for wavelengths much larger than the molecular diameter the radiation undergoes (elastic) *Rayleigh scattering*, named for Lord Rayleigh, Baron J. W. Strutt. The attenuation is quantified by the Rayleigh cross-section [Lip69]

$$\sigma_R(\omega) = \frac{8\pi}{3} \left(\frac{1}{\lambda}\right)^4 \frac{|\alpha_{mol}|^2}{\epsilon_0^2}. \qquad (\text{B.8-16})$$

This scattering is characteristically proportional to the inverse fourth power of the wavelength, $\lambda = 2\pi c/\omega$.

Raman scattering.   Inelastic scattering of photons replaces one light frequency (or photon energy) with another. An important example of inelastic scattering, *Raman* *scattering* [Ram28; Ram28b; Her50b; Blo67; Lal71; Ray81], can either subtract energy from photons, producing lower-frequency Stokes radiation, or add energy to photons, producing higher-frequency *anti-Stokes radiation*; see Figure 5.3. The energy change of the photons comes from the atoms or molecules of the medium. Because their energy states are discrete, only discrete photon energy changes can occur. The Stokes (and anti-Stokes) radiation therefore consists of discrete frequencies (discrete spectral lines) whose values are characteristic of the particular molecules that comprise the medium. The energy deposited in the medium by Stokes radiation can be treated as an excitation wave, sometimes termed a spin wave.

Although the language of photons is often used to describe Raman scattering, the effects can be fully treated with semiclassical theory in which the atom or molecule obeys quantum mechanics and the fields are classical.

Brillouin scattering.   Another example of inelastic scattering occurs when radiation, passing through matter, excites large-scale collective vibrational modes of the media, that carry energy and momentum with wave-particle properties akin to those of photons. The low-frequency sound waves generated in crystals are known as phonons.

Fractal scattering.   Fractals, introduced on page 210, have a connection to optics (and thereby to photons) through interest in light scattering [All86; Akk10] and in novel image-forming optical elements based on *Fresnel zone plates* (FZP): Structures of alternating transmissive and opaque rings used to mimic the focusing action of a lens. A so-called pinhole *photon sieve* is an FZP in which a pattern of pinholes replaces the clear zones [Kip01; Men05; And05]. When these involve fractal structure of the optical element they will produce fractal fields (one might say fractal photons) [Saa03; Gim06].

Appendix C
Coupled atom and field equations

The lack of research results may seem a desolate landscape.

The preceding two appendices described the basic quantum theory of atoms acted upon by prescribed fields and of photons in free space. The presence of matter adds additional fields to the Maxwell equations, discussed in Appendix C.2. No longer are traveling wavelike disturbances identifiable as purely electric and magnetic fields. These waves (and their quantum increments of photons) must incorporate properties of the matter through which they travel. When the radiation is steady and weak, the effects of matter appear as a velocity-altering refractive index in the wave equations, as described in Appendix C.3. The resulting approximation, of fields affected by prescribed bulk-matter behavior, underlies classical, linear optics of lenses and diffraction and leads to the variety of scattering effects. The interface between differing materials causes reflections. Although single photons cannot be further divided, beam splitters are commonplace optical elements. Appendix C.3.5 discusses their mathematics.

A separation of matter and radiation into two systems, as in eqn (5.1-1a), is of most use when the mutual alterations, of atoms by fields and of fields by atoms, can be regarded as a small effect, to be treated by the mathematics of perturbation theory. Appendix C.4 begins the discussion of models in which these two systems are treated as interacting parts of a single system—the field affecting atoms which, in turn, modify the field, as in eqn (5.1-1b). The atoms are regarded as being governed by quantum theory—they have discrete energy states and wavefunctions—but the field may either be classical (termed *semiclassical* models) or, like the atoms, it may fully fit the requirements of quantum theory.

Appendix C.5 discusses the atom-field Hamiltonian in a photon-number representation, as is commonly done in QED and Quantum Optics, and starts the connection with contemporary work that finds practical applications for the quantum nature of electromagnetic-field increments—the photons. When the system of atoms and radiation is enclosed within a cavity the field quanta become standing-wave photons. Appendix C.6 discusses the exactly soluble Jaynes-Cummings model of fully quantum-mechanical atoms and cavity-enclosed (standing-wave) quantized radiation. This model provides a clear demonstration for the existence of discrete increments of electromagnetic field—of *cavity photons*. Appendix C.7 describes one of the procedures for creating single photons on demand. The concluding portion, Appendix C.8, discusses the correlation of coupled atoms and photons, and the general notion of entanglement that underlies contemporary reliance on quantum theory for secure communication.

C.1   Field sources

Although it is useful to consider fields without referencing their sources, as was done in Appendix B, all electromagnetic fields ultimately originate with electric charges and currents. For projects that employ simple laboratory facilities these charges are distributions of electrons and nuclei, organized as atoms and molecules or as lattice structures. In the simplest idealizations these are treated as collections of moving, well-separated, point charges, and the fields of interest are well outside the atoms and molecules in which these are bound—a description suited to the Feynman picture of radiation in Section 8.3. The following paragraph quantifies the connection between radiation and charge acceleration, discussed qualitatively in Section 2.8.

Point-charge radiation.   The electric and magnetic fields in the radiation zone of a point charge q, moving nonrelativistically with velocity $\mathbf{v}$ and acceleration $\dot{\mathbf{v}}$, are [Mil19]:

$$\mathbf{E}(\mathbf{r},t) = \frac{q}{4\pi\epsilon_0}\frac{1}{c^2 r^3}\,\mathbf{r}\times(\mathbf{r}\times\dot{\mathbf{v}}), \qquad \mathbf{B}(\mathbf{r},t) = \frac{q}{4\pi\epsilon_0}\frac{1}{c^3 r^2}\,\dot{\mathbf{v}}\times\mathbf{r}. \qquad \text{(C.1-1)}$$

From the Poynting vector associated with these fields one finds a dipole pattern for the angular distribution of the instantaneous rate, per solid angle, at with the accelerated charge radiates energy. Specifically, the differential power $dP(t)$ being radiated away by an accelerating charge q at time t into differential solid angle $d\Omega$ is [Mil19]

$$\frac{dP(t)}{d\Omega} = \frac{1}{4\pi\epsilon_0}\frac{q^2}{4\pi c^3}\,|\dot{\mathbf{v}}(t)|^2\,\sin^2\theta(t), \qquad \text{(C.1-2)}$$

where θ is the angle between $\mathbf{r}$ and the acceleration $\dot{\mathbf{v}}$. From the integral over solid angles one obtains the *Larmor formula* for the power being radiated,

$$P(t) = \frac{1}{4\pi\epsilon_0}\frac{2q^2\dot{v}(t)^2}{3c^3}. \qquad \text{(C.1-3)}$$

Greater acceleration produces higher power, faster production of radiation energy. The radiated energy must, ultimately, come from work done by whatever force produces the acceleration. Periodic acceleration, say of an oscillator, will radiate power proportional to the fourth power of its frequency (square of the second derivative of position). In classical physics the radiation production continues as long as the acceleration is present—hence a purely classical picture predicts that an electron encircling a nucleus would spiral into it. Only with quantum theory and its imposition of momentum uncertainty and discrete energies came the explanation for atom stability. The lack of spontaneous emission by an atomic electron in its ground state, despite its continual acceleration, is attributable to exact destructive interference between the radiation-reaction field of the accelerated charge and vacuum fluctuations; see [Mil84; Mil19].

Bulk matter. Matter comes in various forms: Gases, in which atoms and molecules move freely between encounters; liquids, in which the particles are packed closely together but slide relatively freely; and solids, in which a rigid framework constrains particles motion; see Section 2.3. The electrons in solids are typically either bound within atoms and other structures (an insulator) or they may be free to move, as a current (a conductor or semiconductor). The bulk matter of primary interest in the present Memoir is material through which photons may travel, meaning primarily neutral gases, liquids, and transparent solids. In all of these the electron charges, and their motions as currents, are localized within atoms; I shall assume that there are no electrons free to move through the material—no free charges or currents.

The wavelengths of optical radiation are much longer than the dimensions of atoms, and so to optical radiation the discrete structure of matter as atoms and molecules appears as a continuous blur of charge and current distributions, averaged over volumes ΔV that contain many atoms but which are much smaller than any wavelength of interest [Str41; Sto63; Van77; Pow80; Cra84]. (Such averaging fails for X-rays and shorter-wavelength radiation.) This idealization, of matter properties distributed as fields—a continuum $\mathbb{R}$ from averaging over discrete points $\mathbb{Q}$—undelies the remainder of this Appendix.

C.2   The Maxwell equations in matter

The Maxwell equations in free space, eqn (B.1-1), used to define Dirac photons, require revision and enlargement in order to treat fields in bulk material. They require an additional vector field, a current density $\mathbf{j}(\mathbf{r}, t)$, and a scalar field, the charge density $^c\rho(\mathbf{r}, t)$, that together express the effect of matter on electromagnetic fields.[1] These fields originate with the distributed atoms and molecules, the electrons and ions, that comprise bulk matter but which are being idealized as spatial averages over unresolvable (subwavelength) continuous distributions of electric charge and current; see Appendix C.2.1. The resulting dynamical partial differential equations read (with suppression of space and time arguments) [Str41; Lan60; Jac75; Sal19; Mil19]

$$\nabla \times \mathbf{E} + \frac{\partial}{\partial t}\mathbf{B} = 0, \qquad \nabla \times \frac{\mathbf{B}}{\mu_0} - \frac{\partial}{\partial t}\epsilon_0 \mathbf{E} = \mathbf{j}. \qquad (\text{C.2-1a})$$

These accompany the static constraints

---

[1] To distinguish charge density $^c\rho$ from the density matrix ρ used elsewhere I here use an unconventional superscript c.

$$\nabla \cdot \mathbf{B} = 0, \qquad \nabla \cdot \epsilon_0 \mathbf{E} = {}^c\rho. \qquad \text{(C.2-1b)}$$

These four vector equations describe temporal and spatial alterations of electric and magnetic fields induced by prescribed charges and currents. Those, in turn, are affected by the fields in accord with the *Lorentz-force* density,

$$\mathbf{F} = {}^c\rho\,\mathbf{E} + \mathbf{j} \times \mathbf{B}. \qquad \text{(C.2-2)}$$

The charge and current density appearing in these equations as sources of the electromagnetic fields satisfy the continuity equation

$$\frac{\partial}{\partial t}{}^c\rho + \nabla \cdot \mathbf{j} = 0. \qquad \text{(C.2-3)}$$

Thus the electric charge appears in the classical theory of electromagnetism as a fluid, continuously distributed in Euclidean space.[2]

To complete the mathematics it is necessary to have not only initial conditions and boundary conditions but some prescription for the matter fields (a *constitutive relationship*), such as the following paragraphs describe.

C.2.1 Polarization and magnetization fields

The spatially averaged electromagnetic properties of electrons and atoms in nonconducting materials are idealized as arising from a continuous distribution of independent individual-atom electric and magnetic dipole moments, distributed as described by a *polarization density* $\mathbf{P}(\mathbf{r}, t)$ and a *magnetization density* $\mathbf{M}(\mathbf{r}, t)$,

$$\mathbf{P}(\mathbf{r}, t) = \sum_\alpha \mathcal{N}_\alpha(\mathbf{r}) \langle \mathbf{d}(\mathbf{r}, t) \rangle_\alpha, \qquad \mathbf{M}(\mathbf{r}, t) = \sum_\alpha \mathcal{N}_\alpha(\mathbf{r}) \langle \mathbf{m}(\mathbf{r}, t) \rangle_\alpha, \qquad \text{(C.2-4)}$$

where $\mathcal{N}_\alpha(\mathbf{r})$ is the number density of atoms of type α at position $\mathbf{r}$ and $\langle \mathbf{d}(\mathbf{r}, t) \rangle_\alpha$ and $\langle \mathbf{m}(\mathbf{r}, t) \rangle_\alpha$ are the instantaneous expectation values at time t of the electric and magnetic dipole moments for single atoms of type α at position $\mathbf{r}$. These macroscopic fields can provide descriptions of individual atoms or other localized sources; here, for description of bulk matter, they are treated as smoothly varying, having discontinuities only at material boundaries. The spatially-averaged dipole moments produce the charge and current densities

$$^c\rho(\mathbf{r}, t) = -\nabla \cdot \mathbf{P}(\mathbf{r}, t), \qquad \mathbf{j}(\mathbf{r}, t) = \frac{\partial}{\partial t}\mathbf{P}(\mathbf{r}, t) + \nabla \times \mathbf{M}(\mathbf{r}, t). \qquad \text{(C.2-5)}$$

The discussion that follows assumes there are no free charges or currents, as occur in plasmas and solid conductors; the charge and current densities of eqn (C.2-5) are the only matter fields of concern.

The D and H fields. When treating the behavior of electromagnetic fields within bulk matter it is often useful to work with fields that combine electromagnetism with

---

[2]It is also possible to treat idealized point charges by using Dirac delta functions for charge distributions.

the polarization and magnetization fields. The auxiliary fields commonly used in the presentation of Maxwell equations are (with SI units)

$$\mathbf{D} = \epsilon_0 \mathbf{E} + \mathbf{P}, \qquad \mathbf{H} = \frac{1}{\mu_0}\mathbf{B} - \mathbf{M}. \tag{C.2-6}$$

The universal constants ϵ_0 (the permittivity of free space) and $\mu_0 = 1/c^2\epsilon_0$ (the permeability of free space) occur as a consequence of using SI units. From these definitions the dynamical *Maxwell equations* eqn (C.2-1a) become

$$\frac{\partial}{\partial t}\mathbf{D} = \nabla \times \mathbf{H}, \qquad \frac{\partial}{\partial t}\mathbf{B} = -\nabla \times \mathbf{E}. \tag{C.2-7a}$$

To these we add the Maxwell equations of constraint,

$$\nabla \cdot \mathbf{D} = 0, \qquad \nabla \cdot \mathbf{B} = 0. \tag{C.2-7b}$$

In the presence of bulk matter it is the field $\mathbf{D}$, rather than $\mathbf{E}$ alone, that must be divergenceless. For quantization of electromagnetic fields in a dielectric medium [Gla91; Hut92; Sip09; Hor14; Ray20b] it is preferable to take the divergenceless $\mathbf{D}$ field, rather than the $\mathbf{E}$ field, as basic when treating nonlinearities as perturbations [Gla91; Sip09; Que17; Ray20].

Discontinuities. Discontinuities of the parameters ϵ and μ at interfaces between different materials produce reflection and partial transmission at mirror interfaces; see Appendix C.3.3. Across any such surface there must be continuity of the component of $\mathbf{B}$ that is normal to the surface and of the component of $\mathbf{E}$ that is tangential to the surface. The conditions can be expressed in terms of a unit vector $\mathbf{e}_\perp$ that is perpendicular to the surface and the fields within the two materials [Str41]:

$$\mathbf{e}_\perp \cdot (\mathbf{B}_2 - \mathbf{B}_1)_{\text{surf}} = 0, \qquad \mathbf{e}_\perp \times (\mathbf{E}_2 - \mathbf{E}_1)_{\text{surf}} = 0. \tag{C.2-8}$$

Here the subscripts identify the two sides of the interface.

There is nothing in such linear relationships to alter the quantum nature of the field and so we can expect that photons arriving at such an interface will remain discrete increments, carrying their quantum heritage even as they alter their course.

C.2.2   Inhomogeneous wave equations in matter

The matter fields $\mathbf{P}$ and $\mathbf{M}$ appear as sources in wave equations derived from the Maxwell equations. In the absence of magnetism density $\mathbf{M}$ the equation for the $\mathbf{E}$ field is

$$\frac{\partial^2}{\partial t^2}\mathbf{E} + c^2\nabla \times \nabla \times \mathbf{E} = -\epsilon_0\frac{\partial^2}{\partial t^2}\mathbf{P}. \tag{C.2-9a}$$

The time variation of the polarization field $\mathbf{P}$ therefore acts as a source for modification of the electric field. The corresponding wave equation for magnetic phenomena in the absence of polarization density $\mathbf{P}$ is

$$\frac{\partial^2}{\partial t^2}\mathbf{B} + c^2\nabla \times \nabla \times \mathbf{B} = \mu_0\nabla \times \nabla \times \mathbf{M}. \tag{C.2-9b}$$

The magnetization $\mathbf{M}$ acts as to alter the $\mathbf{H}$ field to make $\mathbf{B}$, while the spatial variation of magnetization acts as a source for modification of the $\mathbf{B}$ field. In turn, the TDSE governs the changes to the atom-based fields $\mathbf{P}$ and $\mathbf{M}$ caused by the electromagnetic fields. Together, the fields and atoms form a single, unified system, as in eqn (5.1-1b). We might expect that excitations of this system could be called photons (or "*dressed photons*"); but they will not be the photons that occur in Appendix B as increments of "bare" electromagnetic fields.

C.3   Bulk-matter steady response

The dipole-moment expectation values of eqn (C.2-4) that underlie the continuously distributed fields $\mathbf{P}(\mathbf{r}, t)$ and $\mathbf{M}(\mathbf{r}, t)$ originate with independent individual atoms within an averaging volume $\Delta\mathcal{V}$. Each of these atoms is affected by the local electric and magnetic fields and by its surroundings. The local radiation field provides energy for excitation while the neighboring atoms provide a thermal environment of uncontrollable irregular interactions. During brief intervals the excitation can be considered coherent and, if there is a match between the radiation frequency and a Bohr frequency, the resulting atomic excitation will exhibit periodic Rabi oscillations at a frequency fixed by the radiation intensity, modeled by the time-dependent Schrödinger equation, as discussed in Appendix A.10.2. But over longer times the thermal background will eventually prevail, and the individual dipole moments will settle to steady values. The rate at which this relaxation occurs is typically set by a relaxation rate γ and a corresponding relaxation time $\tau = 1/\gamma$ that depend upon the nature of the background bath—see Section 5.1.3. It is in the limit of complete adjustment to the thermal environment and the ongoing radiation, as discussed in Appendix A.14.3, that the regime of classical linear optics occurs, a regime in which the $\mathbf{P}$ and $\mathbf{M}$ fields are typically proportional to the applied electromagnetic fields. It is under these conditions of steady-state atomic response (linear optics [Hec87; Bor99; Sal19]) that susceptibilities and refractive indices are defined.

Susceptibility. A steady, relatively weak, electric field traveling through an idealized isotropic medium (one in which there is no distinction between directions) generally creates a polarization field (a volume density of induced electric-dipole moments) proportional to the instantaneous electric field,

$$\mathbf{P}(\mathbf{r}, t) = \chi^e(\mathbf{r})\, \epsilon_0 \mathbf{E}(\mathbf{r}, t). \tag{C.3-1a}$$

The coefficient $\chi^e(\mathbf{r})$ is the *electric susceptibility*, here taken to be stationary in time but to have spatial variation as would occur with an inhomogeneous medium where composition varies with position $\mathbf{r}$. The magnetization density $\mathbf{M}$ (a volume density of magnetic-dipole moments) is similarly expressible with a *magnetic susceptibility* $\chi^m(\mathbf{r})$, traditionally defined in terms of the magnetic intensity $\mathbf{H} = \mu\mathbf{B}$,

$$\mathbf{M}(\mathbf{r}, t) = \chi^m(\mathbf{r})\, \mathbf{H}(\mathbf{r}, t). \tag{C.3-1b}$$

Through these relationships the steady-state properties of the material are entirely described by the constant electric and magnetic susceptibilities, functions only of position. Here they are treated as scalars, but more generally, in anisotropic media, they are two-index tensors that relate a single component of a vector $\mathbf{P}$ or $\mathbf{M}$ to a super-position of components of the vectors $\mathbf{E}$ and $\mathbf{H}$,

$$P_i = \sum_j \chi_{ij}^e \epsilon_0 E_j, \qquad M_i = \sum_j \chi_{ij}^m H_j. \tag{C.3-2}$$

Although the susceptibilities originate with microscopic dipole moments, and there-fore are linked to quantities that can, in principle, be evaluated from the quantum mechanics of atoms and bulk matter, they appear here as macroscopic parameters that can be determined from experiment on particular materials.

When the fields are weak and steady, so that the bulk material has had time to equilibrate with the distorting effects of the fields, it is possible to express the two auxiliary fields in terms of spatially-varying dielectric constants $\epsilon(\mathbf{r})$ and $\mu(\mathbf{r})$, in the form

$$\mathbf{D}(\mathbf{r}, t) = \epsilon(\mathbf{r})\, \mathbf{E}(\mathbf{r}, t), \qquad \mathbf{H}(\mathbf{r}, t) = \frac{1}{\mu(\mathbf{r})} \mathbf{B}(\mathbf{r}, t), \tag{C.3-3}$$

where the material properties are entirely expressed by the time-independent factors

$$\epsilon(\mathbf{r}) = \epsilon_0[1 + \chi^e(\mathbf{r})], \qquad \mu(\mathbf{r}) = \mu_0[1 + \chi^m(\mathbf{r})]. \tag{C.3-4}$$

By the nature of the averaging procedure that replaces discrete atomic constituents of matter by a continuum distribution of mass, charge, and multipole moment, the spatial variation of $\epsilon(\mathbf{r})$ and $\mu(\mathbf{r})$ is presumed to be slow and slight over a wavelength. Though shown as scalars, in general they are tensors.

When the variation of dielectric properties is slow (over many wavelengths and periods) and independent of direction the Maxwell equations become

$$\frac{n^2}{c^2} \frac{\partial}{\partial t} \mathbf{E} = \nabla \times \mathbf{B}, \qquad \frac{\partial}{\partial t} \mathbf{B} = -\nabla \times \mathbf{E}, \tag{C.3-5a}$$

$$\epsilon \nabla \cdot \mathbf{E} = 0, \qquad \nabla \cdot \mathbf{B} = 0, \tag{C.3-5b}$$

where $n = c\sqrt{\epsilon\mu}$ is the (slowly varying) refractive index, taken here to be a scalar.

C.3.1 Homogeneous wave equations in linearly-responding matter

In regions where ϵ and μ are spatially-uniform scalars both $\mathbf{E}$ and $\mathbf{B}$ (and their vector potentials) satisfy the vector wave equation

$$\nabla \times \nabla \times \mathbf{F}(\mathbf{r}, t) + \frac{1}{\mu\epsilon} \frac{\partial^2}{\partial t^2} \mathbf{F}(\mathbf{r}, t) = 0. \tag{C.3-6}$$

Wavelike solutions to these equations, incorporating properties of bulk matter, are what one may consider as providing photons within matter, counterparts of the vacuum-traveling photons of Feynman.

The relationships of eqn (C.3-3) are too restrictive when we consider the effects of coherent excitation upon the atoms that comprise bulk matter, but they serve as the foundation for the vast enterprise of linear optics, with its apertures, lenses, mirrors, gratings and optical fibers.

Refractive index; Phase velocity. The main effect of a weak, steady radiation field upon atoms is to induce a stationary, mean electric-dipole moment proportional to the electric field. For bulk matter this effect is expressed by means of a polarization field and a constant, or slowly varying, susceptibility. With this steady-state approximation for the induced dipole moment the resulting wave equation for a divergenceless electric field (and spatially uniform or slowly varying, isotropic material) becomes

$$\nabla^2 \mathbf{E}(\mathbf{r}, t) - \frac{n^2}{c^2} \frac{\partial^2}{\partial t^2} \mathbf{E}(\mathbf{r}, t) = 0, \tag{C.3-7}$$

and the free-space propagation speed c is replaced by the *phase velocity* v_{ph}, fixed by the *refractive index* n:

$$v_{\mathrm{ph}} = \frac{c}{n}, \qquad n = \sqrt{\frac{\epsilon}{\epsilon_0} \frac{\mu}{\mu_0}} = \sqrt{(\chi^e + 1)(\chi^m + 1)}. \tag{C.3-8}$$

If the material has no magnetization, then the refractive index is expressible in terms of the dielectric susceptibility, leading to the commonly used expressions

$$v_{\mathrm{ph}} = \frac{c}{\sqrt{1 + \chi^e}}, \qquad n^2 = \frac{\epsilon}{\epsilon_0} = 1 + \chi^e. \tag{C.3-9}$$

For application to optical radiation the spatial variation of the electric vector in eqn (C.3-7) is established primarily by the wavelengths associated with the dominant carrier frequencies of the field. Changes of material properties occurring over a larger distance can be incorporated into the wave equation as slowly changing susceptibilities $\chi(\mathbf{r})$, and the associated refractive index, $n(\mathbf{r})$, see eqn (C.3-22). This relatively slow spatial variation leads to refraction and gradual beam redirection. More abrupt changes of material properties can be treated as a discontinuity, as in eqn (C.2-8). These lead to reflection at surfaces. Reflection is absent at an interface if there is a matching of the *wave impedance* across the boundary,

$$Z = \sqrt{\frac{\mu\mu_0}{\epsilon\epsilon_0}} = \sqrt{\frac{\mu}{\epsilon}} Z_0. \tag{C.3-10}$$

Metamaterials. For ordinary materials both ϵ and μ are positive. However, a variety of *metamaterials* have been created for which one or both of these quantities may be negative [Pen00; Smi04; Pen04; Pen06]. When both these quantities are negative the square root that defines the refractive index in eqn (C.3-8) must be taken as negative [Pen00].

Frequency dependent susceptibility. Material cannot polarize instantaneously in response to an applied field. A more general formulation describes the polarization field as a convolution of the earlier electric field history,

$$\mathbf{P}(\mathbf{r}, t) = \epsilon_0 \int_{-\infty}^{t} dt' \, \chi^e(t - t') \mathbf{E}(\mathbf{r}, t'). \tag{C.3-11}$$

The Fourier transform of this equation leads to a complex-valued frequency-dependent susceptibility relating the positive-frequency parts of the $\mathbf{P}$ and $\mathbf{E}$ fields

$$\mathbf{P}^{(+)}(\mathbf{r}, \omega) = \epsilon_0 \chi^e(\mathbf{r}, \omega) \mathbf{E}^{(+)}(\mathbf{r}, \omega). \tag{C.3-12}$$

The real and imaginary parts of this susceptibility,

$$\chi^e(\omega) = \chi^R(\omega) + i\chi^I(\omega), \tag{C.3-13}$$

are not independent: They are related by the *Kramers-Kronig relations* [Sha64; Sch98; Sch08; Mil10; Boh10] (an example of a *Hilbert transform* [Jac75]),

$$\chi^R(\omega) = \frac{1}{\pi} \mathcal{P} \int_{-\infty}^{\infty} d\omega' \frac{\chi^I(\omega')}{\omega' - \omega}, \qquad \chi^I(\omega) = -\frac{1}{\pi} \mathcal{P} \int_{-\infty}^{\infty} d\omega' \frac{\chi^R(\omega')}{\omega' - \omega}. \tag{C.3-14}$$

Here $\mathcal{P}$ denotes the *Cauchy principal value*—the integral omitting the singular part at $\omega' = \omega$ [Bar03b].

Lorentz atoms. A dilute collection of Lorentz oscillators (or *Lorentz atoms*), provides a particularly useful model of the frequency dependence of optical properties found in many gases. The individual oscillator is an electron, with mass m_e, bound to a fixed center by a restoring force that leads to harmonic motion with frequency ω_0, damped at a rate γ. These oscillators are regarded as distributed with a number density $\mathcal{N}$. The frequency-dependent electric susceptibility is [Mil88]

$$\chi^e(\omega) = \frac{\omega_p^2}{\omega_0^2 - \omega^2 - i\omega\gamma}, \qquad \omega_p^2 = \frac{\mathcal{N}e^2}{\epsilon_0 m_e}. \tag{C.3-15}$$

Here ω_p is the plasma frequency of the electron density.

Microscopic description: Polarizability. The two macroscopic fields **P** and **M** have their origin in spatial distributions of induced (or permanent) individual atomic moments and so, in principle, they can be calculated from any model of individual atoms. The main effect of a weak steady radiation field is to induce in atoms and molecules a stationary, mean dipole moment proportional to the electric field. This induced equilibrium moment is commonly quantified with a single-atom frequency-dependent *polarizability*[3] $\mathsf{X}(\omega)$ associated with the response produced by a monochromatic electric field,

$$\langle \mathbf{d}(\mathbf{r}) \rangle = \mathsf{X}(\omega) \, \epsilon_0 \, \mathbf{E}^{\mathrm{loc}}(\mathbf{r}, \omega). \tag{C.3-16}$$

The field appearing here is the electric field directly around the dipole moment—the *local field*. When the dipole is part of a solid it will be affected not only by the field $\mathbf{E}^{\mathrm{ext}}$ that an experimenter applies but also by the electric field produced by all of the neighboring atoms. Feynman shows that for a homogeneous isotropic medium the relationship between the local field and the experimentally controlled field is

$$\mathbf{E}^{\mathrm{loc}}(\mathbf{r}, \omega) = \mathbf{E}^{\mathrm{ext}}(\mathbf{r}, \omega) + \frac{\mathbf{P}(\mathbf{r}, \omega)}{3\epsilon_0}. \tag{C.3-17}$$

This Lorentz correction to the local field leads to the expression (the *Clausius-Mossotti equation* for mixed materials):

---

[3]The usual symbol is α or $\boldsymbol{\alpha}$. Here I use X to avoid confusion with other uses of α.

$$3\frac{n^2-1}{n^2+2} = \sum_{\alpha} \mathcal{N}_{\alpha} X_{\alpha}(\omega).$$ (C.3-18)

For atoms in a vapor the field from neighbors is negligible, and it is appropriate to take $\mathbf{E}^{\text{local}} = \mathbf{E}^{\text{ext}} = \mathbf{E}$. The connection of the single-atom polarizability with the bulk electric susceptibility is then, for a single species of atom, the commonly used formula

$$\chi^e(\mathbf{r}, \omega) = \mathcal{N}(\mathbf{r}) X(\omega).$$ (C.3-19)

The group velocity of pulses. For a monochromatic plane wave that has frequency ω and wavevector magnitude $k = |\mathbf{k}|$ the space-time variation of the field traveling through a uniform medium can be taken as $\exp[i\mathbf{k} \cdot \mathbf{r} - i\omega t]$ and the wave equation eqn (C.3-7) becomes a *dispersion relation*,

$$- |\mathbf{k}|^2 + \frac{n^2}{c^2}\omega^2 = 0, \quad \text{or} \quad ck = n\omega.$$ (C.3-20)

Though not shown explicitly, the refractive index n appearing here is frequency dependent. The phase fronts of peak and valley amplitudes of this monochromatic wave move together along the propagation axis with phase velocity v_{ph}, inversely proportional to the static refractive index. Both of the quantities v_{ph} and n depend upon the frequency of steady monochromatic radiation. When this carrier frequency is near a Bohr-transition frequency (a resonance condition) the refractive index can vary significantly with frequency. There the phase velocity can be less than, equal to, or greater than the speed of light in vacuum, c.

Fields that are pulsed, rather than monochromatic, are expressible by Fourier transform as an integral over frequencies; see Appendix B.5.1. From such a superposition of monochromatic waves of differing propagation vectors $\mathbf{k}$ one can form a wavepacket whose constructive interference creates a traveling pulse. Each component of the wavepacket, having a slightly different refractive index, travels at a different speed and so the pulse, defined by regions of constructive and destructive interference, becomes shifted relative to its motion at speed c in a vacuum. For sufficiently short distances, when the pulse distortion is not too great, the one-dimensional pulse motion can be described by a *group velocity* v_g involving a group refractive index n_g,

$$v_g = \frac{c}{n_g}, \quad n_g = n + \omega\frac{dn}{d\omega}.$$ (C.3-21)

This velocity, like the phase velocity, may be less than, equal to, or greater than c.

Signal velocity. It is generally accepted that, in keeping with the special theory of relativity, information cannot travel at a speed exceeding c. There has consequently been much study of the *signal velocity* of light pulses. Careful analysis of how signals are to be recognized has led to the conclusion that c is the upper bound on their velocity [Web54; Bri60; Smi70; Boy09; Oug19].

Anisotropy. When the induced dipole-moment is along the field direction the polarizability X and the susceptibility χ are scalars. More generally these are tensors, matrix operators that alter the direction of **P** from that of **E** in eqn (C.3-16); see eqn (C.3-2). These effects are described most simply with a Cartesian coordinate system in which one axis is aligned with the propagation direction. The electric vector then typically has a negligible component along that direction, and only two transverse components of the polarization field are affected.

Birefringence. Birefringent crystals have two distinct axes perpendicular to the propagation axis. For each of these axes a linearly polarized electric vector induces a different steady response within the crystal. The two polarizations therefore have different refractive indices and undergo different phase shifts as they propagate. The two transverse directions for which the difference is greatest are known as the fast and slow axes, and these labels can be used in place of x, y or H, V. In traveling through such a birefringent medium the orthogonal polarization components of the beam acquire different phase delays but the magnitude of the field components are not affected. The result affects the projection of the angular momentum along the propagation axis, and can convert polarization between, say, linear and circular modes or two linear-polarization modes.

C.3.2 Steady paraxial beams; Guided waves

When fields act steadily on bulk matter the transients of coherent excitation (i.e. the Rabi frequencies) die away and the bulk polarization and magnetization take values determined by the fields that drive them. In the simplest situations the dependence is first order, with proportionality of real-valued electric and magnetic susceptibilities as given in eqn (C.3-1). Then the wave equation eqn (C.3-7) for the **E** field incorporates the presence of polarizable media by means of the spatially-varying refractive index,

$$\left[\nabla^2 - \frac{n(\mathbf{r})^2}{c^2} \frac{\partial^2}{\partial t^2} \right] \mathbf{E}(\mathbf{r}, t) = 0. \tag{C.3-22}$$

Abrupt changes of the refractive index cause reflections, either to confine the field within an enclosure or to redirect a traveling wave.

A common application of this equation occurs with waveguide structures, in which transverse variation of the refractive index confines a propagating field to a region along an axis, where the index is n_0 and there is negligible variation of the refractive index along the axis direction. To model the propagation of laser radiation through such a structure we regard the radiation as a steady, transversely localized beam, idealized as a monochromatic plane-wave carrier field propagating along the waveguide (taken as the z axis) and modulated by an amplitude $\mathcal{E}(\mathbf{r})$ that varies only slowly along z:

$$\mathbf{E}^{(+)}(\mathbf{r}, t) = \mathbf{e}\, e^{ikz - i\omega t} \mathcal{E}(\mathbf{r}), \qquad ck = n_0 \omega. \tag{C.3-23}$$

As shown, we take the relationship between carrier frequency ω and wavevector k to be the refractive-index value n_0. Because the envelope $\mathcal{E}(\mathbf{r})$ is assumed to vary slowly along the propagation axis, z, we ignore the second derivative with respect to z (thereby making the *slowly-varying envelope approximation*, SVEA) and obtain the resulting equation

$$\left[\nabla^2 - \frac{n(\mathbf{r})^2}{c^2} \frac{\partial^2}{\partial t^2} \right] \mathbf{E}^{(+)}(\mathbf{r}, t)$$

$$= \mathbf{e}\, e^{ikz - i\omega t} \left[\nabla_\perp^2 + i2k \frac{\partial}{\partial z} - k^2 + n(\mathbf{r})^2\lambdabar^2 \right] \mathcal{E}(\mathbf{r}), \qquad \text{(C.3-24)}$$

where $\lambdabar = \lambda/2\pi = \omega/c$ and $\nabla_\perp^2$ is the two-dimensional Laplacian in the transverse (x, y) plane. The result is the *paraxial wave equation* for propagation in matter, involving a first-order derivative along the propagation direction [Lon05; Lon06; Lon06b; Lon07],

$$i\lambdabar \frac{\partial}{\partial z} \mathcal{E}(\mathbf{r}) = \left[-\frac{\lambdabar^2}{2n_0} \left(\frac{\partial^2}{\partial x^2} + \frac{\partial^2}{\partial y^2} \right) + V(\mathbf{r}) \right] \mathcal{E}(\mathbf{r}). \qquad \text{(C.3-25)}$$

Here the effective potential $V(\mathbf{r})$ incorporates the wave-guiding effect of the refractive index,

$$V(\mathbf{r}) = \frac{n_0^2 - n(\mathbf{r})^2}{2n_0} \approx n_0 - n(\mathbf{r}). \qquad \text{(C.3-26)}$$

It is this equation that governs the propagation of monochromatic light through optical fibers. Hence any interpretation of photons traveling through fibers must be based on such a paraxial equation (or on the equivalent for pulsed radiation, eqn (C.4-9)).

Equation (C.3-25) has the structure of a two-dimensional time-dependent Schrödinger equation but with the independent spatial variable z in place of time t and $\lambdabar$ in place of $\hbar$. Just as with the TDSE we introduce a set of basis wavefunctions $\psi_n(\mathbf{r})$ from which to build a wavefunction $\Psi(\mathbf{r}, t)$, so also can we here introduce a set of electromagnetic mode fields $f_n(\mathbf{r})$ from which to build solutions to the paraxial wave equation, a general procedure for solving the Maxwell wave equations known as *coupled-mode* or *coupled-wave* theory [Sny72; Yar73; Sny83; Per99; Hua09].

Field confinement. When a field is confined in one or two directions perpendicular to the direction of energy flow, as happens when it is formed into a beam or travels through a dielectric waveguide, there must occur at the beam edges some longitudinal field components along the propagation direction. This longitudinal component of the field vectors becomes significant when the beam is strongly focused or is confined by refractive-index variation to a waveguide whose cross section has a dimension comparable to a wavelength [Lod15].

C.3.3 Reflection and refraction

When a beam of light encounters an abrupt change in the refractive index (a discontinuity of material properties) the Maxwell equations require that the electric and magnetic fields on each side of the matter discontinuity should maintain the continuity expressed by eqn (C.2-8). As a consequence, a portion of the beam energy may pass on through the interface (*refraction*), appearing as a transmitted beam, while the remainder returns, as a reflected beam, in the original material. The refracted and reflected fields depend upon the frequency, direction, and polarization of the incident beam, and on the adjacent refractive indices, in a manner that follows from application of eqn (C.2-8).

Bending rays and photons. The behavior of light at an interface is treated by considering its constituent frequencies, propagation directions, and field-vector directions. The simplest model is that of a monochromatic plane wave idealized as a *ray* of light defined by a propagation vector $\mathbf{k}(\mathbf{n})$ whose magnitude $k(\mathbf{n})$ is set by a frequency ω and a refractive index $\mathbf{n}$, in accord with the dispersion relation of eqn (C.3-20),

$$k(\mathbf{n}) \equiv \sqrt{\mathbf{k}(\mathbf{n})^2} = \frac{\mathbf{n}}{c}\omega. \tag{C.3-27}$$

This ray (a mathematical line) intersects an idealized flat interface of matter discontinuity. Let us take this interface to be the horizontal plane, that of Cartesian coordinates x, z. The incident ray thus lies in a vertical plane, the *plane of incidence*, having coordinates x, y, with y being the vertical direction (i.e. the direction of the normal to the discontinuity surface). By definition, the propagation vector has no z component, only components $k_x(\mathbf{n})$ and $k_y(\mathbf{n})$ in the plane of incidence. The angle of incidence θ_{in} for the ray, as it passes from index $\mathbf{n}_1$ to $\mathbf{n}_2$, measured from the vertical direction in this example, is defined by

$$\cos\theta_{\text{in}} = k_y(\mathbf{n}_1)/k(\mathbf{n}_1). \tag{C.3-28}$$

The abrupt redirection of a ray at an interface, or its more gradual change when passing through material that has a gradually changing refractive index, underlies the design of lenses. The rays can be regarded as comprising *bent photons*.

Fields at interfaces. In free space the electric and magnetic vectors are perpendicular to the direction of energy flow (the propagation vector $\mathbf{k}$). Under these conditions a linearly polarized electric vector associated with a given wavevector is commonly expressed by two components: A field labeled p that lies in the plane of incidence (here the x, y plane) but is perpendicular to the vector $\mathbf{k}$, and a field labeled s that is parallel to the interface surface and perpendicular to the incidence plane (here the z direction). The positive frequency part of such a field, constructed to be transverse to the propagation direction, has the structure

$$\mathbf{E}^{(+)}(\mathbf{n}; \mathbf{r}, t) = [\mathcal{E}_p \mathbf{e}_p + \mathcal{E}_s \mathbf{e}_s] \exp[i\mathbf{k}(\mathbf{n}) \cdot \mathbf{r} - i\omega t], \tag{C.3-29a}$$

$$\mathbf{B}^{(+)}(\mathbf{n}; \mathbf{r}, t) = [\mathcal{B}_p \mathbf{e}_p + \mathcal{B}_s \mathbf{e}_s] \exp[i\mathbf{k}(\mathbf{n}) \cdot \mathbf{r} - i\omega t], \tag{C.3-29b}$$

in which the exponential phase is expressible as

$$\mathbf{k}(\mathbf{n}) \cdot \mathbf{r} = k_p \mathbf{e}_p \cdot \mathbf{r} + k_s \mathbf{e}_s \cdot \mathbf{r} + k_k \mathbf{e}_k \cdot \mathbf{r}. \tag{C.3-29c}$$

The ray is termed "s polarized" or transverse electric (TE) when $\mathcal{E}_p = 0$ and is "p polarized" or transverse magnetic (TM) if $\mathcal{B}_p = 0$. Circularly (or general elliptically) polarized radiation is expressible as a phased coherent superposition of the s and p fields.

From the continuity relationships imposed by the Maxwell equations at the interface between two dielectrics one obtains the *Fresnel equations* for the field amplitudes in region 2 from their values in region 1 [Hec87; Bor99; Sal19]. These equations provide expressions for the fraction of radiation energy (and the field amplitudes) that is transmitted or reflected as a function of frequency, angle of incidence, and polarization; see Appendix C.3.5 for examples of their application.

Multilayer mirrors and filters. A succession of parallel interfaces, separated by uniform dielectric material (or vacuum), acts upon a monochromatic plane wave to introduce interference that can, for a given frequency, incidence angle, and polarization, produce any designed combination of transmission and reflection [Mac86; Wei87; Sal19]. In particular, a multilayer design can act as a mirror without introducing the loss that occurs with a metal mirror. With such an optical element the location of the reflection is not sharply defined, as it is with a metallic mirror: It is distributed over the thickness of the multilayer and therefore does not provide a sharply defined cavity. However, for any calculated stationary-field distribution that fits between two such mirrors it is possible to define an effective cavity length,

$$L_{\text{eff}} = \frac{\int dx \ |\mathcal{E}(x)|^2}{\max |\mathcal{E}|^2},$$ (C.3-30)

that characterizes the spatial structure of the mode.

C.3.4   Evanescent fields

The interface conditions upon fields adjacent to a discontinuity in bulk properties provide, for TE radiation in which the electric vector lies in the plane of the interface, *Snell's law* [Hec87; Bor99; Sal19] for the connection between the propagation angle in region 2 (a refracted ray) in terms of the angle in region 1 (an incident ray) and the two refractive indices [Hec87; Bor99; Sal19],

$$\sin \theta_2 = \frac{n_1}{n_2} \sin \theta_1.$$ (C.3-31)

Transmission becomes impossible, and a ray undergoes total internal reflection (TIR) [Van16; Bek18] when it passes into a region of lower refractive index (say from glass to air) and the angle of incidence θ_1 exceeds the critical angle defined by

$$\sin \theta_{\text{crit}} = \frac{n_2}{n_1}.$$ (C.3-32)

At this angle, and higher, there will be in (low-index) region 2 an *evanescent field* whose magnitude falls exponentially with distance away from the interface [Bli12; Bli14b; Van16; Bek18]. This field has evanescent waves, localized in the vicinity of the surface discontinuity, that propagate in a direction dictated by the incident and reflected waves. These superpositions of electromagnetic and polarization fields are known as *surface polaritons* [Bli12; Bli17].

Multiple-fiber waveguides. The refractive-index variation across an optical fiber can direct propagating waves along the fiber axis (an optical waveguide), but in the transverse direction there exist non-propagating evanescent fields whose amplitude diminishes exponentially with distance. When two fibers are adjacent to each other the evanescent fields provide a linkage between two regions of propagation, and propagating energy can flow transversely between the two fibers. The distribution of pulse intensity with distance along a fiber is mathematically equivalent to the population distribution with time amongst quantum states of an atom or similar structure: Both intensity and probability are squares of interfering amplitudes. Thus a system of N fibers has similar dynamics to an N-state quantum system, with propagation distance z along the fiber axes replacing the time variable t of the time-dependent Schrödinger equation [Lon06; Lon07; Lon09; Sho17]. The physical arrangement of the fibers—their transverse separation—defines the couplings between the fiber fields—the counterparts of Rabi frequencies. Just as suitably designed optical pulses can transfer atomic population between discrete quantum states, so too can suitably designed arrangements of fibers lead to pulse transfer between fibers or to beam splitting amongst fibers. The pulses traveling through such a system have properties attributable to atomic-excitation photons, but these entities comprise linked electromagnetic fields and matter, moving with localized velocity that is adjustable by construction. When the pulses originate with quantum transitions, then they can retain some identifiable quantum characteristics in fibers.

Whispering-gallery modes. Curving walls along which sound waves travel with little attenuation have been historically interesting. The noteworthy experimental and theoretical studies of the whispering gallery in the dome of St Pauls cathedral by Lord Rayleigh showed that the sound "crept" along the wall. An analog of such acoustical structures, in which evanescent waves are crucial, is found in the electromagnetic field modes of small dielectric spheres surrounded by vacuum [Bra89; Vah03; For15; Lod17; Sch18b]. The modes are factorizable in spherical coordinates, with Bessel functions for the radial part inside the sphere, outgoing-wave Hankel functions in the exterior and Legendre polynomials $P_n^m(\cos\theta)$ for the azimuthal angular part. A *whispering-gallery mode* (WGM) is a wave for which the Bessel function has no roots inside the sphere and which has identical and large indices n and m for the azimuthal variation, giving concentration in an equatorial band. There are an infinite number of such modes, with frequencies that alter with sphere radius and with distortions of the spherical shape. Their quality factor can be very high: A frequency-matching photon injected into one of these modes, via coupling to an external source, will be sustained with little loss.

Transverse spin. indexspin!transverse The presence of an evanescent field has a significant effect upon wave propagation, and hence upon the field modes that serve as photons. The review [Van16] presents a simple discussion of the essentials, followed below.

Consider the phase of a monochromatic plane wave of frequency $\omega = ck$ that is incident upon an interface beyond which there is an evanescent field. The propagation vector in that material is complex, say

$$\mathbf{k} = \boldsymbol{\kappa} + i\boldsymbol{\eta}, \tag{C.3-33}$$

where $\boldsymbol{\kappa}$ sets the direction of phase (and momentum) propagation and $\boldsymbol{\eta}$ sets the direction of evanescent decay, away from the surface of refractive-index discontinuity. From the free-space Maxwell equations we have the dispersion relation

$$\mathbf{k} \cdot \mathbf{k} = (\omega/c)^2, \tag{C.3-34}$$

and so, from the real and imaginary parts of this scalar product, we have the conditions

$$\kappa^2 - \eta^2 = (\omega/c)^2, \qquad \boldsymbol{\kappa} \cdot \boldsymbol{\eta} = 0. \tag{C.3-35}$$

These state that the vectors $\boldsymbol{\kappa}$ and $\boldsymbol{\eta}$ must be orthogonal: The propagation direction of an evanescent wave is perpendicular to its direction of decay.

A useful complete set of unit vectors to describe this scenario [Van16] starts with the direction of the wave vector $\mathbf{k}$, from which we obtain two additional orthogonal unit vectors in the plane transverse to this (see Appendix B.1.6),

$$\mathbf{e}_k = \frac{\mathbf{k}}{|\mathbf{k}|} \qquad \mathbf{e}_s = \frac{\boldsymbol{\kappa} \times \boldsymbol{\eta}}{|\boldsymbol{\kappa} \times \boldsymbol{\eta}|} = \mathrm{i}\frac{\mathbf{k} \times \mathbf{k}^*}{|\mathbf{k} \times \mathbf{k}^*|}, \qquad \mathbf{e}_p = \mathbf{e}_k \times \mathbf{e}_s. \tag{C.3-36}$$

These three unit vectors form an orthogonal triad,

$$\mathbf{e}_k \cdot \mathbf{e}_s = \mathbf{e}_k \cdot \mathbf{e}_p = \mathbf{e}_s \cdot \mathbf{e}_p = 0. \tag{C.3-37}$$

When expressed with such vectors the electric-field direction (its polarization) is entirely defined by its momentum relative to the interface.

Because the $\mathbf{E}$ field is transverse, $\mathbf{k} \cdot \mathbf{E} = 0$, it must be a superposition of the s and p unit vectors. Its positive-frequency part can therefore be written

$$\mathbf{E}^{(+)}(\mathbf{r}, t) = [\mathcal{E}_s \mathbf{e}_s + \mathcal{E}_p \mathbf{e}_p] \exp(\mathrm{i}\boldsymbol{\kappa} \cdot \mathbf{r} - \boldsymbol{\eta} \cdot \mathbf{r} - \mathrm{i}\omega t). \tag{C.3-38}$$

The s vector is orthogonal to the decay direction and so it is in the plane of discontinuity, transverse to the evanescent-wave propagation in this plane. It is real, and so the orthogonal field component remains constant—it is linearly polarized. The p vector appearing here is complex and is expressible as [Van16]

$$\mathbf{e}_p = \mathrm{i}\frac{\eta}{k}\left(\frac{\boldsymbol{\kappa}}{\kappa}\right) - \frac{\kappa}{k}\left(\frac{\boldsymbol{\eta}}{\eta}\right). \tag{C.3-39}$$

The corresponding electric vector therefore has two phase-related components, one in the direction of evanescent decay, the other in the propagation direction. With each cycle of the field the electric vector traces out an ellipse—it is elliptically polarized. This behavior is a rotation—a spinning—and the s vector that defines the axis of this spin is transverse to the phase-propagation direction.

To picture this scenario, let the evanescent-decay direction be vertical (y), so that the dielectric discontinuity is in a horizontal plane (x, z). Let the phase-propagation vector $\boldsymbol{\kappa}$ define the z direction, so that the plane of incidence is (y, z). An s-polarized field has horizontal (H) linear polarization, while a p-polarized field is a coherent superposition of a vertical (V) field and a field along the propagation direction. Whereas the electric vector of a traditional circular polarization field is akin to the blades of a spinning airplane propeller (with spin axis along the propagation direction), the p field here is akin to the spokes of a bicycle wheel, whose spin axis is transverse to the travel direction – a so-called "photonic wheel" [Ban13; Aie15b; Aie16; Bli17].

We define "spin" to be the intrinsic handedness (left-right circular) of the p basis vector. It is noteworthy that a change of propagation direction $\boldsymbol{\kappa} \to -\boldsymbol{\kappa}$ for fixed decay direction $\boldsymbol{\eta}$ changes the handedness of p polarization, and flips the direction of spin $\mathbf{e}_s$. The sign of this spin is locked to the direction of propagation (so-called spin-momentum locking [Bli15; Van16; Lod17; Bek18]): A forward-traveling wave is associated with positive spin, a backward-traveling wave has negative spin. This "handedness" of the p-field wave is akin to the handedness of neutrinos.

C.3.5 Beam splitters

Beam splitters, along with lenses and mirrors, are commonplace elements of optical workplaces. They allow not only splitting a single beam into two redirected beams but also combining two beams into a single beam. One may well ask how it is possible to reconcile such splitting of a beam with the notion of a photon as an indivisible entity. For the answer we need to regard the radiation beam as a mode that has a presence along multiple paths. This is also how we must regard various interference effects in which a beam illuminates several slits, from which it recombines to form an interference pattern. In so doing, we find confirmation that, indeed, there exist field increments (photons) that are indivisible.

Classical beam splitter. The simplest form of a beam splitter, historically available as a partially silvered mirror, reflects a fraction R of an incident beam intensity while transmitting the remainder, T. When the two fractions are equal the device is termed a 50:50 beam splitter. Figure C.1a shows a schematic representation of an idealized beam splitter in which there occurs neither loss nor gain of energy. When the input is a single photon it must either be detected by detector D1 or detector D2: There will be no simultaneous (coincident) signals from these. More general beam splitters combine two input fields to produce two output fields, an example of a four-port device, as shown schematically in frame b.

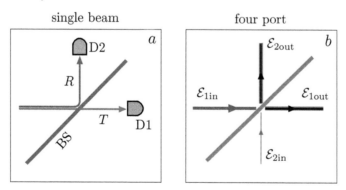

Fig. C.1 *Beam splitter as (a) single mode or (b) four-port device.*

For classical fields the output fields are related to the input fields by the relations [Man84; Ou87; Gar08; Bar09]

$$\mathcal{E}_{1\text{out}} = \mathsf{t}_1 \, \mathcal{E}_{1\text{in}} + \mathsf{r}_1 \, \mathcal{E}_{2\text{in}}, \qquad \mathcal{E}_{2\text{out}} = \mathsf{t}_2 \, \mathcal{E}_{2\text{in}} + \mathsf{r}_2 \, \mathcal{E}_{1\text{in}}, \tag{C.3-40}$$

where the coefficients of transmission t_i and reflection r_i satisfy the relationships

$$|t_1|^2 = |t_2|^2 = T, \qquad |r_1|^2 = |r_2|^2 = R, \qquad |t_i|^2 + |r_i|^2 = R + T = 1. \qquad \text{(C.3-41)}$$

There are two popular choices for the phases of these coefficients [Bar09]:

Choice 1: $t_1 = t_2 = \sqrt{T}, \quad r_1 = r_2 = i\sqrt{R},$

Choice 2: $t_1 = t_2 = \sqrt{T}, \quad r_1 = -r_2 = \sqrt{R}.$

Quantum beam splitter.   As discussed in [Man84; Bar09; Gar08] the quantized version of a beam splitter replaces each of the classical field amplitudes $\mathcal{E}_i$ by a photon annihilation operator $\hat{a}_i$, so that the classical relationship eqn (C.3-40) of output amplitudes to input amplitudes becomes

$$\begin{bmatrix} \hat{a}_{1\text{out}} \\ \hat{a}_{2\text{out}} \end{bmatrix} = \begin{bmatrix} t_1\hat{a}_{1\text{in}} + r_1\hat{a}_{2\text{in}} \\ t_2\hat{a}_{2\text{in}} + r_2\hat{a}_{1\text{in}} \end{bmatrix} = \begin{bmatrix} t_1 & r_1 \\ r_2 & t_2 \end{bmatrix} \begin{bmatrix} \hat{a}_{1\text{in}} \\ \hat{a}_{2\text{in}} \end{bmatrix}. \qquad \text{(C.3-42a)}$$

To describe the input of the beam splitter that produces a specified output we invert these formulas,

$$\begin{bmatrix} \hat{a}_{1\text{in}} \\ \hat{a}_{2\text{in}} \end{bmatrix} = \begin{bmatrix} t_1\,\hat{a}_{1\text{out}} + r_2\,\hat{a}_{2\text{out}} \\ t_2\,\hat{a}_{2\text{out}} + r_1\,\hat{a}_{1\text{out}} \end{bmatrix} = \begin{bmatrix} t_1 & r_2 \\ r_1 & t_2 \end{bmatrix} \begin{bmatrix} \hat{a}_{1\text{out}} \\ \hat{a}_{2\text{out}} \end{bmatrix}. \qquad \text{(C.3-42b)}$$

Although a classical beam splitter has no constraint on the magnitude of any of the field amplitudes, the quantum version must maintain the photon commutators [Gar08; Bar09],

$$[\hat{a}_i , \hat{a}_j^\dagger] = \delta_{ij}, \qquad \text{(C.3-43)}$$

and give the results for decreasing and increasing the number of photons,

$$\hat{a}_i|n_i\rangle = \sqrt{n_i}|n_i - 1\rangle, \qquad \hat{a}_i^\dagger|n_i\rangle = \sqrt{n_i + 1}|n_i + 1\rangle, \qquad \text{(C.3-44)}$$

found in the eigenstates of the mode photon-number operators

$$\hat{N}_i|n_i\rangle = n_i|n_i\rangle, \qquad \hat{N}_i = \hat{a}_i^\dagger\hat{a}_i. \qquad \text{(C.3-45)}$$

Each mode has a zero-photon vacuum state $|0\rangle$ from which further energy (further photons) cannot be removed,

$$\hat{a}_i|0_i\rangle = 0. \qquad \text{(C.3-46)}$$

Example.   As an example, the photon-number operators associated with output beams relates to the input fields as [Gar08]

$$\begin{aligned} \hat{n}_{1\text{out}} &= \hat{a}_{1\text{out}}^\dagger\hat{a}_{1\text{out}} \\ &= (t_1^*\hat{a}_{1\text{in}}^\dagger + r_1^*\hat{a}_{2\text{in}}^\dagger)(t_1\hat{a}_{1\text{in}} + r_1\hat{a}_{2\text{in}}) \\ &= t_1^* t_1\hat{a}_{1\text{in}}^\dagger\hat{a}_{1\text{in}} + t_1^* r_1\hat{a}_{1\text{in}}^\dagger\hat{a}_{2\text{in}} + r_1^* t_1\hat{a}_{2\text{in}}^\dagger\hat{a}_{1\text{in}} + r_1^* r_1\hat{a}_{2\text{in}}^\dagger\hat{a}_{2\text{in}} \\ &= T\,\hat{n}_{1\text{in}} + R\,\hat{n}_{2\text{in}} + \sqrt{RT}\left(\hat{a}_{1\text{in}}^\dagger\hat{a}_{2\text{in}} + \hat{a}_{2\text{in}}^\dagger\hat{a}_{1\text{in}}\right), \qquad \text{(C.3-47a)} \end{aligned}$$

$$\begin{aligned} \hat{n}_{2\text{out}} &= \hat{a}_{2\text{out}}^\dagger\hat{a}_{2\text{out}} \\ &= \left(t_2^*\hat{a}_{2\text{in}}^\dagger + r_2^*\hat{a}_{1\text{in}}^\dagger\right)\left(t_2\hat{a}_{2\text{in}} + r_2\hat{a}_{1\text{in}}\right) \\ &= R\,\hat{n}_{1\text{in}} + T\,\hat{n}_{2\text{in}} - \sqrt{RT}\left(\hat{a}_{1\text{in}}^\dagger\hat{a}_{2\text{in}} + \hat{a}_{2\text{in}}^\dagger\hat{a}_{1\text{in}}\right). \qquad \text{(C.3-47b)} \end{aligned}$$

Because the model has introduced no loss, the sum of photon numbers in both output channels is unchanged [Gar08],

$$\hat{n}_{2\text{out}} + \hat{n}_{1\text{out}} = \hat{n}_{2\text{in}} + \hat{n}_{1\text{in}}. \tag{C.3-48}$$

When these act on photon-number states, the values of expectation values are integers. If there is only a single input photon, then it will be found in only one of the output channels: It is not possible to detect half a photon, nor is it possible to detect, simultaneously, a photon in each output channel. The electric field amplitudes passing through linear devices such as beam splitters may follow concurrent pathways to their eventual detection. Unlike classical fields, the detection step is related to quantum operators of photon creation and annihilation that alter detection amounts by discrete increments. In particular, photons cannot be removed from the vacuum.

Example. As another example [Hon87; Ou87], the output of a beam splitter that has single photons in each of the input ports is, with notation $|n_1, n_2\rangle$,

$$\begin{aligned}
\hat{a}^\dagger_{1\text{in}} \hat{a}^\dagger_{2\text{in}} |\text{vac}\rangle &\Rightarrow \left(\mathfrak{t}_1^* \hat{a}^\dagger_{1\text{out}} + \mathfrak{r}_2^* \hat{a}^\dagger_{2\text{out}} \right) \left(\mathfrak{t}_2^* \hat{a}^\dagger_{2\text{out}} + \mathfrak{r}_1^* \hat{a}^\dagger_{1\text{out}} \right) |\text{vac}\rangle \\
&= \mathfrak{t}_1^* \mathfrak{t}_2^* |1_1, 1_2\rangle + \mathfrak{r}_2^* \mathfrak{r}_1^* |1_1, 1_2\rangle + \mathfrak{t}_1^* \mathfrak{r}_1^* \sqrt{2}|2_1, 0_2\rangle + \mathfrak{r}_2^* \mathfrak{t}_2^* \sqrt{2}|0_1, 2_2\rangle \\
&= (T - R) |1_1, 1_2\rangle + \sqrt{2RT} \left(|2_1, 0_2\rangle - |0_1, 2_2\rangle \right). \tag{C.3-49}
\end{aligned}$$

Figure C.2 illustrates this behavior, showing the four possible connections from inputs a and b of a 50:50 beam splitter (with $T = R$) [Lou05; Bar09b]. When the input fields are those of identical photons, only the left-hand connections are possible: There are no simultaneous counts (coincidences) in the two detectors. This behavior is evidence for the quantum nature of the fields—for photons.

Coincident photons. Discussions of beam splitter action readily offer proposals for testing the existence of photons or, more precisely, for testing the quantum nature of a field. However, experiments inevitably encounter the fact that a field whose frequency is sharply defined must have an extensive duration. The processes that are acknowledged to create photons produce fields that have some temporal extent, and so the notion of simultaneous detection of photon parts requires attention. The approach used in quantum optics is through consideration of correlations, discussed in the following section.

C.3.6 Nonlinear optics

The most common effect of an optical-frequency radiation field upon an atom is an alteration of populations by means of the interaction energy of an atomic dipole-transition moment in the electric field of the radiation. In altering populations of states 1 and 2 the radiation alters also the expectation values of the dipole-transition moment between these two states. The spatial distribution of this expectation value is the polarization field $\mathbf{P}$ that controls the propagation of the $\mathbf{E}$ field; see eqn (C.2-9a) and eqn (C.2-4). When the illumination is steady and prolonged, so that coherent transients have diminished and susceptibility can be defined, the change in the propagating field is attributable to an altered refractive index and attenuation coefficient.

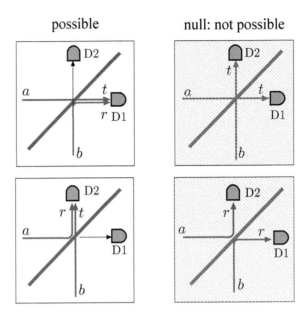

Fig. C.2 *Beam splitter with identical-photon inputs.*

The field-induced alteration of the steady-state dipole-moment expectation value—an *induced* moment—is strongest for resonant excitation, when the carrier frequency of the field ω is close to matching the Bohr transition frequency ω_{12},

$$\hbar\omega \approx \hbar\omega_{12} \equiv |E_2 - E_1|. \tag{C.3-50}$$

But situations exist in which, even without this resonant condition, the smaller field-induced changes have significant effects on other fields, even fields that are not initially present.

For example, two radiation fields, acting between three quantum states, create possibly three pairs of transitions and three Bohr transition frequencies, the pairs 1-2, 2-3, and 3-1, and three corresponding Rabi frequencies; see Figure 5.3 on page 126. Consider a situation in which a controlled pump field, having carrier frequency ω_{pump}, links states 1 and 2 resonantly, while a controlled second (idler) field, of carrier frequency ω_{idler}, links states 2 and 3 resonantly. Each of these transitions is associated with a dipole-transition moment, d_{12} and d_{23} respectively, whose spatially-distributed expectation value is responsible for a polarization field that alters the pump and idler fields as they propagate. Each electric field not only alters the populations affecting the dipole-transition moment (and polarization field) associated with its own resonant frequency (an example of linear optics) but, because it alters population in state 2 shared with the other field, it produces changes there as well. This is an example of *nonlinear* optics [Blo82; Boy08; Gae06; New11; Dru14], to be contrasted with the various effects of propagation, transmission, refraction, and diffraction that constitute the subject of ordinary linear optics [Hec87; Bor99; Sal19].

The following paragraphs describe two examples in which imposed fields can generate a new field, either steadily or transiently.

Second order. When the system has no center of symmetry then there will be a nonzero dipole-transition moment d_{13} between states 1 and 3 to complete the set of three transitions into a closed loop, as shown in the following expression:

$$|1\rangle \xleftarrow{\;pump\;}{d_{12}} |2\rangle \xleftarrow{\;idler\;}{d_{23}} |3\rangle \xleftarrow{\;signal\;}{d_{31}} |1\rangle. \qquad (C.3\text{-}51)$$

The three quantum-state energies E_n may rank in any order, and alternative situations are possible in which the three fields act in other sequences, such as *idler, pump, signal.* From considerations of photon-energy conservation this third transition (here called the signal field) has a frequency that is either the sum or the difference of the two other frequencies (which need not match any Bohr frequencies):

$$\omega_{signal} = |\omega_{pump} \pm \omega_{idler}|. \qquad (C.3\text{-}52)$$

The minus sign (and energy difference) is to be taken only when the energy E_2 is the largest (a Lambda linkage of the three states, $1-2-3$). The absolute value signs accommodate the possibility of starting from an excited state. This new (signal) field will be created whenever the pump and idler fields are present, an example of a second-order nonlinear effect known as frequency *up-conversion* (with the plus sign) or *down-conversion* (with the minus sign).

Third order. When the system has a center of symmetry, as occurs for an isolated atom in free space, then atomic selection rules forbid the completion of a three-state loop linkage. Instead, an allowed loop can take place if one of the interactions proceeds by a two-photon transition, driven by two pump fields (possibly the same field) involving virtual intermediate states, as in the following scheme

$$|1\rangle \xrightarrow{\;pump\,a\;}{d_{1v}} |v\rangle \xrightarrow{\;pump\,b\;}{d_{v2}} |2\rangle \xleftarrow{\;idler\;}{d_{23}} |3\rangle \xleftarrow{\;signal\;}{d_{31}} |1\rangle. \qquad (C.3\text{-}53)$$

Again, as with eqn (C.3-51), the fields may occur with any ordering along the loop linkage, and the energies E_n may also rank in any order. The individual carrier frequencies need not match any Bohr frequencies but overall energy conservation leads to the requirement

$$\omega_{signal} = |\omega_{pump\,a} \pm \omega_{pump\,b} \pm \omega_{idler}|. \qquad (C.3\text{-}54)$$

This is a third-order nonlinear process, known as *four-wave mixing*, in which the signal field has a frequency fixed by the other frequencies and the arrangement of the energy levels.

Nonlinear susceptibilities. Much of the literature of nonlinear optics [Blo82; Boy08; Gae06; New11; Dru14] deals with steady illumination in which the dipole-moment expectation values that produce the polarization field **P** have equilibrated with their environment. Typically the controlled carrier frequencies do not match Bohr frequencies, so that the procedure of adiabatic elimination is applicable; see Appendix A.12.

Under such circumstances the relationship between **P** and **E** is expressible in terms of a nonlinear susceptibility times a product of several **E**-field components. The field amplitudes $\mathcal{E}(\omega)$ appearing in those expressions, for example eqn (C.3-55) and eqn (C.3-56) below, have interpretation either as classical fields or as photon-field creation and annihilation operators.

For the second-order nonlinearity the polarization field for the signal field is proportional to the product of amplitudes for the other two fields and a susceptibility $\chi^{(2)}$, as in the traditional formula relating field amplitudes [Gae06],

$$P_i(\omega) = \epsilon_0 \sum_{jk} \chi_{ijk}^{(2)}(\omega; \omega_m, \omega_n)\, \mathcal{E}_j(\omega_m)\, \mathcal{E}_k(\omega_n). \tag{C.3-55}$$

Here the indices ijk denote Cartesian components of vector fields. Such formulas must be made specific with positive- or negative-frequency amplitudes $\mathcal{E}_j^{(+)}$ or $\mathcal{E}_j^{(-)}$ depending on whether the field is adding or subtracting energy (and photons). Similarly, the traditional form for the third-order nonlinearity is

$$P_i(\omega) = \epsilon_0 \sum_{jkl} \chi_{ijkl}^{(3)}(\omega; \omega_m, \omega_n, \omega_o)\, \mathcal{E}_j(\omega_m)\, \mathcal{E}_k(\omega_n)\, \mathcal{E}_l(\omega_o). \tag{C.3-56}$$

Again, various situations require either $\mathcal{E}_j^{(+)}$ or $\mathcal{E}_j^{(-)}$ in the construction of the polarization field components.

As was noted with the definition of linear response, eqn (C.3-2), the formulas of eqn (C.3-55) and eqn (C.3-56) assume steady, equilibrated matter response to controlled (monochromatic) carrier frequencies (pump and idler). However, contemporary interest in nonlinear optics often involves pulses that are shorter than relaxation times. The traditional nonlinear susceptibilities, such as shown in eqn (C.3-55) and eqn (C.3-56), then do not give a correct description of the system, and explicit time dependence must be evaluated using the time-dependent Schrödinger equation. The following section introduces the needed formalism.

C.4   Bulk-matter transient sources

The preceding portions of the discussion of radiation in matter have made the assumption that the individual dipole-moment expectation values appearing in eqn (C.2-4) were to be taken as equilibrium values appropriate to weak, steady radiation. To treat pulsed radiation having duration shorter than the relaxation time for atoms in the material, we must examine the expectation values as expressed in terms of off-diagonal elements of the density matrix (i.e. coherences)

$$\langle \mathbf{d}(\mathbf{r}, t) \rangle \equiv \langle \Psi(\mathbf{r}, t) | \mathbf{d} | \Psi(\mathbf{r}, t) \rangle = \sum_{mn} \rho_{nm}(\mathbf{r}, t)\, e^{-i\zeta_n + i\zeta_m} \langle \psi_m | \mathbf{d} | \psi_n \rangle. \tag{C.4-1}$$

Like the time t, the position $\mathbf{r}$ appears here as a parameter, not a dynamical variable. The phases ζ_n in the exponentials, chosen in the rotating-wave picture to make the coherences $\rho_{nm}(t)$ slowly varying, involve various carrier frequencies. For example, the three-state system used with the P and S fields in the Lambda linkage has the dipole-moment expectation value

$$\langle \mathbf{d}(\mathbf{r},t)\rangle = e^{-i\varphi_P}\,\rho_{21}(\mathbf{r},t)\,\mathbf{d}_{12} + e^{-i\varphi_S}\,\rho_{23}(\mathbf{r},t)\,\mathbf{d}_{32} + e^{-i\varphi_P+i\varphi_S}\,\rho_{31}(\mathbf{r},t)\,\mathbf{d}_{13}$$

$$+ e^{+i\varphi_P}\,\rho_{12}(\mathbf{r},t)\,\mathbf{d}_{21} + e^{+i\varphi_S}\,\rho_{32}(\mathbf{r},t)\,\mathbf{d}_{23} + e^{+i\varphi_P-i\varphi_S}\,\rho_{13}(\mathbf{r},t)\,\mathbf{d}_{31}.$$

(C.4-2)

Consequently, the positive-frequency part of the polarization field for a single atomic species has the construction

$$\mathbf{P}^{(+)}(\mathbf{r},t) = \mathcal{N}(\mathbf{r})\sum_{mn}\rho_{nm}(\mathbf{r},t)\,\mathbf{d}_{mn}\exp(-i\zeta_n + i\zeta_m).$$

(C.4-3)

The set of exponentials appearing here with paired phases are to be matched with exponentials present in the electric field of the wave equation, eqn (C.2-9a). For classical radiation we express these as a mode expansion of the form

$$\mathbf{E}^{(+)}(\mathbf{r},t) = \tfrac{1}{2}\sum_{\nu}\mathcal{E}_\nu(t)\,\mathbf{U}_\nu(\mathbf{r})\,\exp(-i\omega_\nu t).$$

(C.4-4)

For application to waves traveling along the z axis (the paraxial-field approximation) the phases are

$$\varphi_\nu = \omega_\nu t - k_\nu z, \qquad ck_\nu = \omega_\nu.$$

(C.4-5)

Each of the frequencies ω_ν appearing in the sum of eqn (C.4-3) selects a similar frequency-component of the $\mathbf{E}$ field of eqn (C.4-4).

The direction of the dipole-moment vector $\mathbf{d}$ and the direction of the mode field $\mathbf{U}_\nu$ must match. The availability of the dipole moments $\mathbf{d}_{nm}$ depends on the properties of the linked states, i.e. whether they support allowed-dipole transitions. In order for these to affect the $\mathbf{E}$ field the companion elements $\rho_{mn}(\mathbf{r},t)$ of the density matrix must be nonzero: There must be appropriate coherences.

If the behavior of the density matrix is fully coherent, i.e. governed by the TDSE, then its Rabi oscillations will add sidebands onto the carrier frequencies. If, instead, the relevant behavior of the density matrix is a steady state, introduced by incoherent processes, then the polarization source term of the wave equation only contributes carrier frequency components. That is the regime of classical, linear optics [Bor99], discussed in the preceding paragraphs. The following paragraphs discuss the situation when field and matter must be considered as coupled parts of a single system, one described by a field that superposes field and matter character:

$$\mathbf{F}^{(+)}(\mathbf{r},t) = \alpha(\mathbf{r},t)\,\mathbf{E}^{(+)}(\mathbf{r},t) + \beta(\mathbf{r},t)\,\epsilon_0\mathbf{P}^{(+)}(\mathbf{r},t).$$

(C.4-6)

Two general situations are of particular interest: Either the EM field travels past a distribution of stationary atoms, or individual atoms move into a cavity that supports a standing-wave EM field.

C.4.1   One dimensional short-pulse propagation

The assumption of steady-state response to the field, made in the previous paragraphs, fails when the field is a pulse whose duration is shorter than the relaxation time of the excitable atoms. To treat the propagation of short pulses of directed beams, such as we have with traveling waves from a laser source, we extract a carrier plane-wave propagating along a Cartesian axis, typically taken as the z axis. After separating the field into complex-valued positive- and negative-frequency parts we write

$$\mathbf{E}^{(+)}(\mathbf{r}, t) = \mathbf{e}\, e^{ikz - i\omega t}\mathbf{e}\,\mathcal{E}(z, t). \tag{C.4-7}$$

Then, with $ck = \omega$, the homogeneous portion of the wave equation becomes

$$\left[\frac{\partial^2}{\partial z^2} + \frac{\partial^2}{\partial x^2} + \frac{\partial^2}{\partial y^2} - \frac{1}{c^2}\frac{\partial^2}{\partial t^2}\right]\mathbf{E}^{(+)}(\mathbf{r}, t)$$

$$= \left(\frac{\partial}{\partial z} - \frac{1}{c}\frac{\partial}{\partial t}\right)\left(\frac{\partial}{\partial z} + \frac{1}{c}\frac{\partial}{\partial t}\right)e^{ikz - i\omega t}\mathbf{e}\,\mathcal{E}(z, t) \tag{C.4-8}$$

$$= \mathbf{e}\,e^{ikz - i\omega t}\left[2ik\left(\frac{\partial}{\partial z} + \frac{1}{c}\frac{\partial}{\partial t}\right) + \frac{\partial^2}{\partial z^2} - \frac{1}{c^2}\frac{\partial^2}{\partial t^2}\right]\mathcal{E}(z, t).$$

In making the change of the last line we neglect any reflections and follow only the forward-traveling wave. We make the SVEA and neglect the second derivatives. The resulting equation for the field amplitude for mode ν selects the corresponding term from the sum of dipole moments, and we obtain the one-dimensional inhomogeneous equation for pulse propagation through matter with coordinates z and ct,

$$\left(\frac{\partial}{\partial z} + \frac{\partial}{\partial ct}\right)\mathcal{E}_\nu(z, t) = i\frac{\mathcal{N}\omega_\nu}{2c\epsilon_0}\mathbf{e}^* \cdot \mathbf{d}_{nm}\,\rho_{mn}(z, t). \tag{C.4-9}$$

This equation, for each field mode, accompanies a TDSE for each coherence $\rho_{mn}(z, t)$. When the pulses are short, so that there is no time for decoherence to affect the atomic dynamics, the coherent excitation of the individual atoms (e.g. Rabi oscillations) produce marked modification of the pulses, with effects such as self-induced transparency (SIT), for pulses of temporal area 2π, and soliton formation [All87]. Unlike the steady-state linear optics of eqn (C.3-25), the equation eqn (C.4-9) brings the possibility of coherent, reversible transfer of energy from field to atom. It therefore transfers any quantum nature of the field to quantum excitation. The process can be reversed, to restore a field that recovers its initial quantum properties.

C.4.2   Dark-state polaritons; Spin waves

When a weak probe-field pulse (P) passes into a medium in which it resonantly couples states 1 and 2, and that medium is subjected to a strong *control field* (S) that resonantly couples state 2 to a third state 3 (in either a Lambda or Ladder linkage), the propagation of the P field is strongly affected. A strong and steady S field will alter the resonance condition of the weak P field and lead to electromagnetically-induced transparency (EIT) [Har90; Bol91; Har97; Kuh98; Hau99; Fle05]. With increasing strength of the control field the group velocity v_g of the P-field pulse diminishes and its electric field transforms into a spatial-distribution field of atomic coherences. The required slow adjustment of the S field, and the use of the dark state, has features in common with STIRAP, and so the procedure has been included in reviews of that process [Vit01; Vit17]. The text below repeats that discussion.

The simplest starting point for a description of this behavior is an idealization of the electric-field envelopes as pulses that move in one direction, taken to be z. For the P field, traveling through a constant and uniform number density $\mathcal{N}$, the slowly-varying envelope, expressed as a Rabi frequency $\Omega_P(z,t)$, satisfies the equation

$$\left(\frac{\partial}{\partial z} + \frac{1}{c}\frac{\partial}{\partial t}\right)\Omega_P(z,t) = \mathrm{i}\frac{\alpha_p\Gamma}{2}\,C_2(z,t)C_1^*(z,t), \qquad \text{(C.4-10)}$$

where Γ denotes the decay rate of state 2 and the P-field resonant absorption coefficient,

$$\alpha_p = \frac{\omega_p\mathcal{N}|d_{12}|^2}{2\epsilon_0 c\hbar\Gamma}, \qquad \text{(C.4-11)}$$

parametrizes the effect of the atoms on the P field. A similar equation describes the S field, which is taken to be sufficiently strong to be relatively unaffected during its travel at speed c.

The probability amplitudes that serve in eqn (C.4-10) as sources to spatial changes in Rabi frequency satisfy the usual three-state TDSE: The probability amplitudes C_1 and C_2 are linked by the P field, while the S field links C_2 and C_3. Adiabatic elimination of these atomic probability amplitudes produces the equation [Vit01]

$$\left(\frac{\partial}{\partial z} + \frac{1}{v_g}\frac{\partial}{\partial t}\right)\Omega_P(z,t) = 0, \quad \text{with} \quad v_g = \frac{c}{1+n_g} \quad \text{and} \quad n_g = \frac{\alpha_p\Gamma c}{\Omega_{\mathrm{rms}}^2}. \qquad \text{(C.4-12)}$$

Here Ω_{rms} is the root-mean-square Rabi frequency.

By adjusting the intensity of the S field, and thereby controlling the RMS Rabi frequency, it is possible to control the propagation velocity v_p of the weak P pulse, to bring it to a full stop, and to re-accelerate it on demand. This behavior is associated with the existence of a quasi-particle called a *dark-state polariton* [Fle96; Fle00; Luk01; Vit01; Fle02; Joh04; Fle05] that is a coherent superposition of electric field (i.e. Rabi frequency) and macroscopic atomic-coherence components (termed a *spin wave*, of excitation)

$$F(z,t) = \cos\theta(z,t)\,\Omega_p(z,t) - \sin\theta(z,t)\,\sqrt{\alpha_p\,c\,\Gamma}\,C_3(z,t)C_1^*(z,t). \qquad \text{(C.4-13)}$$

The angle θ that quantifies the relative amounts of field and matter excitation is defined by

$$\tan\theta(z,t) = \frac{\sqrt{\alpha_p\,c\,\Gamma}}{\Omega_S(z,t)}. \qquad \text{(C.4-14)}$$

The dark-state polariton obeys the simple propagation equation

$$\left[\frac{\partial}{\partial t} + c\cos^2\theta(z,t)\frac{\partial}{\partial z}\right]F(z,t) = 0. \qquad \text{(C.4-15)}$$

If Ω_S (and hence θ) is approximately uniform in z then eqn (C.4-15) describes form-preserving propagation of a quasiparticle (a spin wave) having propagation velocity

$$v(t) = c\cos^2\theta(t). \tag{C.4-16}$$

Stopping a pulse of radiation. By slowly reducing the S-field Rabi frequency an experimenter can adiabatically rotate $\theta(t)$ from 0 to $\pi/2$ and thereby transform an initially pure electromagnetic polariton ($F = \Omega_P$) into a pure atomic polarization ($F = \sqrt{\alpha c \Gamma}C_3 C_1^*$). Concurrently the polariton propagation velocity slows from the vacuum speed of light to zero. The P pulse is thus "stopped": Its coherent information is transferred to collective atomic excitation distributed in space. The atomic polarization can be extracted by reversing the transfer process, i.e. S-field adjustment that rotates θ back from $\pi/2$ to 0 and thereby recreates the P pulse.

The stopping and release of the incident P pulse makes possible storage of the phase as well as the amplitude of a field, unlike conventional photographic techniques or radiation detectors, which record only intensities. Thus the technique has drawn interest from those who wish to store photons [Gor07; Hei13; Sch16].

C.5   The atom-photon Hamiltonian

When we regard the electromagnetic field as a quantum system its properties, like those of internal excitation of an atom, derive from a statevector constructed from field modes. The undisturbed Hamiltonian H^0 comprises two parts, corresponding to the isolated atom, H^{atom} and the free-space field H^{field}. The full Hamiltonian combines these with the interaction energy between atoms and fields, V

$$H = H^{\text{atom}} + H^{\text{field}} + V. \tag{C.5-1}$$

There is no explicit indication of time dependence for the interaction when it refers to a closed quantum system. (Time dependence does occur when we describe an atom moving into a cavity field.) The atom-field statevector then has a product-space structure

$$|\Psi\rangle = |\text{atom}\rangle \otimes |\text{field}\rangle. \tag{C.5-2}$$

The use of abstract vectors from independent degrees of freedom is here written using the notation $\otimes$ for a *tensor product* (or *outer product* of two vectors); see eqn (C.8-1) in Appendix C.8.

When such combinations of (bare) atom and field states are eigenstates of the full Hamiltonian H of eqn (C.5-1) they are known as *dressed* states. When the Hamiltonian is time dependent its instantaneous eigenstates are termed *adiabatic* states.

The electric field appearing in the dipole interaction becomes an operator, incorporating photon creation and annihilation operators. For each single-mode field, of carrier frequency ω_ν and direction $\mathbf{e}_\nu$, the construction for the electric field at the center of mass, $\mathbf{R} = 0$, becomes the operator

$$\hat{\mathbf{E}}(0,t) = \tfrac{1}{2}\mathbf{e}_\nu\,\mathcal{E}_\nu(t)\hat{a}_\nu\,\mathrm{e}^{-\mathrm{i}\omega_\nu t} + \tfrac{1}{2}\mathbf{e}_\nu^*\mathcal{E}_\nu(t)^*\,\hat{a}_\nu^\dagger\,\mathrm{e}^{\mathrm{i}\omega_\nu t}, \tag{C.5-3}$$

where the label ν identifies the selected mode of the vector Helmholtz equation, eqn (B.3-3); see Appendix B.3.1.

C.5.1   The two-state atom with photons

Consider a single two-state atom, with states labeled g (for ground) and e (for excited) held in (or directed into) a cavity in which one of the allowable field-mode frequencies ω_ν nearly matches the atom Bohr frequency ω_{eg} defined by

$$\hbar\omega_{eg} = E_e - E_g. \tag{C.5-4}$$

In considering the combination of atom states and field states with which to describe the atom-field system we consider pairs such as

$$|1\rangle = |g\rangle_{\text{atom}}|n_\nu + 1\rangle_{\text{cav}}, \quad \text{Energy } E_1 = E_g + (n_\nu + 1)\hbar\omega_\nu, \tag{C.5-5a}$$
$$|2\rangle = |e\rangle_{\text{atom}}|n_\nu\rangle_{\text{cav}}, \quad \text{Energy } E_2 = E_e + n_\nu\hbar\omega_\nu, \tag{C.5-5b}$$

in which the excited atomic state e is paired with an n-photon state and the atomic state g has an additional photon. The energy difference between the two field-atom basis states is

$$E_2 - E_1 = E_e - E_g - \hbar\omega_\nu \equiv \hbar(\omega_{eg} - \omega_\nu) = \hbar\Delta, \tag{C.5-6}$$

and the linkage between these is one in which, with the RWA, the atomic transition $e \to g$ accompanies the creation of a photon,

$$|g\rangle_{\text{atom}}|n_\nu + 1\rangle_{\text{cav}} \xleftarrow{\text{cav}} |e\rangle_{\text{atom}}|n_\nu\rangle_{\text{cav}}. \tag{C.5-7}$$

When the cavity-mode frequency ω_ν is equal to the Bohr frequency ω_{eg} these two states are degenerate; all other states differ by the energy of at least one photon, $\hbar\omega_\nu$. With allowance of some detuning the paired equations for nearly degenerate atom-fields states, the quantized-field version of the two-state TDSE in the rotating-wave approximation, eqn (A.9-23), take the form

$$i\frac{d}{dt}C_1(t) = \tfrac{1}{2}\Omega(n_\nu)^* C_2(t), \tag{C.5-8a}$$

$$i\frac{d}{dt}C_2(t) = \Delta\, C_2(t) + \tfrac{1}{2}\Omega(n_\nu)\, C_1(t), \tag{C.5-8b}$$

where the variation of Rabi frequency $\Omega(n_\nu)$ with photon number n_ν depends on whether the n-photon state is paired with the excited state, as it is above, to give the formula

$$\Omega(n_\nu) = \sqrt{n_\nu + 1}\,\Omega_0 \quad \text{if} \quad \begin{array}{l} |1\rangle = |g\rangle_{\text{atom}}|n_\nu + 1\rangle_{\text{cav}} \\ |2\rangle = |e\rangle_{\text{atom}}|n_\nu\rangle_{\text{cav}} \end{array}, \tag{C.5-9a}$$

or is paired with the ground state, to give the formula

$$\Omega(n_\nu) = \sqrt{n_\nu}\,\Omega_0 \quad \text{if} \quad \begin{array}{l} |1\rangle = |g\rangle_{\text{atom}}|n_\nu\rangle_{\text{cav}} \\ |2\rangle = |e\rangle_{\text{atom}}|n_\nu - 1\rangle_{\text{cav}} \end{array}. \tag{C.5-9b}$$

With either choice the dependence upon field-mode intensity enters through the vacuum Rabi frequency for the transition,

$$\hbar\Omega_0 = -\langle g|\mathbf{d} \cdot \mathbf{e}_\nu|e\rangle\mathcal{E}_\nu. \tag{C.5-10}$$

The field amplitude $\mathcal{E}_\nu$ appearing here is to be evaluated from a vector Helmholtz equation, (B.3-3), appropriate to the particular cavity.

C.5.2 Radiative rates with photons

For comparison with incoherent rate-equation descriptions of radiation-induced change the usual parametrization involves transition rates that average over radiation modes that share a given frequency, the Bohr frequency in the present example. From time-dependent perturbation theory applied to an interaction Hamiltonian the single atom transition rate is to be evaluated from *Fermi's Famous Golden Rule* [Gry10; Mil19] [see eqn (A.16-10) in Appendix A.16.2],

$$\mathcal{R}_{2\to1} = \frac{2\pi}{\hbar} |V_{12}|^2 \, \varpi(\omega_{12}), \tag{C.5-11}$$

where $\varpi(\omega_{12})$ is the density of states[4] for the field modes, evaluated at the Bohr frequency. Applying this approach to an electric dipole transition, with photon states taken from free-space field modes, we obtain the expression (starting from n photons acting on the excited state of the atom),

$$\mathcal{R}_{2\to1} = \frac{2\pi}{\hbar} \varpi(\omega_{12}) \left| \tfrac{1}{2}[n(\omega_{12}) + 1]\hbar\Omega_0 \right|^2. \tag{C.5-12}$$

The photon-number factor $n(\omega_{12}) + 1$ appearing here has two effects. The term $n(\omega_{12})$ describes stimulated emission—change proportional to the existing radiation intensity. In the absence of any radiation (no photons) the unity 1 leads to spontaneous emission. If the atom has a specified orientation then the radiation has a characteristic dipole-radiation pattern. When the rate is averaged over atom orientations the radiation is isotropic.

By placing the atom into a restrictive environment, such as a cavity or a photonic lattice that restricts and enhances particular directions of the density of states, the spontaneous emission can be crafted toward creation of directed photons. If this transition is the only avenue of radiative decay the excitation probability $P_2(t)$ decays exponentially, with rate A_{21} and half life $t_{1/2} = \ln(2)/A_{21}$.

C.6 The Jaynes-Cummings model

By considering only field-atom states that are coupled by (nearly) resonant radiation, and maintaining the RWA, we define the *Jaynes-Cummings model* (JCM) [Jay63; Mil91; Sho93; Sho05; Lar07; Sho07; Gre13; Gro13; Mil19]: an infinite set of paired equations that associate a single Fock state with each of the two atomic states g and e. A representative equation pair reads (with the assumption of real-valued Rabi frequencies and the RWA)

$$i\frac{d}{dt}C_1(n_\nu; t) = \tfrac{1}{2}\Omega(n_\nu)\, C_2(n_\nu; t), \tag{C.6-1a}$$

$$i\frac{d}{dt}C_2(n_\nu; t) = \Delta\, C_2(n_\nu; t) + \tfrac{1}{2}\Omega(n_\nu)\, C_1(n_\nu; t). \tag{C.6-1b}$$

The argument n_ν, on the Rabi frequency and the paired probability amplitudes, identifies the photon number associated with the atomic excited state. When we allow motion of the atom through the cavity the Rabi frequencies become time dependent.

---

[4]To avoid confusion with the symbol for the density matrix I use ϖ rather than the more common choice ρ for the density of states.

C.6.1 JCM solutions

The JCM deals with a single-mode field and an infinite set of independent pairs of coupled ODEs whose RWA Hamiltonians can be written

$$W^{(n)} = \begin{bmatrix} 0 & \frac{1}{2}\Omega(n) \\ \frac{1}{2}\Omega(n) & \Delta \end{bmatrix}. \tag{C.6-2}$$

As with the basic semiclassical excitation of a two-state atom having constant Rabi frequency and constant detuning the JCM equations, (C.6-1), have closed-form oscillatory solutions. Here these solutions involve the n-photon mixing angle θ_n and flopping frequency Υ_n, defined as

$$\tan^2 \theta_n = \Omega(n)/\Delta, \qquad \Upsilon_n = \sqrt{\Omega(n)^2 + \Delta^2}. \tag{C.6-3}$$

When the excited atomic state e is paired with an n-photon state of the field and the ground atomic state g is paired with a field having $n + 1$ photons the Rabi frequency is

$$\Omega(n) = \sqrt{n+1}\,\Omega_0. \tag{C.6-4}$$

If at time $t = 0$ an excited atom encounters a field that has exactly n photons (a Fock state) then the population inversion at time t undergoes oscillations in accord with the formula

$$w(t) = |C_2(n;t)|^2 - |C_1(n;t)|^2 = -1 + \sin^2(\theta_n t)[1 + \cos(\Upsilon_n t)]. \tag{C.6-5}$$

As with the semiclassical two-state model, the populations are purely periodic, with complete inversion possible if and only if the field frequency is resonant with the Bohr frequency. When the initial state is that of an excited atom in the vacuum $(n = 0)$ the Rabi oscillations occur at the vacuum Rabi frequency Ω_0.

Photon creation. An atom, in its excited state, upon entering a cavity that holds exactly n_ν photons in mode ν, will , in accord with eqn (C.6-1), begin a Rabi oscillation that will add a single photon to this cavity field. If the velocity of the atom is properly adjusted (to produce a pi pulse), it will leave its excitation as an additional field increment. In particular, if the cavity initially contains no radiation of mode ν—it is in the vacuum-state of this mode—the passing atom will, with a pi-pulsed Rabi frequency create a single-photon field in the cavity. A subsequent unexcited atom will, with a pi-pulse interaction, carry this excitation away—it will remove a single photon from the cavity.

By monitoring the state of the emerging atoms an experimenter can deduce the state of the fields photons before and after each atom interacts. In principle, an experimenter can produce definite Fock states by adjusting the atom velocity to prepare and then maintain a fixed Rabi frequency and a consequent fixed photon number in the cavity. The JCM thus offers opportunities for very convincing demonstrations that the electromagnetic fields in cavities exist in discrete increments—as standing-wave photons.

This scenario underlies the experimental demonstration of the single-atom maser and related effects of cavity quantum-electrodynamics (*cavity QED*) [San83; Wal88; Har89; Kim98; Vah03; Mil05; Mes06; Har06; Har13].

Photon energy conservation. One of the things that concerned me when I first encountered the JCM, and has concerned subsequent graduate students, is how energy is conserved when a single atom absorbs a non-resonant photon, i.e. when the field carrier frequency differs from the Bohr frequency. The JCM offers a model of just such a situation, when the detuning Δ is nonzero. With this model there are three contributions to the system energy. In addition to the obvious energy of atomic excitation, and the energy of the photons, there is an interaction energy that must be considered. It is to this term that nonresonant excitation contributes.

Photon loss. The basic two-state JCM model requires alteration to account for two mechanisms that produce photon loss. The first, acting on a cavity-enclosed atom as well as on an unconfined atom, is the possibility of spontaneous emission. To treat such decays that carry the atom between the two states that are used for eqn (A.10-1) or eqn (C.6-1) it is necessary to use a density-matrix equation. However, if the predominant decay modes carry the atom into other states than the single ground state that is assumed in those equations, then this action can be simulated by introducing a complex-valued energy for the excited atom states; the imaginary part of this energy is the rate of probability loss to spontaneous emission out of the two-state system.

The other loss rate is that of the cavity itself. Through its walls there occurs, inevitably, some loss. This loss provides an outgoing field that can serve as a controlled source of single photons.

C.6.2   Photon-averaged JCM: Proof of photons

Now suppose the field has some unknown number of photons, specified only by the mean number of photons $\bar{n}$ and the probability p_n for observing n photons. Then the averaged population in state k, being a weighted average over all possible photon numbers, is

$$\bar{P}_k(t) = \sum_n^\infty p_n \, |C_k(n;t)|^2. \tag{C.6-6}$$

Consider, in particular, behavior of an initially excited atom (inversion $w = +1$) in a resonant field ($\Delta = 0$). The formula for the photon-averaged population inversion is a sum over cosines,

$$\bar{w}(t) = \sum_n^\infty p_n \, \cos(\sqrt{n}\,\Omega_1\, t). \tag{C.6-7}$$

Because the oscillation frequencies are proportional to $\sqrt{n}$ they are incommensurable (not integer multiples of a common factor) and so, for any photon distribution and an initially excited atom, the oscillations undergo a "collapse"—a (temporary) lack of inversion; see Figure C.3.

Surprisingly, the discreteness of the photons leads to "revivals" of the inversion oscillations [Ebe80; Nar81; Kni82; Sho93; Mil19]. These are most evident for a coherent-state field, when they produce nearly perfect restorations of Rabi oscillations at a sequence of times. These revivals occur when oscillations become in phase with one another and so they occur only when the Rabi frequencies form a discrete set. They therefore provide definitive evidence for the quantum nature of a cavity field and the discreteness of its increments—evidence for photons as discrete energy increments.

The photon-number distribution in a thermal field is much broader than that in a coherent state, and the resulting time history does not show the distinct revivals found with a coherent state [Kni82]. Nonetheless there is an irregular pattern of brief inversions, consequence of discrete photon numbers, that does not occur when the photon distribution is a continuum of values.

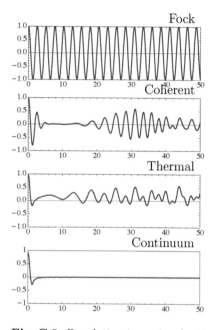

Figure C.3 shows examples of JCM population inversions for four photon distributions that have mean photon number of 6, as shown in Figure B.3:

a **Fock** state of exactly 6 photons, showing Rabi oscillations;

a **coherent** state, showing collapse and a revival of oscillations;

a **thermal** distribution, showing irregularity;

and a **continuum**, obtained with an integral over n rather than a discrete sum.

The continuum inversion history shows none of the features of discrete-photon distributions.

Fig. C.3 *Population inversion for $\overline{n} = 6$*

C.7 Cavity STIRAP

The STIRAP procedure for transferring atom population (see page 129) accompanies transfer of a photon from the P field to the S field. This field change has been used to place a single S photon into a cavity. With a change of notation for the three atomic states to $g - e - x$ the combined atom+field states, and their linkages for a classical P field, read

$$|g\rangle_{\text{atom}}|0\rangle_S \xleftarrow{\;P(t)\;} |e\rangle_{\text{atom}}|0\rangle_S \xleftarrow{\;S\;} |x\rangle_{\text{atom}}|1\rangle_S. \qquad (C.7\text{-}1)$$

The atom first moves into the empty cavity, where the vacuum Rabi frequency for the S field acts. The atom, still in the cavity, next encounters the P field of a laser. The resulting loss of a single P photon has insignificant effect on the laser, and so it may be treated as a classical field with controlled time dependence. Its energy is divided between the cavity S field and the Raman excitation $g \to s$ of the atom. The paired sequence of interactions S, P produces the change

$$|g\rangle_{\text{atom}}|0\rangle_S \xrightarrow{S,P(t)} |x\rangle_{\text{atom}}|1\rangle_S. \tag{C.7-2}$$

The stored S-field photon subsequently leaks through one of the cavity mirrors, at a fixed rate, to produce a traveling-wave photon and leave the cavity again empty,

$$|x\rangle_{\text{atom}}|1_S\rangle_S \xrightarrow{\text{leak}} |x\rangle_{\text{atom}}|0\rangle_S. \tag{C.7-3}$$

As a final step, the atom population is reset to the initial state by optical pumping,

$$|x\rangle_{\text{atom}}|0\rangle_S \xrightarrow{\text{restore}} |g\rangle_{\text{atom}}|0\rangle_S. \tag{C.7-4}$$

Quantum P field. When both fields are weak, so that a quantum description is needed for each, the linkage pattern is

$$|g\rangle_{\text{atom}}|n_P\rangle_P|0\rangle_S \xleftarrow{P} |e\rangle_{\text{atom}}|n_P - 1\rangle_P|0\rangle_S \xleftarrow{S} |s\rangle_{\text{atom}}|n_P - 1\rangle_P|1_S\rangle_S, \tag{C.7-5}$$

and the STIRAP sequence S, P produces the change

$$|g\rangle_{\text{atom}}|n_P\rangle_P|0\rangle_S \xleftarrow{S,P} |s\rangle_{\text{atom}}|n_P - 1\rangle_P|1_S\rangle_S. \tag{C.7-6}$$

The STIRAP sequence is seen to produce the transfer of a single photon from the P field to the S field. Accompanying this field change is a two-photon atom transition from state g to state s.

C.8 Product spaces; Entanglement

For a single bound electron the product form of the wavefunction of eqn (A.5-13) implies independent, uncorrelated motion of radial and angular variables. For the field example of eqn (B.1-48) the space-time amplitudes are correlated with field-vector direction—with polarization. With two independent vector spaces the construction (a *bipartite* statevector) reads

$$|V\rangle = |A\rangle_1 \otimes |B\rangle_2, \tag{C.8-1}$$

where the subscripts 1 and 2 label the distinct vector spaces (hereafter referred to as the A space and the B space) of the two independent degrees of freedom. The notation $\otimes$ means that to get an example of a state $|V\rangle$ we pick a particular state from the A-space set $|A_1\rangle, |A_2\rangle, \ldots$, and a particular state from the B-space set $|B_1\rangle, |B_2\rangle, \ldots$, to form a particular composite C-space state that is identified by the two labels of its components, for example the single *product state*,

$$|V_{ij}\rangle = |A_i\rangle_1|B_j\rangle_2. \tag{C.8-2}$$

Such products occur in two contexts. In one form, both degrees of freedom—both Hilbert subspaces A and B—refer to independent attributes of a single indivisible entity, such as the radial and azimuthal degrees of freedom for a bound electron or the polarization and transverse position of a single classical beam of light. Alternatively, the two subspaces may refer to entities that can be spatially separated and distinguished, such as two atoms or two spins or two photons or a photon and an electron.

More generally one has multiple products of independent subspaces, in the form

$$|V\rangle = |A\rangle_1 \otimes |B\rangle_2 \otimes |C\rangle_3 \cdots . \tag{C.8-3}$$

Here, and in the bipartite example eqn (C.8-2) the subscripts $1, 2, 3, \ldots$ that identify the separate degrees of freedom are redundant: These are implicit in the ordering of the kets, and need not be shown explicitly. An even more compact notation is

$$|A_i\rangle_1 \otimes |B_j\rangle_2 \otimes |C_k\rangle_3 \cdots = |A_i B_j C_k \cdots\rangle. \tag{C.8-4}$$

The indices i and j in this example pick out particular vectors from the two sets.

The individual subspaces $A, B, \cdots$ appearing in these constructions may themselves be composites of more elementary spaces—they may be *reducible*. The subspaces and descriptors are *irreducible* if they cannot be expressed as products of subspaces associated with simpler degrees of freedom.

Examples of products. The separable functions mentioned in Appendix A.5 provide examples of the preceding formalism for multipartite composites. Write the spin orbital of eqn (A.5-14) for a single electron in Dirac form as

$$\psi(\check{\mathbf{s}}, \mathbf{r}) = \sum_{\mu n \ell m} C_{\mu n \ell m} \, \chi^\mu(\check{\mathbf{s}}) \, R_{n\ell}(r) \, \Theta_{\ell m}(\vartheta) \, \Phi_m(\varphi), \tag{C.8-5a}$$

or, with Dirac notation,

$$\langle \check{\mathbf{s}}, \mathbf{r}|\psi\rangle = \sum_{\mu n \ell m} C_{\mu n \ell m} \, \langle \check{\mathbf{s}}|\chi_\mu\rangle \, \langle \mathbf{r}|\psi_{n\ell m}\rangle \equiv \sum_{\mu n \ell m} C_{\mu n \ell m} \, \langle \check{\mathbf{s}}, \mathbf{r}|\chi_\mu \psi_{n\ell m}\rangle. \tag{C.8-5b}$$

These expressions give the space-spin representation of the product kets $|\chi_\mu \psi_{n\ell m}\rangle$ for a single electron. Multielectron states can be constructed from summed products of such single-electron states.

A similar Dirac-notation expression for a classical electric field, expressed in terms of unit vectors $\mathbf{e}_j(\mathbf{r})$ and discrete spatial field modes $U_{j\nu}(\mathbf{r})$, might read

$$\mathbf{E}^{(+)}(\mathbf{r}, t) \equiv \langle \mathbf{r}, t|E^{(+)}\rangle = \sum_{j\nu} a_{j\nu}(t)\langle \mathbf{r}|e_j\rangle\langle \mathbf{r}|U_{j\nu}\rangle \equiv \sum_{j\nu} a_{j\nu}(t)\langle \mathbf{r}|e_j U_{j\nu}\rangle. \tag{C.8-6}$$

Here the product space involves the kets $|e_j U_{j\nu}\rangle$ for a single classical field mode. Although the notation is what we find with quantum theory, the field hereby described can be completely classical.

C.8.1   Separability and correlation

The most general example of a bipartite state fitting the construction of eqn (C.8-1) has the form of a sum over subspace descriptors, as in eqn (A.5-12),

$$|V\rangle = \sum_{ij} c_{ij}|V_{ij}\rangle \equiv \sum_{ij} a_i b_j |A_i\rangle|B_j\rangle. \tag{C.8-7}$$

Here the positioning of the kets is significant: They refer to degrees of freedom for space A and space B, respectively. A very special construction, termed *separable*, occurs when the sum contains only a single term, as in eqn (C.8-2). Such a quantum system is recognizable as being in a definite quantum state of each separate subsystem. No measurement is required to gain this information. Any other situation, known as a *nonseparable* state, involves at least two terms in the sum. Nonseparable states are those that cannot be written as a single product of two states from separate subspaces, no matter how we organize the basis states of the subspaces. When a system is described by a nonseparable state then a measurement on one part (say that of the A subspace) immediately gives previously unknown information about the other part (the B subspace). In such a nonseparable state the two constituents are said to be *correlated* (or entangled, see Appendix C.8.4 below).

Correlation examples. A variety of degrees of freedom find use in describing bipartite systems. The notation for the individual unit vectors $|A_i\rangle$ and $|B_j\rangle$ is simplest when only two values are possible for a degree of freedom. A two-state atom is often presented with basis vectors $|g\rangle, |e\rangle$ ("ground" and "excited" state) or $|a\rangle, |b\rangle$. It is common to regard a two-state system as an example of a spin-half particle, whose descriptive vector space is spanned by two unit vectors denoted as $|\uparrow\rangle, |\downarrow\rangle$ (or "up" and "down") or $|+\rangle, |-\rangle$. When dealing with quantum information the typical notation for two orthogonal states is $|1\rangle, |0\rangle$ (or "yes" and "no").

Similar notation is found in descriptions of radiation. When describing a radiation beam the two elementary polarization states can be those of circular polarization $|R\rangle, |L\rangle$ or linear polarization $|H\rangle, |V\rangle$ or $|X\rangle, |Y\rangle$. The connection between radiation fields and Hilbert space is made more evident by using the Dirac notation $|e_i\rangle$ for the Cartesian unit vectors $\mathbf{e}_i$ and rewriting eqn (A.5-16) as

$$\mathbf{E}(\mathbf{r}, t) = \sum_i \mathcal{E}_{i\nu} \langle \mathbf{r}, t | e_i U_\nu \rangle. \tag{C.8-8}$$

A discretized-frequency radiation beam might be described by just two colors, with unit vectors $|b\rangle, |r\rangle$ (or "blue" and "red"). Or the states might refer to a field in which two paths, described by two discrete rays, $|r_1\rangle, |r_2\rangle$, arrive at a position.

Any of these examples of paired variables can occur as the A space or the B space. Thus one might find, for the first product state $|A_1 B_1\rangle$ of a list such notation as

$$|\uparrow\uparrow\rangle, \quad |11\rangle, \quad |++\rangle, \quad |\uparrow 1\rangle, \quad |1+\rangle, \quad |\uparrow +\rangle, \quad |bb\rangle, \quad \cdots, \tag{C.8-9}$$

in which symbol position within a ket $|\cdots\rangle$ identifies the subspace. Any of these independent degrees of freedom, whether associated with a single particle (or beam) or with two distinct particles (or beams), may appear in a statevector as correlated.

Two limiting cases of the possible composite states are of particular interest: [Ebe15; Ebe16; Qua17]

$$|V^{\text{unc}}\rangle = \frac{1}{2}\left[\left(|A_1\rangle + |A_2\rangle\right)_A \otimes \left(|B_1\rangle + |B_2\rangle\right)_B\right], \qquad \text{(C.8-10a)}$$

$$|V^{\text{cor}}\rangle = \frac{1}{\sqrt{2}}\left[\left(|A_1\rangle \otimes |B_1\rangle\right)_{AB} + \left(|A_2\rangle \otimes |B_2\rangle\right)_{AB}\right]. \qquad \text{(C.8-10b)}$$

(Either or both of the plus signs could be replaced by minus signs without altering physical consequences discussed in the following.)

Separable, uncorrelated. First, suppose the composite state is the product of two states that are each superpositions in their respective spaces. say as in eqn (C.8-10a). This particular composite state is expressible either as the product of two specific superposition states $|A\rangle$ and $|B\rangle$ in separate subspaces,

$$|V^{\text{unc}}\rangle = |A\rangle|B\rangle, \qquad |A\rangle = \frac{1}{\sqrt{2}}\left[|A_1\rangle + |A_2\rangle\right], \qquad |B\rangle = \frac{1}{\sqrt{2}}\left[|B_1\rangle + |B_2\rangle\right], \quad \text{(C.8-11a)}$$

or as a specific coherent superposition of all four possible bipartite product states,

$$|V^{\text{unc}}\rangle = \frac{1}{2}\left[|A_1 B_1\rangle + |A_1 B_2\rangle + |A_2 B_1\rangle + |A_2 B_2\rangle\right]. \qquad \text{(C.8-11b)}$$

With such a product state a measurement of the A space, say certainty that it is in the "A_1" state, gives no information about the properties in the B space. Similarly a measurement of the B space, say certainty of the "B_1" state, gives no information about the A space:

$$\langle A_1|V^{\text{unc}}\rangle = \tfrac{1}{2}\left[|B_1\rangle + |B_2\rangle\right], \qquad \langle B_1|V^{\text{unc}}\rangle = \tfrac{1}{2}\left[|A_1\rangle + |A_2\rangle\right]. \qquad \text{(C.8-12)}$$

In the separable simple-product state $|V^{\text{unc}}\rangle$ the two DoFs are uncorrelated.

Nonseparable, uncorrelated. Alternatively, consider a composite state that is constructed by summing two of the possible paired states, that of eqn (C.8-10b):

$$|V^{\text{cor}}\rangle = \frac{1}{\sqrt{2}}\left[|A_1 B_1\rangle + |A_2 B_2\rangle\right]. \qquad \text{(C.8-13)}$$

This construction is not expressible as a simple product; it is not separable. Here a measurement on the A variables provides a complete determination of the B variables, and vice versa. For example

$$\langle A_1|V^{\text{cor}}\rangle = \frac{1}{\sqrt{2}}|B_1\rangle, \qquad \langle B_1|V^{\text{cor}}\rangle = \frac{1}{\sqrt{2}}|A_1\rangle. \qquad \text{(C.8-14)}$$

The A and B parts of the unseparable composite state $|V^{\text{cor}}\rangle$ are said to be correlated.

C.8.2 Quantifying correlations

A variety of measures are available for quantifying the amount of correlation between degrees of freedom that is present in a particular quantum state or in a mixture of states. The procedures are most simply presented when there are only two degrees of freedom and each of these has two possible states. To describe the composite system we require four orthogonal bipartite states.

The simplest choice would be product states, say the so-called standard states,

$$|\psi_1\rangle = |A_1\rangle|B_1\rangle, \quad |\psi_2\rangle = |A_1\rangle|B_2\rangle, \quad |\psi_3\rangle = |A_2\rangle|B_1\rangle, \quad |\psi_4\rangle = |A_2\rangle|B_2\rangle, \quad \text{(C.8-15a)}$$

or, with up-down notation for the two possibilities in A and B,

$$|\psi_1\rangle = |\uparrow\rangle|\uparrow\rangle, \quad |\psi_2\rangle = |\uparrow\rangle|\downarrow\rangle, \quad |\psi_3\rangle = |\downarrow\rangle|\uparrow\rangle, \quad |\psi_4\rangle = |\downarrow\rangle|\downarrow\rangle. \quad \text{(C.8-15b)}$$

Often a more useful set, obtained by combining pairs of the states $|\psi_j\rangle$, is the four orthonormal bipartite *Bell states*,

$$|\phi^\pm\rangle = \frac{1}{\sqrt{2}}[|\uparrow\uparrow\rangle \pm |\downarrow\downarrow\rangle], \quad |\psi^\pm\rangle = \frac{1}{\sqrt{2}}[|\uparrow\downarrow\rangle \pm |\downarrow\uparrow\rangle]. \quad \text{(C.8-16)}$$

A quantification of correlation obtains by defining a basis of phased Bell states [Hil97],

$$|e_1\rangle = |\phi^+\rangle, \quad |e_2\rangle = i|\phi^-\rangle, \quad |e_3\rangle = i|\psi^+\rangle, \quad |e_4\rangle = |\psi^-\rangle. \quad \text{(C.8-17)}$$

Using this set we can write any bipartite statevector as a four-state superposition

$$|\Psi\rangle = \sum_{i=1,4} \alpha_j |e_j\rangle. \quad \text{(C.8-18)}$$

A measure of correlation is the *concurrence* [Ben96; Hil97; Woo98; Run01; Run03],

$$\mathcal{C} = \left| \sum_j \alpha_j^2 \right|. \quad \text{(C.8-19)}$$

This concurrence varies from $\mathcal{C} = 0$ for a separable state to $\mathcal{C} = 1$ for a maximally correlated state that cannot be expressed as a product of two states.

C.8.3   Reduced density matrices

The relationships between distinct Hilbert spaces can be examined by extraction from the two-space density matrices smaller matrices that refer to the separate degrees of freedom. With any set of basis states we define a Hermitian coherency matrix (a composite density matrix) with elements M_{ij}:

$$\mathsf{M} = \begin{bmatrix} |\psi_1\rangle\langle\psi_1| & |\psi_2\rangle\langle\psi_2| \\ |\psi_3\rangle\langle\psi_3| & |\psi_4\rangle\langle\psi_4| \end{bmatrix}, \quad M_{ij} = \langle\psi_i|\mathsf{M}|\psi_j\rangle = M_{ji}^*. \quad \text{(C.8-20)}$$

From any such coherency matrix, or any density matrix of a composite system, say of parts A and B, we obtain a reduced density matrix by taking the trace over the states of one of the subsystems. For example, the state $|\Psi_{AB}\rangle = |A\rangle \otimes |B\rangle$ has the pure-state density matrix

$$\mathsf{M}(AB) = |\Psi_{AB}\rangle\langle\Psi_{AB}|. \tag{C.8-21}$$

From this we can form two reduced density matrices by taking traces,

$$\boldsymbol{\rho}(A) = \mathrm{Tr}_B\mathsf{M}(AB) = \sum_j \langle B_j|\Psi_{AB}\rangle\langle\Psi_{AB}|B_j\rangle, \tag{C.8-22a}$$

$$\boldsymbol{\rho}(B) = \mathrm{Tr}_A\mathsf{M}(AB) = \sum_i \langle A_i|\Psi_{AB}\rangle\langle\Psi_{AB}|A_i\rangle. \tag{C.8-22b}$$

Applied to the pure-state coherency matrix $\mathsf{M}(AB)$ of eqn (C.8-20) the trace Tr_B gives the reduced density matrix $\boldsymbol{\rho}(A)$. This has the dimension of the A subspace and contains all the information needed to describe measurements involving the degree of freedom associated with part A:

$$\boldsymbol{\rho}(A) = \mathrm{Tr}_B\,\mathsf{M}(AB) = \begin{matrix} |A_1\rangle & |A_2\rangle \\ \begin{bmatrix} M_{11} + M_{33} & M_{12} + M_{34} \\ M_{21} + M_{43} & M_{22} + M_{44} \end{bmatrix} & \begin{matrix} \langle A_1| \\ \langle A_2| \end{matrix} \end{matrix}. \tag{C.8-23a}$$

Similarly $\boldsymbol{\rho}(B)$ informs about part B:

$$\boldsymbol{\rho}(B) = \mathrm{Tr}_A\,\mathsf{M}(AB) = \begin{matrix} |B_1\rangle & |B_2\rangle \\ \begin{bmatrix} M_{11} + M_{22} & M_{13} + M_{24} \\ M_{31} + M_{42} & M_{33} + M_{44} \end{bmatrix} & \begin{matrix} \langle B_1| \\ \langle B_2| \end{matrix} \end{matrix}. \tag{C.8-23b}$$

The subscripts appearing here on the composite matrix M are those of the original bipartite basis states ψ_i.

C.8.4   Polarization, concurrence, and correlation

Given a 2×2 density matrix, such as $\boldsymbol{\rho}(A)$ or $\boldsymbol{\rho}(B)$, we can apply all of the several quantitative descriptors of coherence such as the Stokes parameters discussed in Appendix B.1.7. The traces of the two reduced density matrices are identical,

$$\mathrm{Tr}\ \mathsf{M} = \mathrm{Tr}\ \boldsymbol{\rho}(A) = \mathrm{Tr}\ \boldsymbol{\rho}(B) = M_{11} + M_{22} + M_{33} + M_{44}. \tag{C.8-24}$$

Both the polarization $\mathcal{P}$ and the concurrence $\mathcal{C}$ defined in Appendix B.1.8 have counterparts for the subsystem density matrices. It is noteworthy that, as pointed out by Eberly [Ebe16], the notion of Polarization can be defined for degrees of freedom other than the spin that is associated with unit vectors in Euclidean space.

Starting from the bipartite coherency matrix $\mathsf{M}(A)$ we obtain measures of the polarization for each of the subspaces, obtainable as generalizations of eqn (B.3-63):

$$\mathcal{P}(A) = \sqrt{1 - \frac{4\mathrm{Det}\ \boldsymbol{\rho}(A)}{[\ \mathrm{Tr}\ \boldsymbol{\rho}(A)\ ]^2}}, \qquad \mathcal{P}(B) = \sqrt{1 - \frac{4\mathrm{Det}\ \boldsymbol{\rho}(B)}{[\ \mathrm{Tr}\ \boldsymbol{\rho}(B)\ ]^2}}. \tag{C.8-25}$$

The corresponding concurrence, generalizing eqn (B.3-66), is obtainable as

$$C(A) = \sqrt{2}\sqrt{1 - \text{Tr } [\rho(A)^2]}, \qquad C(B) = \sqrt{2}\sqrt{1 - \text{Tr } [\rho(B)^2]}. \qquad \text{(C.8-26)}$$

The two versions yield identical numerical values, $C(A) = C(B) \equiv C$. As with the earlier discussion, these polarizations and concurrences are complementary aspects of coherence: For pure bipartite states they satisfy the relationships

$$\mathcal{P}(A)^2 + C(A)^2 = 1, \qquad \mathcal{P}(B)^2 + C(B)^2 = 1. \qquad \text{(C.8-27)}$$

These are mathematical versions of the statement that maximally polarized states are disentangled and maximally entangled states are unpolarized [Fed14].

C.8.5   Entanglement

Increasingly the term "entanglement" is used to describe correlations, without requiring physical separation of the parts [Ben96; Hil97; Ved97; Kni98; Nie00; Rai01; Bar03; Min05; Mes06; Har06; Van07; Ami08; Hor09; Ebe15; Mil19]. Entanglement of spatially separated entities is often regarded as a "quintessential quantum attribute", as *the* essential "weirdness" trait of quantum theory [Kni98; Qia18], but when it is defined as a measure of inseparability of independent vector spaces it can be found and studied in classical optics, as pointed out by Robert Spreeuw in 1998 [Spr98] and in numerous recent works [Spr98; Lee04; Lui09; Qia11; Gho14; Kar15; Ebe16; Qua17; Sch17]. The use of Hilbert spaces has long been a staple of quantum theory, but they have application to classical systems as well. Correlations that occur between the degrees of freedom for a single particle, or a single field mode, has been called *local*, or *classical* entanglement [Spr98; Kar15]. By contrast, correlations that occur between entities that can be spatially separated and distinguished, such as two spins or two photons or a photon and an atom, are termed *nonlocal*, or *quantum* entanglement.

Parametric downconversion: Photon-pair entanglement. One of the useful frequency-changing techniques provided by nonlinear optics is the parametric downconversion of a monochromatic beam of frequency ω into two lower-frequency beams whose frequencies ω_1 and ω_2 sum to the original frequency; see page 155. Although the frequencies and directions of the various beams may be regarded as fixed by photon-energy and photon-momentum considerations there remains a possibility of controlling the polarizations of the two photons, each of which may be expressed in terms of unit polarization vectors, say those of horizontal (H) and vertical (V) polarization. The two fields that emerge from the nonlinear process form a *biphoton* (see [Fed14]) whose statevector can be expressed as

$$|\Psi\rangle = C_1|2_H\rangle + C_2|1_H, 1_V\rangle + C_3|2_V\rangle. \qquad \text{(C.8-28)}$$

This relationship exhibits several entanglements between photon polarizations [Rar87; Shi88; Rub94; Law04].

Schmidt decomposition. In general an arbitrary statevector describing a bipartite system requires a sum over all possible elements of the tensor product of the vectors associated with the two subsystems,

$$|\Psi\rangle = \sum_{ij} c_{ij} |a_i\rangle_1 |b_j\rangle_2. \tag{C.8-29}$$

The set of amplitudes a_i are associated with components of unit vectors of space $|A\rangle$ while the amplitudes b_j are associated with unit vectors that span the space of $|B\rangle$. The dimension of the two vector spaces $|A\rangle$ and $|B\rangle$ need not have the same dimension, so the matrix of elements c_{ij} need not be square.

It turns out that if the system is in a pure state, described by a statevector $|\Psi\rangle$, then it is possible to find an orthonormal set of basis vectors $|u_i\rangle$ in $|A\rangle$ and $|v_i\rangle$ in $|B\rangle$ such that we can express the full system statevector as a single sum [Fed14]

$$|\Psi\rangle = \sum_n \sqrt{\lambda_n} |u_n\rangle_1 |v_n\rangle_2. \tag{C.8-30}$$

Here the summation goes over the smaller dimension of the two Hilbert spaces. This is the *Schmidt decomposition* [Eke95; Qia11; Fed14]. Each statevector has such a decomposition, but different statevectors will have different decompositions. Each decomposition has the normalization

$$\sum_n \lambda_n = 1. \tag{C.8-31}$$

The probability of finding the system in state $|u_n\rangle_1 |v_n\rangle_2$ is λ_n.

The importance of the Schmidt decomposition is that if we determine that the subsystem of $|A\rangle$ is with certainty in state $|u_n\rangle$ then we know that the subsystem of $|B\rangle$ is in the state $|v_n\rangle$. A system is completely *disentangled* if the Schmidt decomposition has only a single term. The more terms in the sum, the greater the entanglement.

Coda

Coda: A passage appended to form a satisfactory ending.

Photons are ubiquitous today. Photons traveling through optical fibers bring us the sounds and images of our news and entertainment. Photons from laser scanners tally our retail purchases. Photons from earth satellites give us global positioning information as we drive our autos. Thermal photons from distant stars inform us of our universe and its history; those from our sun give us the warmth to make life possible on our earth.

Throughout this Memoir I have aimed to show that the notion of photon has had a variety of useful interpretations. Evidence suggests that no single, simple definition of a photon will fit the needs of all researchers—that, unlike in retailing, there is no "one size fits all" photon. The colorful land of Oz is assuredly "not Kansas", but each of those two locales offers satisfactory opportunity for reader-satisfying activities. So too can different views of photons be satisfactory in differing surroundings.

The preceding three appendices have discussed, in some technical detail, the various notions of photons that have been, and which continue to be, part of the enterprises of physics and technology—of Photonics. Ongoing developments, made possible by new devices, are taking the notion of photons still further, calling attention to not only unfamiliar aspects of classical electromagnetic theory but to aspects of quantum theory that are only now being explored [Rub17]. It takes no imagination to suggest that the details presented in these appendices will underlie even more remarkable developments in the decades ahead. I expect that some writers will continue find results to be strange and paradoxical, even weird, while others will recognize continuing examples of familiar concepts. Something for everyone.

Just as the physics of photons has continued to advance during the last few decades, so too has Mathematics, influenced by theoretical interest in topology and quantum gravity and in practical needs of statistics, communication, and encryption—but that is another story [Ell15].

The annual Xmas cartoons

For more than three score years I have created an annual holiday-greeting cartoon. Those from my time in grad school depicted the changing views of a graduate student toward his academic environment. They are collected here as a parallel to the changing views toward photons that theme the rest of the book.

The first year I was busy with studies, catching up on topics that had not been part of my undergraduate curriculum. Amongst the books by MIT faculty still on my bookshelves are [Bla52; Cot62; Eva55; Gui49; Hil49; Hil52; Mor53; Mor64; Sla60; Sla77; Str41; Str48; Wie48].

The second year I was busy with exams, in a small room in what had been built as the Riverside Hotel and was known in my day as The Graduate House, "A gentlemen's club and a home away from home". as headmaster Avery Ashdown would tell us.

The third year I was immersed in experimental work. Running a cyclotron, chemically separating reaction products, recording radioactivity, computing with Hollerith punched cards.

The fourth year I had time for some recreation: Learning to ice skate, singing Bach and Purcell with the Choral Society and bringing waterpolo to MIT (and later to Harvard)—I played against Army at West Point.

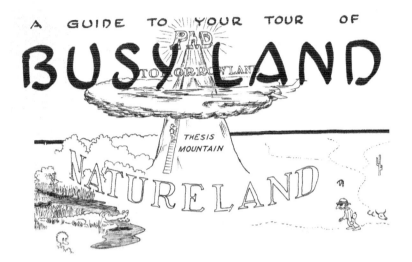

The ultimate goal.

References

[Aar65]   Arons A.B. and Peppard M.B. "Einstein's proposal of the photon concept—a translation of the Annalen der Physik Paper of 1905". *Am J Phys*, **33**, 367–374 (1965).

[Abb85]   Abbott T.A. and Griffiths D.J. "Acceleration without radiation". *Am J Phys*, **53**, 1203–1211 (1985).

[Abo17]   Abouraddy A.F. "What is the maximum attainable visibility by a partially coherent electromagnetic field in Young's double-slit interference?" *Opt Exp*, **25**, 18331–18342 (2017).

[Ack73]   Ackerhalt J.R., Knight P.L. and Eberly J.H. "Radiation reaction and radiative frequency shifts". *Phys Rev Lett*, **30**, 456–460 (1973).

[Ack74]   Ackerhalt J.R. and Eberly J.H. "Quantum electrodynamics and radiation reaction: Nonrelativistic atomic frequency shifts and lifetimes". *Phys Rev D*, **10**, 3350–3375 (1974).

[Ack76]   Ackerhalt J.R. and Eberly J.H. "Coherence versus incoherence in stepwise laser excitation of atoms and molecules". *Phys Rev A*, **14**, 1705–1710 (1976).

[Ack84]   Ackerhalt J.R. and Milonni P.W. "Interaction Hamiltonian of quantum optics". *J Opt Soc Am B*, **1**, 116–120 (1984).

[Adl87]   Adler C.G. "Does mass really depend on velocity, dad?" *Am J Phys*, **55**, 739–743 (1987).

[Aga90]   Agarwal G.S. "Dressed-state lasers and masers". *Phys Rev A*, **42**, 686–688 (1990).

[Aha04]   Aharonian F.A. *Very High Energy Cosmic Gamma Radiation: A Crucial Window on the Extreme Universe* (World Scientific, Singapore, 2004).

[Aha08]   Aharonov Y. and Rohrlich D. *Quantum Paradoxes: Quantum Theory for the Perplexed* (Wiley-VCH, Weinheim, 2008).

[Aie10]   Aiello A., Marquardt C. and Leuchs G. "Transverse angular momentum of photons". *Phys Rev A*, **81**, 053838 (2010).

[Aie15]   Aiello A. and Berry M.V. "Note on the helicity decomposition of spin and orbital optical currents". *J Opt*, **17**, 062001 (2015).

[Aie15b] Aiello A., Banzer P., Neugebauer M. and Leuchs G. "From transverse angular momentum to photonic wheels". *Nature Phot*, **9**, 789–795 (2015).

[Aie16]   Aiello A. and Banzer P. "The ubiquitous photonic wheel". *J Opt*, **18**, 1–8 (2016).

[Ait85]   Aitchison I.J.R. "Nothing's plenty. The vacuum in modern quantum field theory". *Contemp Phys*, **26**, 333–391 (1985).

[Ait91]   Aitchison I.J.R. and Mavromatos N.E. "Anyons". *Contemp Phys*, **32**, 219–233 (1991).

[Akk10]   Akkermans E., Dunne G.V. and Teplyaev A. "Thermodynamics of photons on fractals". *Phys Rev Lett*, **105**, 230407 (2010).

[Alf80]   Alferness R.C. "Efficient waveguide electro-optic TE ↔ TM mode converter–wavelength filter". *App Phys Lett*, **36**, 513–515 (1980).

[All16]   Allen L. "Orbital angular momentum: a personal memoir". *Phil Trans Roy Soc A*, **375**, 20160280 (2016).

[All62]   Allen C.W. *Astrophysical Quantities* (Athlone Press, London, 1962), 2nd edition.

[All86]   Allain C. and Cloitre M. "Optical diffraction on fractals". *Phys Rev B*, **33**, 3566–3569 (1986).

[All87]   Allen L. and Eberly J.H. *Optical Resonance and Two Level Atoms* (Dover, New York, 1987).

[All92]   Allen L., Beijersbergen M.W., Spreeuw R.J. and Woerdman J.P. "Orbital angular momentum of light and the transformation of Laguerre-Gaussian laser modes". *Phys Rev A*, **45**, 8185–8189 (1992).

[All99]   Allen L., Padgett M. and Babiker M. "The orbital angular momentum of light". *Prog Opt*, **39**, 291–372 (1999).

[Alo11]   Alonso M.A. "Wigner functions in optics: describing beams as ray bundles and pulses as particle ensembles". *Adv Opt Phot*, **3**, 272–365 (2011).

[Alt05]   Altmann S.L. *Rotations, Quaternions, and Double Groups* (Dover, Mineola, NY, 2005).

[Alt89]   Altmann S.L. "Hamilton, Rodrigues, and the quaternion scandal: What went wrong with mathematical discoveries in the nineteenth century?" *Mathematics Mag*, **62**, 291–308 (1989).

[Alt96]   Altmann S.L. "Clifford algebra, symmetries, and vectors". *Int J Quant Chem*, **60**, 359–372 (1996).

[Ame19]   Amenomori M. and 90 others. "First detection of photons with energy beyond 100 TeV from an astrophysical source". *Phys Rev Lett*, **123**, 051101 (2019).

[Ami08]   Amico L., Fazio R., Osterloh A. and Vedral V. "Entanglement in many-body systems". *Rev Mod Phys*, **80**, 517–576 (2008).

[And04]   Andrews D.L. and Bradshaw D.S. "Virtual photons, dipole fields and energy transfer: A quantum electrodynamical approach". *Eur J Phys*, **25**, 845–858 (2004).

[And05]   Andersen G. "Large optical photon sieve". *Opt Lett*, **30**, 2976–2978 (2005).

[And12]   Andrews D.L. and Coles M.M. "Measures of chirality and angular momentum in the electromagnetic field." *Opt Lett*, **37**, 3009–3011 (2012).

[And14]   Andrews D.L. and Bradshaw D.S. "The role of virtual photons in nanoscale photonics". *Ann d Physik*, **526**, 173–186 (2014).

[And18]   Andrews D.L., Jones G.A., Salam A. and Woolley R.G. "Perspective: Quantum Hamiltonians for optical interactions". *J Chem Phys*, **148**, 040901 (2018).

[Aol15]   Aolita L., de Melo F. and Davidovich L. "Open-system dynamics of entanglement: A key issues review". *Rep Prog Phys*, **78**, 042001 (2015).

[Arl00]   Arlt J. and Dholakia K. "Generation of high-order Bessel beams by use of an axicon". *Opt Comm*, **177**, 297–301 (2000).

[Arl01] Arlt J., Garces-Chavez V., Sibbett W. and Dholakia K. "Optical micromanipulation using a Bessel light beam". *Opt Comm*, **197**, 239–245 (2001).

[Arm62] Armstrong J.A., Bloembergen N., Ducuing J. and Pershan P.S. "Interactions between light waves in a nonlinear dielectric". *Phys Rev*, **127**, 1918–1939 (1962).

[Arn04] Arnoldus H.F. and Foley J.T. "The dipole vortex". *Opt Comm*, **231**, 115–128 (2004).

[Arn05] Arnoldus H.F. "Vortices in multipole radiation". *Opt Comm*, **252**, 253–261 (2005).

[Arn16] Arnoldus H.F., Li X. and Xu Z. "The giant dipole vortex". *J Mod Opt*, **63**, 1068–1072 (2016).

[Arr17] Arrayás M., Bouwmeester D. and Trueba J.L. "Knots in electromagnetism". *Phys Rep*, **667**, 1–61 (2017).

[Art14] Artal P. "Optics of the eye and its impact in vision: a tutorial". *Adv Opt Phot*, **6**, 340–367 (2014).

[Art16] Artin E. *Geometric Algebra* (Dover, Mineola, NY, 2016).

[Aru17] Arun K., Gudennavar S. and Sivaram C. "Dark matter, dark energy, and alternate models: A review". *Adv Space Res*, **60**, 166 – 186 (2017).

[Asp01] Asplund R. and Björk G. "Reconstructing the discrete Wigner function and some properties of the measurement bases". *Phys Rev A*, **64**, 012106 (2001).

[Ath72] Athay R.G. *Radiative Transfer in Spectral Lines* (Reidel, Dordrecht, 1972).

[Aut55] Autler S.H. and Townes C.H. "Stark effect in rapidly varying fields". *Phys Rev*, **100**, 703–722 (1955).

[Auz09] Auzinsh M., Budker D. and Rochester S.M. "Light-induced polarization effects in atoms with partially resolved hyperfine structure and applications to absorption, fluorescence, and nonlinear magneto-optical rotation". *Phys Rev A*, **80**, 053406 (2009).

[Auz10] Auzinsh M., Budker D. and Rochester S.M. *Optically Polarized Atoms* (Oxford Univ Press, Oxford, 2010).

[Bab73] Babiker M., Power E.A. and Thirunamachandran T. "Atomic field equations for Maxwell fields interacting with non-relativistic quantal sources". *Proc Roy Soc (Lond) A*, **332**, 187–197 (1973).

[Bad10] Badescu V. "Tables of Rosseland mean opacities for candidate atmospheres of life hosting free-floating planets". *Cent Eur J Phys*, **8**, 463–479 (2010).

[Bae01] Baez J.C. "The octonions". *Bull Am Math Soc*, **39**, 145–205 (2001).

[Bai07] Baierlein R. "Does nature convert mass into energy?" *Am J Phys*, **75**, 320–325 (2007).

[Bal14] Ball P. *Serving the Reich: The Struggle for the Soul of Physics under Hitler* (Univ Chicago, Chicago, 2014).

[Bal70] Ballentine L.E. "The statistical interpretation of quantum mechanics". *Rev Mod Phys*, **42**, 358–381 (1970).

[Ban13] Banzer P., Neugebauer M., Aiello A., Marquardt C., Lindlein N., Bauer T. and Leuchs G. "The photonic wheel—demonstration of a state of light with purely transverse angular momentum". *J Eur Opt Soc Rapid Pub*, **8**, 13031 (2013).

[Bar01]   Bartelmann M. and Schneider P. "Weak gravitational lensing". *Phys Rep*, **340**, 291–472 (2001).

[Bar02]   Barnett S.M. and Radmore P.M. *Methods in Quantum Optics* (Oxford Univ Press, Oxford, 2002).

[Bar02b] Barnett S.M. "Optical angular-momentum flux". *J Opt B*, **4**, S7–S16 (2002).

[Bar03]   Barnum H., Knill E., Ortiz G. and Viola L. "Generalizations of entanglement based on coherent states and convex sets". *Phys Rev A*, **68**, 032308 (2003).

[Bar03b] Barnett S.M. and Radmore P.M. *Methods in Theoretical Quantum Optics* (Oxford Univ Press, Oxford, 2003).

[Bar06]   Bartschat K. "7. Density matrices". In G. Drake, (ed.) *Springer Handbook of Atomic, Molecular and Optical Physics*, chapter 7, 123–134 (Springer, New York, 2006), 2nd edition.

[Bar09]   Barnett S.M. *Quantum Information* (Oxford Univ Press, Oxford, 2009).

[Bar09b] Barrett S. "Quantum optics: Single photons shape up". *Nature Phot*, **3**, 430–432 (2009).

[Bar09c] Barros H.G., Stute A., Northup T.E., Russo C., Schmidt P.O. and Blatt R. "Deterministic single-photon source from a single ion". *New J Phys*, **11**, 103004 (2009).

[Bar10]   Barnett S.M. "Resolution of the Abraham-Minkowski dilemma". *Phys Rev Lett*, **104**, 070401 (2010).

[Bar10b] Barnett S.M. and Loudon R. "The enigma of optical momentum in a medium." *Phil Trans Roy Soc A*, **368**, 927–39 (2010).

[Bar10c] Barnett S.M. "Rotation of electromagnetic fields and the nature of optical angular momentum". *J Mod Opt*, **57**, 1339–1343 (2010).

[Bar11]   Barnett S.M. "On the six components of optical angular momentum". *J Opt*, **13**, 064010 (2011).

[Bar12]   Barnett S.M. "Measuring the wave function of light". *Phys World*, **6**, 28–29 (2012).

[Bar12b] Barnett S.M., Cameron R.P. and Yao A.M. "Duplex symmetry and its relation to the conservation of optical helicity". *Phys Rev A*, **86**, 013845 (2012).

[Bar14]   Barnett S.M. "Optical Dirac equation". *New J Phys*, **16**, 093008 (2014).

[Bar15]   Barz S. "Quantum computing with photons: Introduction to the circuit model, the one-way quantum computer, and the fundamental principles of photonic experiments". *J Phys B*, **48**, 083001 (2015).

[Bar16]   Barnett S.M., Allen L., Cameron R.P., Gilson C.R., Padgett M.J., Speirits F.C. and Yao A.M. "On the natures of the spin and orbital parts of optical angular momentum". *J Opt*, **18**, 064004 (2016).

[Bar77]   Barone S.R., Narcowich M.A. and Narcowich F.J. "Floquet theory and applications". *Phys Rev*, **15**, 1109–1125 (1977).

[Bar86]   Barnett S.M. and Pegg D.T. "Phase in quantum optics". *J Phys A*, **19**,

3849–386 (1986).

[Bar88]   Barnett S.M. and Knight P.L. "Squeezing the vacuum in atom-field interactions". *Phys Scripta*, **T21**, 5–10 (1988).

[Bar89]   Barnett S.M. and Pegg D.T. "On the Hermitian optical phase operator". *J Mod Opt*, **36**, 7–19 (1989).

[Bar91]   Barnett S.M. and Phoenix S.J.D. "Information theory, squeezing, and quantum correlations". *Phys Rev A*, **44**, 535–545 (1991).

[Bar94]   Barnett S.M. and Allen L. "Orbital angular momentum and nonparaxial light beams". *Opt Comm*, **110**, 670–678 (1994).

[Bar94b]   Barnett S.M. "A backward look at the future". *Contemp Phys*, **35**, 409–411 (1994).

[Bat05]   Batchelor D.B., Berry L.A., Bonoli P.T., Carter M.D., Choi M., D'Azevedo E., D'Ippolito D.A., Gorelenkov N., Harvey R.W., Jaeger E.F., Myra J.R., Okuda H., Phillips C.K., Smithe D.N. and Wright J.C. "Electromagnetic mode conversion: Understanding waves that suddenly change their nature". *J Phys Conf Ser*, **16**, 35–39 (2005).

[Bau18]   Bäuerle C., Christian Glattli D., Meunier T., Portier F., Roche P., Roulleau P., Takada S. and Waintal X. "Coherent control of single electrons: A review of current progress". *Rep Prog Phys*, **81**, 056503 (2018).

[Bay06]   Baylis W.E. "12. Atomic multipoles". In G. Drake, (ed.) *Springer Handbook of Atomic, Molecular and Optical Physics*, chapter 12, 221–226 (Springer, New York, 2006), 2nd edition.

[Bay92]   Baylis W.E., Huschilt J. and Wei J. "Why i ?" *Am J Phys*, **60**, 788–797 (1992).

[Bay93]   Baylis W.E., Bonenfant J., Derbyshire J. and Huschilt J. "Light polarization: A geometric – algebra approach". *Am J Phys*, **61**, 534–545 (1993).

[Bec10]   Beckley A.M., Brown T.G. and Alonso M.A. "Full Poincaré beams". *Opt Exp*, **18**, 10777–10785 (2010).

[Beg84]   Begelman M.C., Blandford R.D. and Rees M.J. "Theory of extragalactic radio sources". *Rev Mod Phys*, **56**, 255–351 (1984).

[Bek11]   Bekshaev A., Bliokh K.Y. and Soskin M. "Internal flows and energy circulation in light beams". *J Opt*, **13**, 053001 (2011).

[Bek18]   Bekshaev A.Y. "Dynamical characteristics of an electromagnetic field under conditions of total reflection". *J Opt*, **20**, 045604 (2018).

[Bel02]   Beloborodov A.M. "Gravitational bending of light near compact objects". *Ap J Lett*, **566**, L85–L88 (2002).

[Bel07]   Beloussov L.V., Voeikov V.L. and Martynyuk V.S. *Biophotonics and coherent systems in biology* (Springer Science & Business Media, New York, 2007).

[Bel66]   Bell J.S. "On the problem of hidden variables in quantum mechanics". *Rev Mod Phys*, **38**, 447–452 (1966).

[Ben00]   Bennett C.H. and DiVincenzo D.P. "Quantum information and computation". *Nature*, **404**, 247–255 (2000).

[Ben16]   Bennett R., Barlow T.M. and Beige A. "A physically motivated quantization of the electromagnetic field". *Eur J Phys*, **37**, 014001 (2016).

[Ben92]   Benner R., Pakulski J.D., Mccarthy M., Hedges J.I. and Hatcher P.G. "Bulk chemical characteristics of dissolved organic matter in the ocean". *Science*, **255**, 1561–1564 (1992).

[Ben93]   Bennett C.H., Crepeau C., Jozsa R. and Peres A. "Teleporting an unknonwn quantum state via dual classical and Einstein-Podolsky-Rosen channels". *Phys Rev Lett*, **70**, 1895–1898 (1993).

[Ben96]   Bennett C.H., DiVincenzo D.P., Smolin J.A. and Wootters W.K. "Mixed-state entanglement and quantum error correction." *Phys Rev A*, **54**, 3824–3851 (1996).

[Ber00]   Berry M.V. "Making waves in physics". *Nature*, **403**, 21–21 (2000).

[Ber01]   Berry M.V. and Dennis M.R. "Knotting and unknotting of phase singularities: Helmholtz waves, paraxial waves and waves in $2+1$ spacetime". *J Phys A*, **34**, 8877–8888 (2001).

[Ber05]   Bernstein J. "Max Born and the quantum theory". *Am J Phys*, **73**, 999–1008 (2005).

[Ber05b] Bertone G., Hooper D. and Silk J. "Particle dark matter: Evidence, candidates and constraints". *Phys Rep*, **405**, 279–390 (2005).

[Ber09]   Berry M.V. "Optical currents". *J Opt A*, **11**, 094001 (2009).

[Ber11]   Berman P.R. and Malinovsky V.S. *Principles of Laser Spectroscopy and Quantum Optics* (Princeton Univ Press, Princeton, NJ, 2011).

[Ber15]   Bergmann K., Vitanov N.V. and Shore B.W. "Perspective: Stimulated Raman adiabatic passage: The status after 25 years". *J Chem Phys*, **142**, 170901 (2015).

[Ber54]   Berankek L. *Acoustics* (McGraw Hill, New York, 1954).

[Ber64]   Berglund C.N. and Spicer W.E. "Photoemission studies of copper and silver: Theory". *Phys Rev*, **136**, A1030–A1044 (1964).

[Ber64b] Berglund C.N. and Spicer W.E. "Photoemission studies of copper and silver: Experiment". *Phys Rev*, **136**, A1044–A1064 (1964).

[Ber75]   Berman P.R. "Theory of collision effects on atomic and molecular line shapes". *App Phys*, **6**, 283–296 (1975).

[Ber82]   Berestetskii V.B., Lifshitz E. and Pitaevskii L.P. *Quantum Electrodynamics* (Butterworth Heinemann, Oxford, 1982), 2nd edition.

[Ber87]   Bertrand J. and Bertrand P. "A tomographic approach to Wigner's function". *Found Phys*, **17**, 397–405 (1987).

[Ber95]   Bergmann K. and Shore B.W. "Coherent population transfer". In H.C. Dai and R.W. Field, (eds.) *Molecular Dynamics and Spectroscopy by Stimulated Emission Pumping*, chapter 9, 315–374 (World Scientific, Singapore, 1995).

[Ber96]   Bernstein P.L. *Against the Gods. The Remarkable Story of Risk* (Wiley, New York, 1996).

[Ber97]   Berman P.R. *Atom Interferometry* (Academic, New York, 1997).

[Ber98]   Bergmann K., Theuer H. and Shore B.W. "Coherent population transfer among quantum states of atoms and molecules". *Rev Mod Phys*, **70**, 1003–1025 (1998).

[Bet05]   Beth T. and Leuchs G., (eds.) *Quantum Information Processing* (Wiley, New York, 2005), 2nd edition.

[Bet57]   Bethe H.A. and Salpeter E.E. *Quantum Mechanics of One- and Two-Electron Atoms* (Academic, New York, 1957).

[Beu35]   Beutler H. "Über Absorptionsserien von Argon, Krypton und Xenon zu Termen zwischen den beiden Ionisierungsgrenzen $^2P^0_{3/2}$ und $^2P^0_{1/2}$ ". *Zs Phys*, **93**, 177–196 (1935).

[Bia00]   Białynicki-Birula I. "Wigner functional of the electromagnetic field". *Opt Comm*, **179**, 237–246 (2000).

[Bia03]   Bialynicki-Birula I. and Bialynicka-Birula Z. "Vortex lines of the electromagnetic field". *Phys Rev A*, **67**, 062114 (2003).

[Bia06]   Bialynicki-Birula I. and Bialynicka-Birula Z. "Beams of electromagnetic radiation carrying angular momentum: The Riemann–Silberstein vector and the classical–quantum correspondence". *Opt Comm*, **264**, 342–351 (2006).

[Bia06b]   Bialynicki-Birula I. and Bialynicka-Birula Z. "Exponential beams of electromagnetic radiation". *J Phys B*, **39**, S545–S553 (2006).

[Bia06c]   Bialynicki-Birula I. "Photon as a quantum particle". *Acta Phys Polonica B*, **37**, 935–946 (2006).

[Bia09]   Bialynicki-Birula I. and Bialynicka-Birula Z. "Why photons cannot be sharply localized". *Phys Rev A*, **79**, 032112 (2009).

[Bia11]   Bialynicki-Birula I. and Bialynicka-Birula Z. "Canonical separation of angular momentum of light into its orbital and spin parts". *J Opt*, **13**, 064014 (2011).

[Bia12]   Bialynicki-Birula I. and Bialynicka-Birula Z. "Uncertainty relation for photons". *Phys Rev Lett*, **108**, 140401 (2012).

[Bia12b]   Bialynicki-Birula I. and Bialynicka-Birula Z. "Heisenberg uncertainty relations for photons". *Phys Rev A*, **86**, 022118 (2012).

[Bia13]   Bialynicki-Birula I. and Bialynicka-Birula Z. "The role of the Riemann-Silberstein vector in classical and quantum theories of electromagnetism". *J Phys A*, **46**, 053001 (2013).

[Bia13b]   Bialynicki-Birula I. and Bialynicka-Birula Z. "Uncertainty relation for focal spots in light beams". *Phys Rev A*, **88**, 052103. (2013).

[Bia14]   Bialynicki-Birula I. "Local and nonlocal observables in quantum optics". *New J Phys*, **16**, 113056 (2014).

[Bia17]   Bialynicki-Birula I. and Bialynicka-Birula Z. "Quantum-mechanical description of optical beams". *J Opt*, **19**, 125201 (2017).

[Bia18]   Bialynicki-Birula I. "Quantum numbers and spectra of structured light". *ArXiv Optics*, **1805.00678**, 1–8 (2018).

[Bia75]   Bialynicki-Birula I. and Bialynicka-Birula Z. *Quantum Electrodynamics* (Pergamon, New York, 1975).

[Bia77]   Bialynicka-Birula Z., Bialynicki-Birula I., Eberly J.H. and Shore B.W. "Coherent dynamics of N-Level atoms and molecules. II. Analytic solutions". *Phys Rev A*, **16**, 2048–2054 (1977).

[Bia91]   Bialynicka-Birula Z., Bialynicki-Birula I. and Salamone G.M. "Spatial antibunching of photons". *Phys Rev A*, **43**, 3696–3703 (1991).

[Bia93]   Białynicki-Birula I., Freyberger M. and Schleich W.P. "Various measures of quantum phase uncertainty: A comparative study". *Phys Scripta*, **1993**, 113–118 (1993).

[Bia96]   Bialynicki-Birula I. "Photon wave function". *Prog Opt*, **36**, 245–294 (1996).

[Bia98]   Białynicki-Birula I. "Exponential localization of photons". *Phys Rev Lett*, **80**, 5247–5250 (1998).

[Bic85]   Bickel W.S. and Bailey W.M. "Stokes vectors, Mueller matrices, and polarized scattered lightr". *Am J Phys*, **53**, 468–478 (1985).

[Bie81]   Biedenharn L.C. and Louck J.D. *Angular Momentum in Quantum Physics* (Addison-Wesley, Reading, MA, 1981).

[Bjo10]   Björk G., Söderholm J., Sánchez-Soto L.L., Klimov A.B., Ghiu I., Marian P. and Marian T.A. "Quantum degrees of polarization". *Opt Comm*, **283**, 4440–4447 (2010).

[Bjo98]   Björk G. and Karlsson A. "Complementarity and quantum erasure in welcher Weg experiments". *Phys Rev A*, **58**, 3477–3483 (1998).

[Bla09]   Bland-Hawthorn J. and Kern P. "Astrophotonics: A new era for astronomical instruments". *Opt Exp*, **17**, 1880–1884 (2009).

[Bla17]   Bland-Hawthorn J. and Leon-Saval S.G. "Astrophotonics: molding the flow of light in astronomical instruments". *Op Exp*, **25**, 15549–15557 (2017).

[Bla52]   Blatt J.M. and Weisskopf V.F. *Theoretical Nuclear Physics* (Wiley, New York, 1952).

[Bla88]   Blatt R., Lafyatis G., Phillips W.D., Stenholm S. and Wineland D.J. "Cooling in Traps". *Phys Scripta*, **T22**, 216–223 (1988).

[Bli11]   Bliokh K.Y. and Nori F. "Characterizing optical chirality". *Phys Rev A*, **83**, 021803 (2011).

[Bli12]   Bliokh K.Y. and Nori F. "Transverse spin of a surface polariton". *Phys Rev A*, **85**, 061801 R (2012).

[Bli13]   Bliokh K.Y., Bekshaev A.Y. and Nori F. "Dual electromagnetism: Hhelicity, spin, momentum, and angular momentum". *New J Phys*, **15**, 033026 (2013).

[Bli14]   Bliokh K.Y., Dressel J. and Nori F. "Conservation of the spin and orbital angular momenta in electromagnetism". *New J Phys*, **16**, 093037 (2014).

[Bli14b]   Bliokh K.Y., Bekshaev A.Y. and Nori F. "Extraordinary momentum and spin in evanescent waves". *Nature Comm*, **5**, 1–8 (2014).

[Bli15]   Bliokh K.Y. and Nori F. "Transverse and longitudinal angular momenta of light". *Phys Rep*, **592**, 1–38 (2015).

[Bli17]   Bliokh K.Y., Bekshaev A.Y. and Nori F. "Optical momentum and angular momentum in complex media: from the Abraham–Minkowski debate to unusual properties of surface plasmon-polaritons". *New J Phys*, **19**, 123014 (2017).

[Blo46]   Bloch F., Hansen W.W. and Packard M. "Nuclear induction". *Phys Rev*, **69**, 127–127 (1946).

[Blo46b]   Bloch F. "Nuclear induction". *Phys Rev*, **70**, 460–474 (1946).

[Blo46c]   Bloch F., Hansen W.W. and Packard M. "The nuclear induction experiment". *Phys Rev*, **70**, 474–485 (1946).

[Blo67]   Bloembergen N. "The stimulated Raman effect". *Am J Phys*, **35**, 989–1023

(1967).

[Blo82] Bloembergen N. "Nonlinear optics and spectroscopy". *Rev Mod Phys*, **54**, 685–695 (1982).

[Blo90] Blow K.J., Loudon R. and Phoenix S.J.D. "Continuum fields in quantum optics". *Phys Rev A*, **42**, 4102–4114 (1990).

[Blu81] Blum K. *Density Matrix Theory and Applications* (Plenum, New York, 1981).

[Bob16] Bobrov V.B., van Heijst G.J., Schram P.P. and Trigger S.A. "Radiation and matter: Electrodynamics postulates and Lorenz gauge". *J Phys Conf Ser*, **774**, 012124 (2016).

[Boh10] Bohren C.F. "What did Kramers and Kronig do and how did they do it?" *Eur J Phys*, **31**, 573–577 (2010).

[Boh23] Bohr N. "The structure of the atom". *Nature*, **112**, 29–44 (1923).

[Boh24] Bohr N., Kramers H.A. and Slater J.C. "LXXVI. The quantum theory of radiation". *Phil Mag*, **47**, 785–802 (1924).

[Bol91] Boller K.J., Imamoğlu A. and Harris S.E. "Observation of electromagnetically induced transparency". *Phys Rev Lett*, **66**, 2593–2596 (1991).

[Bon65] Bonch-Bruevich A.M. and Khodovoi V.A. "Multiphoton processes". *Sov Phys Usp*, **85**, 3–64 (1965).

[Bon68] Bonch-Bruevich A.M. and Khodovoi V.A. "Current methods for the study of the Stark effect in atoms". *Physics-Uspekhi*, **93**, 71–110 (1968).

[Bor09] Boruah B.R. "Dynamic manipulation of a laser beam using a liquid crystal spatial light modulator". *Am J Phys*, **77**, 331–336 (2009).

[Bor10] Boradjiev I.I. and Vitanov N.V. "Stimulated Raman adiabatic passage with unequal couplings: Beyond two-photon resonance". *Phys Rev A*, **81**, 053415 (2010).

[Bor17] Boriskina S.V., Zandavi H., Song B., Huang Y. and Chen G. "Heat is the new light". *Opt Phot News*, **28**, 26–33 (2017).

[Bor25] Born M. and Jordan P. "Zur Quantenmechanik". *Zs Phys*, **34**, 858–888 (1925).

[Bor26] Born M., Heisenberg W. and Jordan P. "Zur Quantenmechanik II". *Zs Phys*, **35**, 557–615 (1926).

[Bor54] Born M. and Huang K. *Dynamical Theory of Crystal Lattices* (Oxford Univ Press, Oxford, 1954).

[Bor66] Bork A.M. ""Vectors versus quaternions"—The letters in Nature". *Am J Phys*, **34**, 202–211 (1966).

[Bor99] Born M. and Wolf E. *Principles of Optics* (Pergamon, New York, 1999), 7th edition.

[Bos02] Boscain U., Charlot G., Gauthier J.P., Guérin S. and Jauslin H.R. "Optimal control in laser-induced population transfer for two- and three-level quantum systems". *J Math Phys*, **43**, 2107–2132 (2002).

[Bos24] Bose S.N. "Planck's Gesetz und Lichtquantenhypothese". *Zs Phys*, **26**, 178–181 (1924).

[Bou97] Bouwmeester D., Pan J.W., Mattle K., Eibl M., Weinfurter H. and Zeilinger A. "Experimental quantum teleportation". *Nature*, **390**, 575–579

(1997).

[Bow02] Bowen W.P., Schnabel R., Bachor H.A. and Lam P.K. "Polarization squeezing of continuous variable Stokes parameters". *Phys Rev Lett*, **88**, 093601 (2002).

[Bow80] Bowen M. and Coster J. "Born's discovery of the quantum-mechanical matrix calculus". *Am J Phys*, **48**, 491–492 (1980).

[Boy08] Boyd R.W. *Nonlinear Optics* (Elsevier, Burlington, MA, 2003), 3rd edition.

[Boy09] Boyd R.W. and Gauthier D.J. "Controlling the velocity of light pulses". *Science*, **326**, 1074–1077 (2009).

[Boy11] Boyer T.H. "Any classical description of nature requires classical electromagnetic zero-point radiation". *Am J Phys*, **79**, 1163–1167 (2011).

[Boy18] Boyer T.H. "Blackbody radiation in classical physics: A historical perspective". *Am J Phys*, **86**, 495–509 (2018).

[Boy19] Boyer T.H. "Stochastic electrodynamics: The closest classical approximation to quantum theory". *Atoms*, **7**, 29–39 (2019).

[Boy69] Boyer T.H. "Derivation of the blackbody radiation spectrum without quantum assumptions". *Phys Rev*, **182**, 1374–1383 (1969).

[Boy75] Boyer T.H. "Random electrodynamics: The theory of classical electrodynamics with classical electromagnetic zero-point radiation". *Phys Rev D*, **11**, 790–808 (1975).

[Bra01] Brattke S., Varcoe B.T.H. and Walther H. "Preparing Fock states in the micromaser". *Opt Exp*, **8**, 4963–4966 (2001).

[Bra03] Bransden B.H. and Joachain C.J. *Physics of Atoms and Molecules* (Longmans, London, 2003), 2nd edition.

[Bra04] Bramon A., Garbarino G. and Hiesmayr B.C. "Quantitative complementarity in two-path interferometry". *Phys Rev A*, **2**, 022112 (2004).

[Bra11] Braak D. "Integrability of the Rabi Model". *Phys Rev Lett*, **107**, 100401 (2011).

[Bra16] Braak D., Chen Q.H., Batchelor M.T. and Solano E. "Semi-classical and quantum Rabi models: in celebration of 80 years". *J Phys A*, **49**, 300301 (2016).

[Bra89] Braginsky V.B., Gorodetsky M. and Ilchenko V. "Quality-factor and nonlinear properties of optical whispering-gallery modes". *Phys Lett A*, **137**, 393–397 (1989).

[Bra96] Braginsky V.B. and Khalili F.Y. "Quantum nondemolition measurements: The route from toys to tools". *Rev Mod Phys*, **68**, 1–11 (1996).

[Bre02] Breuer H.P. and Petruccione F. *The Theory of Open Quantum Systems* (Oxford Univ Press, Oxford, 2002).

[Bre61] Breene R.G. *The Shift and Shape of Spectral Lines* (Pergamon, New York, 1961).

[Bre75] Brewer R.G. and Hahn E.L. "Coherent two-photon processes: Transient and steady-state cases". *Phys Rev A*, **11**, 1641–1649 (1975).

[Bri01] Brixner T. and Gerber G. "Femtosecond polarization pulse shaping." *Opt Lett*, **26**, 557–559 (2001).

[Bri10]   Brif C., Chakrabarti R. and Rabitz H. "Control of quantum phenomena: past, present and future". *New J Phys*, **12**, 075008 (2010).

[Bri60]   Brillouin L. *Wave Propagation and Group Velocity* (Academic, New York, 1960).

[Bri68]   Brink D.M. and Satchler G.R. *Angular Momentum* (Clarendon Press, Oxford, 1968).

[Bro03]   Brown J.M. and Carrington A. *Rotational Spectroscopy of Diatomic Molecules* (Cambridge Univ Press, Cambridge, 2003).

[Bro06]   Brodin G., Eriksson D. and Marklund M. "Graviton mediated photon-photon scattering in general relativity". *Phys Rev D*, **74**, 124028 (2006).

[Bru09]   Brunner D., Gerardot B.D., Dalgarno P.A., Wüst G., Karrai K., Stoltz N.G., Petroff P.M. and Warburton R.J. "A coherent single-hole spin in a semiconductor". *Science*, **325**, 70–72 (2009).

[Buc01]   Buchwald J.Z. and Warwick A. *Histories of the Electron: The Birth of Microphysics* (MIT Press, Cambridge, MA, 2001).

[Buc12]   Buckley S., Rivoire K. and Vučković J. "Engineered quantum dot single-photon sources." *Rep Prog Phys*, **75**, 126503 (2012).

[Buc86]   Buckle S.J., Barnett S.M., Knight P.L., Lauder M.A. and Pegg D.T. "Atomic interferometers". *Optica Acta*, **33**, 1129–1140 (1986).

[Bul10]   Buller G.S. and Collins R.J. "Single-photon generation and detection". *Meas Sci Technol*, **21**, 012002 (2010).

[Buo16]   Buonaura B. and Giuliani G. "Wave and photon descriptions of light: Historical highlights, epistemological aspects and teaching practices". *Eur J Phys*, **37**, 055303 (2016).

[Bur02]   Burlakov A.V. and Chekhova M.V. "Polarization optics of biphotons". *J Exp Theo Phys Lett*, **75**, 432–438 (2002).

[Bur14]   Burton D.A. and Noble A. "Aspects of electromagnetic radiation reaction in strong fields". *Contemp Phys*, **55**, 110–121 (2014).

[Bur14b]   Burhop E.H.S. *The Auger Effect and Other Radiationless Transitions* (Cambridge Univ Press, Cambridge, 2014).

[Buz96]   Bužek V. and Hillery M. "Quantum copying: Beyond the no-cloning theorem." *Phys Rev A*, **54**, 1844–1852 (1996).

[Buz98]   Bužek V., Derka R., Adam G. and Knight P.L. "Reconstruction of quantum states of spin systems: From quantum Bayesian inference to quantum tomography". *Ann Phys*, **266**, 454–496 (1998).

[Cab03]   Caban P. and Rembieliński J. "Photon polarization and Wigner's little group". *Phys Rev A*, **68**, 042107 (2003).

[Cah69]   Cahill K.E. and Glauber R.J. "Ordered expansion in boson amplitude operators". *Phys Rev*, **177**, 1857–1881 (1969).

[Cah69b]   Cahill K.E. and Glauber R.J. "Density operators and quasiprobability distributions". *Phys Rev*, **177**, 1882–1902 (1969).

[Cah80]   Cahill T.A. "Proton microprobes and particle-induced X-ray analytical systems". *Ann Rev Nuc Particle Sci*, **30**, 211–252 (1980).

[Cah81]   Cahill T.A., Kusko B. and Schwab R.N. "Analyses of inks and papers in historical documents through external beam PIXE techniques". *Nuc Inst*

Meth, **181**, 205–208 (1981).

[Cal06]   Calvo G., Picón A. and Bagan E. "Quantum field theory of photons with orbital angular momentum". *Phys Rev A,* **73**, 013805 (2006).

[Cam12]   Cameron R.P., Barnett S.M. and Yao A.M. "Optical helicity, optical spin and related quantities in electromagnetic theory". *New J Phys,* **14**, 053050 (2012).

[Cam12b]   Cameron R.P. and Barnett S.M. "Electric-magnetic symmetry and Noether's theorem". *New J Phys,* **14**, 123019 (2012).

[Cam13]   Cameron R.P., Barnett S.M., Yao A.M., Drummond P.D., Bliokh K.Y., Bekshaev A.Y., Kofman A.G. and Nori F. "Dual electromagnetism: helicity, spin, momentum, and angular momentum". *New J Phys,* **15**, 033026 (2013).

[Cam15]   Cameron R.P., Speirits F.C., Gilson C.R., Allen L. and Barnett S.M. "The azimuthal component of Poynting's vector and the angular momentum of light". *J Opt,* **17**, 125610 (2015).

[Cam18]   Cameron R.P. "Monochromatic knots and other unusual electromagnetic disturbances: light localised in 3D". *J Phys Commun,* **2**, 015024 (2018).

[Can85]   Cannon C.J. *The Transfer of Spectral Line Radiation* (Cambridge Univ Press, New York, 1985).

[Car13]   Carusotto I. and Ciuti C. "Quantum fluids of light". *Rev Mod Phys,* **85**, 299–366 (2013).

[Car68]   Carruthers P. and Nieto M.M. "Phase and angle variables in quantum mechnics". *Rev Mod Phys,* **40**, 411–440 (1968).

[Car71]   Carroll L. "The Walrus and the Carpenter". in: *Through the Looking-Glass and What Alice Found There* (1871).

[Car74]   Carter W.H. "Electromagnetic beam fields". *Optica Acta,* **21**, 871–892 (1974).

[Car81]   Cartan É. *The Theory of Spinors* (Dover, New York, 1967).

[Car89]   Carmichael H.J., Singh S., Vyas R. and Rice P.R. "Photoelectron waiting times and atomic state reduction in resonance fluorescence". *Phys Rev A,* **39**, 1200–1218 (1989).

[Cas08]   Case W.B. "Wigner functions and Weyl transforms for pedestrians". *Am J Phys,* **76**, 937–946 (2008).

[Cas13]   Cassel K.W. *Variational Methods with Applications in Science and Engineering* (Cambridge Univ Press, Cambridge, 2013).

[Cat99]   Catanese M. and Weekes T.C. "Very high energy gamma-ray astronomy". *Pub Astron Soc Pacific,* **111**, 1193–1222 (1999).

[Cav07]   Cavalieri A.L. and 15 others. "Attosecond spectroscopy in condensed matter". *Nature,* **449**, 1029–1032 (2007).

[Cav80]   Caves C.M., Thorne K.S., Drever R.W.P., Sandberg V.D. and Zimmermann M. "On the measurement of a weak classical force coupled to a quantum-mechanical oscillator. I. Issues of principle". *Rev Mod Phys,* **52**, 341–392 (1980).

[Cer11]   Cerjan A. and Cerjan C. "Orbital angular momentum of Laguerre – Gaussian beams beyond the paraxial approximation". *J Opt Soc Am A,* **28**,

2253–2260 (2011).

[Cha02]   Chan K.W., Law C.K. and Eberly J.H. "Localized single-photon wave functions in free space". *Phys Rev Lett*, **88**, 100402 (2002).

[Cha08]   Chang D.E., Gritsev V., Morigi G., Vuletić V., Lukin M.D. and Demler E.A. "Crystallization of strongly interacting photons in a nonlinear optical fibre". *Nature Phys*, **4**, 884–889 (2008).

[Cha10]   Chanyal B.C., Bisht P.S. and Negi O.P.S. "Generalized octonion electrodynamics". *Int J Theo Phys*, **49**, 1333–1343 (2010).

[Cha14]   Chang D., Vuletić V. and Lukin M. "Quantum nonlinear optics—photon by photon". *Nature Phot*, **8**, 685–694 (2014).

[Cha14b] Chanyal B.C., Bisht P.S. and Negi O.P.S. "Octonion electrodynamics in isotropic and chiral medium". *Int J Mod Phys A*, **29**, 1450008 (2014).

[Cha15]   Chang Y.P., Horke D.A., Trippel S. and Küpper J. "Spatially-controlled complex molecules and their applications". *Int Rev Phys Chem*, **34**, 557–590 (2015).

[Cha15b] Chanyal B.C. "Octonion generalization of Pauli and Dirac matrices". *Int J Geom Meth Mod Phys*, **12**, 1550007 (2015).

[Cha15c] Chanyal B.C., Sharma V.K. and Negi O.P.S. "Octonionic gravi-electro-magnetism and dark matter". *Int J Theo Phys*, **54**, 3516–3532 (2015).

[Cha60]   Chandrasekhar S. *Radiative Transfer* (Dover, New York, 1960).

[Che16]   Chekhova M.V. and Ou Z.Y. "Nonlinear interferometers in quantum optics". *Adv Opt Phot*, **8**, 104–155 (2016).

[Che46]   Chevalley C. *Theory of Lie Groups* (Princeton Univ Press, Princeton, NJ, 1946).

[Cho10]   Choi C.L. and Alivisatos A.P. "From artificial atoms to nanocrystal molecules: Preparation and properties of more complex nanostructures". *Ann Rev Phys Chem*, **61**, 369–389 (2010).

[Cho75]   Chow W.W., Scully M.O. and Stoner Jr. J.O. "Quantum-beat phenomena described by quantum electrodynamics and neoclassical theory". *Phys Rev A*, **11**, 1380–1388 (1975).

[Cla72]   Clauser J.F. "Experimental limitations to the validity of semiclassical radiation theories". *Phys Rev A*, **6**, 49–54 (1972).

[Cla74]   Clauser J.F. "Experimental distinction between the quantum and classical field-theoretic predictions for the photoelectric effect". *Phys Rev D*, **9**, 853–860 (1974).

[Cle16]   Clegg B. *Are Numbers Real?* (St Martin's Press, New York, 2016).

[Clo06]   Clowe D., Bradač M., Gonzalez A.H., Markevitch M., Randall S.W., Jones C. and Zaritsky D. "A direct empirical proof of the existence of dark matter". *Ap J Lett*, **648**, L109–L113 (2006).

[Coh77]   Cohen-Tannoudji C. and Reynaud S. "Dressed-atom description of resonance fluorescence and absorption spectra of a multi-level atom in an intense laser beam". *J Phys B*, **10**, 345–363 (1977).

[Coh97]   Cohen-Tannoudji C., Dupont-Roc J. and Grynberg G. *Photons and Atoms: Introduction to Quantum Electrodynamics* (Wiley-Interscience, New York, 1997).

[Coh98]   Cohen-Tannoudji C. "Nobel lecture: Manipulating atoms with photons". *Rev Mod Phys*, **70**, 707–719 (1998).

[Coi93]   Coifman R., Rokhlin V. and Wandzura S. "The fast multipole method for the wave equation: A pedestrian prescription". *IEEE Anten Prop*, **35**, 7–12 (1993).

[Col12]   Coles M.M. and Andrews D.L. "Chirality and angular momentum in optical radiation". *Phys Rev A*, **85**, 063810 (2012).

[Col14]   Coles P.J., Kaniewski J. and Wehner S. "Equivalence of wave-particle duality to entropic uncertainty". *Nature Comm*, **5**, 1–8 (2014).

[Col90]   Colthup N.B., Daly L.H. and Wiberley S.E. *Introduction to Infrared and Raman Spectroscopy* (Academic, London, 1990), 3rd edition.

[Con03]   Conway J.H. and Smith D.A. *On Quaternions and Octonions* (AK Peters/CRC Press, Boca Raton, FL, 2003).

[Con53]   Condon E.U. and Shortley G.H. *The Theory of Atomic Spectra* (Cambridge Univ Press, Cambridge, 1953).

[Coo77]   Cooper M.J. "Compton scattering and electron momentum". *Contemp Phys*, **18**, 489–517 (1977).

[Coo79]   Cook R.J. and Shore B.W. "Coherent dynamics of N-level atoms and molecules. III. An analytically soluble periodic case". *Phys Rev A*, **20**, 539–544 (1979).

[Coo82]   Cook R.J. "Photon dynamics". *Phys Rev A*, **25**, 2164–2167 (1982).

[Cor97]   Cornwall J.F. *Group Theory in Physics: An Introduction* (Academic, New York, 1997).

[Cos19]   Costa e Silva W., Goulart E. and Ottoni J.E. "On spacetime foliations and electromagnetic knots". *J Phys A*, **52**, 265203 (2019).

[Cot62]   Cotton F.A. and Wilkinson G. *Advanced Inorganic Chemistry* (Interscience, New York, 1962).

[Cot90]   Cotton F.A. *Chemical Applications of Group Theory* (Wiley, New York, 1990), 3rd edition.

[Cow81]   Cowan R.D. *The Theory of Atomic Structure and Spectra* (Univ California Press, Berkeley, CA, 1981).

[Cox46]   Cox R.T. "Probability, frequency and reasonable expectation". *Am J Phys*, **14**, 1–13 (1946).

[Cra82]   Cray M., Shih M.L. and Milonni P.W. "Stimulated emission, absorption, and interference". *Am J Phys*, **50**, 1016–1021 (1982).

[Cra84]   Craig D.P. and Thirunamachandran T. *Molecular Quantum Electrodynamics* (Academic, London, 1984).

[Cri19]   Crimin F., Mackinnon N., Götte J.B. and Barnett S.M. "On the conservation of helicity in a chiral medium". *J Opt*, **21**, 094003 (2019).

[Cri69]   Crisp M.D. and Jaynes E.T. "Radiative effects in semiclassical theory". *Phys Rev*, **179**, 1253–1261 (1969).

[Cri73]   Crisp M.D. "Adiabatic-following approximation". *Phys Rev A*, **8**, 2128–2135 (1973).

[Cri96]   Crisp M.D. "Relativistic neoclassical radiation theory". *Phys Rev A*, **54**, 87–92 (1996).

[Cro09]   Cronin A., Schmiedmayer J. and Pritchard J.D. "Optics and interferome-
            try with atoms and molecules". *Rev Mod Phys*, **81**, 1051–1129 (2009).
[Cub05]   Cubel T., Teo B.K., Malinovsky V.S., Guest J.R., Reinhard A., Knuffman
            B., Berman P.R. and Raithel G. "Coherent population transfer of ground-
            state atoms into Rydberg states". *Phys Rev A*, **72**, 023405 (2005).
[Cur12]   Curceanu C., Gillaspy J.D. and Hilborn R.C. "Resource Letter SS – 1:
            The spin-statistics connection". *Am J Phys*, **80**, 561–577 (2012).
[Dah90]   Dahleh M.A., Peirce A.P. and Rabitz H. "Optimal control of uncertain
            quantum systems". *Phys Rev A*, **42**, 1065–1079 (1990).
[Dal12]   Dalhuisen J.W. and Bouwmeester D. "Twistors and electromagnetic
            knots". *J Phys A*, **45**, 135201 (2012).
[Dal82]   Dalibard J., Dupont-Roc J. and Cohen-Tannoudji C. "Vacuum fluctu-
            ations and radiation reaction: identification of their respective contribu-
            tions". *J Physique*, **43**, 1617–1638 (1982).
[Dal85]   Dalton B.J., Mcduff R. and Knight P.L. "Coherent population trapping.
            Two unequal phase fluctuating laser fields". *Optica Acta*, **32**, 61–70 (1985).
[Dal99]   Dalton B.J., Ficek Z. and Swain S. "Atoms in squeezed light fields". *J
            Mod Opt*, **46**, 379–474 (1999).
[Dam03]   Damascelli A., Hussain Z. and Shen Z.X. "Angle-resolved photoemission
            studies of the cuprate superconductors". *Rev Mod Phys*, **75**, 473–541
            (2003).
[Dar32]   Darwin C.G. "Notes on the theory of radiation". *Proc Roy Soc (Lond) A*,
            136, 36–52 (1932).
[Dav16]   Davis J.A., Moreno I., Badham K., Sánchez-López M.M. and Cottrell D.M.
            "Nondiffracting vector beams where the charge and the polarization state
            vary with propagation distance". *Opt Lett*, **41**, 2270–2273 (2016).
[Dav19]   Davis M.L. *Lost Gutenberg. The Astounding Story of One Book's Five-
            Hundred-Year Odyssey* (Penguin Random House, New York, 2019).
[Dav70]   Davies E.B. and Lewis J.T. "An operational approach to quantum prob-
            ability". *Comm Math Phys*, **17**, 239–260 (1970).
[Dav79]   Davis L. "Theory of electromagnetic beams". *Phys Rev A*, **19**, 1177–1179
            (1979).
[Dav79b]   Davidson K. and Netzer H. "The emission lines of quasars and similar
            objects". *Rev Mod Phys*, **51**, 715–765 (1979).
[Daw86]   Dawkins R. *The Blind Watchmaker* (W W Norton, New York, 1986).
[DeB02]   de Bernardis P. and 35 others. "A flat universe from high-resolution maps
            of the cosmic microwave background radiation". *Nature*, **404**, 955–959
            (2000).
[DeS74]   de Shalit A. and Feshbach H. *Theoretical Nuclear Physics* (Wiley, New
            York, 1974).
[DeV44]   DeVault D. "A method of teaching the electronic structure of the atom.
            1. Elementary presentation". *J Chem Ed*, **21**, 526–534 (1944).
[DeZ18]   De Zela F. "Optical approach to concurrence and polarization". *Opt Lett*,
            43, 2603–2606 (2018).
[DeZ18b]   De Zela F. "Hidden coherences and two-state systems". *Optica*, **5**, 243–250

(2018).

[Deh54]   Dehmelt H.G. "Nuclear quadrupole resonance". *Am J Phys*, **93**, 110–120 (1954).

[Del06]   Dell'Anno F., De Siena S. and Illuminati F. "Multiphoton quantum optics and quantum state engineering". *Phys Rep*, **428**, 53–168 (2006).

[Dem05]   Demtröder W. *Molecular Physics: Theoretical Principles and Experimental Methods* (Wiley-VCH, Weinheim, 2005).

[Dem10]   Demtröder W. *Atoms, Molecules, and Photons: An Introduction to Atomic-, Molecular-, and Quantum-Physics* (Springer, Berlin, 2010), 2nd edition.

[Dem27]   Dempster A. and Batho H. "Light quanta and interference". *Phys Rev*, **30**, 644–648 (1927).

[Dem96]   Demtröder W. *Laser Spectroscopy* (Springer-Verlag, Berlin, 1996), 2nd edition.

[Dem98]   Demtröder W. *Experimentalphysik 4* (Springer, Berlin, 1998).

[Den09]   Dennis M., O'Holleran K. and Padgett M. "Singular optics: optical vortices and polarization singularities". *Prog Opt*, **53**, 293–363 (2009).

[Den10]   Dennis M.R., King R.P., Jack B., Oholleran K. and Padgett M.J. "Isolated optical vortex knots". *Nature Phys*, **6**, 118–121 (2010).

[Der04]   Derbyshire J. *Prime Obsession. Bernhard Riemann and the Greatest Unsolved Problem in Mathematics* (Penguin Group, New York, 2004).

[Des77]   Desloge E.A. and Karch R.I. "Noether's theorem in classical mechanics". *Am J Phys*, **45**, 336–339 (1977).

[Dev04]   Devoret M.H., Wallraff A. and Martinis J.M. "Superconducting qubits: A short review". *ArXiv preprint cond-mat/0411174*, 1–41 (2004).

[Dev12]   Devlin K. *The Joy of Sets: Fundamentals of Contemporary Set Theory* (Springer Science & Business Media, New York, 2012).

[Dev74]   Devaney A.J. and Wolf E. "Multipole expansions and plane wave representations of the electromagnetic field". *J Math Phys*, **15**, 234–244 (1974).

[DiP12]   Di Piazza A., Müller C., Hatsagortsyan K.Z. and Keitel C.H. "Extremely high-intensity laser interactions with fundamental quantum systems". *Rev Mod Phys*, **84**, 1177–1228 (2012).

[DiV00]   DiVincenzo D.P. "The physical implementation of quantum computation". *Fortsch Phys*, **78**, 771–783 (2000).

[Die03]   Dietz K. "On the parametrization of Lindblad equations". *J Phys A*, **36**, 5595–5603 (2003).

[Dim08]   Dimitrova T.L. and Weis A. "The wave-particle duality of light: A demonstration experiment". *Am J Phys*, **76**, 137–142 (2008).

[Dir24]   Dirac P.A.M. "Note on the Doppler principle and Bohr's frequency condition". *Math Proc Camb Phil Soc*, **22**, 432–433 (1924).

[Dir27]   Dirac P.A.M. "The quantum theory of the emission and absorption of radiation". *Proc Roy Soc (Lond) A*, **114**, 243–265 (1927).

[Dir28]   Dirac P.A.M. "The quantum theory of the electron". *Proc Roy Soc (Lond) A*, **117**, 610–624 (1928).

[Dir58]   Dirac P.A.M. *The Principles of Quantum Mechanics* (Clarendon Press,

Oxford, 1958), 4th edition.

[Dir65] Dirac P.A.M. "Quantum electrodynamics without dead wood". *Phys Rev*, **139**, B684–B690 (1965).

[Dix94] Dixon G.M. *Division Algebras: Octonions, Quaternions, Complex Numbers and the Algebraic Design of Physics* (Kluwer, Dordrech, 1994).

[Dod02] Dodonov V.V. "'Nonclassical' states in quantum optics: A 'squeezed ' review of the first 75 years". *J Opt B*, **4**, R1–R33 (2002).

[Dod83] Dodd J.N. "The Compton effect-a classical treatment". *Eur J Phys*, **4**, 205–211 (1983).

[Dom02] Domokos P., Horak P. and Ritsch H. "Quantum description of light-pulse scattering on a single atom in waveguides". *Phys Rev A*, **65**, 033832 (2002).

[Dom15] Dömötör P., Földi P., Benedict M.G., Shore B.W. and Schleich W.P. "Scattering of a particle with internal structure from a single slit: exact numerical solutions". *New J Phys*, **17**, 023044 (2015).

[Don11] Donovan A., Beltrani V. and Rabitz H. "Quantum control by means of Hamiltonian structure manipulation". *Phys Chem Chem Phys*, **13**, 7348–7362 (2011).

[Dor03] Doran C. and Lasenby A. *Geometric Algebra for Physicists* (Cambridge Univ Press, Cambridge, 2003).

[Dra06] Drake G.W.F. *Springer Handbook of Atomic, Molecular, and Optical Physics* (Springer, New York, 2006).

[Dre15] Dressel J., Bliokh K.Y. and Nori F. "Spacetime algebra as a powerful tool for electromagnetism". *Phys Rep*, **589**, 1–71 (2015).

[Dri78] Driscoll W.G. and Vaughan W. *Handbook of Optics* (McGraw-Hill, New York, 1978).

[Dru14] Drummond P.D. and Hilary M. *The Quantum Theory of Nonlinear Optics* (Cambridge Univ Press, New York, 2014).

[Dru90] Drummond P.D. "Electromagnetic quantization in dispersive inhomogeneous nonlinear dielectrics". *Phys Rev A*, **42**, 6845–6857 (1990).

[Dua01] Duan L.M., Lukin M.D., Cirac J.I. and Zoller P. "Long-distance quantum communication with atomic ensembles and linear optics". *Nature*, **414**, 413–418 (2001).

[Dua10] Duan L.M. "Colloquium: Quantum networks with trapped ions". *Rev Mod Phys*, **82**, 1209–1224 (2010).

[Duc98] Duck I. and Sudarshan E.C.G. "Toward an understanding of the spin-statistics theorem". *Am J Phys*, **66**, 284–303 (1998).

[Dud12] Dudin Y.O. and Kuzmich A. "Strongly interacting Rydberg excitations of a cold atomic gas." *Science*, **336**, 887–889 (2012).

[Dud96] Dudley J.M. and Kwan A.M. "Richard Feynman's popular lectures on quantum electrodynamics: The 1979 Robb lectures at Auckland University". *Am J Phys*, **64**, 694–698 (1996).

[Dun03] Dunsby C. and French P.M.W. "Techniques for depth-resolved imaging through turbid media including coherence-gated imaging". *J Phys D*, **36**, R207–R227 (2003).

[Dup09] Duparc O.H. "Pierre Auger – Lise Meitner: Comparative contributions to the Auger effec". *Int J Mat Res*, **100**, 1162–1166 (2009).

[Dur00]   Dürr S. and Rempe G. "Can wave–particle duality be based on the uncertainty relation?" *Am J Phys*, **68**, 1021–1024 (2000).

[Dur87]   Durnin J. "Exact solutions for nondiffracting beams. I. The scalar theory". *J Opt Soc Am A*, **4**, 651–654 (1987).

[Dur87b] Durnin J., Miceli, J. J. Jr. and Eberly J.H. "Diffraction-free beams". *Phys Rev Lett*, **58**, 1499–1501 (1987).

[Dur88]   Durnin J., Miceli, J. J. Jr. and Eberly J.H. "Comparison of Bessel and Gaussian beams". *Opt Lett*, **13**, 79–80 (1988).

[Dur98]   Dürr S., Nonn T. and Rempe G. "Fringe visibility and which-way information in an atom interferometer". *Phys Rev Lett*, **81**, 5705–5709 (1998).

[Dys06]   Dyson F. *The Scientist as a Rebel* (New York Review of Books, New York, 2006).

[Dys49]   Dyson F.J. "The radiation theories of Tomonaga, Schwinger, and Feynman". *Phys Rev*, **75**, 486–502 (1949).

[Ebe15]   Eberly J.H. "Shimony–Wolf states and hidden coherences in classical light". *Contemp Phys*, **56**, 407–416 (2015).

[Ebe16]   Eberly J.H. "Correlation, coherence and context". *Laser Phys*, **26**, 084004 (2016).

[Ebe16b] Eberly J.H., Qian X.F., Qasimi A.A., Ali H., Alonso M.A., Gutiérrez-Cuevas R., Little B.J., Howell J.C., Malhotra T. and Vamivakas A.N. "Quantum and classical optics – emerging links". *Phys Scripta*, **91**, 063003 (2016).

[Ebe17]   Eberly J.H., Qian X.F. and Vamivakas A.N. "Polarization coherence theorem". *Optica*, **4**, 1113–1114 (2017).

[Ebe68]   Eberly J.H. "Coherent radiation and two-level electric dipole transitions, a quantitative classical analogy". *Phys Lett*, **26A**, 499–500 (1968).

[Ebe77]   Eberly J.H., Shore B.W., Bialynicka-Birula Z. and Bialynicki-Birula I. "Coherent dynamics of N-level atoms and molecules I. Numerical experiments". *Phys Rev A*, **16**, 2038–2047 (1977).

[Ebe77b] Eberly J.H. and Wódkiewicz K. "The time-dependent physical spectrum of light". *J Opt Soc Am*, **67**, 1252–1261 (1977).

[Ebe80]   Eberly J.H., Narozhny N.B. and Sanchez-Mondragon J.J. "Periodic spontaneous collapse and revival in a simple quantum model". *Phys Rev Lett*, **44**, 1323–1326 (1980).

[Ebe84]   Eberly J.H., Wódkiewicz K. and Shore B.W. "Noise in strong laser-atom interactions: Phase telegraph noise". *Phys Rev A*, **30**, 2381–2389 (1984).

[Ebe87]   Eberly J.H. and Milonni P.W. "Quantum optics". In vol 11, (ed.) *Encycl. of Physical Science and Technology*, (Academic, New York, 1987).

[Edd59]   Eddington A.S. *The Internal Constitution of the Stars* (Dover, New York, 1959).

[Edm57]   Edmonds A.R. *Angular Momentum in Quantum Mechanics* (Princeton Univ Press, Princeton, NJ, 1957).

[Edm74]   Edmonds, J. D. Jr. "Quaternion quantum theory: New physics or number mysticism. 2. Curved space". *Am J Phys*, **220**, 220–223 (1974).

[Edm78]   Edmonds J.D. "Maxwell's eight equations as one quaternion equation".
           Am J Phys, **46**, 430–431 (1978).

[Ehl97]   Ehlers J. and Rindler W. "Local and global light bending in Einstein's
           and other gravitational theories". *Gen Relativ Grav*, **29**, 519–529 (1997).

[Ein05]   Einstein A. "On a heuristic point of view about the creation and conversion
           of light". *Ann Phys (Leipz)*, **17**, 132–148 (1905).

[Ein35]   Einstein A., Podolsky B. and Rosen N. "Can quantum-mechanical descrip-
           tion of physical reality be considered complete?" *Phys Rev*, **47**, 777–780
           (1935).

[Ein79]   Einwohner T.H., Wong J. and Garrison J.C. "Effects of alternative tran-
           sition sequences on coherent photoexcitation". *Phys Rev A*, **20**, 940–947
           (1979).

[Eis06]   Eisaman M.D., Fleischhauer M. and Zibrov A.S. "Electromagnetically
           induced transparency: Toward quantum control of single photons". *Opt
           Phot News*, **17**, 23–27 (2006).

[Eis11]   Eisaman M.D., Fan J., Migdall A. and Polyakov S. "Invited review article:
           Single-photon sources and detectors". *Rev Sci Inst*, **82**, 071101 (2011).

[Eke91]   Ekert A.K. "Quantum cryptography based on Bell's theorem". *Phys Rev
           Lett*, **67**, 661–663 (1991).

[Eke94]   Ekert A. and Palma G.M. "Quantum cryptography with interferometric
           quantum entanglement". *J Mod Opt*, **41**, 2413–2423 (1994).

[Eke95]   Ekert A.K. and Knight P.L. "Entangled quantum systems and the Schmidt
           decomposition". *Am J Phys*, **63**, 415–423 (1995).

[Ell15]   Ellenberg J. *How Not to be Wrong. The Power of Mathematical Thinking*
           (Penguin, New York, 2015).

[Elv00]   Elvis M. "A structure for quasars". *Ap J*, **545**, 63–76 (2000).

[End89]   Tonomura A., Endo J., Matsuda T., Kawasaki T. and Exawa H. "Demon-
           stration of single-electron buildup of an interference pattern". *Am J Phys*,
           57, 117–120 (1989).

[Eng96]   Englert B.G. "Fringe visibility and which-way information: An inequality".
           Phys Rev Lett, **77**, 2154–2157 (1996).

[Erk10]   Erkmen B.I. and Shapiro J.H. "Ghost imaging: from quantum to classical
           to computational". *Adv Opt Phot*, **2**, 405–450 (2010).

[Eva55]   Evans R.D. *The Atomic Nucleus* (McGraw-Hill, New York, 1955).

[Eyr44]   Eyring H., J.Walter and Kimball G.E. *Quantum Chemistry* (Wiley, New
           York, 1944).

[Fal11]   Falkovich G. *Fluid Mechanics (A Short Course for Physicists)* (Cambridge
           Univ Press, Cambridge, 2011).

[Fan35]   Fano U. "Sullo spettro di assorbimento dei gas nobili presso il limite dello
           spettro d'arco". *Nuov Cim*, **12**, 154–161 (1935).

[Fan49]   Fano U. "Remarks on the classical and quantum-mechanical treatment of
           partial polarization". *J Opt Soc Am*, **39**, 859–863 (1949).

[Fan54]   Fano U. "A Stokes-parameter technique for the treatment of polarization
           in quantum mechanics". *Phys Rev*, **93**, 121–123 (1954).

[Fan57]   Fano U. "Description of states in quantum mechanics by density matrix and operator techniques". *Rev Mod Phys*, **29**, 74–93 (1957).

[Fan61]   Fano U. "Effects of configuraton interaction on intensities and phase shifts". *Phys Rev*, **124**, 1866–1878 (1961).

[Fan65]   Fano U. and Cooper J.W. "Line profiles in the far-uv absorption spectra of the rare gases". *Phys Rev*, **137**, A1364–A1379 (1965).

[Far05]   Faraday M. *The Chemical History of a Candle* (Barnes & Noble, New York, 2005).

[Far07]   Farsiu S., Christofferson J., Eriksson B., Milanfar P., Friedlander B., Shakouri A. and Nowak R. "Statistical detection and imaging of objects hidden in turbid media using ballistic photons". *App Opt*, **46**, 5805–5822 (2007).

[Far16]   Farrera P., Heinze G., Albrecht B., Ho M., Chávez M., Teo C., Sangouard N. and de Riedmatten H. "Generation of single photons with highly tunable wave shape from a cold atomic quantum memory". *Nature Comm*, **7**, 1–10 (2016).

[Fea96]   Fearn H., James D.F.V. and Milonni P.W. "Microscopic approach to reflection, transmission, and the Ewald–Oseen extinction theorem". *Am J Phys*, **64**, 986–995 (1996).

[Fed14]   Fedorov M.V. and Miklin N.I. "Schmidt modes and entanglement". *Contemp Phys*, **55**, 94–109 (2014).

[Fei74]   Feibelman P.J. and Eastman D.E. "Photoemission spectroscopy: Correspondence between quantum theory and experimental phenomenology". *Phys Rev B*, **10**, 4932–4947 (1974).

[Fel68]   Feller W. *Introduction to Probability Theory and its Applications*, volume 1 (Wiley, New York, 1968), 3rd edition.

[Fer32]   Fermi E. "Quantum theory of radiation". *Rev Mod Phys*, **4**, 87–132 (1932).

[Fes58]   Feshbach H. and Villars F. "Elementary relativistic wave mehanics of spin 0 and spin 1/2 particles". *Rev Mod Phys*, **30**, 24–45 (1958).

[Fes58b]   Feshbach H. "Unified theory of nuclear reactions". *Ann Phys*, **5**, 357–390 (1958).

[Fes62]   Feshbach H. "A unified theory of nuclear reactions. II". *Ann Phys*, **19**, 287–313 (1962).

[Few97]   Fewell M.P., Shore B.W. and Bergmann K. "Coherent population transfer among three states: Full algebraic solutions and the relevance of non adiabatic processes to transfer by delayed pulses". *Australian J Phys*, **50**, 281–308 (1997).

[Fey48]   Feynman R.P. "Space-time approach to non-relativistic quantum mechanics". *Rev Mod Phys*, **64**, 367–387 (1948).

[Fey57]   Feynman R.P., Vernon, F. L. Jr. and Hellwarth R.W. "Geometrical representation of the Schrödinger equation for solving maser problems". *J App Phys*, **28**, 49–52 (1957).

[Fey63]   Feynman R.P. *Feynman Lectures* (Addison-Wesley, Reading, MA, 1963).

[Fey66]   Feynman R.P. "The development of the space-time view of quantum electrodynamics". *Science*, **153**, 699–708 (1966).

[Fey69]   Feynman R.P. "What is science". *Phys Teacher*, **7**, 313–320 (1969).

[Fey85]   Feynman R.P. *QED: The Strange Theory of Lght and Matter* (Princeton

Univ Press, Princeton, NJ, 1985).

[Fic12] Fickler R., Lapkiewicz R., Plick W.N., Krenn M., Schaeff C., Ramelow S. and Zeilinger A. "Quantum entanglement of high angular momenta." *Science*, **338**, 640–6433 (2012).

[Fic13] Fickler R., Krenn M., Lapkiewicz R., Ramelow S. and Zeilinger A. "Real-time Imaging of quantum entanglement". *Sci Rep*, **3**, 1–5 (2013).

[Fin03] Finkelstein D. "What is a photon?" *OPN Trends of Opt Phot News*, S12–S17 (2003).

[Fin18] Finnegan W. "Kelly Slater's shock wave". *New Yorker*, **December 17**, 44–55 (2018).

[Fin62] Finkelstein D., Jauch J.M., Schiminovich S. and Speiser D. "Foundations of quaternion quantum mechanics". *J Math Phys*, **3**, 207–220 (1962).

[Fir13] Firstenberg O., Peyronel T., Liang Q.Y., Gorshkov A.V., Lukin M.D. and Vuletić V. "Attractive photons in a quantum nonlinear medium." *Nature*, **502**, 71–75 (2013).

[Fir16] Firstenberg O., Adams C.S. and Hofferberth S. "Nonlinear quantum optics mediated by Rydberg interactions". *J Phys B*, **49**, 152003 (2016).

[Fis06] Fischer C.F. "21. Atomic structure: Multiconfiguration Hartree-Fock theories". In G. Drake, (ed.) *Springer Handbook of Atomic, Molecular and Optical Physics*, chapter 21, 307–324 (Springer, New York, 2006), 2nd edition.

[Fiu62] Fiutak J. "The multipole expansion in quantum theory'". *Can J Phys*, **41**, 12–20 (1962).

[Fle00] Fleischhauer M. and Lukin M.D. "Dark-state polaritons in electromagnetically induced transparency". *Phys Rev Lett*, **84**, 5094–50977 (2000).

[Fle02] Fleischhauer M. and Lukin M.D. "Quantum memory for photons: Dark-state polaritons". *Phys Rev A*, **65**, 022314 (2002).

[Fle05] Fleischhauer M., Imamoğlu A. and Marangos J.P. "Electromagnetically induced transparency: Optics in coherent media". *Rev Mod Phys*, **77**, 633–673 (2005).

[Fle96] Fleischhauer M. and Manka A.S. "Propagation of laser pulses and coherent population transfer in dissipative three-level systems: An adiabatic dressed-state picture." *Phys Rev A*, **54**, 794–803 (1996).

[For08] Forterre Y. and Pouliquen O. "Flows of dense granular media". *Ann Rev Fluid Mech*, **40**, 1–24 (2008).

[For15] Foreman M.R., Swaim J.D. and Vollmer F. "Whispering gallery mode sensors". *Adv Opt Phot*, **7**, 168–240 (2015).

[For19] Forn-Díaz P., Lamata L., Rico E., Kono J. and Solano E. "Ultrastrong coupling regimes of light-matter interaction". *Rev Mod Phys*, **91**, 025005 (2019).

[For60] Forsyth A. *Calculus of Variations* (Dover, New York, 1960).

[Fox72] Fox R.F. "Contributions to the theory of multiplicative stochastic processes". *J Math Phys*, **13**, 1196–1207 (1972).

[Fox87] Fox C. *An Introduction to the Calculus of Variations* (Dover, New York, 1987).

[Fox88]   Fox R.F., Gatland I.R., Roy R. and Vemuri G. "Fast, accurate algorithm for numerical simulation of exponentially correlated colored noise". *Phys Rev A*, **38**, 5938–5940 (1988).

[Fra11]   Franson J. "Entanglement from longitudinal and scalar photons". *Phys Rev A*, **84**, 033809 (2011).

[Fra17]   Franke-Arnold S. "Optical angular momentum and atoms ". *Phil Trans Roy Soc A*, **375**, 20150435 (2017).

[Fra97]   Franklin A. "Are there really electrons? Experiment and reality". *Phys Today*, **50**, 26–33 (1997).

[Fre18]   Frezza F., Mangini F. and Tedeschi N. "Introduction to electromagnetic scattering: Tutorial". *J Opt Soc Am A*, **35**, 163–173 (2018).

[Fri09]   Frisch O.R. "Take a photon...". *Contemp Phys*, **50**, 59–67 (2009).

[Fri65]   Friedman F.L. and Sartori L. *The Classical Atom* (Addison-Wesley, Reading, MA, 1965).

[Gae06]   Gaeta A.L. and Boyd R.W. "72. Nonlinear optics". In G. Drake, (ed.) *Springer Handbook of Atomic, Molecular and Optical Physics*, chapter 72, 1051–1064 (Springer, New York, 2006), 2nd edition.

[Gal05]   Galvez E.J., Holbrow C.H., Pysher M.J., Martin J.W., Courtemanche N., Heilig L. and Spencer J. "Interference with correlated photons: Five quantum mechanics experiments for undergraduates". *Am J Phys*, **73**, 127–140 (2005).

[Gal06]   Gallagher T.F. "14. Rydberg atoms". In G. Drake, (ed.) *Springer Handbook of Atomic, Molecular and Optical Physics*, chapter 14, 235–246 (Springer, New York, 2006), 2nd edition.

[Gal06b]   Gallagher A. "19. Line shapes and radiation transfer". In G. Drake, (ed.) *Springer Handbook of Atomic, Molecular and Optical Physics*, chapter 19, 279–294 (Springer, New York, 2006), 2nd edition.

[Gal08]   Gallagher T.F. and Pillet P. "Dipole-dipole interactions of Rydberg atoms". *Adv At Mol Opt Phys*, **56**, 161–218 (2008).

[Gal12]   Galvez E.J., Khadka S., Schubert W.H. and Nomoto S. "Poincaré-beam patterns produced by nonseparable superpositions of Laguerre–Gauss and polarization modes of light". *App Opt*, **51**, 2925–2934 (2012).

[Gal14]   Galvez E.J. "Resource Letter SPE-1: Single-Photon experiments in the undergraduate laboratory". *Am J Phys*, **82**, 1018–1028 (2014).

[Gan05]   Gantmacher F.R. and Brenner J.L. *Applications of the Theory of Matrices* (Dover, Mineola, NY, 2005).

[Gan60]   Gantmacher F.R. *Theory of Matrices* (Chelsea, New York, 1960).

[Gar01]   Gardner M. "A skeptical look at Karl Popper". *Skeptical Inquirer*, **25**, 13–14 (2001).

[Gar04]   Garrison J.C. and Chiao R.Y. "Canonical and kinetic forms of the electromagnetic momentum in an ad hoc quantization scheme for a dispersive dielectric". *Phys Rev A*, **70**, 053826 (2004).

[Gar08]   Garrison J.C. and Chaio R.Y. *Quantum Optics* (Oxford Univ Press, Oxford, 2008).

[Gar85]   Gardiner C.W. *Handbook of Stochastic Methods* (Springer, New York,

1985).

[Gas18] Gasser T.M., Thoeny A.V., Plaga L.J., Köster K.W., Etter M., Böhmer R. and Loerting T. "Experiments indicating a second hydrogen ordered phase of ice VI". *Chem Sci*, **9**, 4224–4234 (2018).

[Gau90] Gaubatz U., Rudecki P., Schiemann S. and Bergmann K. "Population transfer between molecular vibrational levels by stimulated Raman scattering with partially overlapping laser fields. A new concept and experimental results". *J Chem Phys*, **92**, 5363–5376 (1990).

[Gaz09] Gazeau J.P. *Coherent States in Quantum Physics* (Wiley, New York, 2009).

[Gea02] Gearhart C.A. "Planck, the Quantum, and the Historians". *Physics in Perspective*, **4**, 170–215 (2002).

[Gen11] Genov G.T., Torosov B.T. and Vitanov N.V. "Optimized control of multistate quantum systems by composite pulse sequences". *Phys Rev A*, **84**, 063413 (2011).

[Gen14] Genov G.T., Schraft D., Halfmann T. and Vitanov N.V. "Correction of arbitrary field errors in population inversion of quantum systems by universal composite pulses". *Phys Rev Lett*, **113**, 043001 (2014).

[Ger01] Gerry C.C. "Remarks on the use of group theory in quantum optics." *Opt Exp*, **8**, 76–85 (2001).

[Ger04] Gerry C.C. and Knight P.L. *Introductory Quantum Optics* (Cambridge Univ Press, Cambridge, 2004).

[Gho14] Ghose P. and Mukherjee A. "Entanglement in classical optics". *Rev Theo Sci*, **2**, 274–288 (2014).

[Gib73] Gibbs H.M., Churchill G.G. and Salamo G.J. "Contradictions with the Neoclassical theory of radiation in weakly excited multilevel systems". *Phys Rev A*, **7**, 1766–1770 (1973).

[Gim06] Giménez F., Monsoriu J.A., Furlan W.D. and Pons A. "Fractal photon sieve". *Opt Exp*, **14**, 11958–11963 (2006).

[Gio04] Giovannetti V., Lloyd S. and Maccone L. "Quantum-enhanced measurements: Beating the standard quantum limit". *Science*, **306**, 1330–1336 (2004).

[Gir09] Girvin S.M., Devoret M.H. and Schoelkopf R.J. "Circuit QED and engineering charge-based superconducting qubits". *Phys Scripta T*, **T137**, 014012 (2009).

[Gir11] Girvin S.M. "Circuit QED: Superconducting Qubits Coupled to Microwave Photons". *Proc 2011 Les Houches Summer School* (2011).

[Gis02] Gisin N., Ribordy G., Tittel W. and Zbinden H. "Quantum cryptography". *Rev Mod Phys*, **74**, 145–195 (2002).

[Giu13] Giuliani G. "Experiment and theory: The case of the Doppler effect for photons". *Eur J Phys*, **34**, 1035–1047 (2013).

[Gla06] Glauber R.J. "Nobel Lecture: One hundred years of light quanta". *Rev Mod Phys*, **78**, 1267–1278 (2006).

[Gla63] Glauber R.J. "The quantum theory of optical coherence". *Phys Rev*, **130**, 2529–2539 (1963).

[Gla63b] Glauber R.J. "Coherent and incoherent states of the radiation field". *Phys*

Rev, **131**, 2766–2788 (1963).

[Gla91] Glauber R.J. and Lewenstein M. "Quantum optics of dielectric media". *Phys Rev A*, **43**, 467–491 (1991).

[Gla95] Glauber R.J. "Dirac's famous dictum on interference: One photon or two?" *Am J Phys*, **63**, 12–12 (1995).

[Gog06] Gogberashvili M. "Octonionic electrodynamics". *J Phys A*, **39**, 7099–7104 (2006).

[Gol50] Goldstein H. *Classical Mechanics* (Addison-Wesley, Cambridge, MA, 1950).

[Gol64] Goldberger M. and Watson K.M. *Collision Theory* (Wiley, New York, 1964).

[Gol65] Gold A. and Bebb H.B. "Theory of multiphoton ionization". *Phys Rev Lett*, **14**, 60–63 (1965).

[Gop31] Göppert-Mayer M. "Über Elementarakte mit zwei Quantensprüngen". *Ann d Physik*, **401**, 273–294 (1931).

[Gor07] Gorshkov A.V., André A., Fleischhauer M., Sørensen A.S. and Lukin M.D. "Universal approach to optimal photon storage in atomic media". *Phys Rev Lett*, **98**, 123601 (2007).

[Gor99] Gordon R.J., Zhu L. and Seideman T. "Coherent control of chemical reactions". *Accounts Chem Res*, **32**, 1007–1016 (1999).

[Gra06] Grant I.P. "22. Relativistic atomic structure". In G. Drake, (ed.) *Springer Handbook of Atomic, Molecular and Optical Physics*, chapter 22, 325–358 (Springer, New York, 2006), 2nd edition.

[Gra78] Gray H.R. and Stroud, C. R. Jr. "Autler-Townes effect in double optical resonance". *Opt Comm*, **25**, 359–362 (1978).

[Gre09] Greentree A.D., Beausoleil R.G., Hollenberg L.C.L., Munro W.J., Nemoto K., Prawer S. and Spiller T.P. "Single photon quantum non-demolition measurements in the presence of inhomogeneous broadening". *New J Phys*, **11**, 093005 (2009).

[Gre13] Greentree A.D., Koch J. and Larson J. "Fifty years of Jaynes–Cummings physics". *J Phys B*, **46**, 220201 (2013).

[Gre88] Greenland P.T. "Laser isotope separation". *Contemp Phys*, **31**, 405–424 (1988).

[Gre98] Greene P.L. and Hall D.G. "Properties and diffraction of vector Bessel–Gauss beams". *J Opt Soc Am A*, **15**, 3020–3027 (1998).

[Gri09] Grigoryan G.G., Nikoghosyan G., Halfmann T., Pashayan-Leroy Y., Leroy C. and Guérin S. "Theory of the bright-state stimulated Raman adiabatic passage". *Phys Rev A*, **80**, 033402 (2009).

[Gri12] Griffiths D.J. "Resource Letter EM-1: Electromagnetic Momentum". *Am J Phys*, **80**, 7–18 (2012).

[Gri12b] Grinstead C.M. and Snell J.L. *Introduction to Pprobability* (American Mathematical Soc., 2012).

[Gri75] Grischkowsky D., Loy M.M.T. and Liao P.F. "Adiabatic following model for two-photon transitions: nonlinear mixing and pulse propagation". *Phys Rev A*, **12**, 2514–2532 (1975).

[Gro13]   Groves E., Clader B.D. and Eberly J.H. "Jaynes–Cummings theory out of the box". *J Phys B*, **46**, 224005 (2013).

[Gru01]   Gruebele M. "Fully quantum coherent control". *Chem Phys*, **267**, 33–46 (2001).

[Gry10]   Grynberg G., Aspect A. and Fabre C. *Introduction to Quantum Optics: From the Semi-Classical Approach to Quantized Light* (Cambridge Univ Press, Cambridge, 2010).

[Gu17]   Gu X., Kockum A.F., Miranowicz A., Xi Liu Y. and Nori F. "Microwave photonics with superconducting quantum circuits". *Phys Rep*, **718-719**, 1–102 (2017).

[Gue03]   Guérin S. and Jauslin H.R. "Control of quantum dynamics by laser pulses: Adiabatic Floquet theory". *Adv Chem Phys*, **125**, 147–268 (2003).

[Gue97]   Guérin S., Monti F., Dupont J.M. and Jauslin H.R. "On the relation between cavity-dressed states, Floquet states, RWA and semiclassical models". *J Phys A*, **30**, 7193–7215 (1997).

[Gui49]   Guillemin E.A. *The Mathematics of Circuit Analysis* (Technology Press - John Wiley, New York, 1949).

[Gul93]   Gull S., Lasenby A. and Doran C. "Imaginary numbers are not real - the geometric algebra of spacetime". *Found Phys*, **23**, 1175–1201 (1993).

[Gun73]   Günaydin M. and Gürsey F. "Quark structure and octonions". *J Math Phys*, **14**, 1651–1667 (1973).

[Gun74]   Günaydin M. and Gürsey F. "Quark statistics and octonions". *Phys Rev D*, **9**, 3387–3391 (1974).

[Gun96]   Günaydin M. "Octonionic Hilbert spaces, the Poincaré group and SU(3)". *J Math Phys*, **1875**, 1875–1883 (1996).

[Guo97]   Guo L. and Santsch P.H. "Composition and cycling in marine environments". *Rev Geophys*, **35**, 17–40 (1997).

[Haa73]   Haake F. *Statistical Treatment of Open Systems by Generalised Master Equations* (Springer, Berlin, 1973).

[Had66]   Hadley H.G. "Predictions of the gravitational bending of light before Einstein". *Am J Phys*, **34**, 162–163 (1966).

[Haf08]   Häffner H., Roos C.F. and Blatt R. "Quantum computing with trapped ions". *Phys Rep*, **469**, 155–203 (2008).

[Hal98]   Halfmann T., Yatsenko L.P., Shapiro M., Shore B.W. and Bergmann K. "Population trapping and laser-induced continuum structure in helium: Experiment and theory". *Phys Rev A*, **58**, R46–R49 (1998).

[Ham62]   Hamermesh M. *Group Theory and Its Application to Physical Problems* (Addison-Wesley, Reading, MA, 1962).

[Ham84]   Dwayne Hamilton J. "The Dirac equation and Hestenes' geometric algebra". *J Math Phys*, **25**, 1823–1832 (1984).

[Han35]   Hansen W.W. "A new type of expansion in radiation problems". *Phys Rev*, **47**, 139–143 (1935).

[Han56]   Hanbury Brown R. and Twiss R.Q. "A test of a new type of stellar interferometer on Sirius". *Nature*, **178**, 1046–1048 (1956).

[Han81]   Han D. and Kim Y.S. "Little group for photons and gauge transformations". *Am J Phys*, **49**, 348–351 (1981).

[Hap72]   Happer W. "Optical pumping". *Rev Mod Phys*, **44**, 169–238 (1972).

[Har06]   Haroche S. and Raimond J. *Exploring the Quantum: Atoms, Cavities and Photons* (Oxford Univ Press, Oxford, 2006).

[Har13]   Haroche S. "Nobel Lecture: Controlling photons in a box and exploring the quantum to classical boundary". *Rev Mod Phys*, **85**, 1083–1102 (2013).

[Har13b] Haroche S. and Raimond J.M. *Exploring the Quantum. Atoms, Cavities, and Photons* (Oxford Univ Press, Oxford, 2013).

[Har89]   Haroche S. and Kleppner D. "Cavity quantum electrodynamics". *Phys Today*, **42**, 24–30 (1989).

[Har90]   Harris S.E., Field J.E. and Imamoğlu A. "Nonlinear optical processes using electromagnetically induced transparency". *Phys Rev Lett*, **64**, 1107–1110 (1990).

[Har92]   Hardy G.H. *A Mathematician's Apology* (Cambridge Univ Press, Cambridge, 1992).

[Har96]   Hariharan P. and Sanders B.C. "Quantum phenomena in optical interferometry". *Prog Opt*, **36**, 49–128 (1996).

[Har97]   Harris S.E. "Electromagnetically induced transparency". *Phys Today*, **50**, 36–44 (1997).

[Has08]   Häseler H., Moroder T. and Lütkenhaus N. "Testing quantum devices: Practical entanglement verification in bipartite optical systems". *Phys Rev A*, **77**, 032303 (2008).

[Hau99]   Hau L.V., Harris S.E., Dutton Z. and Behroozi C.H. "Light speed reduction to 17 metres per second in an ultracold atomic gas". *Nature*, **397**, 594–598 (1999).

[Haw01]   Hawton M. and Baylis W.E. "Photon position operators and localized bases". *Phys Rev A*, **64**, 012101 (2001).

[Haw07]   Hawton M. "Photon wave mechanics and position eigenvectors". *Phys Rev A*, **75**, 062107 (2007).

[Haw19]   Hawton M. and Debierre V. "Photon position eigenvectors, Wigner's little group, and Berry's phase". *J Math Phys*, **60**, 052104 (2019).

[Haw99]   Hawton M. "Photon position operator with commuting components". *Phys Rev A*, **59**, 954–959 (1999).

[Hea77]   Healy W.P. "Centre of mass motion in non-relativistic quantum electrodynamics". *J Phys A*, **10**, 279–298 (1977).

[Hec70]   Hecht E. "Note on an operational definition of the Stokes parameters". *Am J Phys*, **38**, 1156–1158 (1970).

[Hec87]   Hecht E. *Optics* (Addison-Wesley, Reading, MA, 1987), 2nd edition.

[Hei10]   Heinze G., Rudolf A., Beil F. and Halfmann T. "Storage of images in atomic coherences in a rare-earth-ion-doped solid". *Phys Rev A*, **81**, 011401 (2010).

[Hei13]   Heinze G., Hubrich C. and Halfmann T. "Stopped light and image storage by electromagnetically induced transparency up to the regime of one minute". *Phys Rev Lett*, **111**, 033601 (2013).

[Hei29]   Heisenberg W. and Pauli W. "Zur Quantendynamik der Wellenfelder". *Zs Phys*, **56**, 1–61 (1929).

[Hei30]   Heisenberg W. and Pauli W. "Zur Quantendynamik der Wellenfelder II". *Zs Phys*, **59**, 168–90 (1930).

[Hei54]   Heitler W. *Quantum Theory of Radiation* (Clarendon Press, Oxford, 1954).

[Hel07]   Hell S.W. "Far-field optical nanoscopy". *Science*, **316**, 1153–1158 (2007).

[Hel18]   Heller E.J. *The Semiclassical Way to Dynamics and Spectroscopy* (Princeton Univ Press, Princeton, NJ, 2018).

[Hel63]   Hellwarth R.W. "Theory of stimulated Raman scattering". *Phys Rev*, **130**, 1850–1852 (1963).

[Hel99]   Hell S.W. and Wichmann J. "Breaking the diffraction resolution limit by stimulated emission: stimulated-emission-depletion fluorescence microscopy". *Opt Lett*, **19**, 780–782 (1994).

[Hen03]   Hennrich M., Legero T., Kuhn A. and Rempe G. "Counter-intuitive vacuum-stimulated Raman scattering". *J Mod Opt*, **50**, 935–942 (2003).

[Hen05]   Hennrich M., Kuhn A. and Rempe G. "Transition from antibunching to bunching in cavity QED". *Phys Rev Lett*, **94**, 053604 (2005).

[Her13]   Herrmann H.J., Hovi J.P. and Luding S. *Physics of Dry Granular Media*, volume 350 (Springer Science & Business Media, New York, 2013).

[Her50]   Herzberg G. *Molecular Spectra and Molecular Structure I. Spectra of Diatomic Molecules* (Van Nostrand, New York, 1950).

[Her50b]   Herzberg G. *Molecular Spectra and Molecular Structure II. Infrared and Raman Spectra of Polyatomic Molecules* (Van Nostrand, New York, 1950).

[Hes03]   Hestenes D. "Oersted Medal Lecture 2002: Reforming the mathematical language of physics". *Am J Phys*, **71**, 104–121 (2003).

[Hes03b]   Hestenes D. "Spacetime physics with geometric algebra". *Am J Phys*, **71**, 691–714 (2003).

[Hes12]   Hestenes D. and Sobczyk G. *Clifford Algebra to Geometric Calculus: A Unified Language for Mathematics* (Reidel, Dordrecht, 2012).

[Hes66]   Hestenes D. *Spacetime Algebra* (Gordon and Breach, New York, 1966).

[Hes67]   Hestenes D. "Real spinor fields". *J Math Phys*, **8**, 798–808 (1967).

[Hes71]   Hestenes D. "Vectors, spinors, and complex numbers in classical and quantum physics". *Am J Phys*, **39**, 1013–1027 (1971).

[Hes75]   Hestenes D. "Observables, operators, and complex numbers in the Dirac theory". *J Math Phys*, **16**, 556–572 (1975).

[Hes86]   Hestenes D. *New Foundations for Classical Mechanics* (Reidel, Dordrecht, 1986).

[Hes86b]   Hestenes D. "A unified language for mathematics and physics". In *Clifford Algebras and their Applications in Mathematical Physics*, 1–23 (Springer, New York, 1986).

[Hes87]   Hestenes D. and Sobczyk G. *Clifford Algebra to Geometric Calculus: A Unified Language for Mathematics and Physics* (Reidel, Dordrecht, 1987).

[Hew68]   Hewish A., Bell S.J., Pilkington J.D.H., Scott P.F. and Collins R.A. "Observation of a rapidly pulsating radio source". *Nature*, **217**, 709–713 (1968).

[Hij07]   Hijlkema M., Weber B., Specht H.P., Webster S.C., Kuhn A. and Rempe

G. "A single-photon server with just one atom". *Nature Phys*, **3**, 253–255 (2007).

[Hil09]   Hillery M. "An introduction to the quantum theory of nonlinear optics". *Acta Phys Slovaca*, **59**, 1–80 (2009).

[Hil49]   Hildebrand F.B. *Advanced Calculus for Engineers* (Prentice-Hall, New York, 1949).

[Hil52]   Hildebrand F.B. *Methods of Applied Mathematics* (Prentice-Hall, Englewood Cliffs, NJ, 1952).

[Hil82]   Hilborn R.C. "Einstein coefficients, cross sections, f values, dipole moments, and all that". *Am J Phys*, **50**, 982–986 (1982).

[Hil97]   Hill S. and Wootters W.K. "Entanglement of a pair of quantum bits". *Phys Rev Lett*, **78**, 5022–5025 (1997).

[Hin67]   Hindmarsh W.R. *Atomic Spectra* (Pergamon, London, 1967).

[Hio81]   Hioe F.T. and Eberly J.H. "N-level coherence vector and higher conservation laws in quantum optics and quantum mechanics". *Phys Rev Lett*, **47**, 838–841 (1981).

[Hio82]   Hioe F.T. and Eberly J.H. "New conservation laws restricting the density matrix of 3-level quantum systems". *App Phys B*, **1**, 105–106 (1982).

[Hio83]   Hioe F.T. "Theory of generalized adiabatic following in multilevel systems". *Phys Lett*, **99A**, 150–155 (1983).

[Hio83b]   Hioe F.T. "Dynamic symmetries in quantum electronics". *Phys Rev A*, **28**, 879–886 (1983).

[Hio84]   Hioe F.T. and Eberly J.H. "Multiple-laser excitation of multilevel atoms". *Phys Rev A*, **29**, 1164–1167 (1984).

[Hio85]   Hioe F.T. "Gellmann dynamic symmetry for N-level quantum systems". *Phys Rev A*, **32**, 2824–2835 (1985).

[Hio87]   Hioe F.T. "N-level quantum systems with SU(2) dynamic symmetry". *J Opt Soc Am B*, **4**, 1327–1332 (1987).

[Hio88]   Hioe F.T. "N-level quantum systems with Gell-Mann dynamic symmetry". *J Opt Soc Am B*, **5**, 859–862 (1988).

[Ho83]   Ho T.S., Chu S.I. and Tietz J.V. "Semiclassical many-mode Floquet theory". *Chem Phys Lett*, **96**, 464–471 (1983).

[Hob05]   Hobson A. "Electrons as field quanta: A better way to teach quantum physics in introductory general physics courses". *Am J Phys*, **73**, 630–634 (2005).

[Hob13]   Hobson A. "There are no particles, there are only fields". *Am J Phys*, **81**, 211–223 (2013).

[Hof12]   Hoffmann N. "Photochemical reactions of aromatic compounds and the concept of the photon as a traceless reagent". *Photochem & Photobio Sci*, **11**, 1613–1641 (2012).

[Hol04]   Holstein B.R. "Effective interactions and the hydrogen atom". *Am J Phys*, **72**, 333–344 (2004).

[Hon11]   Honer J., Löw R., Weimer H., Pfau T. and Büchler H.P. "Artificial atoms can do more than atoms: Deterministic single photon subtraction from arbitrary light fields". *Phys Rev Lett*, **107**, 093601 (2011).

[Hon87]   Hong C.K., Ou Z.Y. and Mandel L. "Measurement of subpicosecond time intervals between two photons by interference". *Phys Rev Lett*, **59**, 2044–2046 (1987).

[Hor09]   Horodecki R., Horodecki M. and Horodecki K. "Quantum entanglement". *Rev Mod Phys*, **81**, 865–942 (2009).

[Hor09b] Horvath H. "Gustav Mie and the scattering and absorption of light by particles: Historic developments and basics". *J Quant Spec Rad Transf*, **110**, 787–799 (2009).

[Hor12]   Horváth I.T. and Joó F., (eds.) *Aqueous Organometallic Chemistry and Catalysis*, volume 5 (Springer Science & Business Media, Dordrecht, 2012).

[Hor12b] Horváth Z.G. "Beyond the beam: A history of multidimensional lasers". *Optic Phot News*, **23**, 36–41 (2012).

[Hor14]   Horsley S.A. and Philbin T.G. "Canonical quantization of electromagnetism in spatially dispersive media". *New J Phys*, **16**, 013030 (2014).

[Hos18]   Hossenfelder S. *Lost in Math. How Beauty Leads Physics Astray* (Basic Books, New York, 2018).

[Hou66]   Houston W.V. "Are electrons real?" *Am J Phys*, **34**, 351–357 (1966).

[Hoy15]   Hoyos C., Sircar N. and Sonnenschein J. "New knotted solutions of Maxwell's equations". *J Phys A*, **48**, 255204 (2015).

[Hra90]   Hradil Z. "Phase in quantum optics and number-phase minimum uncertainty states". *Phys Lett A*, **146**, 1–5 (1990).

[Hu01]   Hu W., Fukugita M., Zaldarriaga M. and Tegmark M. "Cosmic microwave background observables and their cosmological implications". *Ap J*, **549**, 669–680 (2001).

[Hu02]   Hu W. and Dodelson S. "Cosmic microwave background anisotropies". *Ann Rev Astron Astrophys*, **40**, 171–216 (2002).

[Hua09]   Huang W.P. and Mu J. "Complex coupled-mode theory for optical waveguides". *Opt Exp*, **17**, 19134–19152 (2009).

[Hub95]   Hubel D.H. *Eye, Brain, and Vision.* (Scientific American Books, New York, 1995).

[Hut92]   Huttner B. and Barnett S.M. "Quantization of the electromagnetic field in dielectrics". *Phys Rev A*, **46**, 4306–4322 (1992).

[Ila81]   Ilamed Y. and Salingaros N. "Algebras with three anticommuting elements. I. Spinors and quaternions". *J Math Phys*, **22**, 2091–2095 (1981).

[Ima03]   Imamoğlu A., Knill E., Tian L. and Zoller P. "Optical pumping of quantum-dot nuclear spins". *Phys Rev Lett*, **91**, 017402 (2003).

[Irv08]   Irvine W.T.M. and Bouwmeester D. "Linked and knotted beams of light". *Nature Phys*, **4**, 716–720 (2008).

[Irv10]   Irvine W.T.M. "Linked and knotted beams of light, conservation of helicity and the flow of null electromagnetic fields". *J Phys A*, **43**, 385203 (2010).

[Isa07]   Isaacson W. *Einstein. His Life and Universe* (Simon & Schuster, New York, 2007).

[Isi17]   Isinger M., Squibb R.J., Busto D., Zhong S., Harth A., Kroon D., Nandi S., Arnold C.L., Miranda M., Dahlström J.M., Lindroth E., Feifel R., Gisselbrecht M. and L'Huillier A. "Photoionization in the time and frequency

	domain". *Science*, **896**, 893–896 (2017).
[Itz66]	Itzykson C. and Nauenberg M. "Unitary groups: Representations and decompositions". *Rev Mod Phys*, **38**, 95–120 (1966).
[Jac01]	Jackson J.D. and Okun L.B. "Historical roots of gauge invariance". *Rev Mod Phys*, **73**, 663–680 (2001).
[Jac02]	Jackson J.D. "From Lorenz to Coulomb and other explicit gauge transformations". *Am J Phys*, **70**, 917–928 (2002).
[Jac75]	Jackson J.D. *Classical Electrodynamics* (Wiley, New York, 1975), 2nd edition.
[Jae92]	Jaeger H.M. and Nagel S.R. "Physics of the granular state". *Science*, **255**, 1523–1531 (1992).
[Jae96]	Jaeger H.M., Nagel S.R. and Behringer R.P. "Granular solids, liquids, and gases". *Rev Mod Phys*, **68**, 1259–1273 (1996).
[Jaf18]	Jaffe C. and Brumer P. "Local and normal modes: A classical perspective". *J Chem Phys*, **73**, 5646–5658 (2018).
[Jak78]	Jaki S.L. "Johann Georg von Soldner and the gravitational bending of light, with an English translation of his essay on it published in 1801". *Found Phys*, **8**, 927–950 (1978).
[Jam01]	James D.F.V., Kwiat P.G., Munro W.J. and White A.G. "Measurement of qubits". *Phys Rev A*, **64**, 052312 (2001).
[Jam92]	James M.B. and Griffiths D.J. "Why the speed of light is reduced in a transparent medium". *Am J Phys*, **60**, 309–313 (1992).
[Jan57]	Janossy L. and Náray Z. "The interference phenomena of light at very low intensities". *Acta Phys Acad Sci Hungaricae*, **7**, 403–425 (1957).
[Jau55]	Jauch J.M. and Rohrlich F. *The Theory of Photons and Electrons* (Addison-Weslely, Cambridge, MA, 1955).
[Jav06]	Javanainen J. "75. Cooling and trapping". In G. Drake, (ed.) *Springer Handbook of Atomic, Molecular and Optical Physics*, chapter 75, 1091–1107 (Springer, New York, 2006), 2nd edition.
[Jay63]	Jaynes E.T. and Cummings F.W. "Comparison of quantum and semiclassical radiation theories with application to the beam maser". *Proc IEEE*, **51**, 89–109 (1963).
[Jay65]	Jaynes E.T. "Gibbs vs Boltzmann entropies". *Am J Phys*, **33**, 391–398 (1965).
[Jay73]	Jaynes E.T. "Survey of the present status of neoclassical radiation theory". In *Coherence and quantum optics*, 35–81 (Springer, 1973).
[Jel90]	Jeleńska-Kuklińska M. and Kuś M. "Exact solution in the semiclassical Jaynes-Cummings model without the rotating-wave approximation". *Phys Rev A*, **41**, 2889–2891 (1990).
[Joa11]	Joannopoulos J.D., Johnson S.G., Winn J.N. and Meade R.D. *Photonic Crystals: Molding the Flow of Light* (Princeton Univ Press, Princeton, NJ, 2011).
[Joh02]	Johnson S.C. and Gutierrez T.D. "Visualizing the phonon wave function". *Am J Phys*, **70**, 227–237 (2002).
[Joh04]	Johnsson M. and Mølmer K. "Storing quantum information in a solid

using dark-state polaritons". *Phys Rev A*, **70**, 032320 (2004).

[Joh55]   Johnson S. *A Dictionary of the English Language* (Project Gutenberg, 1755).

[Jon94]   Jones D.G.C. "Two slit interference—classical and quantum pictures". *Eur J Phys*, **15**, 170–178 (1994).

[Jud75]   Judd B.R. *Angular Momentum Theory for Diatomic Molecules* (Academic, New York, 1975).

[Kai05]   Kaiser D. "Physics and Feynman's diagrams". *Am Scientist*, **93**, 156–165 (2005).

[Kan71]   Kantor W. "Equivalence of Compton effect and Doppler effect". *Spectr Lett*, **4**, 59–60 (1971).

[Kar09]   Karimi E., Piccirillo B., Nagali E., Marrucci L. and Santamato E. "Efficient generation and sorting of orbital angular momentum eigenmodes of light by thermally tuned q-plates". *App Phys Lett*, **94**, 231124 (2009).

[Kar15]   Karimi E. and Boyd R.W. "Classical entanglement?" *Science*, **350**, 1172–1173 (2015).

[Kas57]   Kastler A. "Optical methods of atomic orientation and of magnetic resonance". *J Opt Soc Am*, **47**, 460–465 (1957).

[Kas93]   Kastner M.A. "Artificial atoms". *Phys Today*, **46**, 24–31 (1993).

[Kat12]   Katz O., Small E. and Silberberg Y. "Looking around corners and through thin turbid layers in real time with scattered incoherent light". *Nat Phot*, **6**, 549–553 (2012).

[Ked13]   Kedia H., Bialynicki-Birula I., Peralta-Salas D. and Irvine W.T.M. "Tying knots in light fields". *Phys Rev Lett*, **111**, 150404 (2013).

[Kei16]   Keiser G. *Biophotonics* (Springer, 2016).

[Kel05]   Keller O. "On the theory of spatial localization of photons". *Phys Rep*, **411**, 1–232 (2005).

[Ken11]   Kenyon I. *The Light Fantastic. A Modern Introduction to Classical and Quantum Optics* (Oxford Univ Press, Oxford, 2011), 2nd edition.

[Ker17]   Kerbstadt S., Timmer D., Englert L., Bayer T. and Wollenhaupt M. "Ultrashort polarization-tailored bichromatic fields from a CEP-stable white light supercontinuum". *Opt Exp*, **25**, 12518 (2017).

[Kha02]   Khalil S. and Munoz C. "The enigma of the dark matter". *Contemp Phys*, **43**, 51–61 (2002).

[Khu10]   Khurgin J.B. "Slow light in various media: a tutorial". *Adv Opt Phot*, **2**, 287–318 (2010).

[Kib65]   Kibble T.W. "Conservation laws for free fields". *J Math Phys*, **6**, 1022–1026 (1965).

[Kid89]   Kidd R., Ardini J. and Anton A. "Evolution of the modern photon". *Am J Phys*, **57**, 27–35 (1989).

[Kim08]   Kimble H.J. "The quantum internet." *Nature*, **453**, 1023–1030 (2008).

[Kim77]   Kimble H.J., Dagenais M. and Mandel L. "Photon antibunching in resonance fluorescence". *Phys Rev Lett*, **39**, 691–695 (1977).

[Kim88]   Kim Y.S. and Wigner E.P. "Cylindrical group and massless particles". In *Special Relativity and Quantum Theory*, 387–391 (Springer, 1988).

[Kim90] Kim Y.S. and Wigner E.P. "Space-time geometry of relativistic particles". *J Math Phys*, **31**, 55–60 (1990).

[Kim98] Kimble H.J. "Strong interactions of single atoms and photons in cavity QED". *Phys Scripta*, **T76**, 127–137 (1998).

[Kin03] Kinney W.H. "Cosmology, inflation and the physics of nothing". In *Techniques and Concepts of High-Energy Physics XII*, 189–243 (Springer, 2003).

[Kip01] Kipp L., Skibowski M., Johnson R., Berndt R., Adelung R., Harm S. and Seemann R. "Sharper images by focusing soft X-rays with photon sieves". *Nature*, **414**, 184–188 (2001).

[Kir04] Kiraz A., Atature M. and Imamoğlu A. "Quantum-dot single-photon sources: Prospects for applications in linear optics quantum-information processing". *Phys Rev A*, **69**, 032305 (2004).

[Kis04] Kis Z., Karpati A., Shore B.W. and Vitanov N.V. "Stimulated Raman adiabatic passage among degenerate-level manifolds". *Phys Rev A*, **70**, 053405 (2004).

[Kis05] Kis Z., Vitanov N.V., Karpati A., Barthel C. and Bergmann K. "Creation of arbitrary coherent superposition states by stimulated Raman adiabatic passage". *Phys Rev A*, **72**, 033403 (2005).

[Kla01] Klar T.A., Engel E. and Hell S.W. "Breaking Abbe's diffraction resolution limit in fluorescence microscopy with stimulated emission depletion beams of various shapes". *Phys Rev E*, **64**, 066613 (2001).

[Kle05] Kleppner D. "Rereading Einstein on radiation". *Phys Today*, **58**, 30–33 (2005).

[Kle08] Klein J., Beil F. and Halfmann T. "Experimental investigations of stimulated Raman adiabatic passage in a doped solid". *Phys Rev A*, **78**, 033416 (2008).

[Kle08b] Klein F. and Sommerfeld A. *The Theory of the Top. Volume I: Introduction to the Kinematics and Kinetics of the Top* (Springer Science & Business Media, New York, 2008).

[Kle65] Klein M.J. "Einstein, specific heats, and the early quantum theory". *Science*, **148**, 173–180 (1965).

[Kle73] Klebesadel R.W., Strong I.B. and Olson R.A. "Observations of gamma-ray bursts of cosmic origin". *Ap J*, **182**, L85–L88 (1973).

[Kle81] Kleppner D. "Inhibited spontaneous emission". *Phys Rev Lett*, **47**, 233–236 (1981).

[Kli15] Klink W.H. and Wickramasekara S. *Relativistic Implications of the Quantum Phase Relativity, Symmetry and the Structure of Quantum Theory I: Galilean Quantum Theory* (Morgan &Claypool, San Rafael, CA, 2015).

[Kni06] Knight P.L. and Scheel S. "81. Quantum information". In G. Drake, (ed.) *Springer Handbook of Atomic, Molecular, and Optical Physics*, chapter 81, 1215–1234 (Springer-Verlag, New York, 2006).

[Kni82] Knight P.L. and Radmore P.M. "Quantum revivals of a two-level system driven by chaotic radiation". *Phys Lett A*, **90**, 342–346 (1982).

[Kni89] Knight P.L. "Quantum Optics". In P. Davies, (ed.) *The New Physics*,

(Cambridge Univ Press, Cambridge, 1989).

[Kni90] Knight P.L., Lauder M.A. and Dalton B.J. "Laser-induced continuum structure". *Phys Rep*, **190**, 1–61 (1990).

[Kni98] Knight P. "Quantum mechanics - Where the weirdness comes from". *Nature*, **395**, 12–13 (1998).

[Koc12] Koch C.P. and Shapiro M. "Coherent control of ultracold photoassociation." *Chem Rev*, **112**, 4928–4948 (2012).

[Koc19] Kockum A.F., Miranowicz A., De Liberato S., Savasta S. and Nori F. "Ultrastrong coupling between light and matter". *Nature Rev Phys*, **1**, 19–40 (2019).

[Koc92] Koch S.W., Knorr A., Binder R. and Lindberg M. "Microscopic theory of Rabi flopping, photon echo, and resonant pulse propagation in semiconductors". *Phys Stat Sol (B)*, **173**, 177–187 (1992).

[Koe96] Koechner W. *Solid-State Laser Engineering* (Springer-Verlag, New York, 1996), 4th edition.

[Kok02] Kok P., Lee H. and Dowling J.P. "Single-photon quantum-nondemolition detectors constructed with linear optics and projective measurements". *Phys Rev A*, **66**, 063814 (2002).

[Kok10] Kok P. and Lovett B.W. *Introduction to Optical Quantum Information Processing* (Cambridge Univ Press, Cambridge, 2010).

[Kok16] Kok P. "Photonic quantum information processing". *Contemp Phys*, **57**, 526–544 (2016).

[Kol99] Kolobov M.I. "The spatial behavior of nonclassical light". *Rev Mod Phys*, **71**, 1539–1589 (1999).

[Kop13] Kopferman H. and Schneider E.E. *Nuclear Moments*, volume 2 (Academic, New York, 2013).

[Kor02] Korolkova N., Leuchs G., Loudon R., Ralph T.C. and Silberhorn C. "Polarization squeezing and continuous-variable polarization entanglement". *Phys Rev A*, **65**, 523061 (2002).

[Kor02b] Korsunsky E.A. and Fleischhauer M. "Resonant nonlinear optics in coherently prepared media: Full analytic solutions". *Phys Rev A*, **66**, 033808 (2002).

[Kor05] Korolkova N. and Loudon R. "Nonseparability and squeezing of continuous polarization variables". *Phys Rev A*, **71**, 032343 (2005).

[Kor05b] Korotkova O. and Wolf E. "Generalized Stokes parameters of random electromagnetic beams". *Opt Lett*, **30**, 198–200 (2005).

[Kot12] Kotlyar V.V., Kovalev A.A. and Soifer V.A. "Hankel–Bessel laser beams". *J Opt Soc Am A*, **29**, 741–747 (2012).

[Kou01] Kouwenhoven L.P., Austing D.G. and Tarucha S. "Few-electron quantum dots". *Rep Prog Phys*, **64**, 701–736 (2001).

[Kox13] Kox A.J. "Hendrik Antoon Lorentz's struggle with quantum theory". *Arch Hist Exact Sci*, **67**, 149–170 (2013).

[Kox97] Kox A.J. "The discovery of the electron: II. The Zeeman effect". *Eur J Phys*, **18**, 139–144 (1997).

[Kra00] Kragh H. "Max Planck: The reluctant revolutionary". *Phys World*, **13**,

31–35 (2000).

[Kra02]   Kragh H. *Quantum Generations: A history of Physics in the Twentieth Century* (Princeton Univ Press, Princeton, NJ, 2002).

[Kra03]   Kragh H. "Magic number: A partial history of the fine-structure constant". *Archive Hist Exact Sci*, **57**, 395–431 (2003).

[Kra07]   Král P., Thanopulos I. and Shapiro M. "Colloquium: Coherently controlled adiabatic passage". *Rev Mod Phys*, **79**, 53–77 (2007).

[Kra14]   Kragh H. "Photon: New light on an old name". arXiv:1401.0293 [physics.hist-ph] (2014).

[Kra19]   Krasnok A., Baranov D., Li H., Miri M.A., Monticone F. and Alú A. "Anomalies in light scattering". *Adv Opt Phot*, **11**, 892–951 (2019).

[Kra92]   Kragh H. "A sense of history: History of science and the teaching of introductory quantum theory". *Science and Education*, **1**, 349–363 (1992).

[Kre16]   Krenn M., Malik M., Erhard M. and Zeilinger A. "Orbital angular momentum of photons and the entanglement of Laguerre-Gaussian modes". *Phil Trans Roy Soc A*, **375**, 20150442 (2016).

[Kru19]   Krüger Y., Mercury L., Canizares A., Marti D. and Simon P. "Metastable phase equilibria in the ice II stability field. A Raman study of synthetic high-density water inclusions in quartz". *Phys Chem Chem Phys*, **21**, 19554–19566 (2019).

[Kuh02]   Kuhn A., Hennrich M. and Rempe G. "Deterministic single-photon source for distributed quantum networking". *Phys Rev Lett*, **89**, 067901 (2002).

[Kuh10]   Kuhn A. and Ljunggren D. "Cavity-based single-photon sources". *Contemp Phys*, **51**, 289–313 (2010).

[Kuh84]   Kuhn T.S. "Revisiting Planck". *Hist Stud Phys Sciences*, **14**, 231–252 (1984).

[Kuh87]   Kuhn T.S. *Black-Body Theory and the Quantum Discontinuity, 1894-1912* (Univ Chicago Press, Chicago, 1987).

[Kuh98]   Kuhn A., Steuerwald S. and Bergmann K. "Coherent population transfer in NO with pulsed lasers: the consequences of hyperfine structure , Doppler broadening and electromagnetically induced absorption". *Eur Phys J D*, **1**, 57–70 (1998).

[Kuh99]   Kuhn A., Hennrich M., Bondo T. and Rempe G. "Controlled generation of single photons from a strongly coupled atom-cavity system". *App Phys B*, **69**, 373–377 (1999).

[Kuk89]   Kuklinski J.R., Gaubatz U., Hioe F.T. and Bergmann K. "Adiabatic population transfer in a three-level system driven by delayed laser pulses". *Phys Rev A*, **40**, 6741–6744 (1989).

[Kum11]   Kumar P., Malinovskaya S.A. and Malinovsky V.S. "Optimal control of population and coherence in three-level Λ systems". *J Phys B*, **44**, 154010 (2011).

[Kum15]   Kumar P. and Zhang B. "The physics of gamma-ray bursts & relativistic jets". *Phys Rep*, **561**, 1–109 (2015).

[Kur00]   Kurtsiefer C., Mayer S., Zarda P. and Weinfurter H. "Stable solid-state source of single photons". *Phys Rev Lett*, **85**, 290–293 (2000).

[Kyo06]   Kyoseva E.S. and Vitanov N.V. "Coherent pulsed excitation of degenerate multistate systems: Exact analytic solutions". *Phys Rev A*, **73**, 023420 (2006).

[Lai89]   Lai Y. and Haus H.A. "Quantum theory of solitons in optical fibers. I.Time-dependent Hartree approximation". *Phys Rev A*, **40**, 844–853 (1989).

[Lai89b]   Lai Y. and Haus H.A. "Quantum theory of solitons in optical fibers, II. Exact solutions". *Phys Rev A*, **40**, 854–866 (1989).

[Lal01]   Laloë F. "Do we really understand quantum mechanics? Strange correlations, paradoxes, and theorems". *Am J Phys*, **69**, 655–701 (2001).

[Lal71]   Lallemand P. "The stimulated Raman effect". In A. Anderson, (ed.) *The Raman Effect*, 287–342 (Dekker, Utrecht, 1971).

[Lam06]   Lambropoulos P. and Petrosyan D. *Fundamentals of Quantum Optics and Quantum Information* (Springer, Berlin, 2006).

[Lam45]   Lamb H. *Hydrodynamics* (Dover, New York, 1945).

[Lam95]   Lamb W.E. "Anti-photon". *App Phys B*, **60**, 77–84 (1995).

[Lan32]   Landau L.D. "Zur Theorie der Energieübertragung bei Stössen". *Phys Zs Sowjet*, **1**, 88–98 (1932).

[Lan32b]   Landau L.D. "Zur Theorie der Energieübertragung II". *Phys Zs Sowjet*, **2**, 46–51 (1932).

[Lan51]   Landau L. and Lifshitz E. *The Classical Theory of Fields* (Addison-Wesley, Cambridge, MA, 1951).

[Lan60]   Landau L.D. and Lifshitz E.M. *Electrodynamics of Continuous Media* (Pergamon, New York, 1960).

[Lan60b]   Landau L.D. and Lifshitz E.M. *Principles of Dynamics* (Pergamon, New York, 1960).

[Lan80]   Lang K.R. *Astrophysical Quantities* (Springer-Verlag, New York, 1980).

[Lan87]   Landau L.D. and Lifshitz E.M. *Fluid Mechanics. Course of Theoretical Physics* (Pergamon, New York, 1987), 2nd edition.

[Lap31]   Laporte O. and Uhlenbeck G.E. "Application of Spinor Analysis to the Maxwell and Dirac Equations". *Phys Rev*, **37**, 1380–1397 (1931).

[Lar07]   Larson J. "Dynamics of the Jaynes–Cummings and Rabi models: Old wine in new bottles". *Phys Scripta*, **76**, 146–160 (2007).

[Lar83]   Larsson M. "Conversion formulas between radiative lifetimes and other dynamical variables for spin-allowed electronic transitions in diatomic molecules". *Astron Astrophys*, **128**, 291–298 (1983).

[Las93]   Lasenby A., Doran C. and Gull S. "Grassmann calculus, pseudoclassical mechanics, and geometric algebra". *J Math Phys*, **34**, 3683–3712 (1993).

[Las98]   Lasenby A., Doran C. and Gull S. "Gravity, gauge theories and geometric algebra". *Phil Trans Roy Soc A*, **356**, 487–582 (1998).

[Law04]   Law C.K. and Eberly J.H. "Analysis and interpretation of high transverse entanglement in optical parametric down conversion". *Phys Rev Lett*, **92**, 127903 (2004).

[Law96]   Law C.K. and Eberly J.H. "Arbitrary control of a quantum electromagnetic field." *Phys Rev Lett*, **76**, 1055–1058 (1996).

[Lax51]   Lax M. "Multiple scattering of waves". *Rev Mod Phys*, **23**, 287–310 (1951).

[Lea02]   Leach J., Padgett M.J., Barnett S.M., Franke-Arnold S. and Courtial J. "Measuring the orbital angular momentum of a single photon". *Phys Rev Lett*, **88**, 257901 (2002).

[Lea05]   Leach J., Dennis M.R., Courtial J. and Padgett M.J. "Vortex knots in light". *New J Phys*, **7** (2005).

[Lea14]   Leader E. and Lorcé C. "The angular momentum controversy: What's it all about and does it matter?" *Phys Rep*, **541**, 163–248 (2014).

[Lee04]   Lee K.F. and Thomas J.E. "Entanglement with classical fields". *Phys Rev A*, **69**, 052311 (2004).

[Lee05]   Lee H.W. "Theory and application of the quantum phase-space distribution functions". *Phys Rep*, **259**, 147–211 (1995).

[Leg80]   Leggett A.J. "Macroscopic quantum systems and the quantum theory of measurement". *Prog Theo Phys Supp*, **69**, 80–100 (1980).

[Lem13]   Lemeshko M., Krems R.V., Doyle J.M. and Kais S. "Manipulation of molecules with electromagnetic fields". *Mol Phys*, **111**, 1648–1682 (2013).

[Leo04]   Leone M., Paoletti A. and Robotti N. "A simultaneous discovery: The case of Johannes Stark and Antonino Lo Surdo". *Physics in Perspective*, **6**, 271–294 (2004).

[Leo95]   Leonhardt U. "Quantum-state tomography and discrete Wigner function". *Phys Rev Lett*, **74**, 4101–4105 (1995).

[Let07]   Letokhov V. *Laser Control of Atoms and Molecules* (Oxford Univ Press, New York, 2007).

[Let79]   Letokhov V.S. "Laser isotope separation". *Nature*, **277**, 605–610 (1979).

[Leu10]   Leuchs G., Villar A.S. and Sánchez-Soto L.L. "The quantum vacuum at the foundations of classical electrodynamics". *App Phys B*, **100**, 9–13 (2010).

[Lev15]   Levy U. and Silberberg Y. "Free-space nonperpendicular electric–magnetic fields". *J Opt Soc Am A*, **32**, 647–653 (2015).

[Lev16]   Levy U. and Silberberg Y. "Weakly diverging to tightly focused Gaussian beams: A single set of analytic expressions". *J Opt Soc Am A*, **33**, 1999–2009 (2016).

[Lev16b]   Levy U., Derevyanko S. and Silberberg Y. "Light modes of free space". *Prog Opt*, **61**, 237–281 (2016).

[Lev79]   Levitt M.H. and Freeman R. "NMR population inversion using a composite pulse". *J Mag Res*, **33**, 473–476 (1979).

[Lev86]   Levitt M.H. "Composite pulses". *Prog Nuc Mag Res Spect*, **18**, 61–122 (1986).

[Lew26]   Lewis G.N. "The conservation of photons". *Nature*, **118**, 874–875 (1926).

[Li08]   Li X., Shu J. and Henk F Arnoldus. "Far-field detection of the dipole vortex". *Opt Lett*, **33**, 2269–2271 (2008).

[Li10]   Li X. and Arnoldus H.F. "Macroscopic far-field observation of the sub-wavelength near-field dipole vortex". *Phys Lett A*, **374**, 1063–1067 (2010).

[Li12]   Li J., Paraoanu G.S., Cicak K., Altomare F., Park J.I., Simmonds R.W., Sillanpää M.A. and Hakonen P.J. "Dynamical Autler-Townes control of a phase qubit". *Sci Rep*, **2**, 645–652 (2012).

[Lig02] Ligare M. and Oliveri R. "The calculated photon: Visualization of a quantum field". *Am J Phys*, **70**, 58–66 (2002).

[Lin03] Lindner N.H., Peres A. and Terno D.R. "Wigner's little group and Berry's phase for massless particles". *J Phys A*, **36**, L449–L454 (2003).

[Lin12] Lindberg J. "Mathematical concepts of optical superresolution". *J Opt*, **14**, 083001 (2012).

[Lin76] Lindblad G. "On the generators of quantum dynamical semigroups". *Comm Math Phys*, **48**, 119–130 (1976).

[Lip64] Lipkin D.M. "Existence of a new conservation law in electromagnetic theory". *J Math Phys*, **5**, 696–700 (1964).

[Lip65] Lipkin H.J. *Lie Groups for Pedestrians* (North Holland, Amsterdam, 1965).

[Lip65b] Lipeles M., Novick R. and Tolk N. "Direct detection of two-photon emission from the metastable state of singly ionized helium". *Phys Rev Lett*, **15**, 690–693 (1965).

[Lip69] Lipson S.G. and Lipson H. *Optical Physics* (Cambridge Univ Press, Cambridge, 1969).

[Lip79] Lippmann B.A. and Schwinger J. "Variational principles for scattering processes. I". *Phys Rev*, **79**, 469–480 (1950).

[Liu10] Liu R.B., Yao W. and Sham L.J. "Quantum computing by optical control of electron spins". *Adv Phys*, **59**, 70–802 (2010).

[Llo17] Lloyd S.M., Babiker M., Thirunavukkarasu G. and Yuan J. "Electron vortices: Beams with orbital angular momentum". *Rev Mod Phys*, **89**, 035004 (2017).

[Lod15] Lodahl P., Mahmoodian S. and Stobbe S. "Interfacing single photons and single quantum dots with photonic nanostructures". *Rev Mod Phys*, **87**, 347–400 (2015).

[Lod17] Lodahl P., Mahmoodian S., Stobbe S., Rauschenbeutel A., Schneeweiss P., Volz J., Pichler H. and Zoller P. "Chiral quantum optics". *Nature*, **541**, 473–480 (2017).

[Lon05] Longdell J.J., Fraval E., Sellars M.J. and Manson N.B. "Stopped light with storage times greater than one second using electromagnetically induced transparency in a solid". *Phys Rev Lett*, **95**, 063601 (2005).

[Lon06] Longhi S. "Adiabatic passage of light in coupled optical waveguides". *Phys Rev E*, **73**, 026607 (2006).

[Lon06b] Longhi S. "Optical realization of multilevel adiabatic population transfer in curved waveguide arrays". *Phys Lett A*, **359**, 166–170 (2006).

[Lon07] Longhi S., Valle G.D., Ornigotti M. and Laporta P. "Coherent tunneling by adiabatic passage in an optical waveguide system". *Phys Rev B*, **76**, 201101 (R) (2007).

[Lon09] Longhi S. "Quantum-optical analogies using photonic structures". *Laser Photon Rev*, **3**, 243–261 (2009).

[Lor52] Lorentz H.A. *The Theory of Electrons* (Dover, New York, 1952).

[Lou00] Loudon R. *The Quantum Theory of Light* (Oxford Univ Press, Oxford,

2000), 3rd edition.

[Lou01] Lounesto P. *Clifford Algebras and Spinors* (Cambridge Univ Press, Cambridge, 2001).

[Lou03] Loudon R. "What is a photon?" *OPN Trends of Opt Phot News*, **14**, S6–S11 (2003).

[Lou05] Lounis B. and Orrit M. "Single-photon sources". *Rep Prog Phys*, **68**, 1129–1179 (2005).

[Lou06] Louck J.D. "2. Angular momentum theory". In G. Drake, (ed.) *Springer Handbook of Atomic, Molecular and Optical Physics*, chapter 2, 9–74 (Springer, New York, 2006), 2nd edition.

[Lou73] Loudon R. *The Quantum Theory of Light* (Clarendon, Oxford, 1973).

[Lou73b] Louisell W.H. *Quantum Statistical Properties of Radiation* (Wiley, New York, 1973).

[Lou83] Loudon R. "Quantum theory of combined first order/second order optical interference experiments". *Opt Comm*, **45**, 361–366 (1983).

[Lou87] Loudon R. and Knight P.L. "Squeezed Light". *J Mod Opt*, **34**, 709–759 (1987).

[Lou92] Lounis B. and Cohen-Tannoudji C. "Coherent population trapping and Fano profiles". *J Physique II*, **2**, 579–592 (1992).

[Loy74] Loy M.M.T. "Observation of population inversion by optical adiabatic rapid passage". *Phys Rev Lett*, **32**, 814–17 (1974).

[Lud91] Ludwig A.C. "The generalized multipole technique". *Comp Phys Comm*, **68**, 306–314 (1991).

[Lui02] Luis A. "Degree of polarization in quantum optics". *Phys Rev A*, **66**, 013806 (2002).

[Lui06] Luis A. and Korolkova N. "Polarization squeezing and nonclassical properties of light". *Phys Rev A*, **74**, 043817 (2006).

[Lui09] Luis A. "Coherence, polarization, and entanglement for classical light fields". *Opt Comm*, **282**, 3665–3670 (2009).

[Luk01] Lukin M.D. and Imamoğlu A. "Controlling photons using electromagnetically induced transparency". *Nature*, **413**, 273–276 (2001).

[Luk03] Lukin M.D. "Colloquium: Trapping and manipulating photon states in atomic ensembles". *Rev Mod Phys*, **75**, 457–472 (2003).

[Lvo09] Lvovsky A.I. "Continuous-variable optical quantum-state tomography". *Rev Mod Phys*, **81**, 299–332 (2009).

[Lyn95] Lynch R. "The quantum phase problem: a critical review". *Phys Rep*, **256**, 367–436 (1995).

[Mac03] Mack H. and Schleich W.P. "A photon viewed from Wigner phase space". *OPN Trends*, 29–36 (2003).

[Mac86] MacLeod H.A. *Thin-Film Optical Filters* (MacMillan, New York, 1986).

[Mah70] Mahan G.D. "Theory of photoemission in simple metals". *Phys Rev B*, **2**, 4334–4350 (1970).

[Maj32] Majorana E. "Atomi orientati in campo magnetico variabile". *Nuov Cim*, **9**, 43–50 (1932).

[Maj76] Majernik V. and Nagy M. "Quaternionic form of Maxwell's equations with sources". *Lett Nuov Cim*, **16**, 265–268 (1976).

[Mal04] Malinovsky V.S. and Sola I.R. "Quantum control of entanglement by phase manipulation of time-delayed pulse sequences. I". *Phys Rev A*, **70**, 042304 (2004).

[Man16] Mannhart J., Boschker H., Kopp T. and Valentí R. "Artificial atoms based on correlated materials". *Rep Prog Phys*, **79**, 084508 (2016).

[Man17] Mansuripur M. "Force, torque, linear momentum, and angular momentum in classical electrodynamics". *App Phys A*, **123**, 065501 (2017).

[Man65] Mandel L. and Wolf E. "Coherence properties of optical fields". *Rev Mod Phys*, **37**, 231–287 (1965).

[Man67] Mandelbrot B. "How long Is the coast of Britain? Statistical self-similarity and fractional dimension". *Science*, **156**, 636–638 (1967).

[Man77] Mandelbrot B.B. *Fractals: Form, Chance, and Dimension* (W H Freeman, San Francisco, 1977).

[Man79] Mandel L. "Sub-Poissonian photon statistics in resonance fluorescence". *Opt Lett*, **4**, 205–207 (1979).

[Man84] Mandel L. "The problem of non-locality and simultaneous photon detection behind a beam splitter". *Phys Lett*, **103**, 416–418 (1984).

[Man86] Mandel L. "Non-classical states of the electromagnetic field". *Phys Scripta*, **1986**, 34–42 (1986).

[Man91] Mandel L. "Coherence and indistinguishability." *Opt Lett*, **16**, 1882–1883 (1991).

[Man95] Mancini S. and Tombesi P. "Quantum dynamics of a dissipative Kerr medium with time-dependent parameters". *Phys Rev A*, **52**, 2475–2478 (1995).

[Man99] Mandel L. "Quantum effects in one-photon and two-photon interference". *Rev Mod Phys*, **71**, S274–S282 (1999).

[Mar06] Martin W.C. and Wiese W.L. "10. Atomic spectroscopy". In G. Drake, (ed.) *Springer Handbook of Atomic, Molecular and Optical Physics*, chapter 10, 175–198 (Springer, New York, 2006), 2nd edition.

[Mar12] Martin J.Ô. "Everything you always wanted to know about the cosmological constant problem (but were afraid to ask)". *Compt Rend Phys*, **13**, 566–665 (2012).

[Mar72] Marrus R. and Schmieder R.W. "Forbidden decays of hydrogenlike and heliumlike argon". *Phys Rev A*, **5**, 1160–1183 (1972).

[Mar92] Marchiafava S. and Rembieliński J. "Quantum quaternions". *J Math Phys*, **33**, 171–173 (1992).

[Mar95] Martin J., Shore B.W. and Bergmann K. "Coherent population transfer in multilevel systems with magnetic sublevels. II. Algebraic analysis". *Phys Rev A*, **52**, 583–593 (1995).

[Mar96] Martin J., Shore B.W. and Bergmann K. "Coherent population transfer in multilevel systems with magnetic sublevels. III. Experimental results". *Phys Rev A*, **54**, 1556–1569 (1996).

[Mar96b] Marburger J.H. "What is a photon?" *Phys Teacher*, **34**, 482–486 (1996).

[Mat67] Mattuck R.D. *A Guide to Feynman Diagrams in The Many-Body Problem*

(McGraw-Hill, New York, 1967).

[Mau04] Maurer C., Becher C., Russo C., Eschner J. and Blatt R. "A single-photon source based on a single Ca⁺ ion". *New J Phys*, **6**, 1–19 (2004).

[Max81] Maxwell J.C. *A Treatise on Electricity and Magnetism* (Clarendon press, Oxford, 1881).

[McC57] McConnell H.M. "Nuclear and electron magnetic resonance". *Ann Rev Phys Chem*, **8**, 105–128 (1957).

[McC67] McCormmach R. "J. J. Thomson and the structure of light". *Brit J Hist Sci*, **3**, 362–387 (1967).

[McD00] McDonald K.T. "Slow light". *Am J Phys*, **68**, 293–294 (2000).

[McD11] McDermott S.D., Yu H.B. and Zurek K.M. "Turning off the lights: How dark is dark matter?" *Phys Rev D*, **83**, 063509 (2011).

[McG05] McGloin D. and Dholakia K. "Bessel beams: Diffraction in a new light". *Contemp Phys*, **46**, 15–28 (2005).

[McM54] McMaster W.H. "Polarization and the Stokes parameters". *Am J Phys*, **22**, 351–362 (1954).

[Meh07] Mehta A. *Granular Physics* (Cambridge Univ Press, Cambridge, 2007).

[Men05] Menon R., Gil D., Barbastathis G. and Smith H.I. "Photon-sieve lithography". *J Opt Soc Am A*, **22**, 342–345 (2005).

[Men12] Menzel R., Puhlmann D., Heuer A. and Schleich W.P. "Wave-particle dualism and complementarity unraveled by a different mode". *Proc Nat Acad Sci*, **109**, 9314–9319 (2012).

[Men61] Menzel D.H. *Mathematical Physics* (Dover, New York, 1961).

[Mes06] Meschede D. and Schenzle A. "79. Entangled atoms and fields: Cavity QED". In G. Drake, (ed.) *Springer Handbook of Atomic, Molecular, and Optical Physics*, chapter 79, 1167–1184 (Springer-Verlag, New York, 2006).

[Mes06b] Meschede D. and Rauschenbeutel A. "Manipulating Single Atoms". *Adv At Mol Opt Phys*, **53**, 75–104 (2006).

[Mes07] Meschede D. *Optics, Light and Lasers: The Practical Approach to Modern Aspects of Photonics and Laser Physics* (Wiley-VCH, Weinheim, 2007), 2nd edition.

[Mes61] Messiah A. *Quantum Mechanics. Volume 1* (North Holland, Amsterdam, 1961).

[Mes85] Meschede D., Walther H. and Müller G. "One-atom maser". *Phys Rev Lett*, **54**, 551–554 (1985).

[Met12] Metcalf H.J. and Van der Straten P. *Laser Cooling and Trapping* (Springer Science & Business Media, New York, 2012).

[Mey07] Meystre P. and Sargent, M. III. *Elements of Quantum Optics* (Springer Science & Business Media, New York, 2007), 3rd edition.

[Mic47] Michels W.C. "The Doppler effect as a photon phenomenon". *Am J Phys*, **15**, 449–450 (1947).

[Mie08] Mie G. "Beiträge zur Optik trüber Medien speziell kolloidaler Goldlösungen". *Ann Phys*, **25**, 377–445. (1908).

[Mil05] Miller R., Northup T.E., Birnbaum K.M., Boca A., Boozer A.D. and Kimble H.J. "Trapped atoms in cavity QED: Coupling quantized light and matter". *J Phys B*, **38**, S551–S565 (2005).

[Mil10]   Milonni P.W. and Boyd R.W. "Momentum of light in a dielectric medium". *Adv Opt Phot*, **2**, 519–553 (2010).

[Mil17]   Milonni P.W. "Review of: "Void: The Strange Physics of Nothing. James Owen Weatherall. 196 pp. Yale U.P"". *Am J Phys*, **85**, 637–638 (2017).

[Mil19]   Milonni P.W. *An Introduction to Quantum Optics and Quantum Fluctuations* (Oxford Univ Press, Oxford, 2019).

[Mil19b] Millot M., Coppari F., Rygg J.R., Barrios A.C., Hamel S., Swift D.C. and Eggert J.H. "Nanosecond X-ray diffraction of shock-compressed superionic water ice". *Nature*, **569**, 251–255 (2019).

[Mil19c] Miller D.A.B. "Waves, modes, communications, and optics: a tutorial". *Adv Opt Phot*, **11**, 679–825 (2019).

[Mil60]   Milne-Thomson L.M. *Theoretical Hydrodynamics* (MacMillan, New York, 1960).

[Mil73]   Milonni P.W., Ackerhalt J.R. and Smith W.A. "Interpretation of radiative corrections in spontaneous emission". *Phys Rev Lett*, **31**, 958–960 (1973).

[Mil75]   Milonni P.W. and Smith W.A. "Radiation reaction and vacuum fluctuations in spontaneous emission". *Phys Rev A*, **11**, 814–824 (1975).

[Mil76]   Milonni P.W. "Semiclassical and quantum-electrodynamical approaches in nonrelativistic radiation theory". *Phys Rep*, **25**, 1–81 (1976).

[Mil81]   Milonni P.W. "Quantum mechanics of the Einstein–Hopf model". *Am J Phys*, **49**, 177–184 (1981).

[Mil82]   Milonni P.W. and Hardies M.L. "Photons cannot always be replicated". *Phys Lett A*, **92**, 321–322 (1982).

[Mil84]   Milonni P.W. "Why spontaneous emission?" *Am J Phys*, **52**, 340–343 (1984).

[Mil88]   Milonni P.W. and Eberly J.H. *Lasers* (Wiley, New York, 1988).

[Mil91]   Milonni P.W. and Singh S. "Some recent developments in the fundamental theory of light". *Adv At Mol Opt Phys*, **28**, 75–142 (1991).

[Mil94]   Milonni P.W. *The Quantum Vacuum. An Introduction to Quantum Electrodynamics* (Academic, New York, 1994).

[Mil97]   Milonni P.W. "Answer to Question # 45 " What (if anything) does the photoelectric effect teach us"". *Am J Phys*, **64**, 11–12 (1997).

[Mil99]   Miller A.D., Caldwell R., Devlin M.J., Dorwart W.B., Herbig T., Nolta M.R., Page L.A., Puchalla J., Torbet E. and Tran H.T. "A measurement of the angular power spectrum of the cosmic microwave background from l = 100 to 400". *Ap J*, **524**, L1–L4 (1999).

[Min05]   Mintert F., Carvalho A.R., Kuś M. and Buchleitner A. "Measures and dynamics of entangled states". *Phys Rep*, **415**, 207–259 (2005).

[Mir95]   Mirman R. *Group Theory: An Intuitive Approach* (World Scientific, Singapore, 1995).

[Mis09]   Mishchenko M.I. "Gustav Mie and the fundamental concept of electromagnetic scattering by particles: A perspective". *J Quant Spec Rad Transf*, **110**, 1210–1222 (2009).

[Mis10]   Mishchenko M.I., Travis L.D. and Mackowski D.W. "T-matrix method and its applications to electromagnetic scattering by particles: A current perspective". *J Quant Spec Rad Transf*, **111**, 1700–1703 (2010).

[Mis91] Mishra S.R. "A vector wave analysis of a Bessel beam". *Opt Comm*, **85**, 159–161 (1991).

[Mit92] Mitalas R. and Sills K.R. "On the photon diffusion time scale for the sun". *Ap J*, **401**, 759–760 (1992).

[Moe04] Moerner W.E. "Single-photon sources based on single molecules in solids". *New J Phys*, **6**, 88–88 (2004).

[Mol69] Mollow B.R. "Power spectrum of light scattered by two-level systems". *Phys Rev*, **188**, 1969–1975 (1969).

[Mol72] Mollow B.R. "Absorption and emissiion line-shape funcitons for driven atoms". *Phys Rev A*, **5**, 1522–1527 (1972).

[Mol93] Molmer K. "Monte Carlo wave-function method in quantum optics". *J Opt Soc Am B*, **10**, 524–538 (1993).

[Mom00] Mompart J. and Corbalan R. "Lasing without inversion". *J Opt B*, **2**, R7–R24 (2000).

[Mon10] Monmayrant A., Weber S. and Chatel B. "A newcomer's guide to ultrashort pulse shaping and characterization". *J Phys B*, **43**, 103001 (2010).

[Mor36] Morse P.M. *Vibrations and Sound* (McGraw Hill, New York, 1936).

[Mor53] Morse P.M. and Feshbach H. *Methods of Theoretical Physics* (McGraw-Hill, New York, 1953).

[Mor64] Morse P.M. *Thermal Physics* (W A Benjamin, New York, 1964).

[Mor83] Morris J.R. and Shore B.W. "Reduction of degenerate two-level excitation to independent two-state systems". *Phys Rev A*, **27**, 906–912 (1983).

[Mos08] Mosley P.J., Lundeen J.S., Smith B.J. and Walmsley I.A. "Conditional preparation of single photons using parametric downconversion: A recipe for purity". *New J Phys*, **10**, 093011 (2008).

[Muc13] Mücke M., Bochmann J., Hahn C., Neuzner A., Nölleke C., Reiserer A., Rempe G. and Ritter S. "Generation of single photons from an atom-cavity system". *Phys Rev A*, **87**, 063805 (2013).

[Muk82] Mukamel S. "Collisional broadening of spectral line shapes in two-photon and multiphoton processes". *Phys Rep*, **93**, 1–60 (1982).

[Mur17] Murray C.R. and Pohl T. "Coherent photon manipulation in interacting atomic ensembles". *Phys Rev X*, **7**, 011007 (2017).

[Mut03] Muthukrishnan A., Scully M.O. and Zubairy M.S. "The concept of the photon – revisited". *OPN Trends*, **October**, S18–S27 (2003).

[Mut05] Muthukrishnan A., Scully M.O. and Zubairy M.S. "The photon wave function". *Proc SPIE*, **5866**, 287–292 (2008).

[Nai03] Nairz O., Arndt M. and Zeilinger A. "Quantum interference experiments with large molecules". *Am J Phys*, **71**, 319–325 (2003).

[Nak07] Nakar E. "Short-hard gamma-ray bursts". *Phys Rep*, **442**, 166–236 (2007).

[Nar81] Narozhny N.B., Sanchez-Mondragon J.J. and Eberly J.H. "Coherence versus incoherence: Collapse and revival in a simple quantum model". *Phys Rev A*, **23**, 236–247 (1981).

[Nau16] Nauenberg M. "Max Planck and the birth of the quantum hypothesis".

Am J Phys, **84**, 709–720 (2016).

[Nay08] Nayak C., Simon S.H., Stern A., Freedman M. and Das Sarma S. "Non-Abelian anyons and topological quantum computation". *Rev Mod Phys*, **80**, 1083–1159 (2008).

[Ned05] Nedderman R.M. *Statics and Kinematics of Granular Materials* (Cambridge Univ Press, Cambridge, 2005).

[Nel17] Nelson P. *From Photon to Neuron: Light, Imaging, Vision* (Princeton Univ Press, Princeton, NJ, 2017).

[Neu17] Neuenschwander D.E. *Emmy Noether's Wonderful Theorem* (Johns Hopkins Univ Press, Baltimore, MD, 2017), revised edition.

[Nev15] Neves A.A.R., Jones P.H., Luo L. and Marago O.M. "Optical cooling and trapping: Introduction". *J Opt Soc Am A*, **32**, OCT1–OCT5 (2015).

[New11] New G.H.C. "Nonlinear optics: The first 50 years". *Contemp Phys*, **52**, 281–292 (2011).

[New49] Newton T.D. and Wigner E.P. "Localized states for elementary wystems". *Rev Mod Phys*, **21**, 400–406 (1949).

[New66] Newton R.G. *Scattering Theory of Waves and Particles* (McGraw-Hill, New York, 1966).

[Nie00] Nielsen M.A. and Chuang I.L. *Quantum Computation and Quantum Information* (Cambridge Univ Press, New York, 2000).

[Nie66] Nielsen A. and Olsen J. "Formal Analogy between Compton Scattering and Doppler Effect". *Am J Phys*, **34**, 621–622 (1966).

[Nis11] Nisbet-Jones P.B.R., Dilley J., Ljunggren D. and Kuhn A. "Highly efficient source for indistinguishable single photons of controlled shape". *New J Phys*, **13**, 103036 (2011).

[Nol13] Nölleke C., Neuzner A., Reiserer A., Hahn C., Rempe G. and Ritter S. "Efficient teleportation between remote single-atom quantum memories". *Phys Rev Lett*, **110**, 140403 (2013).

[Nor14] Northup T. and Blatt R. "Quantum information transfer using photons". *Nature Phot*, **8**, 356–363 (2014).

[Nor17] Norton J.D. "Thermodynamically reversible processes in statistical physics". *Am J Phys*, **85**, 135–145 (2017).

[Now12] Nowack R.L. "A tale of two beams: An elementary overview of Gaussian beams and Bessel beams". *Studia Geophysica et Geodaetica*, **56**, 355–372 (2012).

[OLe64] O'Leary A.J. "Redshift and deflection of photons by gravitation: a comparison of relativistic and Newtonian treatments". *Am J Phys*, **32**, 52–55 (1964).

[ONe63] O'Neil E.L. *Introduction to Statistical Optics* (Addison-Wesley, Reading, MA, 1963).

[Oha86] Ohanian H.C. "What is spin?" *Am J Phys*, **54**, 500–505 (1986).

[Ohm09] Ohmori K. "Wave-packet and coherent control dynamics." *Ann Rev Phys Chem*, **60**, 487–511 (2009).

[Oku89] Okun L.B. "The concept of mass". *Phys Today*, **42**, 31–36 (1989).

[Ong98] Ong R.A. "Very high-energy gamma-ray astronomy". *Phys Rep*, **305**,

93–202 (1998).

[Orb16]   Orbes A.N.F. and Udley A.N.D. "Creation and detection of optical modes with spatial light modulators". *Adv Opt Phot*, **8**, 200–227 (2016).

[Ore84]   Oreg J., Hioe F.T. and Eberly J.H. "Adiabatic following in multilevel systems". *Phys Rev A*, **29**, 690–697 (1984).

[Oro00]   Oron R., Blit S., Davidson N., Friesem A.A., Bomzon Z. and Hasman E. "The formation of laser beams with pure azimuthal or radial polarization". *App Phys Lett*, **77**, 3322–3324 (2000).

[Ort18]   Ortigoso J. "Twelve years before the quantum no-cloning theorem". *Am J Phys*, **86**, 201–205 (2018).

[Ota11]   Ota Y., Iwamoto S., Kumagai N. and Arakawa Y. "Spontaneous two-photon emission from a single quantum dot". *Phys Rev Lett*, **107**, 233602 (2011).

[Ou87]   Ou Z.Y., Hong C.K. and Mandel L. "Relation between input and output states for a beam splitter". *Opt Comm*, **63**, 118–122 (1987).

[Oug19]   Oughstun K.E. *Electromagnetic and Optical Pulse Propagation* (Springer, Cham, Switzerland, 2019), 2nd edition.

[Oug88]   Oughstun K.E. and Sherman G.C. "Propagation of electromagnetic pulses in a linear dispersive medium with absorption (the Lorentz medium)". *J Opt Soc Am B*, **5**, 817–849 (1988).

[Oxb05]   Oxborrow M. and Sinclair A.G. "Single-photon sources". *Contemp Phys*, **46**, 173–206 (2005).

[Pac93]   Pace A., Collett M. and Walls D. "Quantum limits in interferometric detection of gravitational radiation". *Phys Rev A*, **47**, 3173–3189 (1993).

[Pad00]   Padgett M.J. and Allen L. "Light with a twist in its tail". *Contemp Phys*, **41**, 275–285 (2000).

[Pad03]   Padgett M.J., Barnett S.M. and Loudon R. "The angular momentum of light inside a dielectric". *J Mod Opt*, **50**, 1555–1562 (2003).

[Pad03b] Padmanabhan T. "Cosmological constant—the weight of the vacuum". *Phys Rep*, **380**, 235–320 (2003).

[Pad04]   Padgett M., Courtial J. and Allen L. "Light's orbital angular momentum". *Phys Today*, **57**, 35–40 (2004).

[Pad11]   Padgett M.J., O'Holleran K., King R.P. and Dennis M. "Knotted and tangled threads of darkness in light beams". *Contemp Phys*, **52**, 265–279 (2011).

[Pad17]   Padgett M.J. "Orbital angular momentum 25 years on [Invited]". *Opt Exp*, **25**, 11265 (2017).

[Pad95]   Padgett M.J. and Allen L. "The Poynting vector in Laguerre-Gaussian laser modes". *Opt Comm*, **121**, 36–40 (1995).

[Pai79]   Pais A. "Einstein and the quantum theory". *Rev Mod Phys*, **51**, 863–914 (1979).

[Pai82]   Pais A. *Subtle is the Lord: The Science and the Life of Albert Einstein* (Oxford Univ Press, Oxford, 1982).

[Pal73]   Paldus J. and Wong H. "Computer generation of Feynman diagrams for perturbation theory I. General algorithm". *Comp Phys Comm*, **6**, 1–7

(1973).

[Pan55] Panofsky W.H.K.H. and Phillips M. *Classical Electricity and Magnetism* (Addison-Wesley, Cambridge, MA, 1955).

[Pat18] Pathak A. and Ghatak A. "Classical light vs. nonclassical light: Characterizations and interesting applications". *J Elec Waves Appl*, **32**, 229–264 (2018).

[Pau04] Paul H. *Introduction to Quantum Optics: From Light Quanta to Qquantum Teleportation* (Cambridge Univ Press, Cambridge, 2004).

[Pau82] Paul H. "Photon antibunching". *Rev Mod Phys*, **54**, 1061–1102 (1982).

[Pav08] Pavia D.L., Lampman G.M., Kriz G.S. and Vyvyan J.A. *Introduction to Spectroscopy* (Brooks Cole, Belmont, CA, 2008), 4th edition.

[Pav08b] Pavesi L. and Fauchet P.M. *Biophotonics* (Springer Science & Business Media, New York, 2008).

[Paz15] Pazourek R., Nagele S. and Burgdörfer J. "Attosecond chronoscopy of photoemission". *Rev Mod Phys*, **87**, 765–802 (2015).

[Pea06] Peach G. "59. Collisional broadening of spectral lines". In G. Drake, (ed.) *Springer Handbook of Atomic, Molecular and Optical Physics*, chapter 59, 875–890 (Springer, New York, 2006), 2nd edition.

[Pea10] Pearson B.J. and Jackson D.P. "A hands-on introduction to single photons and quantum mechanics for undergraduates". *Am J Phys*, **78**, 471–484 (2010).

[Pee03] Peebles P.J.E. and Ratra B. "The cosmological constant and dark energy". *Rev Mod Phys*, **75**, 559–606 (2003).

[Peg88] Pegg D.T. and Barnett S.M. "Unitary phase operators in quantum mechanics". *Europhys Lett*, **6**, 483–487 (1988).

[Peg89] Pegg D.T. and Barnett S.M. "Phase properties of the quantized single-mode electromagnetic field". *Phys Rev A*, **50**, 1665–1675 (1989).

[Peg97] Pegg D.T. and Barnett S.M. "Tutorial review. Quantum optical phase". *J Mod Opt*, **44**, 225–264 (1997).

[Pen00] Pendry J.B. "Negative refraction makes a perfect lens". *Phys Rev Lett*, **85**, 3966–3969 (2000).

[Pen04] Pendry J.B. "Negative refraction". *Contemp Phys*, **45**, 191–202 (2004).

[Pen06] Pendry J.B., Schurig D. and Smith D.R. "Controlling electromagnetic fields." *Science*, **312**, 1780–1782 (2006).

[Pen65] Penzias A.A. and Wilson R.W. "A measurement of excess antenna temperature at 4080 Mc/s." *Ap J*, **142**, 419–421 (1965).

[Pen76] Pendry J.B. "Theory of photoemission". *Surf Sci*, **57**, 679–705 (1976).

[Per04] Pereira E., Martinho J.M.G. and Berberan-Santos M.N. "Photon trajectories in incoherent atomic radiation trapping as Lévy flights". *Phys Rev Lett*, **93**, 120201 (2004).

[Per09] Perwass C. *Geometric Algebra with Applications in Engineering* (Springer, Berlin, 2009).

[Per18] Pérez J.P. and Lamine B. "Phase velocity and light bending in a gravitational potential". *Eur J Phys*, **39**, 055602 (2018).

[Per72] Perelomov A. "Coherent states for arbitrary Lie group". *Comm Math*

Phys, **26**, 222–236 (1972).

[Per99]   Peral E. and Yariv A. "Supermodes of grating-coupled multimode waveguides and application to mode conversion between copropagating modes mediated by backward Bragg scattering". *J Lightw Techn*, **17**, 942–947 (1999).

[Pet08]   Petrosyan D. and Fleischhauer M. "Quantum information processing with single photons and atomic ensembles in microwave coplanar waveguide resonators". *Phys Rev Lett*, **100**, 170501 (2008).

[Pet10]   Petrosyan D., Nikolopoulos G.M. and Lambropoulos P. "State transfer in static and dynamic spin chains with disorder". *Phys Rev A*, **81**, 042307 (2010).

[Pet11]   Petrosyan D., Otterbach J. and Fleischhauer M. "Electromagnetically induced transparency with Rydberg atoms". *Phys Rev Lett*, **107**, 213601 (2011).

[Pey12]   Peyronel T., Firstenberg O., Liang Q.Y., Hofferberth S., Gorshkov A.V., Pohl T., Lukin M.D. and Vuletić V. "Quantum nonlinear optics with single photons enabled by strongly interacting atoms." *Nature*, **488**, 57–60 (2012).

[Pfe07]   Pfeifer R.N., Nieminen T.A., Heckenberg N.R. and Rubinsztein-Dunlop H. "Colloquium: Momentum of an electromagnetic wave in dielectric media". *Rev Mod Phys*, **79**, 1197–1216 (2007).

[Phi98]   Phillips W.D. "Nobel Lecture: Laser cooling and trapping of neutral atoms". *Rev Mod Phys*, **70**, 721–741 (1998).

[Pho00]   Phoenix S.J.D., Barnett S.M. and Chefles A. "Three-state quantum cryptography". *J Mod Opt*, **47**, 507–516 (2000).

[Pic14]   Picton Drake S. and Purvis A. "Everyday relativity and the Doppler effect". *Am J Phys*, **82**, 52–59 (2014).

[Pie06]   Pierrat R., Greffet J.J. and Carminati R. "Photon diffusion coefficient in scattering and absorbing media". *J Opt Soc Am A*, **23**, 1106–1110 (2006).

[Pie88]   Pierce A.P., Dahleh M.A. and Rabitz H. "Optimal control of quantum-mechanical systems: Existence, numerical approximaiton, and applications". *Phys Rev A*, **37**, 4950–4964 (1988).

[Pir10]   Piro N., Rohde F., Schuck C., Almendros M., Huwer J., Ghosh J., Haase A., Hennrich M., Dubin F. and Eschner J. "Heralded single-photon absorption by a single atom". *Nature Phys*, **7**, 17–20 (2010).

[Pla16]   Plato A.D.K., Hughes C.N. and Kim M.S. "Gravitational effects in quantum mechanics". *Contemp Phys*, **57**, 477–495 (2016).

[Pli13]   Plick W.N., Krenn M., Fickler R., Ramelow S. and Zeilinger A. "Quantum orbital angular momentum of elliptically symmetric light". *Phys Rev A*, **87**, 033806 (2013).

[Por01]   Porras M.A., Borghi R. and Santarsiero M. "Relationship between elegant Laguerre-Gauss and Bessel-Gauss beams." *J Opt Soc Am A*, **18**, 177–184 (2001).

[Pou59]   Pound R.V. and Rebka, G.A. Jr. "Gravitational red-shift in nuclear resonance". *Phys Rev Lett*, **3**, 439–441 (1959).

[Pou60]   Pound R.V. and Rebka, G.A. Jr. "Apparent weight of photons". *Phys Rev Lett*, **4**, 337–341 (1960).

[Pow64]   Power E.A. *Introductory Quantum Electrodynamics* (Longmans, London, 1964).

[Pow78]   Power E.A. and Thirunamachandran T. "On the nature of the Hamiltonian for the interaction of radiation with atoms and molecules: (e/mc) p.A, -μ.E, and all that". *Am J Phys*, **46**, 370–378 (1978).

[Pow80]   Power E.A. and Thirunamachandran T. "The multipolar Hamiltonian in radiation theory". *Proc Roy Soc (Lond) A*, **372**, 265–273 (1980).

[Pow85]   Power E.A. and Thirunamachandran T. "Further remarks on the Hamiltonian of quantum optics". *J Opt Soc Am B*, **2**, 1100–1105 (1985).

[Pra03]   Präkelt A., Wollenhaupt M., Assion A., Horn C., Sarpe-Tudoran C., Winter M. and Baumert T. "Compact, robust, and flexible setup for femtosecond pulse shaping". *Rev Sci Inst*, **74**, 4950–4953 (2003).

[Pra03b] Prasad P.N. *Introduction to Biophotonics* (Wiley Online Library, New York, 2003).

[Pra05]   Pratt S.T. "Vibrational autoionization in polyatomic molecules". *Ann Rev Phys Chem*, **56**, 281–308 (2005).

[Pre98]   Preskill J. "Lecture Notes for Physics 229: Quantum Information and Computation". *Lecture Notes* (1998).

[Pry04]   Pryde G.J., Brien J.L.O., White A.G., Bartlett S.D. and Ralph T.C. "Measuring a photonic qubit without destroying It". *Phys Rev Lett*, **92**, 190402 (2004).

[Pur01]   Puri R.R. *Mathematical Methods of Quantum Optics* (Springer, New York, 2001).

[Pur46]   Purcell E.M., Torrey H.C. and Pound R.V. "Resonance absorption by nuclear magnetic moments in a solid". *Phys Rev*, **69**, 37–38 (1946).

[Pur46b] Purcell E.M. "Spontaneous emission probabilities at radio frequencies". *Phys Rev*, **69**, 681–681 (1946).

[Qia11]   Qian X.F. and Eberly J.H. "Entanglement and classical polarization states". *Opt Lett*, **36**, 4110–4112 (2011).

[Qia13]   Qian X.F. and Eberly J.H. "Entanglement is sometimes enough". *ArXiv*, **14627**, 1–5 (2013).

[Qia15]   Qian X.F., Little B., Howell J.C. and Eberly J.H. "Shifting the quantum-classical boundary: Theory and experiment for statistically classical optical fields". *Optica*, **2**, 611–615 (2015).

[Qia16]   Qian X.F., Malhotra T., Vamivakas A.N. and Eberly J.H. "Coherence constraints and the last hidden optical coherence". *Phys Rev Lett*, **117**, 153901 (2016).

[Qia18]   Qian X.F., Vamivakas A.N. and Eberly J.H. "Entanglement limits duality and vice versa". *Optica*, **5**, 942–947 (2018).

[Qia18b] Qian X.F., Vamivakas A.N. and Eberly J.H. "Bohr's complementarity: Completed with entanglement". *ArXiv:1803.04611*, **04611**, 1–8 (2018).

[Qua17]   Quan X.F., Vamivakas A.N. and Eberly J.H. "Emerging connections: Classical and quantum optics". *Opt Phot News*, **28**, 34–41 (2017).

[Que17]   Quesada N. and Sipe J.E. "Why you should not use the electric field to quantize in nonlinear optics". *Opt Lett*, **42**, 3443–3446 (2017).

[Qui06]   Quimby R.S. *Photonics and Lasers: An Introduction* (John Wiley & Sons, Hoboken, NJ, 2006).

[Qur16]   Qureshi T. "Quantitative wave-particle duality". *Am J Phys*, **84**, 517–521 (2016).

[Rab00]   Rabitz H. and Zhu W. "Optimal control of molecular motion: design, implementation, and inversion." *Accounts Chem Res*, **33**, 572–578 (2000).

[Rab02]   Rabitz H. "Optimal control of quantum systems: Origins of inherent robustness to control field fluctuations". *Phys Rev A*, **66**, 063405 (2002).

[Rab16]   Raboy M. *Marconi. The Man Who Networked the World* (Oxford Univ Press, New York, 2016).

[Rab36]   Rabi I.I. "On the process of space quantization". *Phys Rev*, **49**, 324–328 (1936).

[Rab37]   Rabi I.I. "Space quantization in a gyrating magnetic field". *Phys Rev*, **51**, 652–654 (1937).

[Rai01]   Raimond J.M., Brune M. and Haroche S. "Manipulating quantum entanglement with atoms and photons in a cavity". *Rev Mod Phys*, **73**, 565–582 (2001).

[Ral06]   Ralph T.C., Bartlett S.D., O'Brien J.L., Pryde G.J. and Wiseman H.M. "Quantum nondemolition measurements for quantum information". *Phys Rev A*, **73**, 012113 (2006).

[Ram28]   Raman C.V. and Krishnan K.S. "The optical analogue of the Compton effect". *Nature*, **121**, 377–378 (1928).

[Ram28b] Raman C.V. "A new type of secondary radiation". *Nature*, **121**, 501–502 (1928).

[Ran05]   Rangacharyulu C. "What is a photon?" *Proc SPIE*, **5866**, 8–16 (2005).

[Ran06]   Rangelov A.A., Vitanov N.V. and Shore B.W. "Extension of the Morris-Shore transformation to multilevel ladders". *Phys Rev A*, **74**, 053402 (2006).

[Ran08]   Rangelov A.A., Vitanov N.V. and Shore B.W. "Population trapping in three-state quantum loops revealed by Householder reflections". *Phys Rev A*, **77**, 033404 (2008).

[Ran10]   Rangelov A.A., Vitanov N.V. and Shore B.W. "Rapid adiabatic passage without level crossing". *Opt Comm*, **283**, 1346–1350 (2010).

[Ran18]   Randall J., Lawrence A.M., Webster S.C., Weidt S., Vitanov N.V. and Hensinger W.K. "Generation of high-fidelity quantum control methods for multilevel systems". *Phys Rev A*, **98**, 043414 (2018).

[Ran90]   Rañada A.F. "Knotted solutions of the Maxwell equations in vacuum". *J Phys A*, **23**, L815–L820 (1990).

[Ran92]   Rañada A.F. "Topological electromagnetism". *J Phys A*, **25**, 1621–1641 (1992).

[Ran95]   Rañada A.F. and Trueba J.L. "Electromagnetic knots". *Phys Lett A*, **202**, 337–342 (1995).

[Ran97]   Rañada A.F. and Trueba J.L. "Two properties of electromagnetic knots".

Phys Lett A, **232**, 25–33 (1997).

[Rao17]   Rao K.R.K. and Suter D. "Nonlinear dynamics of a two-level system of a single spin driven beyond the rotating-wave approximation". *Phys Rev A*, **95**, 053804 (2017).

[Rar87]   Rarity J.G., Tapster P.R. and Jakeman E. "Observation of sub-poissonian light in parametric downconversion". *Opt Comm*, **62**, 201–206 (1987).

[Ray12]   Raymer M.G. and Srinivasan K. "Manipulating the color and shape of single photons". *Phys Today*, **65**, 32–37 (2012).

[Ray20]   Raymer M.G. "Quantum theory of light in a dispersive structured linear dielectric: a macroscopic Hamiltonian tutorial treatment". *J Mod Opt*, 1–17 (2020).

[Ray20b] Raymer M.G. and Walmsley I.A. "Temporal modes in quantum optics: then and now". *Phys Scripta*, **95**, 064002 (2020).

[Ray81]   Raymer M.G. and Mostowski J. "Stimulated Raman scattering: Unified treatment of spontaneous initiation and spatial propagation". *Phys Rev A*, **24**, 1980–1993 (1981).

[Ray90]   Raymer M.G. "Observations of the modern photon". *Am J Phys*, **58**, 11–11 (1990).

[Ray97]   Raymer M.G. "Measuring the quantum mechanical wave function". *Contemp Phys*, **38**, 343–355 (1997).

[Red13]   Redzić D.V. "The case of the Doppler effect for photons revisited". *Eur J Phys*, **34**, 1355–1366 (2013).

[Rei02]   Reimann S.M. and Manninen M. "Electronic structure of quantum dots". *Rev Mod Phys*, **74**, 1283–1342 (2002).

[Rei13]   Reiserer A., Ritter S. and Rempe G. "Nondestructive detection of an optical photon". *Science*, **342**, 1349–1351 (2013).

[Rem91]   Rempe G., Scully M.O. and Walther H. "The one-atom maser and the generation of nonclassical light". *Phys Scripta*, **1991**, 5–13 (1991).

[Rey69]   Reynolds G.T., Spartalian K. and Scarl D.B. "Interference effects produced by single photons". *Nuov Cim B*, **61**, 355–364 (1969).

[Ric00]   Rice S.A. and Zhao M. *Optical Control of Molecular Dynamics* (Wiley, New York, 2000).

[Rie14]   Rieländer D., Kutluer K., Ledingham P.M., Gündogan M., Fekete J., Mazzera M. and Riedmatten H.D. "Quantum storage of heralded single photons in a praseodymium-doped crystal". *Phys Rev Lett*, **112**, 040504 (2014).

[Rie16]   Rieländer D., Lenhard A., Mazzera M. and de Riedmatten H. "Cavity enhanced telecom heralded single photons for spin-wave solid state quantum memories". *New J Phys*, **18**, 123013 (2016).

[Rie98]   Rieke F. and Baylor D.A. "Single-photon detection by rod cells of the retina". *Rev Mod Phys*, **70**, 1027–1036 (1998).

[Rig86]   Rigden J.S. "Quantum states and precession: The two discoveries of NMR". *Rev Mod Phys*, **58**, 433–448 (1986).

[Roc05]   Roche J. "What is mass?" *Eur J Phys*, **26**, 225–242 (2005).

[Roc13]   Rocci A. "On first attempts to reconcile quantum principles with gravity".

J Phys Conf Ser, **470**, 012004 (2013).

[Rod67]   Rodberg L.S. and Thaler R.M. *Introduction to the Quantum Theory of Scattering* (Academic, New York, 1967).

[Ros55]   Rose M.E. *Multipole Fields* (Wiley, New York, 1955).

[Ros61]   Rose D.J. and Clark, Melville Jr. *Plasmas and Controlled Fusion* (MIT Press - John Wiley, New York, 1961).

[Roy89]   Royer A. "Measurement of quantum states and the Wigner function". *Found Phys*, **19**, 3–32 (1989).

[Rub17]   Rubinsztein-Dunlop H., Forbes A., Berry M.V., Dennis M.R. and Andrews D.L. "Roadmap on structured light". *J Opt*, **19**, 13001 (2017).

[Rub94]   Rubin M.H., Klyshko D.N., Shih Y.H. and Sergienko A.V. "Theory of two-photon entanglement in type-II optical parametric down-conversion". *Phys Rev A*, **50**, 5122–5133 (1994).

[Rue96]   Rueckner W. and Titcomb P. "A lecture demonstration of single photon interference,". *Am J Phys*, **64**, 184–188 (1996).

[Run01]   Rungta P., Buzek V., Caves C., Hillery M. and Milburn G.J. "Universal state inversion and concurrence in arbitrary dimensions". *Phys Rev A*, **64**, 042315 (2001).

[Run03]   Rungta P. and Caves C.M. "Concurrence-based entanglement measures for isotropic states". *Phys Rev A*, **67**, 012307 (2003).

[Rus03]   Russell P. "Photonic crystal fibers". *Science*, **299**, 358–362 (2003).

[Ryb85]   Rybicki G.B. and Lightman A.P. *Radiative Processes in Astrophysics* (Wiley, New York, 1985).

[Saa03]   Saavedra G., Furlan W.D. and Monsoriu J.A. "Fractal zone plates". *Opt Lett*, **28**, 971–973 (2003).

[Sac19]   Sacks O. "Personal History: The machine stops". *New Yorker*, **Feb 11**, 28–29 (2019).

[Saf10]   Saffman M., Walker T.G. and Mø lmer K. "Quantum information with Rydberg atoms". *Rev Mod Phys*, **82**, 2313–2363 (2010).

[Sag98]   Saghafi S. and Sheppard C.J.R. "Near field and far field of elegant Hermite-Gaussian and Laguerre-Gaussian modes". *J Mod Opt*, **45**, 1999–2009 (1998).

[Sal10]   Saleh M.F., Di Giuseppe G., Saleh B.E. and Teich M.C. "Photonic circuits for generating modal, spectral, and polarization entanglement". *IEEE Photonics J*, **2**, 736–752 (2010).

[Sal11]   Saldanha P.L. and Monken C.H. "Interaction between light and matter: A photon wave function approach". *New J Phys*, **13**, 073015 (2011).

[Sal19]   Saleh B.E.A. and Teich M.C. *Fundamentals of Photonics* (John Wiley & Sons, New York, 2019), 3rd edition.

[Sal74]   Salzman W.R. "Quantum mechanics of systems periodic in time". *Phys Rev A*, **10**, 461–465 (1974).

[San04]   Sangouard N., Guérin S., Yatsenko L.P. and Halfmann T. "Preparation of coherent superposition in a three-state system by adiabatic passage". *Phys Rev A*, **70**, 013415 (2004).

[San12]   Sanders B.C. "Review of entangled coherent states". *J Phys A*, **45**, 244002

(2012).

[San83]   Sanchez-Mondragon J.J., Narozhny N.B. and Eberly J.H. "Theory of spontaneous-emission line shape in an ideal cavity". *Phys Rev Lett*, **51**, 550–553 (1983).

[Sap06]   Sapirstein J.R. "27. Quantum electrodynamics". In G. Drake, (ed.) *Springer Handbook of Atomic, Molecular and Optical Physics*, chapter 27, 413–428 (Springer, New York, 2006), 2nd edition.

[Sar74]   Sargent, Murray III, Scully M.O. and Lamb, Willis E. Jr. *Laser Physics* (Addison-Wesley, Reading, MA, 1974).

[Sar76]   Sargent, Murray III and Horwitz P. "Three-level Rabi flopping". *Phys Rev A*, **13**, 1962–1964 (1976).

[Sat14]   Sathyamoorthy S.R., Tornberg L., Kockum A.F., Baragiola B.Q., Combes J., Wilson C.M., Stace T.M. and Johansson G. "Quantum nondemolition detection of a propagating microwave photon". *Phys Rev Lett*, **112**, 093601 (2014).

[Sca05]   Scarani V., Iblisdir S., Gisin N. and Acín A. "Quantum cloning". *Rev Mod Phys*, **77**, 1225–1256 (2005).

[Sca11]   Scala M., Militello B., Messina A. and Vitanov N.V. "Stimulated Raman adiabatic passage in a Λ system in the presence of quantum noise". *Phys Rev A*, **83**, 012101 (2011).

[Sca11b] Scala M., Militello B., Messina A. and Vitanov N.V. "Detuning effects in STIRAP processes in the presence of quantum noise". *Optics and Spectroscopy*, **111**, 589–592 (2011).

[Sch01]   Schleich W.P. *Quantum Optics in Phase Space* (Wiley-VCH, Weinheim, 2001).

[Sch03]   Schnabel R., Bowen W.P., Treps N., Ralph T.C., Bachor H.A. and Lam P.K. "Stokes-operator-squeezed continuous-variable polarization states". *Phys Rev A*, **67**, 012316 (2003).

[Sch05]   Schlosshauer M. "Decoherence , the measurement problem , and interpretations of quantum mechanics". *Rev Mod Phys*, **76**, 1267–1305 (2005).

[Sch07]   Scheel S., Florescu M., Häffner H., Lee H., Strekalov D.V., Knight P.L. and Dowling J.P. "Single photons on demand from tunable 3D photonic band-gap structures". *J Mod Opt*, **54**, 409–416 (2007).

[Sch07b] Schaefer B., Collett E., Smyth R., Barrett D. and Fraher B. "Measuring the Stokes polarization parameters". *Am J Phys*, **75**, 163–168 (2007).

[Sch08]   Scheel S. and Buhmann S.Y. "Macroscopic quantum electrodynamics–Concepts and applications". *Acta Phys Slovaca*, **58**, 675–809 (2008).

[Sch09]   Scheel S. "Single-photon sources–an introduction". *J Mod Opt*, **56**, 141–160 (2009).

[Sch10]   Schweitzer G.K. and Pesterfield L.L. *The Aqueous Chemistry of the Elements* (Oxford Univ Press, Oxford, 2010).

[Sch10b] Schneider D.P., Richards G.T., Hall P.B., Strauss M.A., Anderson S.F., Boroson T.A., Ross N.P., Shen Y., Brandt W., Fan X. *et al.* "The Sloan digital sky survey quasar catalog. V. Seventh data release". *Astron J*, **139**, 2360–2373 (2010).

[Sch16]   Schraft D., Hain M., Lorenz N. and Halfmann T. "Stopped light at high storage efficiency in a $Pr^{3+} : Y_2SiO_5$ crystal". *Phys Rev Lett*, **116**, 073602 (2016).

[Sch17]   Schroeder D.V. "Entanglement isn't just for spin". *Am J Phys*, **85**, 812–820 (2017).

[Sch17b]   Schnabel R. "Squeezed states of light and their applications in laser interferometers". *Phys Rep*, **684**, 1–51 (2017).

[Sch18]   Scholz R. and Friege G. "Undergraduate quantum optics: Experimental steps to quantum physics". *Eur J Phys*, **39**, 055301 (2018).

[Sch18b]   Schwefel H.G., Trainor L.S., Rueda A. and Sedlmeir F. "Nonlinear and quantum optics within whispering gallery mode resonators". *J Opt*, **18**, 123002 (2018).

[Sch22]   Schrödinger E. "Dopplerprinzip und Bohrsche Frequenzbedingung". *Phys Zs*, **23**, 301–303 (1922).

[Sch27]   Schrödinger E. "Über den Comptoneffekt". *Ann d Physik*, **82**, 828–832 (1927).

[Sch35]   Schrödinger E. "Die gegenwärtige Situation in der Quantenmechanik". *Naturwiss*, **48**, 807–812, 823–828, 844–849 (1935).

[Sch35b]   Schrödinger E. "Discussion of probability relations between separated systems". *Math Proc Camb Phil Soc*, **31**, 555–563 (1935).

[Sch36]   Schrödinger E. "Probability relations between separated systems". *Math Proc Camb Phil Soc*, **32**, 446–452 (1936).

[Sch37]   Schwinger J. "On nonadiabatic processes in Inhomogeneous fields". *Phys Rev*, **51**, 648–651 (1937).

[Sch48]   Schwinger J. "Quantum electrodynamics. I. A covariant formulation". *Phys Rev*, **74**, 1439–1461 (1948).

[Sch49]   Schwinger J. "Quantum electrodynamics. II. Vacuum polarization and self-energy". *Phys Rev*, **75**, 651–679 (1949).

[Sch49b]   Schwinger J. "Quantum electrodynamics. III. The electromagnetic properties of the electron-radiative corrections to scattering". *Phys Rev*, **76**, 790–817 (1949).

[Sch63]   Schmidt M. "3C 273: a star-like object with large red-shift". *Nature*, **197**, 1040–1043 (1963).

[Sch65]   Schwinger J. "On angular momentum". In L.C. Biedenharn and H.v. Dam, (eds.) *Quantum Theory of Angular Momentum*, 229–279 (Academic, New York, 1965).

[Sch77]   Schadee A. "On the Zeeman effect in electronic transitions". *J Quant Spec Rad Transf*, **19**, 517–531 (1977).

[Sch89]   Schwinger J. "A path to quantum electrodynamics". *Phys Today*, **42**, 42–48 (1989).

[Sch98]   Scheel S., Knöll L. and Welsch D.G. "QED commutation relations for inhomogeneous Kramers-Kronig dielectrics". *Phys Rev A*, **58**, 700–706 (1998).

[Sch98b]   Schwarzschild B. "Physics Nobel prize goes to Tsui, Stormer and Laughlin for the fractional quantum Hall effect". *Phys Today*, **51**, 17–19 (1998).

[Scu18]   Scully M., Sokolov A. and Svidzinsky A. "Virtual photons from the Lamb shift to black holes". *Opt Phot News*, **29**, 34–40 (2018).

[Scu66]   Scully M.O. and Lamb, W. E. Jr. "Quantum theory of an optical maser. I. General theory". *Phys Rev*, **159**, 208–226 (1966).

[Scu72]   Scully M.O. and Sargent, Murray III. "The concept of the photon". *Phys Today*, **25**, 38–47 (1972).

[Scu79]   Scully M.O. "General-relativistic treatment of the gravitational coupling between laser beams". *Phys Rev D*, **19**, 3582–3591 (1979).

[Scu97]   Scully M.O. and Zubairy M.S. *Quantum Optics* (Cambridge Univ Press, Cambridge, 1997).

[Sek09]   Sekatski P., Brunner N., Branciard C., Gisin N. and Simon C. "Towards quantum experiments with human eyes as detectors based on cloning via stimulated emission". *Phys Rev Lett*, **103**, 113601 (2009).

[Sen73]   Senitzky I.R. "Radiation-reaction and vacuum-field effects in Heisenberg-picture quantum electrodynamics". *Phys Rev Lett*, **31**, 955–958 (1973).

[Ser02]   Series G.W. "Radio-frequency spectroscopy of excited atoms". *Rep Prog Phys*, **22**, 280–328 (2002).

[Ser78]   Series G.W. "A semi-classical approach to radiation problems". *Phys Rep*, **43**, 1–41 (1978).

[Ser86]   Series G.W. "A long journey with classical fields". *Phys Scripta*, **T12**, 5–13 (1986).

[Sha03]   Shapiro M. and Brumer P. "Coherent control of molecular dynamics". *Rep Prog Phys*, **66**, 859–942 (2003).

[Sha06]   Shapiro M. and Brumer P. "Quantum control of bound and continuum state dynamics". *Phys Rep*, **425**, 195–264 (2006).

[Sha09]   Shapiro E.A., Milner V. and Shapiro M. "Complete transfer of populations from a single state to a preselected superposition of states using piecewise adiabatic passage: Theory". *Phys Rev A*, **79**, 023422 (2009).

[Sha15]   Sharma K. and Friedrich B. "Directional properties of polar paramagnetic molecules subject to congruent electric, magnetic and optical fields". *New J Phys*, **17**, 045017 (2015).

[Sha48]   Shannon C.E. "A mathematical theory of communication". *Bell Syst Tech J*, **27**, 623–656 (1948).

[Sha64]   Sharnoff M. "Validity Conditions for the Kramers-Kronig Relations". *Am J Phys*, **32**, 40–44 (1964).

[Sha91]   Shapiro J.H. and Shepard S.R. "Quantum phase measurement: A system-theory perspective". *Phys Rev*, **43**, 3795–3818 (1991).

[Sha97]   Shapiro M. and Brumer P. "Quantum control of chemical reactions by laser light". *J Chem Soc, Faraday Trans*, **93**, 1263–1277 (1997).

[She65]   Shen Y.R. and Bloembergen N. "Theory of stimulated Brillouin and Raman scattering". *Phys Rev*, **137**, A1787–A1805 (1965).

[Shi65]   Shirley J.H. "Solution of the Schrödinger equation with a Hamiltonian periodic in time". *Phys Rev*, **138**, B979–B987 (1965).

[Shi88]   Shih Y.H. and Alley C.O. "New type of Einstein-Podolsky-Rosen-Bohm experiment using pairs of light quanta produced by optical parametric down conversion". *Phys Rev Lett*, **61**, 2921–2923 (1988).

[Shi90]   Shi S. and Rabitz H. "Quantum mechanical optimal control of physical observables in microsystems". *J Chem Phys*, **92**, 364–376 (1990).

[Sho00]   Shore B.W. "Universal variational principles and the legacy of Jeremy Bentham". In W.E. Lamb, W. Schleich and H. Walther, (eds.) *Ode to a Quantum Physicist: A Festschrift in Honor of Marlan O. Scully*, 51–58 (North-Holland, Amsterdam, 2000).

[Sho01]   Shore B.W., Johnson M.A., Kulander K.C. and Davis J. "The Livermore experience: Contributions of J. H. Eberly to laser excitation theory." *Opt Exp*, **8**, 28–43 (2001).

[Sho05]   Shore B.W. and Knight P.L. "Surprises in physics: Overturning conventional wisdom". *Laser Phys*, **15**, 1448–1457 (2005).

[Sho06]   Shore B.W. and Vitanov N.V. "Overdamping of coherently driven quantum systems". *Contemp Phys*, **47**, 341–362 (2006).

[Sho07]   Shore B.W. "Sir Peter Knight and the Jaynes–Cummings model". *J Mod Opt*, **54**, 2009–2016 (2007).

[Sho08]   Shore B.W. "Coherent manipulations of atoms using laser light". *Acta Phys Slovaca*, **58**, 243–486 (2008).

[Sho11]   Shore B.W. *Manipulating Quantum Structures Using Laser Pulses* (Cambridge Univ Press, Cambridge, 2011).

[Sho13]   Shore B.W. "Pre-history of the concepts underlying stimulated Raman adiabatic passage (STIRAP)". *Acta Phys Slovaca*, **63**, 361–481 (2013).

[Sho13b]   Shore B.W. "Two-state behavior in N -state quantum systems: The Morris–Shore transformation reviewed". *J Mod Opt*, **61**, 787–815 (2013).

[Sho15]   Shore B.W., Dömötör P., Sadurní E., Süssmann G. and Schleich W.P. "Scattering of a particle with internal structure from a single slit". *New J Phys*, **17**, 013046 (2015).

[Sho17]   Shore B.W. "Picturing stimulated Raman adiabatic passage: A STIRAP tutorial". *Adv Opt Phot*, **9**, 563–719 (2017).

[Sho61]   Shore B.W., Wall N.S. and Irvine, J. W. Jr. "Interactions of 7.5-Mev protons with copper and vanadium". *Phys Rev*, **123**, 276–283 (1961).

[Sho65]   Shore B.W. and Menzel D.H. "Generalized tables for the calculation of dipole transition probabilities". *Ap J Supp*, **12**, 187–213 (1965).

[Sho65b]   Shore B.W. "Method for calculating matrix elements between configurations with several open l shells". *Phys Rev*, **139**, A1042–A1048 (1965).

[Sho67]   Shore B.W. "Scattering theory of absorption-line profiles and refractivity". *Rev Mod Phys*, **39**, 439–462 (1967).

[Sho68]   Shore B.W. and Menzel D.H. *Principles of Atomic Spectra* (Wiley, New York, 1968).

[Sho68b]   Shore B.W. "Parametrization of absorption-line profiles". *Phys Rev A*, **171**, 43–54 (1968).

[Sho69]   Shore B.W. "Dielectronic recombination". *Ap J*, **158**, 1205–1218 (1969).

[Sho77]   Shore B.W. and Ackerhalt J.R. "Dynamics of multilevel laser excitation: Three-level atoms". *Phys Rev A*, **15**, 1640–1646 (1977).

[Sho78]   Shore B.W. "Effects of magnetic sublevel degeneracy on Rabi oscillations".

Phys Rev A, **17**, 1739–1746 (1978).

[Sho79]   Shore B.W. and Cook R.J. "Coherent dynamics of N-level atoms and molecules. IV. Two-and three-level behavior". *Phys Rev A*, **20**, 1958–1964 (1979).

[Sho79b]   Shore B.W. "Definition of virtual levels". *Am J Phys*, **47**, 262–263 (1979).

[Sho81]   Shore B.W. "Two-level behavior of coherent excitation of multilevel systems". *Phys Rev A*, **24**, 1413–1418 (1981).

[Sho81b]   Shore B.W. and Johnson M.A. "Effects of hyperfine structure on coherent excitation". *Phys Rev A*, **23**, 1608–1610 (1981).

[Sho84]   Shore B.W. "Gating of population flow in resonant multiphoton excitation". *Phys Rev A*, **29**, 1578–1582 (1984).

[Sho90]   Shore B.W. *The Theory of Coherent Atomic Excitation* (Wiley, New York, 1990).

[Sho91]   Shore B.W., Meystre P. and Stenholm S. "Is a quantum standing wave composed of two traveling waves?" *J Opt Soc Am B*, **8**, 903–910 (1991).

[Sho93]   Shore B.W. and Knight P.L. "The Jaynes-Cummings model". *J Mod Opt*, **40**, 1195–1238 (1993).

[Sho95]   Shore B.W. "Examples of counter-intuitive physics". *Contemp Phys*, **36**, 15–28 (1995).

[Sho95b]   Shore B.W., Martin J., Fewell M.P. and Bergmann K. "Coherent population transfer in multilevel systems with magnetic sublevels. I. Numerical studies." *Phys Rev A*, **52**, 566–582 (1995).

[Shu08]   Shu J., Li X. and Arnoldus H.F. "Energy flow lines for the radiation emitted by a dipole". *J Mod Opt*, **55**, 2457–2471 (2008).

[Sie73]   Siegman A.E. "Hermite-Gaussian functions of complex argument as optical-beam eigenfunctions". *J Opt Soc Am*, **63**, 1093–1094 (1973).

[Sie86]   Siegman A. *Lasers* (University Science Books, Mill Valley, CA, 1986).

[Sil02]   Silva C.C. and de Andrade Martins R. "Polar and axial vectors versus quaternions". *Am J Phys*, **70**, 958–963 (2002).

[Sil07]   Silberhorn C. "Detecting quantum light". *Contemp Phys*, **48**, 143–156 (2007).

[Sil12]   Silberstein W. "LXXVI. Quaternionic form of relativity". *Phil Mag S 6*, **23**, 790–809 (1912).

[Sil68]   Silk J. "Cosmic black-body radiation and galaxy formation". *Ap J*, **151**, 459–470 (1968).

[Sim00]   Simon C., Weihs G. and Zeilinger A. "Optimal quantum cloning via stimulated emission". *Phys Rev Lett*, **84**, 2993–2996 (2000).

[Sim16]   Simon D.S. *A Guided Tour of Light Beams* (Morgan & Claypool, San Rafael, CA, 2016).

[Sin05]   Singh S. *Big Bang. The Origin of the Universe* (Harper Collins, New York, 2005).

[Sin19]   Sinha U., Sahoo S.N., Singh A., Joarder K., Chatterjee R. and Chakraborti S. "Single-photon sources". *Opt Phot News*, **28**, 34–39 (2019).

[Sin99]   Singh S. *The Code Book. The Evolution of Secrecy from Mary Queen of Scots to Quantum Cryptography* (Doubleday, New York, 1999).

[Sip09]   Sipe J.E. "Photons in dispersive dielectrics". *J Opt A*, **11**, 114006 (2009).

[Sip95]   Sipe J.E. "Photon wave functions". *Phys Rev A*, **52**, 1875–1883 (1995).

[Sir17]   Sirleto L., Vergara A. and Ferrara M.A. "Advances in stimulated Raman scattering in nanostructures". *Adv Opt Phot*, **9**, 169–217 (2017).

[Sla60]   Slater J.C.   *Quantum Theory of Atomic Structure* (McGraw-Hill, New York, 1960).

[Sla75]   Slater J.C.   *Solid-State and Molecular Theory: A Scientific Biography* (Wiley, New York, 1975).

[Sla77]   Slater J.C. *Quantum Theory of Matter* (R. E. Krieger, Huntington, NY, 1977), 2nd edition.

[Slu11]   Slussarenko S., Murauski A., Du T., Chigrinov V., Marrucci L. and Santamato E. "Tunable liquid crystal q-plates with arbitrary topological charge". *Opt Exp*, **19**, 4085–4090 (2011).

[Smi04]   Smith D.R., Pendry J.B. and Wiltshire M.C.K. "Metamaterials and negative refractive index." *Science*, **305**, 788–792 (2004).

[Smi07]   Smith B.J. and Raymer M.G. "Photon wave functions, wave-packet quantization of light, and coherence theory". *New J Phys*, **9**, 414–451 (2007).

[Smi70]   Smith R.L. "The velocities of light". *Am J Phys*, **38**, 978–984 (1970).

[Sny72]   Snyder A.W. "Coupled-mode theory for optical fibers". *J Opt Soc Am*, **62**, 1267–1277 (1972).

[Sny83]   Snyder A.W. and Love J.D. *Optical Waveguide Theory* (Chapman and Hall, New York, 1983).

[Sob12]   Sobel'man I.I., Vainshtein L.A. and Yukov E.A. *Excitation of Atoms and Broadening of Spectral Lines*, volume 15 (Springer Science & Business Media, New York, 2012).

[Sob63]   Sobolev V.V. *A Treatise on Radiative Transfer* (Van Nostrand, New York, 1963).

[Sob72]   Sobel'man I.I. *Introduction to the Theory of Atomic Spectra* (Pergamon, New York, 1972).

[Sob92]   Sobel'man I.I. *Atomic Spectra and Radiative Transitions* (Springer, New York, 1992), 2nd edition.

[Son17]   Sonnleitner M. and Barnett S.M. "The Röntgen interaction and forces on dipoles in time-modulated optical fields". *Eur Phys J D*, **71**, 336–349 (2017).

[Son97]   Song J., Lu C.C. and Chew W.C. "Multilevel fast multipole algorithm for electromagnetic scattering by large complex objects". *IEEE Trans Anten Prop*, **45**, 1488–1493 (1997).

[Sos01]   Soskin M.S. and Vasnetsov M.V. "Singular optics". *Prog Opt*, **42**, 219–276 (2001).

[Spr98]   Spreeuw R.J.C. "A classical analogy of entanglement". *Found Phys*, **28**, 361–374 (1998).

[Sta18]   Staggs S., Dunkley J. and Page L. "Recent discoveries from the cosmic microwave background: a review of recent progress". *Rep Prog Phys*, **81**, 044901 (2018).

[Sta98]   Stalgies Y., Siemers I., Appasamy B. and Toschek P.E. "Light shift and

Fano resonances in a single cold ion". *J Opt Soc Am B*, **15**, 2505–2513 (1998).

[Ste04] Steele J.M. *The Cauchy–Schwarz Master Class: an Introduction to the Art of Mathematical Inequalities.* (Cambridge Univ Press, Cambridge, 2004).

[Ste06] Stenholm S. "69. Absorption and gain spectra". In G. Drake, (ed.) *Springer Handbook of Atomic, Molecular and Optical Physics*, chapter 69, 1009–1022 (Springer, New York, 2006), 2nd edition.

[Ste08] Stern A. "Anyons and the quantum Hall effect — a pedagogical review". *Ann Phys*, **323**, 204–249 (2008).

[Ste66] Stephenson R.J. "Development of vector analysis from quaternions". *Am J Phys*, **34**, 194–201 (1966).

[Ste73] Stenholm S. "Quantum theory of electromagnetic fields interacting with atoms and molecules". *Phys Rep*, **6**, 1–122 (1973).

[Ste78] Stenholm S. "Theoretical foundations of laser spectroscopy". *Phys Rep*, **43**, 151–221 (1978).

[Ste92] Stenholm S. "Simultaneous measurement of conjugate variables". *Ann Phys*, **218**, 233–254 (1992).

[Ste98] Steane A. "Quantum computing". *Rep Prog Phys*, **61**, 117–173 (1998).

[Sto05] Stone N. "Table of nuclear magnetic dipole and electric quadrupole moments". *At Data Nuc Data Tables*, **90**, 75–176 (2005).

[Sto09] Stobińska M., Alber G. and Leuchs G. "Perfect excitation of a matter qubit by a single photon in free space". *Europhys Lett*, **86**, 14007 (2009).

[Sto10] Stoian R., Wollenhaupt M., Baumert T. and Hertel I.V. "Temporal pulse tailoring in laser manufacturing technologies". In *Laser Precision Microfabrication*, 121–144 (Springer-Verlag, Berlin, 2010).

[Sto10b] Stobińska M., Alber G. and Leuchs G. "Quantum electrodynamics of one-photon wave packets". *Adv Quant Chem*, **60**, 457–483 (2010).

[Sto19] Stöhr J. "Overcoming the diffraction limit by multi-photon interference: a tutorial". *Adv Opt Phot*, **11**, 215–313 (2019).

[Sto52] Stokes G.G. "On the change of refrangibility of light". *Phil Trans Roy Soc (Lond)*, **142**, 463–562 (1852).

[Sto63] Stone J.M. *Radiation and Optics. An Introduction to the Classical Theory* (McGraw-Hill, New York, 1963).

[Sto94] Storey P., Tan S., Collett M. and Walls D. "Path detection and the uncertainty principle". *Nature*, **367**, 626–628 (1994).

[Str16] Strauch F.W. "Resource Letter QI-1: Quantum Information". *Am J Phys*, **84**, 495–507 (2016).

[Str17] Streltsov A., Adesso G. and Plenio M.B. "Colloquium: Quantum coherence as a resource". *Rev Mod Phys*, **89**, 041003 (2017).

[Str41] Stratton J.A. *Electromagnetic Theory* (McGraw Hill, New York, 1941).

[Str48] Struik D.J. *A Concise History of Mathematics* (Dover, New York, 1948).

[Str70] Stroud, C. R. Jr. and Jaynes E.T. "Long-term solutions in semiclassical radiation theory". *Phys Rev A*, **1**, 106–121 (1970).

[Str86] Strnad J. "Photons in introductory quantum physics". *Am J Phys*, **54**, 650–652 (1986).

[Str86b] Strnad J. "The Compton effect—Schrödinger's treatment". *Eur J Phys*, **7**, 217–221 (1986).

[Str95] Strekalov D.V., Sergienko A.V., Klyshko D.N. and Shih Y.H. "Observation of two-photon "Ghost" interference and diffraction". *Phys Rev Lett*, **74**, 3600–3603 (1995).

[Stu32] Stueckelberg E.C.G. "Theorie der unelastichen Stösse zwischen Atomen". *Helv Phys Acta*, **5**, 369–423 (1932).

[Stu95] Stumm W. and Morgan J.J. *Aquatic Chemistry* (Wiley-Interscience, New York, 1995), 3rd edition.

[Sty00] Styer D.F. "Insight into entropy". *Am J Phys*, **68**, 1090–1096 (2000).

[Sug04] Sugon Q.M. and McNamara D.J. "A geometric algebra reformulation of geometric optics". *Am J Phys*, **72**, 92–97 (2004).

[Sug18] Sugic D. and Dennis M.R. "Singular knot bundle in light". *J Opt Soc Am A*, **35**, 1987–1999 (2018).

[Sus08] Susskind L. *The Black Hole War* (Little, Brown, New York, 2008).

[Sus64] Susskind L. and Glogower J. "Quantum mechanical phase and time operator". *Physics Physique Fizika*, **1**, 49–61 (1964).

[Svi13] Svidzinsky A.A., Yuan L. and Scully M.O. "Transient lasing without inversion". *New J Phys*, **15**, 053044 (2013).

[Swe08] Swendsen R.H. "Explaining irreversibility". *Am J Phys*, **76**, 643–648 (2008).

[Swe11] Swendsen R.H. "How physicists disagree on the meaning of entropy". *Am J Phys*, **79**, 342–348 (2011).

[Sza01] Szapudi I., Prunet S., Pogosyan D., Szalay A.S. and Bond J.R. "Fast cosmic microwave background analyses via correlation functions". *Ap J*, **548**, L115–L118 (2001).

[Sza92] Szanton A. *The Recollections of Eugene P Wigner* (Springer, Boston, 1992).

[Sze04] Szekeres P. *A Course in Modern Mathematical Physics. Groups, Hilbert Space and Differential Geometry* (Cambridge Univ Press, New York, 2004).

[Tak71] Takatsuji M. "Propagation of a light pulse in a two-photon resonant medium". *Phys Rev A*, **4**, 808–810 (1971).

[Tan11] Tanışlı M. and Kansu M.E. "Octonionic Maxwell's equations for bi-isotropic media". *J Math Phys*, **52**, 053511 (2011).

[Tay09] Taylor G.I. "Interference fringes with feeble light". *Proc Camb Phil Soc*, **15**, 114–115 (1909).

[Tay72] Taylor J.R. *Scattering Theory* (Wiley, New York, 1972).

[Tem06] Temkin A. and Bhatia A.K. "25. Autoionization". In G. Drake, (ed.) *Springer Handbook of Atomic, Molecular and Optical Physics*, chapter 25, 391–400 (Springer, New York, 2006), 2nd edition.

[Tem13] Temkin A. *Autoionization: Rrecent Developments and Applications* (Springer Science & Business Media, New York, 2013).

[Tem16] Tempone-Wiltshire S.J., Johnstone S.P. and Helmerson K. "Optical vortex knots–one photon at a time". *Sci Rep*, **6**, 24463 (2016).

[Ter61] ter Haar D. "Theory and application of the density matrix". *Rep Prog*

Phys, **24**, 304–362 (1961).

[Ter67]   ter Haar D., (ed.) *The Old Quantum Theory* (Pergamon, New York, 1967).

[Tho04]   Thorn J.J., Neel M.S., Donato V.W., Bergreen G.S., Davies R.E. and Beck M. "Observing the quantum behavior of light in an undergraduate laboratory". *Am J Phys*, **72**, 1210–1219 (2004).

[Tho67]   Thomson W. "On vortex atoms". *Proc Roy Soc Edinb*, **6**, 94–105 (1867).

[Tho92]   Thompson R.J., Rempe G. and Kimble H.J. "Observation of normal-mode splitting for an atom in an optical cavity". *Phys Rev Lett*, **68**, 1132–1135 (1992).

[Tid90]   Tidwell S.C., Ford D.H. and Kimura W.D. "Generating radially polarized beams interferometrically". *App Opt*, **29**, 2234–2239 (1990).

[Tol31]   Tolman R.C., Ehrenfest P. and Podolsky B. "On the gravitational field produced by light". *Phys Rev*, **37**, 602–615 (1931).

[Tor11]   Torosov B.T. and Vitanov N.V. "Evolution of superpositions of quantum states through a level crossing". *Phys Rev A*, **84**, 063411 (2011).

[Tor11b]   Torosov B. and Vitanov N.V. "Smooth composite pulses for high-fidelity quantum information processing". *Phys Rev A*, **83**, 053420 (2011).

[Tor15]   Torosov B.T., Kyoseva E.S. and Vitanov N.V. "Composite pulses for ultrabroad-band and ultranarrow-band excitation". *Phys Rev A*, **92**, 033406 (2015).

[Tor15b]   Törmö P. and Barnes W.L. "Strong coupling between surface plasmon polaritons and emitters: A review". *Rep Prog Phys*, **78**, 013901 (2015).

[Tor18]   Torosov B.T. and Vitanov N.V. "Arbitrarily accurate twin composite π-pulse sequences". *Phys Rev A*, **97**, 043408 (2018).

[Tow13]   Townes C.H. and Schawlow A.L. *Microwave Spectroscopy* (Dover, Mineola, NY, 2013).

[Tra01]   Trayling G. and Baylis W.E. "A geometric basis for the standard-model gauge group". *J Phys A*, **34**, 3309–3324 (2001).

[Tri80]   Trimmer J.D. "The present situation in quantum mechanics: A translation of Schrödinger's "Cat paradox" paper". *Proc Am Phil Soc*, **24**, 323–338 (1980).

[Tri88]   Trimble V. "Dark matter in the universe: Where, what, and why?" *Contemp Phys*, **29**, 373–392 (1988).

[Tse19]   Tse J.S. "A twist in the tale of the structure of ice". *Nature*, **569**, 495–496 (2019).

[Tuc75]   Tucker W.H. *Radiation Processes in Astrophysics* (MIT Press, Cambridge, MA, 1975).

[Una97]   Unanyan R.G., Yatsenko L.P., Bergmann K. and Shore B.W. "Laser-induced adiabatic atomic reorientation with control of diabatic losses". *Opt Comm*, **139**, 48–54 (1997).

[Una98]   Unanyan R.G., Fleischhauer M., Shore B.W. and Bergmann K. "Robust creation and phase-sensitive probing of superposition states via stimulated Raman adiabatic passage (STIRAP) with degenerate dark states". *Opt Comm*, **155**, 144–54 (1998).

[Una99]   Unanyan R.G., Shore B.W. and Bergmann K. "Laser-driven population transfer in four-level atoms: Consequences of non-Abelian geometrical adiabatic phase factors". *Phys Rev A*, **59**, 2910–2919 (1999).

[Vah03]   Vahala K.J. "Optical microcavities." *Nature*, **424**, 839–846 (2003).

[Van07]   Van Enk S.J., Lütkenhaus N. and Kimble H.J. "Experimental procedures for entanglement verification". *Phys Rev A*, **75**, 052318 (2007).

[Van08]   Van Kampen N.G. "The scandal of quantum mechanics". *Am J Phys*, **76**, 989–990 (2008).

[Van12]   van de Meerakker S.Y.T., Bethlem H.L., Vanhaecke N. and Meijer G. "Manipulation and control of molecular beams". *Chem Revs*, **112**, 4828–4878 (2012).

[Van13]   Van Enk S.J. "The covariant description of electric and magnetic field lines of null fields: Application to Hopf-Rañada solutions". *J Phys A*, **46**, 175204 (2013).

[Van16]   Van Mechelen T. and Jacob Z. "Universal spin-momentum locking of evanescent waves". *Optica*, **3**, 118–126 (2016).

[Van74]   Van den Doel R. and Kokkedee J.J.J. "Comment on the radiative level shift in the neoclassical radiation theory". *Phys Rev A*, **9**, 1468–1469 (1974).

[Van76]   Van Kampen N.G. "Stochastic differential equations". *Phys Rep*, **24**, 171–228 (1976).

[Van77]   Van Vleck J.H. and Huber D.L. "Absorption, emission, and linebreadths: A semihistorical perspective". *Rev Mod Phys*, **49**, 939–959 (1977).

[Var00]   Varcoe B.T.H., Brattke S., Weidinger M. and Walther H. "Preparing pure photon number states of the radiation field". *Nature*, **403**, 743–746 (2000).

[Var04]   Varcoe B.T.H., Brattke S. and Walther H. "The creation and detection of arbitrary photon number states using cavity QED". *New J Phys*, **6**, 97–109 (2004).

[Vas07]   Vasilev G.S., Ivanov S.S. and Vitanov N.V. "Degenerate Landau-Zener model: Analytical solution". *Phys Rev A*, **75**, 013417 (2007).

[Vas09]   Vasilev G.S., Kuhn A. and Vitanov N.V. "Optimum pulse shapes for stimulated Raman adiabatic passage". *Phys Rev A*, **80**, 013417 (2009).

[Vas10]   Vasilev G.S., Ljunggren D. and Kuhn A. "Single photons made-to-measure". *New J Phys*, **12**, 063024 (2010).

[Vas89]   Vassiliadis D.V. "Wigner's little group and decomposition of Lorentz transformations". *J Math Phys*, **30**, 2177–2180 (1989).

[Ved01]   Vedral V. *Modern Foundations of Quantum Optics* (World Scientific, Singapore, 2001).

[Ved06]   Vedral V. *Introduction to Quantum Information Science* (Oxford Univ Press, New York, 2006).

[Ved97]   Vedral V., Plenio M.B., Rippin M.A. and Knight P.L. "Quantifying entanglement". *Phys Rev Lett*, **78**, 2275–2279 (1997).

[Vel07]   Vellekoop I.M. and Mosk A.P. "Focusing coherent light through opaque strongly scattering media". *Opt Lett*, **32**, 2309–2311 (2007).

[Vel10]   Vellekoop I.M., Lagendijk A. and Mosk A.P. "Exploiting disorder for perfect focusing". *Nature Phot*, **4**, 320–322 (2010).

[Vew03]   Vewinger F., Heinz M., Garcia-Fernandez R., Vitanov N.V. and Bergmann

K. "Creation and measurement of a coherent superposition of quantum states". *Phys Rev Lett*, **91**, 213001 (2003).

[Vew10] Vewinger F., Shore B.W. and Bergmann K. "Superpositions of degenerate quantum states: Preparation and detection in atomic beams". *Adv At Mol Opt Phys*, **58**, 113–172 (2010).

[Vit00] Vitanov N.V. "Measuring a coherent superposition". *Opt Comm*, **179**, 73–83 (2000).

[Vit00b] Vitanov N.V. "Measuring a coherent superposition of multiple states". *J Phys B*, **33**, 2333–2346 (2000).

[Vit01] Vitanov N.V., Fleischhauer M., Shore B.W. and Bergmann K. "Coherent manipulation of atoms and molecules by sequential laser pulses". *Adv At Mol Opt Phys*, **46**, 55–190 (2001).

[Vit01b] Vitanov N.V., Halfmann T., Shore B.W. and Bergmann K. "Laser-induced population transfer by adiabatic passage techniques". *Ann Rev Phys Chem*, **52**, 763–809 (2001).

[Vit03] Vitanov N.V., Kis Z. and Shore B.W. "Coherent excitation of a degenerate two-level system by an elliptically polarized laser pulse". *Phys Rev A*, **68**, 063414 (2003).

[Vit05] Vitanov N.V. and Shore B.W. "Quantum transitions driven by missing frequencies". *Phys Rev A*, **72**, 052507 (2005).

[Vit09] Vitanov N.V., Shore B.W. and Yatsenko L.P. "Atomic absorpton profiles associated with pulsed excitation". *Ukr J Phys*, **54**, 53–62 (2009).

[Vit11] Vitanov N.V. "Arbitrarily accurate narrowband composite pulse sequences". *Phys Rev A*, **84**, 065404 (2011).

[Vit17] Vitanov N.V., Rangelov A.A., Shore B.W. and Bergmann K. "Stimulated Raman adiabatic passage in physics, chemistry and beyond". *Rev Mod Phys*, **89**, 015006 (2017).

[Vit95] Vitanov N.V. and Knight P.L. "Coherent excitation of a two-state system by a train of short pulses". *Phys Rev A*, **52**, 2245–2261 (1995).

[Vit97] Vitanov N.V. and Stenholm S. "Analytic properties and effective two-level problems in stimulated Raman adiabatic passage". *Phys Rev A*, **55**, 648–660 (1997).

[Vit97b] Vitanov N.V. and Stenholm S. "Properties of stimulated Raman adiabatic passage with intermediate-level detuning". *Opt Comm*, **135**, 394–405 (1997).

[Vit98] Vitanov N.V. "Adiabatic population transfer by delayed laser pulses in multistate systems". *Phys Rev A*, **58**, 2295–2309 (1998).

[Vit98b] Vitanov N.V., Shore B.W. and Bergmann K. "Adiabatic population transfer in multistate chains via dressed intermediate states". *Eur Phys J D*, **4**, 15–29 (1998).

[Vit98c] Vitanov N.V. "Analytic model of a three-state system driven by two laser pulses on two-photon resonance". *J Phys B*, **31**, 709–725 (1998).

[Vit98d] Vitanov N.V. "Transition times in the Landau-Zener model". *Phys Rev A*, **59**, 998–994 (1998).

[Vit99] Vitanov N.V., Suominen K.A. and Shore B.W. "Creation of coherent atomic superpositions by fractional stimulated Raman adiabatic passage". *J Phys B*, **32**, 4535–4546 (1999).

[Vog06] Vogel W. and Welsch D.G. *Quantum Optics* (Wiley-VCH, Berlin, 2006).

[Vol93] Vold T.G. "An introduction to geometric calculus and its application to electrodynamics". *Am J Phys*, **61**, 505–513 (1993).

[Vol93b] Vold T.G. "An introduction to geometric algebra with an application in rigid body mechanics". *Am J Phys*, **61**, 491–504 (1993).

[Voo17] Vool U. and Devoret M.H. "Introduction to quantum electromagnetic circuits". *Int J Circ Theory App*, **45**, 897–934 (2017).

[Wal06] Walther H., Varcoe B.T.H., Englert B.G. and Becker T. "Cavity quantum electrodynamics". *Rep Prog Phys*, **69**, 1325–1382 (2006).

[Wal07] Walls D.F. and Milburn G.J. *Quantum Optics* (Springer-Verlag, Berlin, 2007), 2nd edition.

[Wal54] Walker M.J. "Matrix calculus and the Stokes parameters of polarized radiation". *Am J Phys*, **22**, 170–174 (1954).

[Wal70] Walker C.T. and Slack G.A. "Who named the -ON's?" *Am J Phys*, **38**, 1380–1389 (1970).

[Wal83] Walls D.F. "Squeezed states of light". *Nature*, **306**, 141–146 (1983).

[Wal88] Walther H. "The single atom maser and the quantum electrodynamics in a cavity". *Phys Scripta*, **T23**, 165–169 (1988).

[Wan16] Wang J.J., Wriedt T., Lock J.A. and Mädler L. "General description of circularly symmetric Bessel beams of arbitrary order". *J Quant Spec Rad Transf*, **184**, 218–232 (2016).

[Wan16b] Wang H., Xu Y., Genevet P., Jiang J.H. and Chen H. "Broadband mode conversion via gradient index metamaterials". *Sci Rep*, **6**, 24529 (2016).

[Wan91] Wang L., Ho P.P., Liu C., Zhang G. and Alfano R.R. "Ballistic 2-D imaging through scattering walls using an ultrafast optical Kerr gate". *Science*, **253**, 769–771 (1991).

[War93] Warren W.S., Rabitz H. and Dahleh M.A. "Coherent control of quantum dynamics: The dream is alive." *Science*, **259**, 1581–1589 (1993).

[Wat07] Waterman P.C. "The T-matrix revisited." *J Opt Soc Am A*, **24**, 2257–2267 (2007).

[Wea16] Weatherall J.O. *Void: The Strange Physics of Nothing* (Yale Univ Press, New Haven, 2016).

[Web54] Weber J. "Phase, group, and signal velocity". *Am J Phys*, **22**, 618–620 (1954).

[Wee89] Weekes T.C., Cawley M., Fegan D., Gibbs K., Hillas A., Kowk P., Lamb R., Lewis D., Macomb D., Porter N. *et al.* "Observation of TeV gamma rays from the Crab nebula using the atmospheric Cerenkov imaging technique". *Astrophys J*, **342**, 379–395 (1989).

[Wei11] Weiner A.M. "Ultrafast optical pulse shaping: A tutorial review". *Opt Comm*, **284**, 3669–3692 (2011).

[Wei15] Weinberger P. "Arthur Cayley and the 'Gruppen Pest'". *Phil Mag*, **95**, 3039–3051 (2015).

[Wei87] Weis R.S. and Gaylord T.K. "Electromagnetic transmission and reflection characteristics of anisotropic multilayered structures". *J Opt Soc Am A*, **4**, 1720–1740 (1987).

[Wen05]   Wendin G. and Shumeiko V. "Superconducting quantum circuits, qubits and computing". *ArXiv preprint cond-mat/0508729*, 1–58 (2005).

[Wen17]   Wendin G. "Quantum information processing with superconducting circuits: a review". *Rep Prog Phys*, **80**, 106001 (2017).

[Wen27]   Wentzel G. "Über die Richtungsverteilung der Photoelektronen". *Zs Phys*, **41**, 828–832 (1927).

[Wey50]   Weyl H. *The Theory of Groups and Quantum Mechanics* (Dover, New York, 1950).

[Whe14]   Wheeler J.A. and Zurek W.H. *Quantum Theory and Measurement* (Princeton Univ Press, Princeton, NJ, 2014).

[Wie48]   Wiener N. *Cybernetics* (Wiley, New York, 1948).

[Wig39]   Wigner E. "On unitary representations of the inhomogeneous Lorentz group". *Ann Math*, **40**, 149–204 (1939).

[Wig60]   Wigner E.P. and Fano U. "Group theory and its application to the quantum mechanics of atomic spectra". *Am J Phys*, **28**, 408–409 (1960).

[Wig63]   Wigner E.P. "The problem of measurement". *Am J Phys*, **31**, 6–15 (1963).

[Wil07]   Wilk T., Webster S.C., Specht H.P., Rempe G. and Kuhn A. "Polarization-controlled single photons". *Phys Rev Lett*, **98**, 063601 (2007).

[Wil13]   Willner A.E. and 17 others. "Optical communications using orbital angular momentum beams". *Adv Opt Phot*, **5**, 66–91 (2013).

[Wil15]   Wilczek F. *A Beautiful Question: Finding Nature's Deep Design* (Penguin, New York, 2015).

[Wil80]   Wilson E.B., Decius J.C. and Cross P.C. *Molecular Vibrations. The Theory of Infrared and Raman Vibrational Spectra* (Dover, New York, 1980).

[Wil82]   Wilczek F. "Quantum mechanics of fractional-spin particles". *Phys Rev Lett*, **49**, 957–959 (1982).

[Wil88]   Will C.M. "Henry Cavendish, Johann von Soldner, and the deflection of light". *Am J Phys*, **56**, 413–415 (1988).

[Win77]   Winans J.G. "Quaternion physical quantities". *Found Phys*, **7**, 341–349 (1977).

[Wod79]   Wódkiewicz K. "Exact solutions of some multiplicative stochastic processes". *J Math Phys*, **20**, 45–48 (1979).

[Wod84]   Wódkiewicz K., Shore B.W. and Eberly J.H. "Pre-Gaussian noise in strong laser-atom interactions". *J Opt Soc Am B*, **1**, 398–405 (1984).

[Wod84b] Wódkiewicz K., Shore B.W. and Eberly J.H. "Noise in strong laser-atom interactions: frequency fluctuations and nonexponential correlations". *Phys Rev A*, **30**, 2390–2398 (1984).

[Wod87]   Wódkiewicz K., Knight P.L., Buckle S.J. and Barnett S.M. "Squeezing and superposition states". *Phys Rev A*, **35**, 2567–2577 (1987).

[Wod92]   Wódkiewicz K. and Eberly J.H. "Click-counting distributions in optical transitions". *Ann Phys*, **216**, 268–290 (1992).

[Wol03]   Wolf E. "Correlation-induced changes in the degree of polarization, the degree of coherence, and the spectrum of random electromagnetic beams on propagation." *Opt Lett*, **28**, 1078–1080 (2003).

[Wol05] Wollenhaupt M., Engel V. and Baumert T. "Femtosecond laser photo-electron spectroscopy on atoms and small molecules: Prototype studies in quantum control". *Ann Rev Phys Chem*, **56**, 25–56 (2005).

[Wol06] Wollenhaupt M., Assion A. and Baumert T. "Femtosecond laser pulses: linear properties, manipulation, generation and measurement". In F. Träger, (ed.) *Springer Handbook of Lasers and Optics*, 937–983 (Springer, New York, 2006).

[Wol07] Wolf E. *Introduction to the Theory of Coherence and Polarization of Light* (Cambridge Univ Press, Cambridge, 2007).

[Wol59] Wolf E. "Coherence properties ol partially polarized electromagnetic radiation". *Nuov Cim*, **XIII**, 1165–1181 (1959).

[Woo09] Wootters W.K. and Zurek W.H. "The no-cloning theorem". *Phys Today*, **62**, 76–77 (2009).

[Woo61] Wood R.W. *Physical Optics* (Dover, New York, 1961).

[Woo79] Wootters W.K. and Zurek W.H. "Complementarity in the double-slit experiment: Quantum nonseparability and a quantitative statement of Bohr's principle". *Phys Rev D*, **19**, 473–484 (1979).

[Woo82] Wootters W.K. and Zurek W.H. "A single quantum cannot be cloned". *Nature*, **299**, 802–803 (1982).

[Woo98] Wootters W.K. "Entanglement of formation of an arbitrary state of two qubits". *Phys Rev Lett*, **80**, 2245–2248 (1998).

[Wun04] Wünsche A. "Quantization of Gauss-Hermite and Gauss-Laguerre beams in free space". *J Opt B*, **6**, S47–S59 (2004).

[Xia13] Xiang Z.L., Ashhab S., You J.Q. and Nori F. "Hybrid quantum circuits: Superconducting circuits interacting with other quantum systems". *Rev Mod Phys*, **85**, 623–653 (2013).

[Xie17] Xie Q., Zhong H., Batchelor M.T. and Lee C. "The quantum Rabi model: solution and dynamics". *J Phys A*, **50**, 113001 (2017).

[Yab93] Yablonovitch E. "Photonic band-gap structures". *J Opt Soc Am B*, **10**, 283–295 (1993).

[Yam08] Yamazaki R., Kanda K., Inoue F., Toyoda K. and Urabe S. "Robust generation of superposition states". *Phys Rev A*, **78**, 023808 (2008).

[Yam99] Yamamoto Y. and Imamoglu A. *Mesoscopic Quantum Optics* (Wiley, New York, 1999).

[Yan09] Yannopapas V. and Vitanov N.V. "First-principles study of Casimir repulsion in metamaterials". *Phys Rev Lett*, **103**, 120401 (2009).

[Yao11] Yao A.M. and Padgett M.J. "Orbital angular momentum: Origins, behavior and applications". *Adv Opt Phot*, **3**, 161–204 (2011).

[Yar06] Yariv A. and Yeh P. *Photonics: Optical Electronics in Modern Communications* (Oxford Univ Press, Oxford, 2006), 6th edition.

[Yar73] Yariv A. "Coupled-mode theory for guided-wave optics". *IEEE J Quant Elec*, **9**, 919–933 (1973).

[Yat06] Yatsenko L.P., Rangelov A.A., Vitanov N.V. and Shore B.W. "Steering population flow in coherently driven lossy quantum ladders." *J Chem*

Phys, **125**, 014302 (2006).

[Yil05]   Yildiz A. and Selvin P.R. "Fluorescence imaging with one nanometer accuracy, Application to molecular motors". *Accounts Chem Res*, **38**,, 574–582 (2005).

[Yod96]   Yodh A. and Chance B. "Spectroscopy and imaging with diffusing light". *Phys Today*, **48**, 34–41 (1995).

[Yoo90]   Yoo K.M. and Alfano R.R. "Time-resolved coherent and incoherent components of forward light scattering in random media". *Opt Lett*, **15**, 320–322 (1990).

[You05]   You J.Q. and Nori F. "Superconducting circuits and quantum information". *Phys Today*, **58**, 42–47 (2005).

[Yur84]   Yurke B. and Denker J.S. "Quantum network theory". *Phys Rev A*, **29**, 1419–1437 (1984).

[Zab15]   Zaburdaev V., Denisov S. and Klafter J. "Lévy walks". *Rev Mod Phys*, **87**, 483–530 (2015).

[Zar88]   Zare R.N. *Angular Momentum: Understanding Spatial Aspects in Chemistry and Physics* (Wiley, New York, 1988).

[Zav04]   Zavatta A., Viciani S. and Bellini M. "Tomographic reconstruction of the single-photon Fock state by high-frequency homodyne detection". *Phys Rev A*, **70**, 1–6 (2004).

[Zei20]   Zeilinger A. "Quantum teleportation". *Sci Am*, **282**, 50–59 (2000).

[Zen32]   Zener C. "Non-adiabatic crossing of energy levels". *Proc Roy Soc (Lond) A*, **137**, 696–699 (1932).

[Zha09]   Zhan Q. "Cylindrical vector beams: from mathematical concepts to applications". *Adv Opt Phot*, **1**, 1–57 (2009).

[Zha12]   Zhang L., Söller C., Cohen O., Smith B.J. and Walmsley I.A. "Heralded generation of single photons in pure quantum states". *J Mod Opt*, **59**, 1525–1537 (2012).

[Zim60]   Ziman J.M. *Electrons and Phonons: The Theory of Transport Phenomena in Solids* (Oxford Univ Press, Oxford, 1960).

[Zol05]   Zoller P. and 38 others. "Quantum information processing and communication". *Eur Phys J D*, **36**, 203–228 (2005).

[Zol11]   Zoller P. "Prospects of quantum information processing with atoms". *Quant Inf Proc*, **10**, 1061–1063 (2011).

[Zou90]   Zou X.T. and Mandel L. "Photon-antibunching and sub-Poissonian photon statistics". *Phys Rev A*, **41**, 475–476 (1990).

Index